The Ice Age

The Ice Age

Jürgen Ehlers
Witzeeze, Germany

Philip D. Hughes
Reader in Physical Geography, School of Environment,
Education & Development, The University of Manchester,
Manchester, United Kingdom

Philip L. Gibbard
Department of Geography, University of Cambridge,
Cambridge, United Kingdom

WILEY Blackwell

Library of Congress Cataloging-in-Publication Data
Ehlers, Jürgen, 1948-
  [Eiszeitalter. English]
  The Ice Age / Jürgen Ehlers, Philip Hughes. and Philip Gibbard.
      pages cm
  Translated from German: Das Eiszeitalter / by Jürgen Ehlers (Springer-Verlag, GmbH, 2011).
  Includes bibliographical references and index.
    ISBN 978-1-118-50781-0 (cloth) – ISBN 978-1-118-50780-3 (pbk.) 1. Glacial epoch.
2. Climatic changes. 3. Glaciology. 4. Glacial landforms. 5. Eolian processes. 6. Ice sheets.
I. Hughes, Philip D. II. Gibbard, Philip L. (Philip Leonard), 1949- III. Title.
  QE697.E4413 2015
  551.7'92–dc23
                                                          2015015483

A catalogue record for this book is available from the British Library.

Cover image: Briksdalsbreen Glacier in Jostedalsbreen, Norway © Getty Images

Set in 11/14 Adobe Garamond Pro by Aptara, India
Printed and bound in Singapore by Markono Print Media Pte Ltd

1   2016

Contents

ABOUT THE AUTHORS   vii
PREFACE   ix
ACKNOWLEDGEMENTS   xi
ABOUT THE COMPANION WEBSITE   xiii

1.   *Introduction*   3
     1.1   In the Beginning was the Great Flood   4
     1.2   The Ice Ages of the Earth   14
     1.3   Causes of an Ice Age   18

2.   *The Course of the Ice Age*   27
     2.1   When did the Quaternary Period Begin?   27
     2.2   What's in Stratigraphy?   33
     2.3   Traces in the Deep Sea   35
     2.4   Systematics of the Ice Age   43
     2.5   Günz, Mindel, Riss and Würm:
           Do They Still Apply?   46
     2.6   Northern Germany and Adjacent Areas   58
     2.7   The British Pleistocene Succession   76
     2.8   Quaternary History of North America   86
     2.9   The Course of the Ice Ages: A Global View 100

3.   *Ice and Water*   107
     3.1   The Origin of Glaciers   107
     3.2   Recent Glaciers: Small and Large   113
     3.3   Dynamics of Ice Sheets   121
     3.4   Meltwater   129

4.   *Till and Moraines: The Traces of Glaciers* 137
     4.1   Till   137
     4.2   Moraines   172

5.   *Meltwater: From Moulins to the Urstromtal* 191
     5.1   Fjords, Channels and Eskers   191
     5.2   Outwash Plains and Gravel Terraces   202
     5.3   Ice-dammed Lakes   207

     5.4   Kames: Deposits at the Ice Margin   213
     5.5   Urstromtäler   220

6.   *Maps: Where Are We?*   227
     6.1   Digital Maps   229
     6.2   Satellite Images: Basic Data for
           Ice-Age Research   236
     6.3   Projections and Ellipsoids   240

7.   *Extent of the Glaciers*   243
     7.1   Exploring the Arctic by Airship   243
     7.2   Glaciers in the Barents Sea   244
     7.3   Isostasy and Eustasy   246
     7.4   Ice in Siberia?   252
     7.5   Asia: The Mystery of Tibet   258
     7.6   South America: Volcanoes
           and Glaciers   265
     7.7   Mediterranean Glaciations   269
     7.8   Were Africa, Australia and
           Oceania Glaciated?   272
     7.9   Antarctica: Eternal Ice?   273

8.   *Ice in the Ground: The Periglacial Areas* 277
     8.1   Definition and Distribution   277
     8.2   Extent of Frozen Ground during
           the Pleistocene   281
     8.3   Frost Weathering   283
     8.4   Cryoplanation   286
     8.5   Rock Glaciers: Glaciers (Almost)
           Without Ice   288
     8.6   Involutions   291
     8.7   Solifluction   294
     8.8   Periglacial Soil Stripes   296
     8.9   Frost Cracks and Ice Wedges   297
     8.10 Pingos, Palsas and other
           Frost Phenomena   301

9. *Hippos in the Thames: The Warm Stages*   311
  9.1   Tar Pits of Evidence   311
  9.2   Development of Fauna   312
  9.3   Development of Vegetation   316
  9.4   Weathering and Soil Formation   324
  9.5   Water in the Desert: The Shifting of
        Climate Zones   336
  9.6   Changes in the Rainforest   345

10. *The Course of Deglaciation*   349
  10.1 Contribution to Landforms   349
  10.2 Ice Decay   350
  10.3 The Origin of Kettle Holes   354
  10.4 Pressure Release   357
  10.5 A Sudden Transition?   359
  10.6 The Little Ice Age   363

11. *Wind, Sand and Stones: Aeolian Processes*   369
  11.1 Dunes   369
  11.2 Aeolian Sand   378
  11.3 Loess   378

12. *What Happened to the Rivers?*   383
  12.1 River Processes and Landforms   383
  12.2 Dry Valleys   386
  12.3 The Rhine: Influences of Alpine
        and Nordic Ice   387
  12.4 The Elbe: Once Flowed to the Baltic Sea 396
  12.5 The Thames: Influence of British Ice   400

13. *North and Baltic Seas during the Ice Age*   405
  13.1 Development of the North Sea   406
  13.2 Development of the Baltic Sea   414

14. *Climate Models and Reconstructions*   427
  14.1 Ice Cores   427
  14.2 The Marine Circulation   429
  14.3 Modelling the Last Ice Sheets   431
  14.4 Modelling Glaciers and Climate   442

15. *Human Interference*   447
  15.1 Out of Africa: Humans Spread Out   448
  15.2 Neanderthals and *Homo sapiens*   452
  15.3 The Middle Stone Age   452
  15.4 The Neolithic Period: The Beginning
        of Agriculture   453
  15.5 Bronze and Iron   454
  15.6 The Romans   455
  15.7 Middle Ages   457
  15.8 Recent Land Grab   457
  15.9 Drying Lakes, Melting Glaciers and
        other Problems   459
  15.10 The Anthropocene: Defining the
        Human Age?   465

REFERENCES   469
INDEX   541

About the Authors

Jürgen Ehlers worked as a Quaternary geologist at the Hamburg State Geological Survey, where he was responsible for geological mapping of the city area, from 1978 until 2013. He has taken part in international research projects. Together with Phil Gibbard and Phil Hughes, he has coordinated the INQUA project 'Extent and Chronology of Quaternary Glaciations'. He is regarded as an expert in Quaternary geology. He is the author of several books on aspects of the Ice Age and about the history and development of the North Sea. He is also Associate Editor of the German Quaternary Association's (DEUQUA) journal *Eiszeitalter und Gegenwart*.

Phil Hughes is Reader in Physical Geography at The University of Manchester. He studied for his first degree in Geography at the University of Exeter, graduating in 1999. This was followed by a Masters in Quaternary Science and a PhD in Geography, both at the University of Cambridge (Darwin College). His PhD, entitled *Quaternary Glaciation in the Pindus Mountains, Northwest Greece*, was completed during 2001–2004 under the supervision of Phil Gibbard (Cambridge) and Jamie Woodward (then Leeds, now Manchester). Phil has since published widely on glaciations in the Mediterranean mountains and his research has included work in Morocco, Spain, Montenegro, Albania and Greece. He has also published on various aspects of glaciation in the British Isles, as well as several theoretical papers on stratigraphy and glacier–climate modelling. Phil collaborated with both Jürgen Ehlers and Phil Gibbard in the recent edited volume of global glaciations (*Quaternary Glaciations—Extent and Chronology: A Closer Look*). Since 2011, Phil has also been Subject Editor for Quaternary Science and Geomorphology for the *Journal of the Geological Society of London*.

Phil Gibbard is Professor of Quaternary Palaeoenvironments in the Department of Geography, University of Cambridge and was a founder member of the Godwin Institute of Quaternary Research (since January 1995), now known as Cambridge Quaternary. He is concurrently a Docent (Adjunct Professor) in the Department of Geosciences and Geography, University of Helsinki where he contributes to the teaching of Quaternary Geology. Phil Gibbard studied Geology at the University of Sheffield in the 1960s, then took his PhD at the University of Cambridge under the supervision of Richard West FRS. He has published extensively on many aspects of Quaternary science. He has published numerous books with Jürgen Ehlers, including a translation of *Quaternary Geology and Glacial Geology* (John Wiley & Sons, 1996). In recognition of Phil's contribution to

Quaternary Science he has received numerous awards, including an honorary doctorate degree (PhD *honoris causa*) from the University of Helsinki (2010); a Doctor of Science (ScD) degree from the University of Cambridge (2010); the André Dumont Medal (2014) from Geologica Belgica; and the James Croll Medal 2014 from Britain's Quaternary Research Association. He was President of the International Commission on Stratigraphy's Subcommission on Quaternary Stratigraphy during 2002–2012. In 2011 he was elected President of the INQUA Commission on Stratigraphy and Geochronology (SACCOM) in 2011. He is also a member of the editorial boards of several journals.

Preface

The Ice Age is the period in which we live; our present interglacial interval is part of the Ice Age. The Ice Age has been a period of extreme climatic variations which are continuing today. Temporarily vast ice sheets covered major parts of the northern continents. At other times, the Sahara was green and inhabited by humans and Lake Chad, which today is the size of Greater London, once covered an area 20% larger than Britain and Ireland combined.

What happened in the Ice Age can only be reconstructed from the traces left behind. The Ice Age created strata that differ from the deposits of other geological periods. This book describes the processes which formed them and the methods by which they can be investigated. In this sense, an Earth scientist's work resembles (and is as exciting as) that of a detective.

This book builds on the original German language edition, *Das Eiszeitalter* (2011; Spektrum Akademischer Verlag) by Jürgen Ehlers. This was also the title of Paul Woldstedt's classic German language textbook on Quaternary geology, which was published in several editions by Ferdinand Enke Verlag. To emphasize the connection between the dramatic climate changes of the past and our present world (Woldstedt 1950), Paul Woldstedt named the journal of the German Quaternary Association (DEUQUA) *Eiszeitalter und Gegenwart* (Ice Age and Present). Jürgen Ehlers did not experience this early phase of German Quaternary research; Paul Woldstedt died in 1973, the year before Ehlers first participated in a DEUQUA meeting at Hofheim.

Since the publication of Woldstedt's *Eiszeitalter*, much has changed. The Ice Age is the period during which humans began to interfere with nature. The changes brought about by such interference are evident around us, and all relevant data are freely available. The idea of what a textbook should look like in order to appeal to the reader has also changed. This book does not merely aim to add another to the market; it is intended to offer a personal vision on the Quaternary, building on the very different experiences and perspectives of the three authors.

Acknowledgements

Many colleagues and friends have critically checked parts of the book and/or contributed illustrations. They include: Dr Hinrich Bäsemann, Tromsø; Dr Jochen Brandt, Helms-Museum, Harburg; Professor Dr Detlef Busche, Universität Würzburg; Dr Gerhard Doppler, Bayerisches Geologisches Landesamt; Professor Dr Edward Evenson, Lehigh University, Pennsylvania; Professor Dr Peter Felix-Henningsen, Universität Gießen; Professor Dr Markus Fiebig, Universität für Bodenkultur, Wien; Professor Dr Magnús Tumi Guðmundsson, University of Iceland, Reykjavik; Dr Bernd Habermann, Stadtarchäologie Buxtehude; Dr Robert Hebenstreit, Freie Universität Berlin; Professor Dr Dieter Jäkel, Freie Universität Berlin; Dr Adriaan Janszen, London; Professor Dr Kurt Kjær, Natural History Museum, Kopenhagen; Professor Dr Keenan Lee, Colorado School of Mines; Dr Christian Hoselmann, Hessisches Landesamt für Geologie und Umwelt; Professor Wighart von Koenigswald, Universität Bonn; Marcus Linke, Landesbetrieb Geoinformation und Vermessung, Hamburg; Professor Dr Thomas Litt, Universität Bonn; Professor Dr Juha-Pekka Lunkka, University of Oulu; Steve Mathers, British Geological Survey, Nottingham; Professor Dr Andrea Moscariello, University of Geneva; Professor Dr Jan Piotrowski, University of Aarhus; Professor Dr Frank Preusser, Universität Freiburg; Professor Vladimir E. Romanovsky, University of Alaska, Fairbanks; Professor Dr Alexei Rudoy, Tomsk State University; Professor Dr Gerhard Schellmann, Universität Bamberg; Professor Dr Christian Schlüchter, Universität Bern; Dr Petra Schmidt, Witzeeze; Gertrud Seehase, Ratzeburg; Professor Dr John Shaw, University of Alberta, Edmonton; Eva-Maria Stellmacher-Ludwig, Wentorf; Dr Hans-Jürgen Stephan, Kiel; Dr Þröstur Þorsteinsson, University of Iceland, Reykjavik; Professor Dr Roland Vinx, Universität Hamburg; Dr Stefan Wansa, Landesamt für Geologie und Bergbau Sachsen-Anhalt; Gerda and Holger Wolmeyer, Hamburg; Professor Dr Jan Zalasiewicz, University of Leicester; Jacob G. Zandstra, Heemskerk; and Professor Dr Bernd Zolitschka, Universität Bremen. It is a pleasure to thank them all. We would also like to thank Elaine Rowan for her thorough copy-editing of this English edition.

About the Companion Website

This book is accompanied by a companion website:

www.wiley.com/go/ehlers/iceage

The website includes:
- Powerpoints of all figures from the book for downloading
- Pdfs of all tables from the book for downloading

Central part of Gorner Glacier, Switzerland. Source: Agassiz (1841).

Chapter 1
Introduction

The Ice Ages! It is difficult, now, to understand the perplexity and bafflement and sheer disbelief that greeted this idea, over a century and a half ago: the idea of vast walls of ice invading from the north to engulf entire landscapes. This seemed like science fiction, a Gothic fantasy on a par with a belief in dragons and fairies and industrious aliens that built canals on Mars.

Zalasiewicz (2009, p. 68)

The meeting of the Swiss Society of Natural Science on 24 July 1837 in Neuchâtel began with a scandal. The young president of the Association, Louis Agassiz, spoke not about the latest results of his studies on fossil fishes as expected, which had made him famous. Instead, he decided to talk about the erratic blocks in the Jura Mountains (and in the vicinity of Neuchâtel) which he said were the legacy of a major glaciation. This 'Discourse of Neuchâtel' is considered the birth of the Ice Age Theory.

Agassiz was not the first person to have said this, but he was the first high-ranking scientist. His speech met with icy disapproval. On the subsequent field trip on 26 July, during which the participants could actually examine the evidence with their own eyes, Agassiz did not succeed in convincing the other experts. The glacial theory appeared to be a non-starter (Imbrie & Imbrie 1979).

The Ice Age, First Edition. Jürgen Ehlers, Philip D. Hughes and Philip L. Gibbard.
© 2016 John Wiley & Sons, Ltd. Published 2016 by John Wiley & Sons, Ltd.
Companion website: www.wiley.com/go/ehlers/iceage

1.1 In the Beginning was the Great Flood

People always tend to explain incomprehensible natural phenomena by processes they know. The notion of an 'Ice Age' was alien to the scientists of earlier centuries. They did know however that, in the course of the Earth's history, back and again extensive areas of land had been inundated by the sea. It therefore seemed to make sense to interpret the legacy of the Quaternary, especially the erratic blocks, as the results of a great flood. Did not the Bible report a devastating deluge? In many parts of the Earth there were traces of that flood to be found. Johann Friedrich Wilhelm Jerusalem listed some of them. He wrote:

> *The greatest attention deserve the southward pointed shape of Africa and India, and all the great embayments all around Asia, from the Red Sea up to Kamchatka, all open to the south, which are the surest proof that the Earth once suffered a violent flood from the south, which is also confirmed by the large amount of skeletons of large land animals found in Siberia which are derived from a more southern country.*

> Jerusalem (1774)

When Jerusalem published these lines, belief in the literal meaning of biblical texts had ceased. Jerusalem, adviser to Duke Karl I of Brunswick-Wolfenbüttel, was one of the most important theologians of the German Enlightenment. He was an educated man who had spent years in Holland and England. In his interpretation of the flood, he includes the dead mammoths from Siberia. He was well aware that 'petrified sea animals spread over the whole earth, such as the horns of Ammon' could not be related to the biblical flood, but a flood – a very, very big flood – still seemed possible.

That the latter might have been the biblical deluge was only believed by a few at the beginning of the 19th century. One of them, the Reverend William Buckland of Oxford, introduced the term 'Diluvium' to the stratigraphic nomenclature in 1823.

While his contemporary Cuvier was convinced that the traces of the deluge were limited to the lowlands and valleys of the Earth, Buckland wrote:

> *The blocks of granite, which have been transported from the heights of Mont Blanc to the Jura mountains, could not have been moved from their parent mountain, which is the highest in Europe, had not that mountain been below the level of the water by which they were so transported.*

> Buckland (1823, p. 221)

Cuvier also wrote:

> *In certain countries, we find a number of large blocks of primitive substances scattered over the surface of secondary formations, and separated by deep valleys or even by arms of the sea, from the peaks or ridges from which they must have been derived. We must necessarily conclude, therefore, either that these blocks have been ejected by eruptions, or that the valleys (which must have stopped their course) did not exist at the time of their being transported; or, lastly, that the motions of the waters by which they were transported, exceeded in violence anything we can imagine at the present day.*

> Cuvier (1827, p. 23)

This early attempt at a natural explanation for the occurrence of boulders far from their source rocks corresponds to the rolling stone or mud flood theory, mainly advocated by Leopold von Buch (1815), but also by Alexander von Humboldt (1845) and the Swedish physician and scientist Nils Gabriel Sefström (1836). They assumed that the erratics had been transported by huge masses of water, the so-called 'petridelaunic flood'. The reason why such masses of water would have been released and flooded out of the Alps and the mountains of Scandinavia remained open.

In England, Charles Lyell had argued in his *Principles of Geology* (1830–33) against the geological significance of disasters. Von Hoff was the first German scientist to turn against Cuvier's catastrophism (1834). Neptunists quarrelled with Plutonists, and eventually the concept of a smooth transformation of the Earth seemed to prevail.

A new interpretation of the erratic blocks was found at the beginning of the 19th century. In a shallow, cold sea icebergs might have transported the boulders. The supporters of the drift theory (Box 1.1), including Darwin and the physicist Helmholtz, were not completely opposed to a larger extension of the former glaciers, but rejected large-scale glaciation. Even when Lyell (1840) discussed the origin of erratic boulders in northern Europe, he was strongly opposed to the neo-catastrophism envisaged by Agassiz.

BOX 1.1 DRIFT-ICE TRANSPORT

The sandstone block in Figure 1.1 is 185 × 175 × 135 cm in size and its weight is estimated at 8 tons. It was found on a salt marsh covered with *Spartina alterniflora*. When the stone was pushed landward by drift ice, it left behind a distinct furrow in the ground (foreground right).

Figure 1.1 A block of sandstone on the lower saltmarsh at Isle-Verte, St Lawrence Estuary, Canada. Photograph by Jean-Claude Dionne.

(continued)

BOX 1.1 DRIFT-ICE TRANSPORT *(CONTINUED)*

Goethe had also heard that drift ice should have transported rock material from Sweden across the Øresund to Denmark. Was this the method by which the boulders in northern Germany had arrived in their present position?

There is no doubt that drift ice can move large stones. The coastal waters of northern Canada are covered with ice in winter. In the spring the ice cover breaks up, resulting in an ice drift along the coast. In its course, frozen rock and soil material are moved and redeposited. The Canadian geographer Jean-Claude Dionne has studied this phenomenon in numerous publications.

Figure 1.2 shows a melting ice block which eroded a 25–30 cm thick layer of salt marsh and redeposited it further downshore. Icebergs produce significantly deeper scours. Corresponding plough marks from icebergs of the Weichselian glaciation are found on the seafloor, for instance in the North Sea.

Figure 1.2 Stranded ice floe with a thick layer of frozen-on soil in the St Lawrence Estuary, Canada. Photograph by Jean-Claude Dionne.

However, Agassiz did not capitulate. In 1840 he published his *Études sur les Glaciers,* followed a year later by the German edition *Studien über die Gletscher.* Both books were printed at the author's expense, and met a cool reception. Alexander von Humboldt advised Agassiz to return to his fossil fishes. 'In so doing,' he wrote, 'you will render a greater service to positive geology, then by these general considerations (a little icy besides) on the revolutions of the primitive world, considerations which, as you well know, convince only those who give them birth' (quoted in Imbrie & Imbrie 1986).

Nevertheless, Agassiz eventually had success with his book. He provided evidence that the legacy of the glaciers could be traced from the current ice margin over a series of end moraines to the foothills of the Alps, and that the path of the erratic blocks could be followed from their source areas to the outer edge of the former glaciers. He had no hesitation in making his ideas public, not only in word but also in pictures. The lavishly illustrated 'atlas' conveyed the views of the author more persuasively than his words.

The scientific breakthrough came with his trip to Britain, where he finally managed to convince William Buckland of his theory. This in turn persuaded Charles Lyell, the most important geologist of his time, and in November 1840 together they presented their new insights to the professional world in front of the Geological Society of London. There was still much scepticism, but now the triumph of the Ice Age Theory was unstoppable.

Agassiz demanded a considerable imagination of his readers. He wrote:

At the end of the geological epoch that preceded the elevation of the Alps, the earth was covered with an immense crust of ice, which stretched forth from the polar regions over most of the northern hemisphere. The Scandinavian and British Peninsula (sic), the North Sea and Baltic Sea, northern Germany, Switzerland, the Mediterranean Sea to the Atlas, North America and Asian Russia, were just a single vast ice field, from which only the highest peaks of the then existing mountains … emerged.

Agassiz (1841, p. 284, translated from the German edition)

The discussion also aroused a great deal of interest from the public. Switzerland and its peaks were among the favourite destinations at the beginning of tourism (Figs 1.3–1.5). The first tourists were mostly English climbers who ventured into the Alps. The journey was initially difficult until, in the last decades of the 19th century, the railway made access much easier (Hachtmann 2007). The improved infrastructure in the Alps also made it easier for scientists to investigate the evidence of former glaciations in the field.

In northern Germany, the 'Glazialtheorie' still predominated. Charpentier (1842) had already postulated the existence of a former northwest European ice sheet that reached as far as England, the Netherlands, the Hartz Mountains, Saxony, Poland and 'almost to Moscow'. He fared no better than Bernhardi (1832) before him or Morlot (1844, 1847) after him. Bernhard Cotta (1848) wrote:

It surpasses the boundaries of the thinkable, to accept glaciers, which should have reached from the mountains of Norway to the Elbe River and as far as Moscow, and even to the coasts of England, and moved across this level ground, with its rough surface, laden with moraine … However, we know from observations in both polar regions of the Earth another kind of natural stone transport, which takes place continually, and which should be well suited to explain the northern boulders of Europe and America, as well as the erratic blocks in Patagonia. That is the transport by floating ice.

The drift theory remained firmly in place in northern Germany for a few more decades (e.g. Cotta 1867; Box 1.2).

PANORAMA DER GLETSCHER DES MONTE ROSA,
Oestlicher Theil der Kette.

Figure 1.3 View from Gornergrat to Monte Rosa, Switzerland. Above: as seen by Agassiz (source: Agassiz 1841), below: 1979 (photograph by Jürgen Ehlers). The perspective is slightly different, but the decline of the ice on the opposite slope is clearly visible.

ZERMATT - GLETSCHER

Unteres Ende.

Figure 1.4 The lower end of the Gorner glacier (formerly called the Zermatt glacier). Neither of the two ladies at the foot of the glacier or the gentlemen on the adjacent rocks (left) seem to have work to do here; they are probably tourists. Source: Agassiz (1841).

J. J. 5496 *Glacier de Corbassière et le Grand Combin (4317 m*

Figure 1.5 An excursion onto the Glacier de Courbassière, Valais, Switzerland. The nineteenth century saw a strong growth in tourism in Switzerland. The wild nature and the glaciers were not only major tourist attractions, but had also become more accessible for scientific research. From the point where the postcard was taken, the glacier seems almost unchanged today. In truth, its length has shrunk by 800 m since 1889.

BOX 1.2 FINAL PROOF OF THE DRIFT THEORY?

Towards the end of the 1870s, Heinrich Otto Lang decided to address the issue of the North German 'Glazialtheorie' by investigating the local boulder inventory (Lang 1879). Lang was born on 10 September 1846 in Gera-Unterhaus; he received his doctorate in 1874 and then became a professor of mineralogy and geology at the University of Göttingen.

When it was brought to his attention that large gravel deposits had been found in Wellen, near Bremen, he had Professor Buchenau from Bremen and a Mr von der Hellen send him 180 stones. He asked them to collect not only those rocks that looked most interesting, but also those who presented 'the essential constituents of the deposit'.

Having received these stones, Lang faced a seemingly impossible task. His work was complicated by the fact that he had never been to Scandinavia, and that part of the relevant literature was not accessible to him. However, he was able to inspect samples from various geological collections including the petrographic collections of the Royal University of Göttingen, which held erratic boulders from the Coburgs (Hannover), Loitz (Pomerania), Denmark, Sweden and Iceland. The Icelandic rocks would have been of little use to him, just like the rocks that the first German North Pole expedition had collected.

Because of the high printing costs, the thesis could not be illustrated in colour. Lang had to resort to accurate description and did his best:

A brownish-red granite (Sample 156), the primary constituents of which are almost entirely made up of feldspar and quartz; to some extent the feldspars constitute a red matrix, in which, when seen with the naked eye, gray quartz grains are embedded, which, when polished, even appear black; other dark, less shiny, irregularly defined spots on the polished surface are sparse; in a crack that in one place includes the mouth of a cavity, traces of ferric hydroxide are found in places, or also a pale greenish mica-like mineral, and it is especially this fact, with makes the rock resembling a granite boulder from Zeitz in Thuringia (from Liebe's private collection) …

Lang wondered: 'Could those stones have been brought by glacier to the north of Germany?' His answer was 'no'. As everybody knows, a glacier can only transport the rocks that it erodes in its source area. In this collection from Wellen, however, a wide variety of rock types was present which obviously did not all come from the same area. When considering transport by drifting icebergs, however, such mixing became far more likely.

Lang put great effort into his study and, even when his work was already in print, wrote some last-minute additions. To his surprise, he got the opportunity for a trip to Christiania (Oslo) and southern Scandinavia. Everything Lang found there confirmed his views: there had been no Ice Age. He concluded his study jokingly: 'One cannot save Mr. Torell the allegation that he has been playing with ice'.

All efforts were in vain, however. That same year Albrecht Penck's essay on the 'Boulder formation of North Germany' (1879) appeared, putting an end to any doubts concerning the 'Glazialtheorie' in the north of the German Empire.

On a field trip in conjunction with a meeting of the German Geological Society in Berlin (on 3 November 1875), the Swedish geologist Otto Torell (1875) noted the scratches on the Muschelkalk of Rüdersdorf, which Sefström (1838) before him had clearly identified as glacial striations. This observation signalled that a change of doctrine was long overdue. The glacial theory at that time had already been generally accepted for more than ten years in England and North America (Dana 1863; Lyell 1863) and Torell had also published his views on the Ice Age in northern Europe before (in 1865). At first, very few colleagues believed him.

The following year (1880) Felix Wahnschaffe found glacial striae at several points on the northern edge of the German uplands. He wrote: 'In Velpke, some 5 km southwest of Oebisfelde, the surface of NE – SW trending almost horizontal sandstones which are over-lain by boulder clay or glacial sand revealed in several quarries extraordinary glacial striae.' The glacial striae belonged to two different ice advances. The older set, striking at 27°, is crossed by a younger system trending at about 84°. A large flagstone of Rhät sandstone was recovered and included in the collection of Royal Prussian Geological Survey. It can be seen in the collections of the Federal Institute for Geosciences and Natural Resources in Spandau, Berlin (Fig. 1.6).

Figure 1.6 Slab of Rhät sandstone from Velpke (10 km ESE of Wolfsburg, Germany) with striae pointing in two different directions. The slab is located in the Museum of the BGR in Spandau. Photograph by Klaus Steuerwald.

Figure 1.7 Geikie's map showing the extent of the glaciers of the 'Third Glacial Epoch' (i.e. Weichselian) in Europe. The southern boundary of the glaciated area is nearly identical with the present state of knowledge. Source: Geikie (1894).

In Britain, James Geikie was a leading proponent of glacialism. In the 1894 edition of his book (first published 1874), Geikie had already included maps that showed the extent of three major glaciations in northern Europe (Fig. 1.7); the framework for more detailed mapping of the following decades was set. Geikie was in contact with the leading geologists of his time, and books and reprints were exchanged. Of course, one had to maintain friendly relations with foreign colleagues. Geikie wrote: 'Dear Monsieur Boule, Allow me to thank you cordially for the excellent analysis of my "Great Ice Age" which you have given in [the magazine] "L'Anthropologie", and for your friendly recommendation of the book to your compatriots …' Of course, it could not do any harm to send the good man a copy of the fully revised third edition as well (Figs 1.8, 1.9).

At that time, the origin of humankind was also of great general interest. Charles Darwin's *On the Origin of Species by Means of Natural Selection, or The Preservation of Favoured Races*

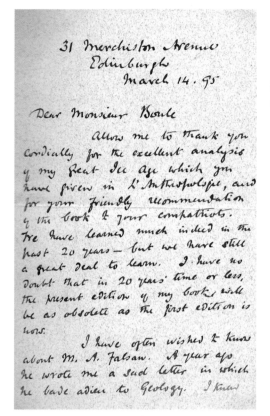

Figure 1.8 Letter in which Geikie thanks Professor Boule. Albert Faslan, mentioned in the letter, was a French natural scientist who had mapped the erratics in the Rhône catchment area. Source: Faslan & Chantre (1877/78).

Figure 1.9 The Wonders of the Primeval World, by Dr W.E.A. Zimmermann. Source: Zimmermann (1885).

in the Struggle for Life, published in 1859, triggered a lively debate among scientists and the public. Parts of his ideas were accepted very quickly (evolution); others, including the selection of species, only decades later. What did man look like in the past? Geikie described the findings, but he drew no picture. Others were less reticent (Fig. 1.10). Dr W.E.A. Zimmermann, for example, presented to his readers 'The miracles of the primeval world' (subtitled 'A popular account of the history of creation and original state of our world as well as the various periods of development of its surface, its vegetation and its inhabitants until the present time'). There were no questions unanswered. Images revealed, for example, 'The Lisbon Earthquake' (smoke, fire, sinking ships) or the 'Erebus volcano in the Southern Ocean' (before the smoking volcano, ice, high waves, sinking ships). A later 'Thirtieth Edition. Supplemented according to the latest state of scientific research' included erratic blocks, but in this 1885 edition these erratics were still accounted for by drift-ice transport.

An excellent illustrated overview of the history of the study of the ice ages is provided in Jamie Woodward's recent publication *The Ice Age: A Very Short Introduction* (2014).

Figure 1.10

Antediluvian man. The
author makes fun of
the artist who dares
to publish 'an image
of our antediluvian
ancestors', but reprints
it all the same. Source:
Zimmermann (1885).

Der vorsintfluthliche Mensch. (Nach Boitard.)

1.2 The Ice Ages of the Earth

It was known internationally by the middle of the nineteenth century that the Ice-Age
glaciation was not an isolated case in Earth's history (Fig. 1.11). When geologists in
northern Germany still believed in drift ice, traces of an older, Permo-Carboniferous ice
age had been identified in the Indian subcontinent in 1856, in Australia in 1859 and
in South Africa in 1868. Later in 1871 scientists were able to detect an even older great
ice age of the Earth, which had taken place during the late Precambrian, the so-called
Vendian (some 600 million years ago). Today an additional period at the end of the
Ordovician has been added (Hirnantian), the ice sheets of which are probably limited
to the Sahara. The first comprehensive overview of the Saharan glaciation was offered by
Deynoux (1980). Moreover, the presence of other even older glacial periods in the Pre-
cambrian, about 950 and 2800–2000 Ma has been proven (Hambrey & Harland 1981;
Harland et al. 1990).

The major glaciations still appear to be exceptions within the Earth's history. The spatial distribution of glacial sediments from these geological eras is by now fairly well known.

The exact location of the poles and the correlation of the scattered occurrences, however, often cannot be established with certainty. One of the few things that is certain is that the ancient glaciations, like their Pleistocene counterparts, had multiple phases.

In the tillite series of Scotland, which date from the latest Precambrian (Port Askeig Formation), numerous layers of rock are found that represent morainic deposits that have been turned into stone (tillite). Glacial deposits from this period have been found in many places all over the globe (e.g. Norway; Figs 1.12, 1.13), leading to the assumption that the Earth at that time might have experienced a long period during which its surface was completely covered by kilometre-thick ice sheets, making all life impossible. The press in particular have embraced this sensational idea. However, today it is known that there never was such a 'Snowball Earth'. Widespread black shales were found in the São Francisco craton in southeastern Brazil, formed during the Neoproterozoic glaciation about 740–700 Ma. The rock contains up to 3 weight percent (wt%) organic carbon, which could only be deposited under the condition that the sea was free of ice (Olcott et al. 2005). Moreover, when the composition of the Port Askaig deposits on Islay (Scotland; Fig. 1.14) is examined, it is found that at least part of the sequence was deposited in open water. Consequently, the glaciers of the Precambrian Varanger ice age were – just like their successors in the later ice ages – limited in extent (Harland 2007).

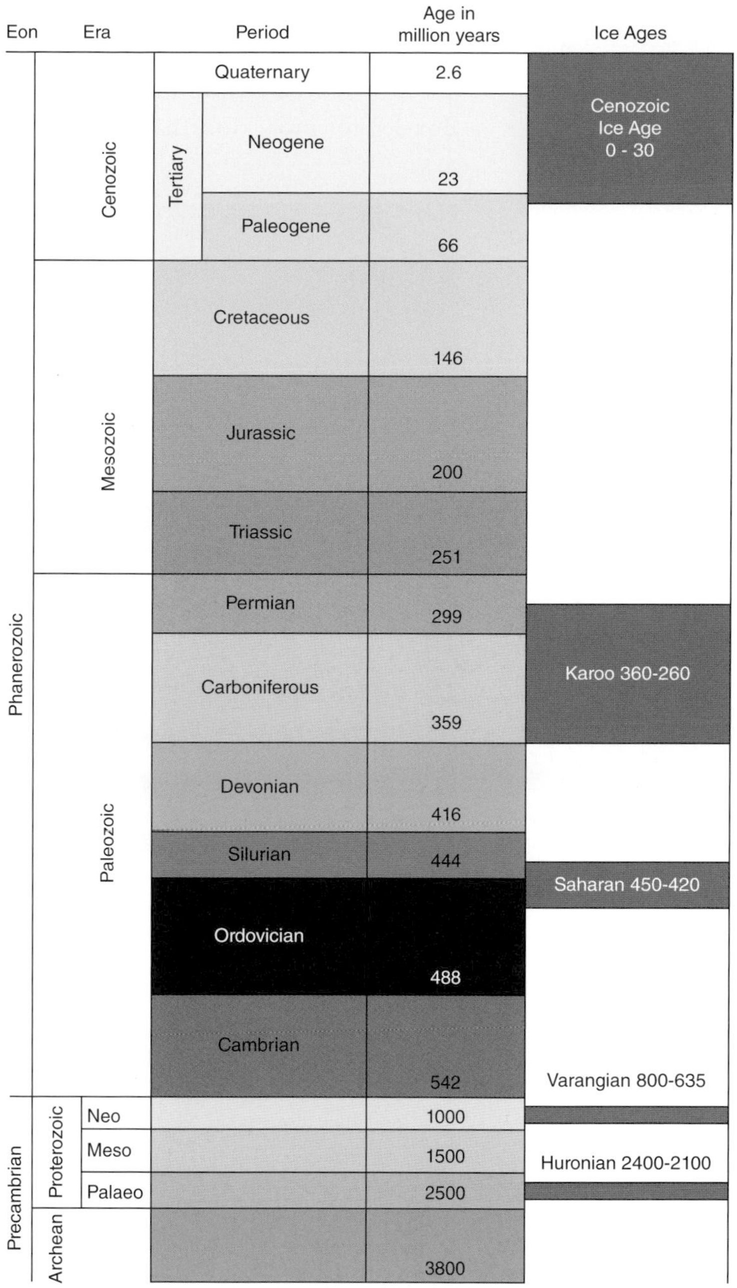

Traces of the Carboniferous glaciation are widespread in the southern continents (the former Gondwanaland), and are particularly well exposed in South Africa. Numerous recent studies have shown that those early glaciations left behind the full inventory of landforms and sediments that we know from the Quaternary glaciations.

Evidence of glaciation at the end of the Ordovician so far has been demonstrated from South Africa and the Sahara. From Europe, glacial deposits from only the latest Precambrian (Neoproterozoic) are known (from Scotland and Norway); corresponding layers are also found in Greenland, Asia, Africa and Australia. The oldest traces of glaciation occur

Eon	Era		Period	Age in million years	Ice Ages
Phanerozoic	Cenozoic	Tertiary	Quaternary	2.6	Cenozoic Ice Age 0 - 30
			Neogene	23	
			Paleogene	66	
	Mesozoic		Cretaceous	146	
			Jurassic	200	
			Triassic	251	
	Paleozoic		Permian	299	Karoo 360-260
			Carboniferous	359	
			Devonian	416	
			Silurian	444	Saharan 450-420
			Ordovician	488	
			Cambrian	542	Varangian 800-635
Precambrian	Proterozoic	Neo		1000	
		Meso		1500	Huronian 2400-2100
		Palaeo		2500	
	Archean			3800	

Figure 1.11 Geological timescale and the occurrence of ice ages in the Earth's history.

in North America (Canadian Shield and Montana), South America (Brazil) and South Africa. However, those old deposits will not be discussed here. The presentation in this book is limited to the most recent Ice Age of the Earth's history: the Quaternary (see Box 1.3 for more information).

Figure 1.12 Neoproterozoic Moelv Tillite at Moelv, Lake Mjøsa, Norway. Above: overview; below: detail. Photographs by Jürgen Ehlers.

Figure 1.13 Neoproterozoic tillite of the Varanger glaciation at Bigganjarga, Karlebotn, Varanger Peninsula, northern Norway. The tillite is part of the Smalfjord Formation, presumably upper Vendian (Varangerian) in age (>640 Ma). Photograph by Juha-Pekka Lunkka.

Figure 1.14 Neoproterozoic Port Askaig Tillite at the Port Askaig ferry terminal, Isle of Islay, Scotland. Photograph by Jürgen Ehlers.

BOX 1.3 QUATERNARY

The Quaternary Period is the youngest period in the Earth's history. It is characterized by cyclic phases of climate change, culminating in some cases in the growth and decay of continental ice sheets separated by warm climate events. In June 2009, the Executive Committee of the International Union of Geological Sciences (IUGS) formally ratified the new definition of the base of the Quaternary System/Period to the Global Stratotype Section and Point (GSSP) of the Gelasian Stage/Age at Monte San Nicola, Sicily, Italy. The base of the Gelasian corresponds to Marine Isotope Stage (MIS) 103, and has an astronomically tuned age of 2.58 Ma.

The Quaternary comprises two series: the Pleistocene and the Holocene. The Pleistocene is the geological series which covers the world's recent period of repeated glaciations. Consequently, the end of the last glaciation marks the end of the Pleistocene. This is easily said, but when exactly did the last glaciation end? Where do we find the geological record that allows us to draw a line at that position?

The Pleistocene–Holocene boundary is not defined in sand or clay or any other type of rock. It is differentiated in the NorthGRIP (NGRIP) core from the Greenland Ice Sheet. The boundary is reflected in an abrupt shift in deuterium excess values, accompanied by more gradual changes in $\delta^{18}O$, dust concentration, a range of chemical species and annual layer thickness. A timescale based on annual layer counting gives an age of 11,700 calendar years b2k (before AD 2000) for the base of the Holocene Series.

1.3 Causes of an Ice Age

We live in an ice age. Even if Europe and North America are recently almost free of substantial ice cover, today's 'warm stage' belongs to one of the colder phases of the Earth's geological history. The polar regions have been free of ice for the majority of the past, and the present temperate latitudes experienced a warmer climate than today.

The climatic fluctuations of the ice age are well known today as a result of investigations of deep-sea sediments, ice cores and sediment cores from inland lakes. The Pleistocene comprises 61 marine oxygen-isotope stages (MIS), representing approximately 30 cold and warm periods. Using palaeomagnetic studies, scientists have been able to date the sequence of cold and warm periods so accurately that the duration of the oscillations is known. During the last 600,000 years (600 ka), climate change was dominated by a glacial–interglacial cycle of approximately 100 ka; before that, shorter cycles of 40 ka prevailed. Today we know that those changes are largely controlled by the interplay of three cyclical variations of the Earth's orbit (Box 1.4; Fig. 1.15), including the eccentricity (100 ka), the angle of inclination of the Earth's axis (c. 41 ka) and the timing of the perihelion (c. 26 ka), which are causing changes in the incoming solar radiation (insolation).

BOX 1.4 EARTH'S ORBIT AROUND THE SUN IS NOT CONSTANT

Eccentricity of the orbit: The orbit of the Earth around the Sun is not a circle but an ellipse, the Sun located at one focus. The Earth's orbital parameters change under the influence of the other planets in our solar system. Sometimes it is almost circular, sometimes more elliptical. The changes occur over a cycle of *c.* 100 ka.

The tilt angle of the Earth's axis: The Earth's axis is currently inclined by 23.4° towards the plane on which the Earth moves around the Sun. The inclination angle varies over a cycle of *c.* 41 ka between 22.1° and 24.5°. A smaller axial inclination leads to cooler summers near the poles, so that the ice formed in the winter may not melt in the summer.

Precession: The position of the Earth's axis in space has not changed much. As the Earth orbits the Sun, its axis always points towards the north in the direction of the North Star. In the long term, however, the axis changes its position. In 12 ka it will point towards Wega (in the constellation Lyra). The precession cycle is *c.* 26 ka. This change means that the Earth reaches its closest point to the Sun in its orbit (the perihelion) in different seasons. The Earth currently reaches perihelion in winter.

The impact of these factors on the overall radiation budget of the Earth is small. A basic requirement of these orbital parameters eventually leading to climate change is that large landmasses are present near the poles; in the Southern Hemisphere, only permanently glaciated Antarctica is in a polar position. When the summers in the Northern Hemisphere are particularly cool (greatest distance to the Sun by the eccentricity, perihelion in summer) and the winters are warmest (minimum tilt of the Earth's axis), the northern continents are covered in snow for a long time. Snow has a greater reflection (albedo) than the ground, which contributes to further cooling.

This finding, which was presented by the Serb mathematician Milankovitch (1941), met initially with much scepticism. The fluctuations in the Earth's orbital parameters occurred throughout the Earth's history, whereas, according to contemporaneous knowledge, there had only been four ice ages. Only much later, when it turned out that the history of climate changes went back far beyond the traditional four ice ages, it became clear that the Milankovitch curve in principle was correct after all.

With the help of sensitivity analysis it has been shown that the astronomical parameters actually act as a pacemaker for the glacial–interglacial cycles. This can cause the climate cycles to be driven by solar radiation, but only at lower CO_2 concentrations. Long-term variations of the atmospheric CO_2 content alone are not sufficient to produce glacial–interglacial cycles. However, if both the orbital and CO_2 variations are considered in the model calculation, both point to the onset of the Ice Age at 2.75 Ma. The Early Pleistocene cycle of 41 ka, the transition to a 100 ka cycle at *c.* 850 ka and the glacial–interglacial cycles of the past 600 ka can all be almost exactly simulated (Berger & Loutre 2004).

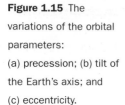

Figure 1.15 The variations of the orbital parameters: (a) precession; (b) tilt of the Earth's axis; and (c) eccentricity.

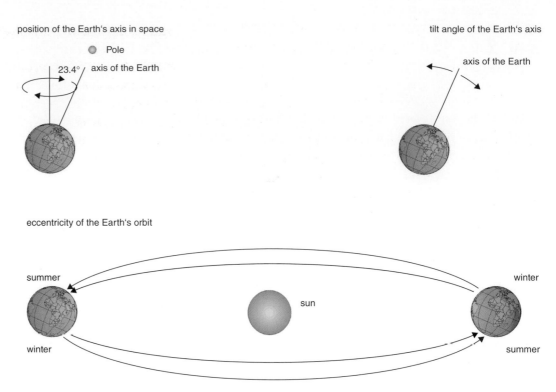

The recent glacial–interglacial cycle of *c.* 100 ka is not only one of the most striking features of the Quaternary climate, but it also determines the future climatic development. Since each of the known climate cycles is characterized by a long cold period followed by a short interglacial (*c.* 10–15 ka), and as our interglacial (the Holocene) has already lasted for 10 ka, it might be suspected that the next glaciation is imminent. However, model calculations have shown that this is not the case. In the next tens of thousands of years, the Earth will have a nearly circular orbit around the Sun. This was also the case, for example, during MIS 11 at *c.* 400 ka before present, but not during the Eemian interglacial (MIS 5e; Berger & Loutre 2002). Accordingly, for the current interglacial a total length of *c.* 30 ka or more can probably be expected. The further increase of the concentration in atmospheric CO_2 by human activities may cause the Greenland Ice Sheet to melt and, within the next 10 ka, disappear completely. Further warming and not cooling is therefore expected.

The climatic cycles are only the 'pacemaker' (Hays et al. 1976), not the causes of the Ice Age. Schwarzbach (1993) suggests large changes in relief (mountain-building phases) as a possible explanation for the initiation of ice ages. Matthias Kuhle (e.g. 1985) was of the opinion that the glaciation of the highlands of Tibet had a significant impact on the global cooling of the Earth. Based on his field studies, he concluded that the snow line in High Asia during the Pleistocene ice ages had been lowered by around 1200–1500 m, so that an ice sheet of *c.* 2.4 million km² formed on the Tibetan Plateau. When High Asia had been uplifted far enough during the Early Pleistocene glacial cycles occurred repeatedly which, in conjunction with the radiation cycles, were sufficient to trigger large-scale global glaciations (Kuhle 1989).

Whilst Kuhle's ideas are controversial (see Sections 2.9.2 and 7.2) it is now undisputed that the relief of the continents has a major influence on global climate. Ruddiman & Kutzbach (1990) highlighted the significant role of the uplift of Tibet and the uplands in western North America on the general circulation of the atmosphere. However, model calculations indicate that this influence on the wind system is not in itself sufficient to trigger ice ages. The reduction of atmospheric CO_2 concentration, which was triggered by the increased chemical weathering of young exposed areas, has also been discussed as a possible cause of the ice ages (Raymo et al. 1988; Saltzman & Maasch 1990).

An important basic condition for the initiation of ice ages appears to be the distribution of large land masses. With the help of modern GIS technology, the former location of the continents and the rough shape of the Earth's surface can be reconstructed quite well. Extensive glaciation can only occur when appropriate land masses are present near the poles. During the Precambrian glaciations, almost all the continents were situated near the South Pole (Blakey 2008). During the Carboniferous glaciations, the southern continent Gondwana was in pole position (Stampfli & Borel 2004). The same applies to the Ordovician glaciation (Stampfli & Borel 2002). However, this also applies to the Devonian, an era for which no traces of glaciation have ever been found (Scotese 2008).

The shifting of the continents in the course of plate tectonics also caused changes in the ocean currents. The closing or opening of important straits has a significant impact on the oceanic circulation. The separation of Australia and South America from Antarctica and the resulting opening of the Tasman and Drake Passage during the Oligocene led to the isolation of Antarctica from warm surface water, forming the basis for the glaciation of that continent. The closure of the Straits of Panama during the early Pliocene stopped the currents running parallel to the equator, resulting in a rapid north–south exchange of water masses in the oceans, favouring glaciation of the northern continents (Smith & Pickering 2003).

Since the Earth has repeatedly undergone ice ages (during the Quaternary, Carboniferous/Permian, Ordovician and several times during the Precambrian), the question arises of whether a common timer can be found for these operations. Since these events seem to have been repeated roughly about every 250 million years, there might be a connection with the rotation of the galaxy. McCrea (1975) assumed that during each rotation the solar system had to pass through dust clouds in the spiral arms of the galaxy, meaning that the total irradiance was reduced; Dennison & Mansfield (1976) disagreed. The question of the common causes of the ice ages currently remains open. An overview of the many factors that may play a role is given in Saltzman's book *Dynamical Paleoclimatology* (2001).

The International Union for Quaternary Science (INQUA) was established in 1928, and exists to encourage and facilitate the research of Quaternary scientists of all disciplines. Five Commissions were established to focus on different fields of research: Coastal & Marine Processes; Humans & the Biosphere; Palaeoclimates; Stratigraphy & Chronology; and Terrestrial Processes, Deposits & History (Box 1.5).

BOX 1.5 INQUA

The International Union for Quaternary Research (INQUA) is the worldwide association of Ice Age research. It was founded on the Geographical Congress 1928 in Copenhagen. The initiative was started by Victor Madsen, then director of Danmarks Geologiske Undersøgelse, taking up a suggestion by Poland. Figure 1.16 shows only the central part of the official photograph of INQUA's founding fathers, which has been preserved in the Natural History Museum in Copenhagen. The names of the participants are listed on the back of the picture; most of the serious-looking men and women are completely forgotten today. The Germans include Paul Woldstedt (front row, third from left) and Rudolf Grahmann (diagonally right behind); on the other edge of the image the Hamburg Professor Gürich can be seen, easily identified by his white goatee. Austrian Gustav Götzinger, who would go on to host the third INQUA Congress in Vienna (1936), is in the centre (behind the man with the cigar). It is however striking that many important Quaternary scientists are missing; not a single North American was present.

Figure 1.16 Foundation of INQUA at the Geographical Congress in Copenhagen in 1928. Reproduced with permission of Kurt Kjær, Natural History Museum, Copenhagen.

INQUA was initially an European organization. Götzinger suggested at the second INQUA Congress in Leningrad to include the Americas and Asia in the future. The extension to a 'global association' was approved at the 16th International Geological Congress in Washington (1933); three years later in Vienna, as well as representatives of the European nations, there were scientists from Japan, the Dutch East Indies, Turkey, Mexico, Argentina and the first five scientists from the United States, a total of 193 participants (Götzinger 1938). The next INQUA conference was planned to be held in England (Cambridge) in 1940, but was sidelined for obvious reasons.

The international conferences have continued since after the Second World War to this day, and take place every four years. They provide scientists with the opportunity to present the latest research and exchange new ideas (Fig. 1.17). The number of participants has increased significantly; the XVIII INQUA Congress in Bern, 2011, saw some 2000 scientists in attendance.

Figure 1.17 Closing ceremony of the XVII INQUA Congress in Cairns in 2007. Photograph by Jürgen Ehlers.

Investigation of the Pleistocene ice age has led to considerable progress in recent decades. This is true not only for the highly technical disciplines of age determination, deepsea exploration and the study of ice cores from Greenland and Antarctica. Even with regard to mapping the limits of the individual glaciations, the changes are enormous. The differences are nicely illustrated in Figure 1.18, which compares Flint's (1971) map of the northern European ice sheets with today's interpretation, based on the INQUA project 'Extent and Chronology of Quaternary Glaciations' (Ehlers et al. 2011a), which have itself been updated for this book.

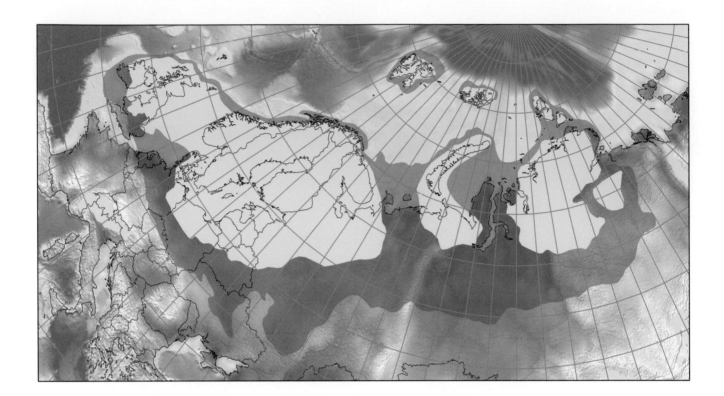

Figure 1.18 Comparison of the glacial limits during the Weichselian glacial maximum (light blue) and the maximum Pleistocene glaciation (dark blue) in northern Europe according to Flint (1971, above) and according to recent interpretations (2011, below). Map from Flint (1971) reproduced with permission of John Wiley & Sons.

Jan Mangerud explains the Eemian Fjøsanger site near Bergen (Norway). Photograph by Jürgen Ehlers.

Chapter 2
The Course of the Ice Age

I n the mid-1970s organic deposits were found at Fjøsanger near Bergen, which were obviously interglacial strata. This caused a sensation, because at that time it was assumed that all such deposits had been eroded by the glaciers of the last ice age. The Quaternary Department of the University of Bergen decided to study the deposits in detail. In the years 1975–76 a 15-m-deep pit was dug down to the bedrock surface, to 1 m below present sea level. The rock was found to be striated and overlain by till, probably of the Saalian Glaciation. The till unit was overlain by sandy beds containing mollusc shells, which were part of a cold-water fauna. These beds were in turn overlain by other marine deposits, and the fauna in these layers changed upwards to include more warmth-loving species until finally a bed yielded fauna indicating that the sea was at least as warm as today. Since this unit was overlain by two more tills, it was concluded that the deposits represented the Eemian (Last) Stage interglacial. This discovery, which Jan Mangerud and his colleagues reported at the 1977 INQUA conference in Birmingham (Mangerud et al. 1977), was later confirmed by further detailed investigations (Mangerud et al. 1981). Even in a landscape so obviously shaped by glacial erosion as the Norwegian coast, in a protected position older deposits were shown to have survived overriding by the Weichselian Glaciation.

2.1 When did the Quaternary Period Begin?

The beginning of the Ice Age is not marked by an abrupt change in climate, but by a gradual, progressive, punctuated transition to cooler climates. Glaciers existed in parts of the world as early as Palaeogene time, while in most other areas glaciation began much later. Consequently, the definition of the boundary between the Tertiary (Palaeogene–Neogene)

The Ice Age, First Edition. Jürgen Ehlers, Philip D. Hughes and Philip L. Gibbard.
© 2016 John Wiley & Sons, Ltd. Published 2016 by John Wiley & Sons, Ltd.
Companion website: www.wiley.com/go/ehlers/iceage

and the Quaternary is more or less arbitrary, with a range of different criteria being used for its delineation.

At about 2.6 Ma, a significant change is found in the sediments of the Lower Rhine area of Germany. At that time, the Rhine catchment area in the south had extended to include the foothills of the Alps, which is seen as a drastic change in the heavy mineral composition of the Rhine sediments (Boenigk 1981). It was also at this time that the thermophilous forest vegetation that had typified the Pliocene was gradually replaced by the cold-tolerant Quaternary plant communities. The gravel composition of the Rhine also changed, and the mollusc associations adapted to the cooler and more variable climate. If these changes did not all occur at precisely the same time, it was at least during this transitional period that a significant floral and faunal adjustment occurred, accompanied by a change in sediment composition. These changes mark the Tertiary–Quaternary boundary in the Lower Rhine area, placed at the Reuverian–Praetiglian stage boundary.

Despite this, the 1948 International Geological Congress decided that the Neogene–Quaternary (Pliocene–Pleistocene) boundary should be placed at the base of the Calabrian Stage (Italy) at 1.64 Ma (Aguirre & Pasini 1985) following the convention that such boundaries should be located in marine sediments. The boundary age was subsequently revised to 1.806 Ma based on astronomical calibration (Lourens et al. 2005). Cold-water indicators are found in Mediterranean sediments for the first time in the Calabrian (among other things, the foraminifer *Hyalinea baltica* was recognized). The international Tertiary–Quaternary boundary was therefore fixed at the upper limit of the Olduvai Subchron, a normal phase within the reversely magnetized Matuyama Chron. Consequently, the boundary could be identified worldwide even in sediments lacking fossils. The position of this boundary was again confirmed at the 1982 INQUA Congress in Moscow.

However, many Quaternary scientists were not satisfied. For example, Shackleton (1987) argued that while the upper limit of the Olduvai event approximates to the onset of the 41 ka climate cycles, it does not represent a significant climatic event. Instead, the base of the Gelasian Stage at *c.* 2.6 Ma best approximates the onset of the intense glacial cycles that characterize the Quaternary. The debate regarding the position of the Pliocene–Pleistocene boundary was summarized by Partridge (1997) who wrote that it is:

> *…questioned by many, including the present writer, whether it is justified to maintain a boundary whose position is rooted in priority and historical usage rather than the current evidence for the considerable earlier onset of a climatic regime which gives an essential unity to the Pleistocene and is clearly mirrored in the global stratigraphic record.*

The issue of the definition of the Quaternary (and by implication, the time-equivalent Pliocene–Pleistocene boundary) was finally settled in June 2009 when the Executive Committee of the International Union of Geological Sciences (IUGS) ratified the definition of the base of the Quaternary being lowered to 2.588 Ma (Gibbard et al. 2010). This followed a tumultuous few years when even the continued acceptance of the term Quaternary as a formal stratigraphical division was threatened (Box 2.1; Bowen & Gibbard 2007).

BOX 2.1 TEARS IN THE EYE

The breeze is rustling the horsetails,
Suspiciously glimmers the sea,
Whilst with tears in his eyes,
An ichthyosaurus swims by.
He laments the corruption of times,
Since a very deplorable tone
Has recently been adopted
In the Lias Formation.

Joseph Victor von Scheffel wrote his poem about the extinct 'fish lizard' in 1866. The poor Ichthyosaurus (Fig. 2.1) today would have even more reason to complain about the 'corruption of times'; his Lias has ceased to be a 'formation' because it did not meet the requirements of modern stratigraphy (Menning & Hendrich 2005).

Figure 2.1 Dinosaurs at close range in *Hagenbecks Tierpark* Zoo in Hamburg. The saurians died out at the end of the Cretaceous. Photograph by Jürgen Ehlers.

(continued)

BOX 2.1 TEARS IN THE EYE *(CONTINUED)*

Even the Quaternary Period, the Ice Age, nearly lost its status as a System in the stratigraphic table by a hair's breadth.

The term Quaternary was coined at a time when nobody knew that there had ever been an Ice Age. It was introduced by Giovanni Arduino in 1760 together with the term Tertiary. Based on his observations of geological strata in northern Italy, he distinguished four orders or large-scale divisions: Primary (basalt, granite, slate); Secondary (fossil calcareous deposits); Tertiary (younger sedimentary deposits); and Fourth Order or Quaternary (recent alluvial deposits). In 1829 Jules Desnoyers applied the terms to his subdivision of the sedimentary sequence in the Paris Basin. He realized that the Quaternary strata were significantly younger than the Tertiary deposits. Although these young layers were very thick in parts of the basin, they were of little geological age, resulting in a very unequal division of the Cenozoic Era (Cenozoic). The extent of how lopsided this subdivision was only became evident when radiometric dating methods became available in the second half of the twentieth century. The 63.7 Ma span of the Tertiary (Palaeogene–Neogene) were followed by only 2.6 Ma of the Quaternary.

Arduino's terminology has long been regarded as no longer appropriate. The terms Primary and Secondary had disappeared from the stratigraphic table in the nineteenth century, and the term Tertiary was excluded from the internationally recognized geological timescale in 2000 (although it remains frequently used). Instead, the Cenozoic was subdivided into the Palaeogene (formerly the lower Tertiary) and Neogene (formerly upper Tertiary).

Was the term 'Quaternary' to become redundant as well? The definition and climatostratigraphical subdivision of this latest unit of the Earth's history fit rather poorly to the other, biostratigraphically defined units. The attempt to fix the beginning of the Quaternary biostratigraphically to 1.805 million years did not match the climatostratigraphical evidence. As a result, the Quaternary disappeared from the International Stratigraphic Table in 2004, at least in the book *Geologic Time Scale 2004*, an official publication of the International Commission on Stratigraphy (ICS) who are responsible for the decision of stratigraphic questions (Gradstein et al. 2005).

This coup triggered strong protests from both INQUA and the national Quaternary Associations. The President of the International Union of Geological Sciences (IUGS), Professor Zhang Hongren, stated that the new *Geological Time Scale* had not been ratified by the IUGS Executive and was therefore not binding. The geological timescale presented in that publication only reflected the personal preferences of some ICS members. The Board of the IUGS stated further that the ICS had acted against the interests and requirements of the IUGS, and harmed the reputation of the ICS and the IUGS. The result was that on 21 May 2009 the ICS not only decided to retain the Quaternary as a formal time division of Period/System status, but also to extend it by moving the base of the Quaternary to 2.6 million years ago (Fig. 2.2). On 29 June 2009 this decision was also ratified by the IUGS Executive.

So everything remains as it was? No. The stratigraphic nomenclature has changed and will change even more in the future. The national stratigraphic committees are redefining and improving their divisions, with the goal of improving the correlation across countries. To be formally recognized, a stratigraphic unit (whether a litho-, bio- or chronostratigraphical division) must meet certain requirements. Its lower boundaries must be established and there must be a type locality, where the stratigraphic position of the formation is illustrated. This approach is being increasingly adopted in many countries and will be further expanded as the need arises.

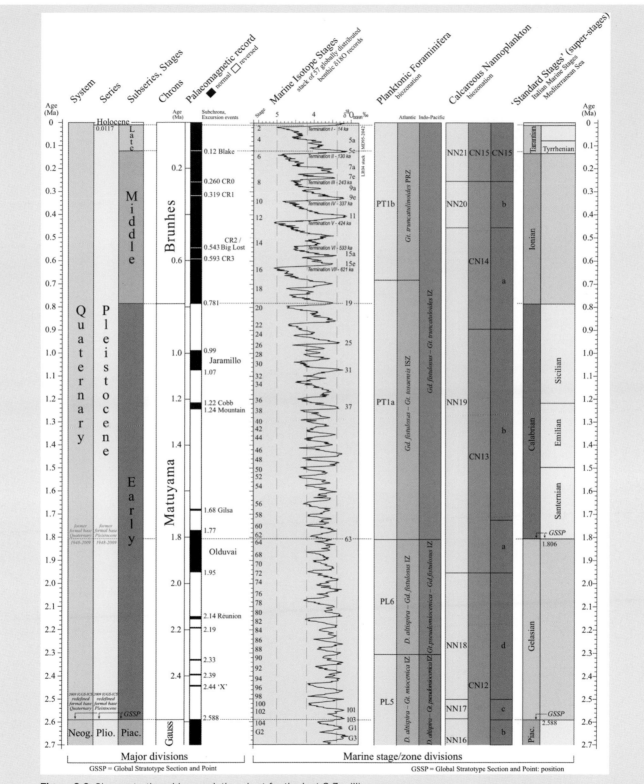

Figure 2.2 Chronostratigraphic correlation chart for the last 2.7 million years.

Source: Cohen & Gibbard (2011), http://quaternary.stratigraphy.org/charts/.

(continued)

BOX 2.1 TEARS IN THE EYE *(CONTINUED)*

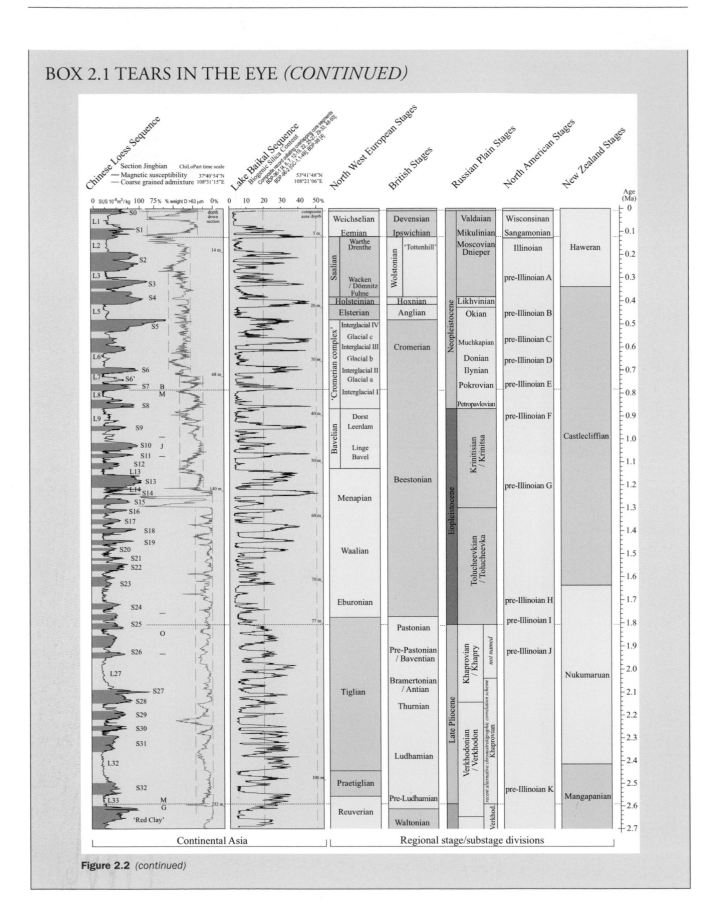

Figure 2.2 *(continued)*

2.2 What's in Stratigraphy?

In bedded rocks the oldest layers are found at the bottom because they were deposited first, and are overlain by progessively younger strata. This rule, known as the 'stratigraphic principle' was formulated for the first time by the Danish scientist Nicolas Steno (Niels Stensen) in 1669. Binding international guidelines on the application of this rule have existed for almost 40 years (Hedberg 1976). However, because many of the stratigraphic divisions were defined before these rules were adopted, the names of many units are less than ideal and sometimes ill-defined. They need to be redefined or replaced by more precise terms.

Stratigraphy includes the classification of rock strata by age. This can be achieved based on the physical composition of the rocks (lithostratigraphy), their fossil content (biostratigraphy), the climate control (climate stratigraphy) or their morphological characteristics (morphostratigraphy), but also on an accurate determination of the rock age (chronostratigraphy; Box 2.2). Today's most commonly used classification system is lithostratigraphy.

BOX 2.2 PALAEOMAGNETISM

An important method of determining the age of rocks is palaeomagnetic polarity. The Earth's magnetic field represents a dipole, which is inclined at about $10°$ to the axis of the Earth. The horizontal component (declination) is the deviation from the north–south direction. The vertical component (inclination) is the angle at which the local magnetic field dips. Direction and strength (intensity) of the magnetic field are dependent on geographic latitude. Near the poles the horizontal component of the magnetic field approaches zero, while the vertical component reaches its highest values (Hambach et al. 2008).

Processes in the liquid outer core control the magnetic field of the Earth. This results in periodic displacements of the magnetic field for periods of days to tens of millions of years. The most dramatic changes are reversals of the magnetic poles, when the North Pole becomes the South Pole and vice versa. Such 'sudden' reversals take place over a period of thousands to tens of thousands of years. The normal or reverse state may then last for hundreds of thousands or millions of years (Table 2.1).

TABLE 2.1 CHANGES OF THE EARTH'S MAGNETIC FIELD.

Type of change	Description	Timescale (a)
Reversal	North Pole to the South Pole reversal (and vice versa)	10^3–10^4
Secular variation	Changes the direction of the geomagnetic field by 10–30°; strength differs by up to 50% of present value	10^3
Excursion	Short-term change in direction of the Earth's magnetic field by more than 30°; strength can decline by up to 10% of today's value	$<10^3$

Source: Hambach et al. (2008).

(continued)

BOX 2.2 PALAEOMAGNETISM (CONTINUED)

How do we measure these changes? Magnetic minerals tend to align themselves corresponding to the current magnetic field of the Earth. In igneous rocks this alignment is maintained permanently, when the rock solidifies. For fast-flowing lava this process takes a few hours to years. Magnetic minerals are also arranged in sediments according to the orientation of the geomagnetic field during their deposition. This type of orientation is weaker and less stable than in the igneous rocks. It can by rearranged, for example, by the action of living organisms (bioturbation), but also by geochemical processes and the formation of new minerals. Moreover, it takes much longer for the mineral grains to reach a stable position. Even in rapidly deposited lake sediments, the magnetic minerals are only oriented after about 150 years (Stockhausen 1998).

The palaeomagnetic timescale is based on studies of the seafloor. At mid-ocean ridges new basalt is constantly being formed. As a result, the bottom of the oceans expands slowly (seafloor spreading). As the Earth does not increase in size, the seafloor eventually has to disappear again somewhere. This happens in the subduction zones in the deep-sea trenches.

The history of the Earth's magnetic field is not only read from the basalts at the bottom of the oceans, but the overlying sediments provide an age control. The basalts can be dated with the K–Ar dating method. It is possible to determine the rate of seafloor spreading, but also to establish a palaeomagnetic timescale reaching back to the Early Jurassic (Nicolas 1995). The geomagnetic timescale of the Quaternary Period consists of two major parts: the present epoch with 'normal' polarity (Brunhes Chron) and the previous period with reverse polarity (Matuyama Chron). The shift occurred about 780 ka ago. In contrast to the Brunhes Chron, the Matuyama Chron contains two major sections with a different (in this case 'normal' polarity): the Jaramillo Subchron and the Olduvai Subchron.

Palaeomagnetic field investigations are also applied to the Asian loess profiles. It could be shown that the earliest loess deposits date back to the late Pliocene. In the future, improved dating methods are likely to allow the use of short-term excursions in the magnetic field for dating purposes.

In lithostratigraphy, the rocks are classified on the basis of the observed lithological properties of the beds and their relative stratigraphic positions. Observable lithological characteristics and reproducible stratigraphic position are the only criteria that can be used in defining lithostratigraphic units. Mappability is another important consideration for the usefulness of such a unit. Taking into account these considerations, the Geological Survey of the Netherlands revised the old lithostratigraphic classification of the Quaternary in the Netherlands. The revision was necessary because (1) the old system from the early 1970s was based heavily on a mixture of bio- and chronostratigraphic assumptions that were partly untestable; and (2) the emphasis of the mapping has moved from purely two-dimensional recording of geological maps towards 2.5D and now to 3D models of the subsurface. The result was a comprehensive reorganization of the Dutch stratigraphic table (Weerts et al. 2005; Westerhoff 2009).

The fundamental unit of lithostratigraphy is the formation. This may be subdivided into subunits (members), further subdivided into beds and, in some cases, smaller-scale units.

2.3 Traces in the Deep Sea

The classical subdivision of the Ice Age was established in the Alps. There, Penck (1882) was not only able to demonstrate that multiple glaciations had occurred, but the classical system of four Alpine glaciations was established (Günz, Mindel, Riss and Würm) (Penck 1899; Penck & Brückner 1901/1909). This was adopted as a first worldwide subdivision of the Ice Age. It was not until the 1970s that this subdivision was considered incomplete. Today, the basic outline of the Quaternary stratigraphy is based on the oxygen-isotope stratigraphy of deep-sea cores, to which all other stratigraphies are adjusted.

The bottom of the oceans can be roughly divided into three main units: shallow-shelf seas; steep continental slopes; and deep-sea floors. The ocean floors are among the deepest areas of the Earth's surface; their depth is only surpassed by that of the deep-sea trenches. The bottoms of the deep-sea areas therefore represent areas of almost continuous sedimentation from which the climate history of the Earth can be reconstructed.

Ideally, the sediments of the ocean floors are the result of slow vertical deposition on an almost flat surface. However, the seafloor sometimes has a strong relief. On steeper slopes landslides occur, mass flow rearranges the sediments and large sediment sheets are accumulated by massive turbidity currents. The stratified sequence of deep-sea sediments is therefore not uniform, and there are likely to be discontinuities. If a relatively complete sequence of layers in a borehole is required, it should preferably be drilled in the range of a plateau or on gently sloping hillsides, where the danger of relocation is relatively low. The boreholes V28-238 and V28-239 of the research vessel *Vema*, upon which the basic deep-sea oxygen-isotope stratigraphy was based, are from the Solomon Plateau at water depths of 3120 m and 3490 m (Shackleton & Opdyke 1973). Hiatuses in the sediment core may be difficult to discover; the sequence of layers encountered is therefore correlated with other cores, and radiometric dating and palaeomagnetic studies in particular provide a basic framework of fixed points to which the layer sequence can be adjusted.

The sediments of the deep sea are a mixture of fine sediments transported from the land, but they also consist of the shells of marine microorganisms. The continental (terrigenous) component consists essentially of clay and partly also silt and fine sand (see Table 2.2) that

TABLE 2.2 GRAIN SIZES ACCORDING TO EUROPEAN STANDARD.

Sediment type	Grain size diameter (mm)
Gravel	2–63
Sand	0.06–2
Silt	0.002–0.06
Clay	<0.002

Source: EN ISO 14688.

the wind has transported into the sea. The latter play a greater role off the West African coast (dust from the Sahara) for example, but also in the Bay of Biscay where they provide an important connection to the northwest European Quaternary sediment stratigraphy of the mainland. Occasionally layers of fine-grained volcanic ash (tephra) are found in the deep-sea sediments. In some areas, such as the North Atlantic, there are also coarser clastic intercalations which are the result of drift-ice transport during cold periods. The marine organic components consist primarily of the housings and hard parts of single-celled animals (mainly foraminifera and radiolaria) or plants (mainly diatoms and coccolithophorids). In the mid-Atlantic, the Pleistocene sediments consist largely of *Globigerina* mud (McIntyre et al. 1972; Dietrich 1992).

Where grain sizes are discussed (Table 2.2), the terminology conforms to the European standard EN ISO 14688. In comparisons with international literature, it should be noted that different classifications may be used in the USA and in Russia, for example.

The proportion of the various components of the deposits of the ocean floor varies greatly, especially depending on the entry of terrigenous components. Since two-thirds of the continental drainage of the Earth is carried into the Atlantic, the sedimentation of terrigenous components there is much higher than in the Pacific. A core through the Quaternary sediments from the Pacific will therefore represent a longer period of time than the same core length in the Atlantic. The sedimentation rates range from less than 1 cm per thousand years in some areas up to about 50 cm per thousand years in small ocean basins. Terrestrial sedimentation in the cold periods was 3–4 times higher than under interglacial conditions (Box 2.3). The reasons for this are increased physical weathering and glacial erosion and

BOX 2.3 GLACIOMARINE SEDIMENTS AND IRD

Occasionally coarse-grained layers, with particles including scattered gravels are found in deep-sea sediments. This is mostly material that has been brought there by iceberg transport. In addition, a minor source of entry is by marine algae (seaweed, e.g. Fig. 2.3), driftwood (mainly in the roots of tree trunks) and held by mammals. Human activities have caused the introduction of recent coarse sediments in the North Atlantic (primarily slag and ash from the days of steamships). While the anthropogenic contaminants can be easily identified, the safe determination of material transported by floating ice (ice-rafted detritus or IRD) is much more difficult. Its identification plays a significant role when it comes to the reconstruction of earlier drift-ice limits. The zone of transportation by algae is largely restricted to warm waters, but it overlaps with the zone of subpolar drift-ice transport.

Transport by floating ice is not only noticed within the pack-ice zone. During the ice ages, the ocean currents transported icebergs far to the south and north. In the North Atlantic drift-ice transport reached south to the latitude of Morocco, and in the South Atlantic from the Antarctic waters north to around Cape Town.

Glaciomarine sediments form by the melting and sediment rain-out from drift ice, by the deposition of fine-grained sediments that have been in suspension, and by turbidity currents. In addition, the sediments may be reworked by iceberg ploughing or bioturbation. These deposits have a greater preservation potential than terrestrial glacial sediments, because the depositional environments of the deep ocean basins lie beyond the limits of glaciations; they are therefore protected from erosion during subsequent glaciations. The seafloor sediment

Figure 2.3 It is not only ice, but also seaweed which is able to transport larger stones from the beach area to the open sea. Photograph by Jürgen Ehlers.

archive is created in this way, which has recorded the past climatic changes in the form of accumulating and melting continental ice sheets and the related changes in the ocean. As a result of glacioisostatic uplift, Quaternary glaciomarine sediments are also often found in areas that are now part of the mainland.

Subaquatic outwash fans have accumulated where the glacier was in contact with the ground. The velocity of meltwater that drained from tunnels under the ice decreased abruptly when entering the sea, so that the sediment was deposited directly in front of the glacier margin.

Where the ice edge remained stable for a long period of time, true morainic ridges accumulated. The extent of these accumulations depends on time and also on sediment supply. In the glaciomarine environment in the Canadian Arctic, morainic deposits have been identified which are less than 1 m thick, while others comprise more than 10-m-thick mounds of clay, silt, sand and mixed-grained sediments.

Where the sedimentation rate is high, alluvial fans can form and morainic ridges may grow to sea level and form so-called ice-contact deltas. Such deltas are often isolated sediment bodies with a flat surface attached to rocky valley slopes (marginal terraces). They are characterized by a steeper proximal ice contact slope and a surface that is dotted with ice holes. In the interior of these deposits, topsets, foresets and bottomsets are found. The topsets are the deposits of an anastomosing river system, and consist of massive gravels which are accordingly subhorizontally stratified. The foresets are dominated by sandy deposits, which have been accumulated under the influence of gravity. Further downslope they grade into bottomsets of massive or graded sand and silt. Beyond the actual delta, the distal bottomsets are composed of finer silt and clay, often containing marine macrofauna. On the side where the glacier was located, the layers may be disturbed by ice pressure and the melting out of buried dead ice or landslides in connection with the glacial retreat. Such ice-contact deltas are widespread in Norway, for example.

increased runoff from meltwater streams. The water bound in glaciers also resulted in a lowering of the sea level and the exposure of more shelf areas, therefore increasing deflation. The stratification of the deep-sea sediments therefore reflects the variations of global climate and can be used to reconstruct the climatic history of the Quaternary. In this context, the oxygen-isotope ratio has been found to be the most appropriate method because it provides reproducible results worldwide.

Oxygen isotopes in sea water are present in the two variants ^{16}O and ^{18}O. In evaporation, the lighter isotope ^{16}O is preferred. Under constant climatic conditions, this fact is irrelevant because the ^{16}O evaporated by precipitation is recycled back into the sea. During cold periods, however, a considerable part of the precipitation does not get back into the sea, but is tied up in glaciers and ice sheets of the mainland. The consequence is that the ^{16}O content of the seawater is reduced. Marine organisms build the two oxygen isotopes in their calcareous shells in the proportion which they find in the sea water. In this way it is possible from the corresponding deposits to reconstruct the composition of sea water, and therefore the approximate climate.

The oxygen-isotope ratio is not only dependent on ice volume, but also on temperature. In the calcite shells ('tests') of foraminifera at lower temperatures, a higher percentage of ^{18}O is incorporated than at higher temperatures. The oceans have a significant temperature stratification. While the near-surface sea water (up to 300 m depth) experiences the thermal fluctuations of the atmosphere (with some delay), the temperature of oceanic deep water (at about 1000 m depth) is often assumed to have changed little throughout the Pleistocene. It is controlled by the temperature of the Antarctic deep water. The calcareous shells of foraminifera which live on the ocean floor (benthic forms) therefore show a distribution of oxygen isotopes which is independent of temperature fluctuations, and only reflects the changing ice volume. The oxygen-isotope curves based on benthic foraminifera are therefore globally reproducible with a high degree of accuracy. This is robust for the 100 ka glacial cycles, but there is evidence that the relationship breaks down at short timescales (*c.* 5 ka), which is explored further below. For now, let us accept the premise that oxygen isotopes are a proxy for global ice volume.

The oxygen-isotope ratio is usually given as the deviation of the percentage of the heavier isotope ^{18}O from the $^{18}O/^{16}O$ ratio specified by a standard determined on a belemnite from the Peedee formation in South Carolina, USA. This value of $\delta^{18}O$ is calculated using the formula:

$$\delta^{18}O = \frac{1000 \times \left[\left(^{18}O/^{16}O \right)_{sample} - \left(^{18}O/^{16}O \right)_{standard} \right]}{\left(^{18}O/^{16}O \right)_{standard}}$$

The dating of the first reference curve of the oxygen-isotope ratio was carried out with the help of five control points. Four of these, within the last 35 ka, were dated using the ^{14}C method. A fifth checkpoint was the Brunhes–Matuyama boundary dated using the K–Ar method (780 ka years before present; Imbrie et al. 1984). The oxygen-isotope chronology and palaeomagnetic timescale are shown in Figure 2.4.

Note that compaction plays a very minor role in Quaternary marine sediments. Studies on a series of cores have shown that density increases with increasing depth from *c.* 1.5 g cm^{-3} to 1.8 g cm^{-3} over several hundred metres. This increase is not continuous, but represents repeated, small-scale changes of density. The changes are less a result of settling rather than

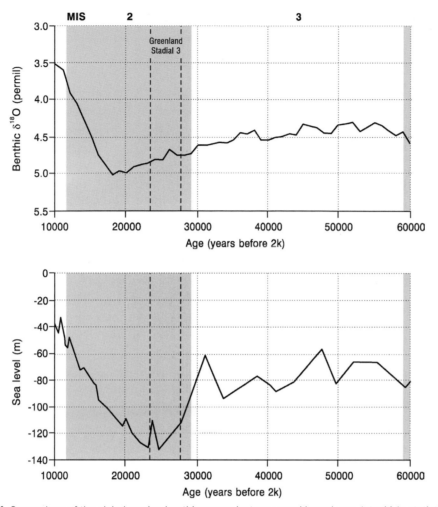

Figure 2.4 Comparison of the global marine benthic oxygen-isotope record based on a 'stack' (a statistical synthesis) of 57 records from around the world (Data from Lisiecki & Raymo 2005 and Thompson & Goldstein 2006, diagram from Hughes & Gibbard 2014). The cold interval represented by Greenland Stadial 3 is indicated by the dashed lines, and Hughes & Gibbard (2014) argued that this interval best represents the global Last Glacial Maximum (c. 25,000 years ago) when global ice volume was at its greatest and global sea levels were at their lowest.

diagenetic processes. They are in total so low that they can be ignored in the processing of the relatively short Quaternary cores for oxygen-isotope studies. However, bioturbation does blur the signal and this limits the utility of the technique for higher-resolution stratigraphy.

Shackleton & Opdyke (1973) therefore applied a linear interpolation between the present and the Brunhes–Matuyama boundary. In core V28-238 this period includes a core length of 12 m, in core V28-239 approximately 7.25 m. For Stage 5e an age of 123 ka BP only for 14C ages results, a value that coincides with results obtained by U–Th dating from the continental area.

Deeper cores now provide an opportunity to provide a framework for the entire Quaternary stratigraphy. Data for the magnetic reversals come from the continental area. In addition to these hard data, calibration of the deep-sea isotope record with the fluctuations in the

Earth's orbital elements (orbital forcing) has been improved repeatedly, such that the sequence and temporal fixation of the events must be regarded as relatively safe (e.g. Shackleton et al. 1990). The time frame of the recent Ice Age has therefore become very much established.

While oxygen-isotope stratigraphy has dominated thinking in Quaternary Science in recent decades, it is not without its problems; there are issues which limit the reliance of the marine oxygen-isotope record as a global time reference. While for glacial cycles (c. 100 ka) the marine oxygen-isotope record is very closely related to global ice volume (e.g. Shackleton 2000; Waelbroeck et al. 2002), it does not offer sufficient resolution to differentiate environmental events at millennial timescales shorter than c. 5 ka. This is partly a consequence of the slow sedimentation rates in the deep oceans and especially bioturbation, which 'is a virtually universal source of degradation for deep-sea records' (Shackleton 1987, p. 183; McCave et al. 1995). In addition, there is the problem of deep-water temperatures which affect fractionation of oxygen isotopes in foraminiferal shells. Shackleton (2000) emphasized that a substantial part of the 100 ka glacial cycle recorded by $\delta^{18}O$ in marine foraminiferal records is a deep-water temperature signal and not an ice volume signal. Skinner & Shackleton (2005) showed that fluctuations in benthic $\delta^{18}O$ and MIS boundaries from different hydrological settings may be significantly diachronous. The use of benthic $\delta^{18}O$ as a proxy for global ice volume as established by Shackleton (1967) begins to 'break down at millennial timescales and in particular across glacial–interglacial transitions'. For relatively short intervals such as the Last Glacial Maximum the marine isotope record is therefore inappropriate for defining its span. Hughes & Gibbard (2014) highlighted this issue and noted that there is an offset of c. 5000 years between global ice volume and the oxygen-isotope records (Fig. 2.4).

The term Last Glacial Maximum (abbreviated to LGM; Box 2.4) refers to the maximum in global ice volume during the last glacial cycle. The original definition of the LGM was described by CLIMAP project members (1976, 1981) and spanned the interval 23–14 ^{14}C ka BP with a

BOX 2.4 LAST GLACIAL MAXIMUM

The term 'Last Glacial Maximum' (LGM) is widely accepted as referring to the maximum global ice volume during the last glacial cycle. In this context, Clark et al. (2009) have argued that most ice sheets around the world reached their maximum between c. 26.5 and 19 ka, preceded by a maximum extent of smaller mountain glaciers between c. 33 and 26.5 ka. This conclusion was reached based on analysis of an extensive dataset covering the period 50–10 ka. However, by limiting their study to the period 50–10 ka, Clark et al. (2009) did not consider the possibility that in many parts of the world the maximum extent of glaciers was reached earlier in the glacial cycle (Ehlers et al. 2011a). The detail has been summarized by Hughes et al. (2014).

However, the LGM has been taken to mean the maximum extent of glacial ice globally, and has been used not only by glacial geologists across the world but by many others who have taken it to be a time-marker for subdivision of the Late Weichselian and its equivalents. As such, it is also used in this book.

In this respect it is a useful term, but it should be properly defined. The question of definition has been considered by Hughes & Gibbard (2014). Such a formalization requires acceptance and ratification by the International Commission on Stratigraphy. At the time of writing, the Subcommission on Quaternary Stratigraphy is in the process of establishing a working group to consider the whole problem of these fine-scale time divisions (Ehlers et al. 2011b).

midpoint at 18 ^{14}C ka BP (Shackleton 1977). Mix et al. (2001) considered that the event should be centred on the radiocarbon calibrated date of 21 cal. ka BP, and should span the period 23–19 or 24–18 cal. ka BP depending on the dating applied (e.g. MARGO project members, 2009). However, other research on global sea-level minima places the global ice maximum slightly earlier at 26–21 ka (Peltier & Fairbanks 2006). The recent paper by Hughes & Gibbard (2014) combined the evidence from marine cores with those from ice cores on land to define the LGM. Unlike marine sediment records which accumulate slowly and can be affected by bioturbation, ice core records have a much higher resolution and can tell us about climate changes on a year-by-year basis. Hughes & Gibbard (2014) identified the period between 27,540 and 23,340 years ago recorded in the Greenland ice core record as the best candidate interval for its formal stratigraphical definition. This interval was characterized by the largest global ice volume, the lowest global sea levels and the greatest atmospheric dust flux in both hemispheres and some of the coldest temperatures of the last ice age (Fig. 2.5). The end of the LGM event is marked by evidence for a large discharge of icebergs into the North Atlantic Ocean (Heinrich Event 2) which coincided with the onset of the collapse of the Laurentide Ice Sheet at *c.* 24 ka, along with other ice sheets in the North Atlantic region (Fig. 2.6).

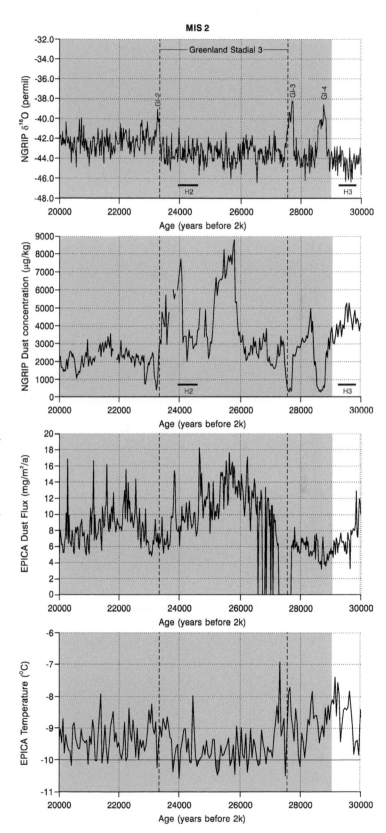

Figure 2.5 Close-up of the Last Glacial Maximum during the time interval 30–20 ka, showing the ice core records from Greenland (NGRIP) and Antarctica (EPICA). The top two diagrams are the oxygen isotope (Andersen et al. 2006) and dust concentration (Ruth et al. 2007) records from the NGRIP core. The NGRIP core is on the GICC05 age model. The bottom two diagrams are the dust flux (Lambert et al. 2012) and deuterium-derived temperature (Jouzel et al. 2007) records from EPICA, Antarctica. The EPICA records are both on the EDC3 age model. H2 and H3 are Heinrich Events when large icebergs were discharged into the North Atlantic Ocean.

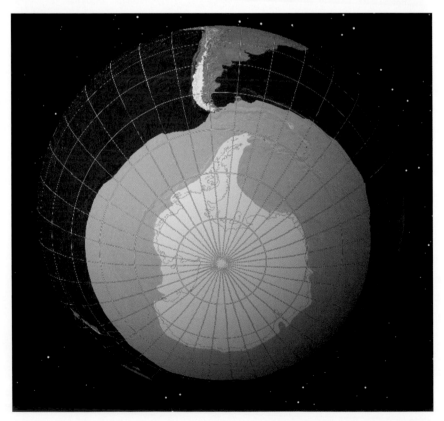

Figure 2.6 Glaciation and sea ice during the Weichselian (last) glacial maximum in the Northern and Southern hemispheres.

2.4 Systematics of the Ice Age

There were several ice ages. An ice age is defined as a period of the Earth's history in which at least one of the poles was glaciated. The ice ages are relatively brief events, which account for only about one-fifth of the Earth's history. A characteristic of an ice age is that the climate changes are much more severe than in other periods of the Earth's history. Within an ice age, we distinguish between cold stages and warm stages or glacials and interglacials.

When focusing on regions such as Central Europe, Britain or North America, we have the special case of cold intervals in which glaciers formed. These cold periods are known as glacials, and the warm periods between the glacials are referred to as interglacials. In an interglacial, the climate was similar to the present or even warmer. Minor thermal fluctuations in which this condition was not met are referred to as interstadials, while equivalent cold phases are known as stadials.

Investigations of deep-sea sediments have shown that this subdivision can be further refined. In sediment cores from the North Atlantic in 1988, the marine geologist Hartmut Heinrich discovered to his surprise a number of coarse sediment layers that even included fine gravel, present in all cores. With such coarse sediment, it could only be sediment transported by icebergs (IRD). Apparently, during the Weichselian cold Stage there were six periods during which icebergs occurred much more frequently. The origin of these 'Heinrich Events', named after their discoverer, is not yet clear. They each coincided with the release of large quantities of fresh water, probably from the Laurentian Ice Sheet area in North America. The Heinrich Events lasted a few hundred years. They have so far not been detected for older glacial periods due to the low resolution of the corresponding cores.

While the Heinrich Events represent short-term cold phases, there is also evidence of equivalent warm phases. Ice-core studies from Greenland suggest that, a series of dramatic climate fluctuations took place during the last glacial period. These short-term fluctuations in temperature, which were only noticed some 20 years ago in cores from the GRIP and GISP research boreholes near the centre of the Greenland Ice Sheet, are named after their discoverers as 'Dansgaard–Oeschger events' (first described by Dansgaard et al. 1993).

The Quaternary is synonymous with extensive glaciation of the Earth's middle and high latitudes. Although there were local precursors, significant glaciation began 35 million years ago in the Oligocene in eastern Antarctica. This was followed by glaciation in mountain areas through the Miocene (in Alaska, Greenland, Iceland and Patagonia), later in the Pliocene (e.g. in the Alps, the Bolivian Andes and possibly in Tasmania) and in the earliest Pleistocene (New Zealand, Iceland and Greenland). Today evidence from both the land and the ocean floors demonstrates that the major continental glaciations outwith the polar regions were markedly restricted to the last 1 Ma to 800 ka or less, rather than occurring throughout the 2.6 Ma of the Quaternary. Marine Isotope Stage (MIS) 22 (c. 870–880 ka) included the first of the major worldwide events with substantial ice volumes that typify the later Pleistocene glaciations (Fig. 2.7).

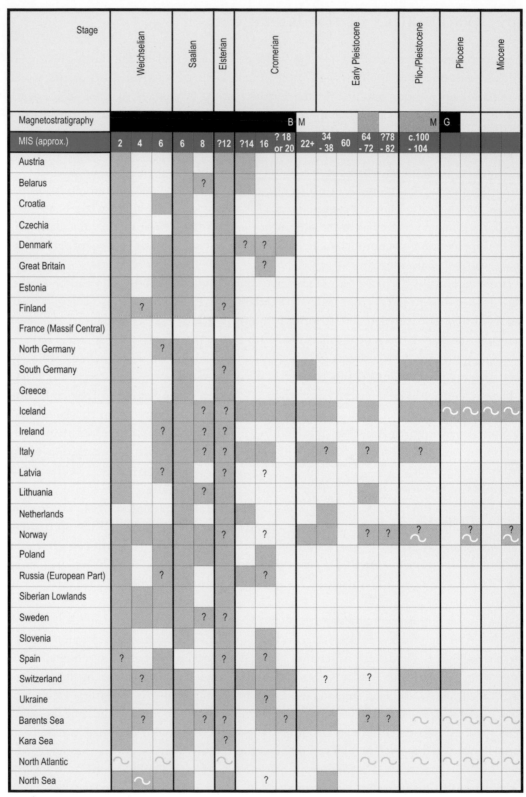

Figure 2.7 Ice-age glaciations in (a) Europe.

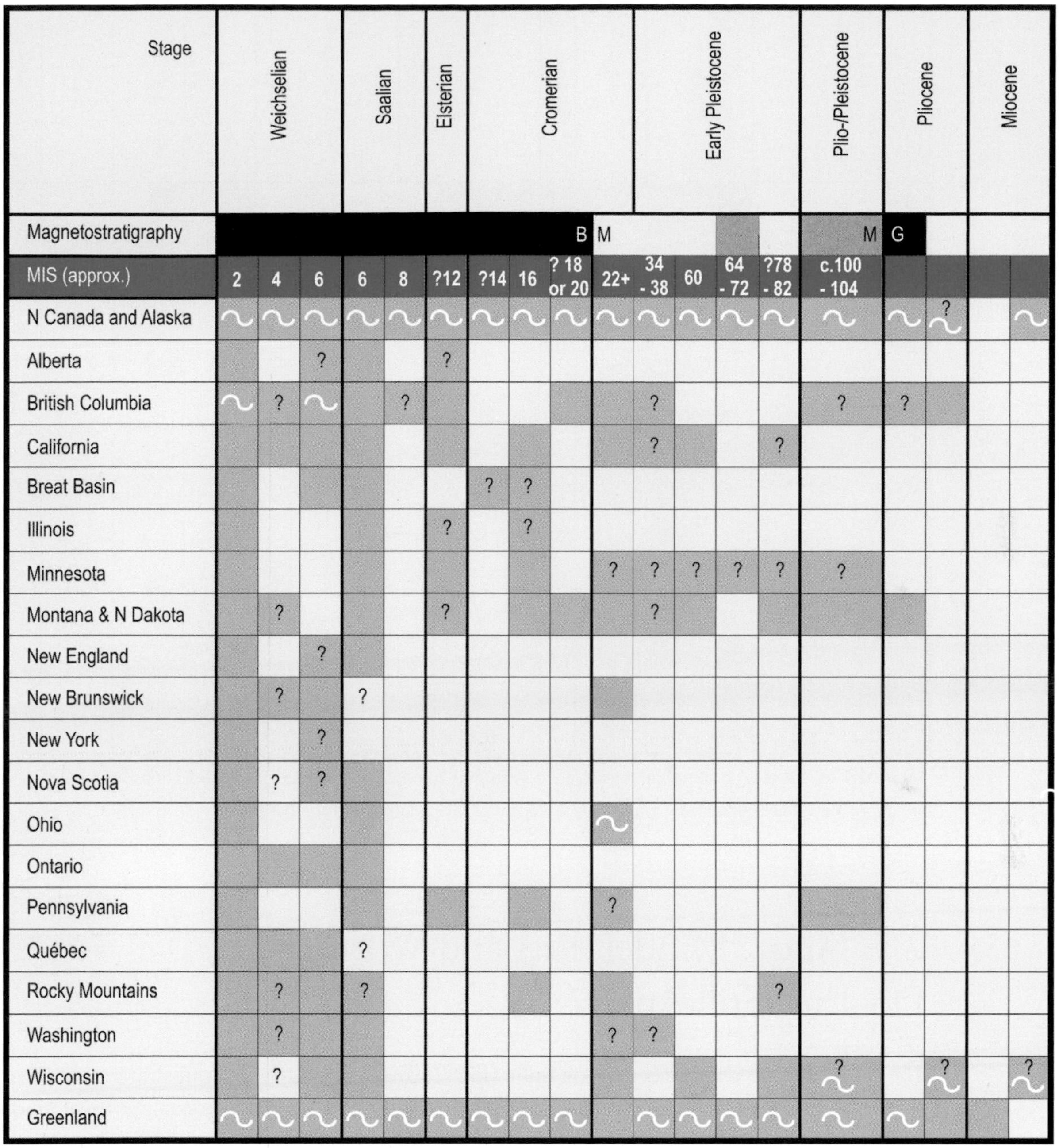

Stage	Weichselian			Saalian		Elsterian	Cromerian			Early Pleistocene					Pilo-/Pleistocene	Pliocene			Miocene
Magnetostratigraphy									B	M					M G				
MIS (approx.)	2	4	6	6	8	?12	?14	16	?18 or 20	22+	34-38	60	64-72	?78-82	c.100-104				
N Canada and Alaska	~	~	~	~	~	~	~	~	~	~	~	~	~	~	~	~	? ~		~
Alberta			?			?													
British Columbia	~	?	~		?					?					?	?			
California										?			?						
Breat Basin							?	?											
Illinois						?		?											
Minnesota										?	?	?	?	?	?				
Montana & N Dakota		?				?				?									
New England			?																
New Brunswick		?		?															
New York			?																
Nova Scotia		?	?																
Ohio										~									
Ontario																			
Pennsylvania										?									
Québec				?															
Rocky Mountains		?		?						?				?					
Washington		?								?	?								
Wisconsin		?													? ~		? ~		~
Greenland	~	~	~	~	~	~	~	~	~	~	~	~	~	~	~	~			

Figure 2.7 *(continued)* (b) North America

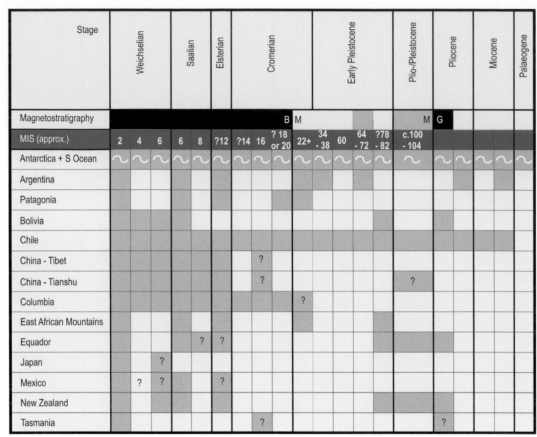

Figure 2.7 *(continued)* (c) South America, Africa, Asia and Australia. Purple squares: glacial deposits; ?: questionable glacial deposits; ~ glaciomarine deposits; MIS: Marine Isotope Stage.

Source: Ehlers & Gibbard (2008).

2.5 Günz, Mindel, Riss and Würm: Do They Still Apply?

The classic Alpine Quaternary stratigraphy (Fig. 2.8) is essentially a morphostratigraphy, that is, it is based on the concept that in each glaciation a 'glacial series' is formed, a series of landform communities consisting of tongue basins with drumlins, morainic ridges and outwash gravel fields. The term was coined by Penck & Brückner (1901–1909). Penck and his student Brückner made a point of proving the direct link between glacial gravels and associated end moraines. They assumed that each glaciation had deposited just one gravel landform unit or terrace.

The morphostratigraphy which forms the basis for the classification of the Alpine glacial deposits is not stratigraphy in a strict sense. It is not based on a series of superimposed strata, but is distinguished from landforms of different age separated from each other by erosional unconformities. Most of these represent glacial deposits. They do not represent the entire Pleistocene sequence of strata, but usually only the maxima of the respective glaciations.

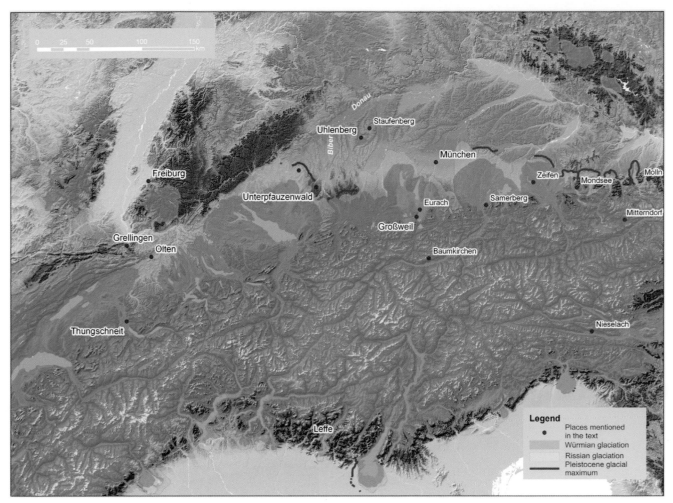

Figure 2.8 The extent of the Alpine glaciations.

Deposits from the early and late phases of glaciation are mostly missing. Additionally, non-glacial or interglacial deposits have rarely been preserved.

For the glacial deposits of each glaciation there are corresponding glaciofluvial (gravel) deposits which, at a greater distance from the terminal moraines, grade into a gravel field that can often be traced over tens of kilometres. Since in the northern Alpine foreland during the Pleistocene both major receiving rivers (Rhine and Danube) and the glacier tongue basins (as starting points of the drainage) were permanently lowered, the youngest gravel fields are located on the deepest and the oldest at the highest parts of the terrain.

Full morphological sequences in the sense of Penck's 'glacial series' are only found in the northern foothills of the Alps for deposits of the Würm, Riss and Mindel glaciations. For the Günz glaciations, this proof was later provided by studies in Upper Austria (Weinberger 1950; Kohl 1958). For even older glaciations, no appropriate landform assemblages have survived.

The morphostratigraphic method has been developed in the northern foothills of the Alps, where it can also be best used. On the southern Alpine margin and in the French Western

Alps, its application is far more difficult. The restrictions apply even more for the Swiss Plateau where the Jura Mountains were also glaciated, impeding free drainage for the glaciers and meltwaters of the Alpine glaciation.

On the basis of glacial morphostratigraphy, the northern foothills of the Alps can be roughly divided into three major units.

- The west is dominated by the Rhine glacier area. In the penultimate glaciations, the ice extended well beyond Lake Constance into the Alpine foreland, crossing the two main receiving waters (the Rhine in the west and the Danube in the east). In this area, older glacial deposits (pre-Riss) are covered by younger strata almost everywhere.
- In the east, the Danube receiving water was separated by a broad belt of Tertiary hills from the Alpine ice. Here the glaciofluvial deposits are restricted to relatively narrow river valleys, where only deposits of more recent glaciations have been preserved.
- In the central area between rivers Riss and Lech however, major spreads of Early Pleistocene gravel have been deposited. Multiple changes of drainage direction have resulted in gravel fields at different heights representing a number of glaciations. Here Penck (1882) could demonstrate the multiplicity of glaciations; further, in the gravel fields of the Riss–Iller–Mindel area the system of four Alpine glaciations (Günz, Mindel, Riss and Würm) was established (Penck 1899; Penck & Brückner 1901–1909).

This was also the area where later the deposits of two earlier glaciations (Donau and Biber glaciations) could be identified (Eberl 1930; Schaefer 1956). The area around Memmingen is still a key area for Alpine Quaternary stratigraphy.

Where the Alpine Quaternary stratigraphy goes beyond the limits of glaciation (because of the strong predominance of meltwater deposits) it is largely based on gravel and sand sequences, each underlying an individual surface or terrace. It was assumed that each gravel terrace corresponds to a glaciation, even if no direct relationship to glacial deposits could be shown. Van Husen (1983) pointed out that the deposition of gravel began during the climatic deterioration long before the actual glaciation. The loss of vegetation resulted in increased erosion and overloading of the rivers, which led to a raising of the entire valley floor. This is indicated by the fact that corresponding gravel terraces were deposited in the non-glaciated Alpine valleys simultaneously (Draxler & van Husen 1989). The latter include the massive terrace sediments of the Inn valley near Innsbruck, which even extend into the tributary valleys. The genesis of most of the gravel bodies has not yet been thoroughly examined (Fiebig & Preusser 2008).

One problem is to assign the individual gravel bodies to certain ice advances. It is not always easy to establish a relationship between a gravel body and a moraine. This also applies to the Inn valley terrace near Innsbruck. Since this gravel accumulation is overlain by till, it cannot have been formed during the last deglaciation. The dating of the over 100-m-thick varved clays at Baumkirchen (east of Innsbruck) has finally clarified that the terrace was deposited in the last glacial period before the Würmian glacial maximum (Fliri 1973).

As long as it was believed that the Alps had been affected by only four major glaciations, morphostratigraphy in most cases could meet the demands. Today, increasingly litho- and biostratigraphical methods are used for the subdivision of the Alpine Quaternary, and palaeomagnetic studies significantly contribute to the age classification of the sediments. However, the latter have only been used in a few places as sufficiently magnetized appropriate fine-grained sediments are comparatively rare. Modern methods of investigation, such as optically stimulated luminescence (OSL) and cosmogenic isotopes, support the dating of the deposits.

2.5.1 TRACES OF OLDER GLACIATIONS

The record of Quaternary deposits of the Alps is nowhere completely preserved. The stratigraphic table must therefore be compiled from a fragmentary record from various subregions, which makes correlation difficult in many cases because of the lack of age determinations. Evidence of early cold stages has long been reported from the Italian and French Alps. In both regions, an abrupt change from fine-grained sediment deposition to coarse gravels and conglomerates took place during the late Pliocene. This change in depositional environment could be explained either by tectonics or as a result of climate change (Billard & Orombelli 1986).

In the intra-montane basin at Leffe, above Bergamo in Lombardy, Taramelli (1898) had found early Quaternary *Schieferkohle*. This is a kind of weakly lithified, strongly compressed peat. The 100-m-thick deposits, which have yielded mammal remains, have been studied in detail. The Leffe Basin sequence was re-investigated by Muttoni et al. (2007) and was shown to contain evidence of 18 climatic cycles, each of which represented an alteration from warm temperate to cool climate. The palaeomagnetically roughly dated sequence ranges from about 1.94 to 0.78 Ma (MIS 19). At that time, the Alpine glaciers in the neighbouring Serio valley advanced for the first time to about 5 km upstream from Leffe. This evidence demonstrates that extensive glaciation had not occurred in the region before the Middle Pleistocene.

In Switzerland however, traces of older glaciations have been reported. First there are the so-called *wanderblöcke*, boulders apparently unrelated to corresponding glacigenic deposits, that have been found in the northwestern Jura Mountains (Hantke 1978). The blocks were derived mainly from the southwestern Black Forest. The southern edge of their distribution follows a line from Grellingen–Fehren–Himmelried–Hölstein–Tenniken–Olten. Hantke suggested that they might represent traces of an extensive Early Pleistocene Black-Forest Glaciation. It is more likely, however, that they are fluvially transported blocks of Pliocene–Early Pleistocene age.

The oldest reliably identified glacial deposits in Switzerland are the so-called 'cover gravels' (*deckenschotter*). These deposits may be even older than was originally assumed (Schlüchter 1988). In northeastern Switzerland, they occur as a relatively high-lying gravel cover that has been dissected subsequently and fragmented to form escarpment-like features. These sedimentary bodies are traditionally subdivided into lower and upper cover gravels and have been associated with the Günz and Mindel divisions of Penck's classical Alpine glacial stratigraphy. They bear traces of glaciofluvial, fluvial and glacigenic depositional environments. The original cover-gravel landscape has largely been destroyed by subsequent erosion and weathering, so that the extent of any early glaciation they might represent is unknown. Quite clearly, however, the ice advances which deposited this sedimentary complex must have extended beyond the Last Glacial Maximum ice limit. On the Belchen plateau in NW Switzerland, the cover gravel is overlain by an alluvial loess sequence with intercalated gravel. This 37-m-thick weathering profile contains a series of palaeosols. Schlüchter (1988) estimates that, in total, a period of more than 1.5 million years was required for this soil formation to have developed. How far pre-weathered material was involved in the deposition cannot be determined, however.

Between the planar deposition of the cover-gravel sheets and the partial filling of overdeepened valleys and troughs in the Swiss Alpine foothills, such as those representing the last four glaciations, there is not only a significant reversal of the morphogenetic conditions but also a time gap. The incision of the deep valleys postdates the formation of the cover gravels. This incision is linked to the uplift of NE Switzerland, and varies from region to region. It was particularly intense and extended in the central and western parts of central Switzerland. A second major developmental phase was the renewed reversal from erosion to accumulation in the main valleys of the Alps. Schlüchter (1989a) calls this a 'Middle Pleistocene revolution'. However, such a marked 'morphological reversal' cannot be detected in the neighbouring southern German and Austrian Alps.

Figure 2.9 The Thalgut exposure in the Aare valley (south of Bern, Switzerland). In the upper part of the exposure: Rotachewald Diamicton above the Upper Münsingen Gravels. Photograph by Christian Schlüchter.

The first indications that there might have been more than the four Alpine glaciations that Penck & Brückner (1901–1909) had reported were found in the northern foothills of the Alps. Eberl (1930) found high-lying gravels in the Iller-Lech region, which he regarded as deposits of a 'Danube Glaciation' ('Donau') which had preceded the Günz Glaciation of Penck and Brückner. Schaefer (1956) added another event, the 'Biber Glaciation' (based on gravel deposits at Staufenberg and in the eastern part of the Aindlingen terrace staircase of the River Lech), which is thought to pre-date the Donau Glaciation. Today both Biber and Donau are now well established as chronostratigraphical terms for the oldest Alpine Quaternary (Figs 2.9, 2.10). That the gravel deposits in question are actually glaciofluvial in origin has not however been demonstrated.

The deposits of the Biber glacial period are gravels, the formation of which included multiple periods of fluvial erosion and accumulation (Becker-Haumann 1998). The next younger deposits of the Donau glacial period were also deposited in several phases. Since interglacial deposits have never been found between the deposits of the Biber and the Donau glacial periods, it seems that the strata grade into one another. By contrast, in the intervening period between the deposition of the Donau gravel and the Günz Glaciation, there was a pronounced erosional phase, however. During this time the Uhlenberg interglacial deposits were laid down. They appear to correspond to one of the interglacials of the Bavelian Stage Complex of the Dutch stratigraphy (late Early Pleistocene).

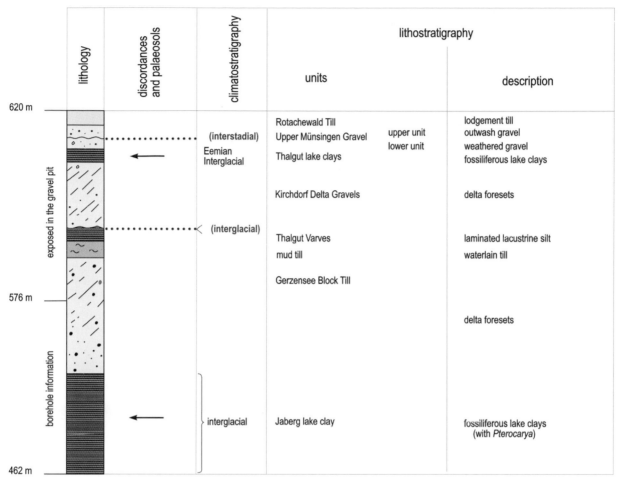

Figure 2.10 Schematic profile of the Thalgut exposure.

Source: Schlüchter (1989b).

2.5.2 GÜNZ

On the basis of recent discoveries the Günz(-ian) includes the deposits of three different glaciations. The older parts of this Günz Complex are reversely magnetized. The majority of the deposits, however, seem to be younger than the Brunhes–Matuyama boundary. Günz moraines have only been found in the border area between the Iller and Wertach glacier. Here, the glaciers of the Günz had apparently advanced further into the Alpine foothills than elsewhere (Habbe 2007).

While the relative order of the gravel terraces is undisputed, the chronological classification of the various early cold periods is extremely problematic. Initial attempts to date the oldest gravel using cosmogenic nuclides have given an age of 2.35 million years for the Günz gravels south of Memmingen, and 680,000 years for the Mindel gravels (Häuselmann et al. 2007).

2.5.3 HASLACH

The precise number of glaciations before the Riss(-ian) is uncertain. As early as in 1952, Schädel, working in the Württemberg Rottal (south of Ulm), had divided the younger (Mindel) 'cover gravel' of Penck & Brückner (1901–09) into two separate bodies. Following detailed investigations, Schreiner & Ebel (1981) found that the 10-m-higher Haslach gravel contains a much lower frequency of crystalline lithologies than the deeper (and therefore younger) Tannheim Gravel. They assigned the newly discovered Haslach Gravel to a Haslach Glaciation, the deposits of which are separated from the Günz deposits by a fossil soil (interglacial). In turn, they are separated from the Mindel Glaciation sediments by the Unterpfauzenwald Interglacial deposits, described by Göttlich & Werner (1974). The pollen diagram from the latter shows primarily coniferous forest with a high proportion of spruce and fir. In addition to beech, *Pterocarya* (wingnut) and, according to new unpublished research, *Ostrya* (hop hornbeam) and *Tsuga* (hemlock) also occur. Together they testify to a relatively warm, humid climate and an age older than Holsteinian (see Section 2.6.2) has to be assumed (Bibus et al. 1996).

So far, the Haslach has only been detected at its type locality in the northeastern Rhine glacier area. However, it is conceivable that Haslach meltwater sediments also occur in the Iller-Lech district (Becker-Haumann 2002). Since the age of the moraine and the gravel is not secured, the temporal relationship of the Haslach deposits to the Mindel Glaciation remains problematic.

2.5.4 MINDEL

The younger cover gravels of the Mindel valley, and their links with moraines north of Obergünzburg, led Penck to designate his second glaciation as the Mindel(-ian). The terminal moraines of this Mindel Glaciation, south of Memmingen, are well developed, and include the end-moraine ridge of Brandholz-Manneberg. While no continuation of the moraine is found to the west, it can be traced eastwards to Obergünzburg (Habbe 1986a).

East of the River Lech, the Mindelian Younger Cover Gravels rarely reach beyond the ancient morainic areas. The Younger Cover Gravels with 'geological organs' (karstic solution pipes within the calcareous gravels) are found, for example, in the Isar valley, south of Munich, in the Mangfall valley, in the Inn glacier area and in the Salzach valley. The sequences in the Isar valley were described by Penck & Brückner (1901–09) and also, for example, by Jerz (1987). The deposits in the Mangfall valley were remapped by Grottenthaler (1985). In the eastern part of the south German Alpine foothills, Mindelian landforms have been mapped in morphological studies (Inn–Chiemsee glacier; Troll 1924). The mapping by Eichler & Sinn (1974) and Grimm et al. (1979) in the western and northwestern part of the Salzach glacier area border on the investigations carried out by Weinberger (1950) in the eastern part of the Salzach glacier area.

The age of the Mindelian is still controversial. In Baden-Württemberg it is – together with the Haslach – assigned to the Matuyama Chron (i.e. older than 0.78 ka), while in Bavaria it is included in the Brunhes Chron (in Bavaria the Günz occurs at the transition of Brunhes–Matuyama). In Baden-Württemberg an additional Hosskirch Cold Stage is reported, which occurs intermediately between Mindelian and Holsteinian and corresponds in age to the Bavarian Mindelian.

2.5.5 MINDEL–RISS INTERGLACIAL

The Mindel–Riss Interglacial, which is equated to the North German Holsteinian Stage, is characterized by the presence of *Pterocarya* (wing nut), *Fagus* (beech) and *Buxus* (box). Two sites that can safely be correlated with the Mindel–Riss Interglacial are the Samerberg 2 profile

(Grüger 1983) and the Thalgut profile in Switzerland (Figs 2.9, 2.10; Welten 1988). The interglacial sequence found at Meikirch is now regarded as younger because of its 'differing vegetational succession' and on the basis of luminescence dating (Preusser et al. 2005).

2.5.6 RISS

When the glaciers advanced once again following the Mindel–Riss Interglacial, they entered a landscape that had been sculpted by several previous glaciations. Tongue basins already existed at the foot of the Alps. However, the Riss Stage glaciers in many areas advanced further into the foothills than ever before. On the basis of U–Th dating of the associated high gravel terrace it is now parallelized with MIS 6 (Preusser 2010). This was the most extensive glaciation in the Emmental area of Switzerland.

In the area of the Rhine glacier, the Riss can be subdivided into two or three ice advances. The deposits of the first Riss advance are characterized by intense weathering. In the Riss valley and in the neighbouring valleys, they underlie middle Riss deposits from which they are separated by gravels. The extent of the intermediate glacier retreat cannot yet be determined. The differences in weathering intensity suggest that the intervening period should at least have had the rank of an interstadial. The till deposition of this first ice advance is limited to single tongue-like advances.

The middle Riss is bounded east of the Rhine glacier area (between Biberach and Leutkirch) by a double-walled end moraine. For this reason, Schreiner (1989) and Ellwanger (1990) termed this the 'double-wall Riss'. In general, the end moraines consist of gravelly push moraines up to 10–30 m in height, and form the eastern margin of the Rhine glacier area 1–3 km behind the middle Riss ice margin.

The extent of the late or younger Riss advance is controversial. Its terminal moraines are relatively incomplete. The eastern periphery of the former Rhine glacier includes moraines found to the north of Ingoldingen and south of Eberhardzell. The associated gravel fields in the Riss and Umlach valleys (up to the north of Warthausen) have largely been subjected to later erosion (Schreiner 1992).

Older moraines have also been reported in Bavaria from the area beyond the classical known Riss terminal moraines. According to investigations of the weathering profiles, it is possible that the two superimposed Riss deposits are separated from each other by a warm climate phase, possibly even an interglacial (Bibus & Kösel 1987; Miara 1995).

Riss moraines have also been mapped in the Salzach and the Traun glacier areas, in Upper Austria's Krems valley. Deposits of the penultimate glaciation have even been preserved in some inner-Alpine areas. For example, deposits of the Riss Glaciation have been identified in the narrow valleys of the rivers Steyr and Enns. However, a separation of these sequences into individual advances or stages has not been attempted. An overview of the state of knowledge is provided by van Husen (2004).

The stratigraphical position of the Riss Alpine Glaciation, at least in some parts of the glaciation area, remains uncertain. Neither the exact extent of the glacial ice advances nor even their number can currently be identified with certainty. Luminescence dating of 'Riss gravel' from the Ingolstadt area has shown, for example, that a portion of the deposits is clearly younger than suspected (90–60 ka) and therefore belongs to the early Würmian (Fiebig &

Preusser 2003). It is clear that a correlation of the classic 'Riss', or parts thereof, to the Saalian Stage Glaciation of northern Europe is not yet possible.

2.5.7 RISS–WÜRM INTERGLACIAL (EEMIAN)

Temperate stage deposits have been known in the Alps since the last century. The first warm deposits identified were the so-called *Schieferkohle*. These weakly humified deposits are found in numerous places throughout the Alps. They often form part of a sediment sequence which includes lake deposits (clays, silts, partly lake marls). These eventually led to infilling of the lake basins, depositing a series of slightly sandy clays with intercalated peats (the latter mostly compressed to thin layers of coal). In Grossweil and in Pfefferbichl (Upper Bavaria) these *Schieferkohle* deposits have been shown to represent warm stage accumulations based on their palynological assemblages. Most workers correlate them with the Eemian Stage interglacial in northern Europe.

In southern Germany, Beug (1972) studied the Eemian interglacial sequence at Zeifen, near Laufen on the Salzach River, and later published the pollen profile of Eurach on Lake Starnberg (Beug 1979). Other palynological investigations include the Eemian profile at Füramoos (Müller 2001). The best-studied palynological sequence from the Alpine region, however, which includes the last interglacial period and three Weichselian interstadials, is that from Samerberg in Bavaria, southeast of Rosenheim (Grüger 1979). Here the vegetation succession is basically the same as that in northern German Eemian deposits. The montane character of the locality is expressed by the high proportion of spruce pollen. *Picea* (spruce) appears early, and quickly achieves a dominant share of the tree pollen content. After a short, pronounced maximum of *Taxus* (yew) the immigration of *Abies* (fir) and *Carpinus* (hornbeam) follows. At the end of the interglacial all three taxa disappear and are replaced by *Pinus* (pine). By contrast, in the lower-altitude Zeifen Eemian deposits the montane species *Picea* and *Abies* are less well represented while *Carpinus* dominates throughout (Grüger 1989).

In Switzerland, deposits of the last interglacial period are not uncommon. Among other things, the interglacial is represented by a series of lake deposits. Within the area of the last glaciation these include the sites at Meikirch, Thungschneit, Thalgut and Wildhaus. Beyond the extent of the last glaciation are the sites of Niederweningen, Sulzberg and Gondiswil.

A thick sequence of lacustrine deposits was exposed on the north shore of Lake Mondsee (Austria), both under- and overlain by till. Palynological studies have shown that the lake deposits at this site include the last interglacial as well as four Early Weichselian mild climate interstadials. The vegetation succession of the interglacial here corresponds to that from the Samerberg site. In this sequence, the climatic optimum is characterized by mixed oak forest, which is rich in *Ilex* (holly) and *Taxus* (yew). This suggests that the annual mean temperature in the region during the Eemian was about 2–3°C higher than during the Holocene climatic optimum. At the end of the interglacial, as in northern Germany, there is a return to total deforestation. At this locality the Early Weichselian interstadials are characterized by renewed spread of forest dominated by *Picea* and, to a lesser extent, by *Quercus* (oak) and *Taxus* (Drescher-Schneider 2000).

2.5.8 WÜRM(-IAN)

In contrast to earlier glaciations, the maximum extent of the Würmian–Weichselian ice is rather well known and detailed maps have been drawn for areas such as the Alps or the Pyrenees (Figs 2.11, 2.12). In Meikirch, the Eemian deposits are overlain only by till from a single glacier advance. The same applies to Gossau, near Zürich. However, in Thalgut, Jaberg, Thungschneit and Cossonay, the last glacial period in Switzerland is represented by two major ice advances, the glaciers of which have each extended into the foothills. The older ice advance is less well preserved than the younger. The profiles from Gossau and Lake Walen show that between the two ice advances an ice-free period occurred during 60–28 ka (Schlüchter 2004). At Wangen, on the Aare River, an erratic block on the Würmian moraine has been dated to 20.1±1 ka by cosmic nuclides (Ivy-Ochs et al. 2004). Also in the French Alps, the main advance of the Würmian Glaciation occurred at about 20 ka (Buoncristiani & Campy 2011).

In contrast to the results from Switzerland, the results from Samerberg (Bavaria) and Mondsee (Austria) show an uninterrupted succession of lake sediments between undisputed deposits of the Eemian Interglacial and the overlying till from the Würmian maximum

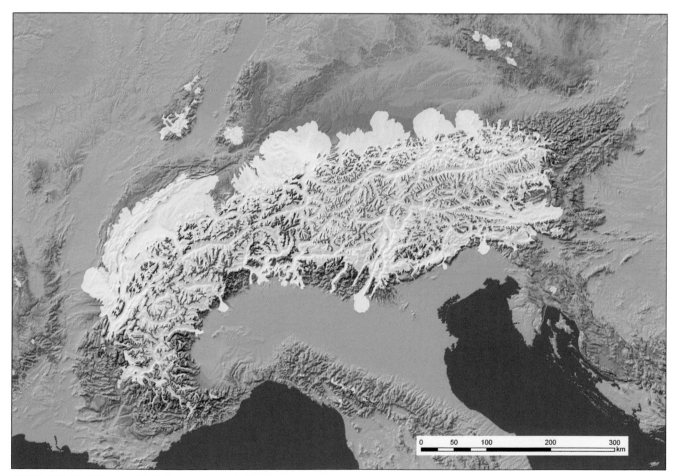

Figure 2.11 The Würmian Glaciation maximum in the Alps.

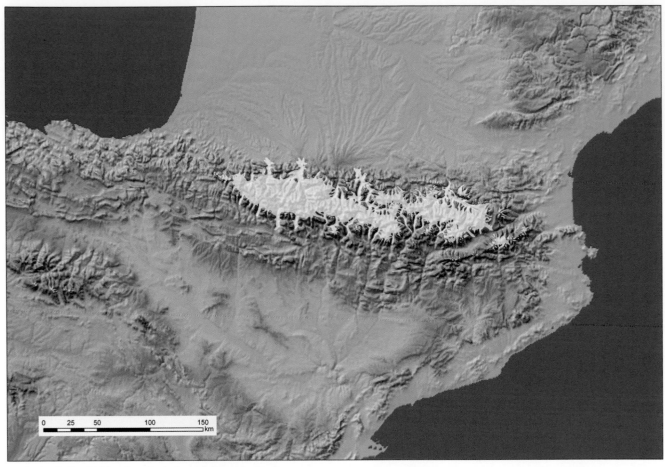

Figure 2.12 The Weichselian Glaciation maximum in the Pyrenees.

(about 20 ka). Moreover, van Husen (1977) has only found traces of a single Würm ice advance in the Trauntal in Austria. There is no evidence of any significant ice advance during the Early Würmian. The main advance during the last glaciation began at about 25–24 ka, and reached its maximum around 20 ka (van Husen 2004).

Within the loess deposits of the Alpine foothills, fossil soils can be frequently encountered. Strong wet soils and humic zones from the Early Würmian occur above the last interglacial soils, with brunification zones and wet soils from the Middle Würmian and poor initial brunification zones from the Upper Würmian. The soil profiles are characterized by gleying and thin decalcified layers. In most cases two–three such gleyed zones can be distinguished, and they often also include ice-wedge casts and cryoturbations (Jerz 1982). Further north towards the Danube, largely decalcified 'brown tundra soils' developed under a drier climate (Brunnacker 1957). New research in the Salzach glacier area has shown that the Middle Würmian loess was deposited under relatively humid conditions as an alluvial to aeolian deposit (Starnberger et al. 2009).

The best-known Würmian deposits from before the glacial maximum are those of the varved clays (*Bändertone*) of Baumkirchen, 15 km east of Innsbruck (Fliri 1973). The climate had already deteriorated significantly when they were deposited, and shrub-tundra prevailed in the surroundings of the lake. These >100-m-thick seasonally stratified clays are overlain here by about 70-m-thick meltwater gravels, which are overlain by a thick till. These laminated clays were deposited under uniform sedimentation conditions in a shallow lake, and include trace fossils of fish that have been found on the surfaces of the individual layers. The sedimentation rate was consistently high; on average it reached about 5 cm a^{-1} (Bortenschlager & Bortenschlager 1978). Radiocarbon dating of wood (*Pinus*, *Alnus* and *Hippophae*) from the *Bändertone* gave ages of between 31.6±13 and 26.8±13 ^{14}C ka BP (Fliri 1973).

Unlike the Scandinavian Ice Sheet, where the ice thickness can only be determined on the basis of a few rare nunataks, in the case of the Alps the elevation of the ice stream network can be reconstructed quite accurately, especially with the help of erratic pebbles.

The advance of the glaciers during the Würmian Glaciation may have been a single ice expansion, but it has left behind a number of retreat-stage series of minor end moraines. Detailed geomorphological studies have been conducted on a number of Alpine glaciation areas. Well-developed glacial series are found in many places, such as south of Winterstettenstadt at the northern end of the former Rhine glacier and at Memmingen on the northern edge of the former Iller glacier. Likewise, the marginal area of the Iller glacier, Würmian ice margins and drainage paths were mapped by Habbe (1986b) and Ellwanger (1988).

During the peak of the Würmian Glaciation multiple transfluences occurred in various parts of the ice-stream network. For example, ice from the Inn valley drained northwards via the Fern Pass and the Seefeld valley, as well as during earlier glaciations. This is reflected in the gravel and clast associations of the corresponding sediments. In the foothills of the Isar–Loisach glacier foreland, up to 35% of central Alpine crystalline components have been identified in the 20–31.5 mm group. Near the eastern edge of the Tölz glacier area, however, where no direct transfluence from the Inn valley occurred the frequencies are only 0–1%, (Dreesbach 1985).

The Salzach glacier marginal area was mapped by Grimm et al. (1979) and Ziegler (1983); no traces of any Early Würmian ice advance were found. In the foothills near Salzburg and Tittmoning, there are two strongly overdeepened basins up to 200 m deep which are completely filled with sediments. The melting of the Salzach glacier left behind a large number of end-moraine ridges, behind some of which moraine-dammed lakes formed. During the Oldest Dryas Stadial the moraine dams were consecutively breached by the meltwaters. 4–6 terraces were formed, the precise chronological classification of which remains unestablished.

Numerous recent studies have also been published from Austria on the developments through the Würmian Stage. For example, van Husen (1977) conducted a detailed investigation of the Würmian deposits in the Traun valley and reconstructed the expansion and morphological expression of the individual glacier fluctuations. The Würmian glacier extents, and deposits in the upper Enns valley, have also been mapped by this author. It was shown that the confluence between the Enns glacier and Traun glacier ceased immediately following the glacial maximum. Mapping the lower Enns valley demonstrated a tripartite division of van Husen's Würmian gravel terraces. These sediment bodies can be connected to those of the Danube terraces near Linz (Kohl 1968). Comparable studies are also available from the Mur and Drau glacial areas. Both north and south of the Alps, the maximum extent of the Würmian glaciers was followed by another advance, during which the ice margin remained several hundred metres behind the maximum limit. This advance left a morphologically much more distinct terminal moraine than the maximum limit, and it was also during this part of the glaciation that the main body of the Würmian gravel terrace was deposited (van Husen 2004).

2.6 Northern Germany and Adjacent Areas

The sea had retreated from the Northwest European Basin in the late Tertiary, and fluvial deposition was predominant since early Miocene time. The rivers of the Baltic Shield and Fennoscandinavian Platform in the north and Variscan mountains in the south transported clastic sediments into the Northwest European and Eastern German–Polish basins. During this time uplift of the surrounding uplands led to increased sedimentation within the basin (Ziegler 1982). The deposits are light-grey- to whitish-coloured quartz sands, with intercalated layers of clay and lignite. The rare gravels consist mainly of well-rounded quartz and quartzite and silicified sedimentary rocks, the latter in particular derived from former limestone areas of the eastern Fennoscandian Shield and the Baltic Platform. This 'Baltic River system' generally followed the course of today's Baltic Sea depression.

The Baltic River system (also called the 'Eridanos System'; Fig. 2.13) continued to flow during the Early Pleistocene. Corresponding deposits are known mainly from the southern margins of the North Sea Basin into which a large delta system prograded (Westerhoff 2009). The alluvial deposits of this system extend from around Bremen well into the Netherlands and beyond. They contain lydite from the uplands as well as large blocks of rock, which may have been transported to their present site by drift ice from the eastern Baltic region (Gripp 1964). The 'Loosen Gravels' of Mecklenburg are also regarded as deposits of this river system (von Bülow 2000).

Evidence of a cold climate and related frost phenomena in northwest Europe has already been found from the Miocene. Sharp-edged silicifications in the Miocene lignite sands at Besenhorst, east of Hamburg may be interpreted as an indication of drift-ice transport (Ehlers et al. 1984). Similarly, the lavender-blue cherts found between the Miocene Lower Lusatian brown coal seams, have also been interpreted accordingly (Ahrens & Lotsch 1976). Ice rafting would also be required for the up to 0.5 m subrounded sandstone blocks and cherts of Scandinavian origin found in the kaolinitic sands on the North Sea island of Sylt (Figs 2.14, 2.15). In addition, sand blocks occur that can only have been moved in a frozen state. Whether the poorly orientated striae on Danian flints and other hard rocks (von Hacht 1987) can be interpreted as glacial striations is questionable. Heavy mineral analysis by Burger (1986) has demonstrated that the Kaolin Sands on Sylt must date back to the Brunssumian Stage (lower Pliocene).

The basic outline of the Quaternary in northern Germany was already established by the end of the nineteenth century. At that time, no-one had any idea of the actual duration of the Ice Age. It had been assumed at an early stage that north Germany (like the Alps) had been glaciated three times (Penck 1879). This view, however, could only be established once Keilhack (1896) in Berlin and Gottsche (1897a) in Hamburg demonstrated the presence of an oldest (third) till in northern Germany.

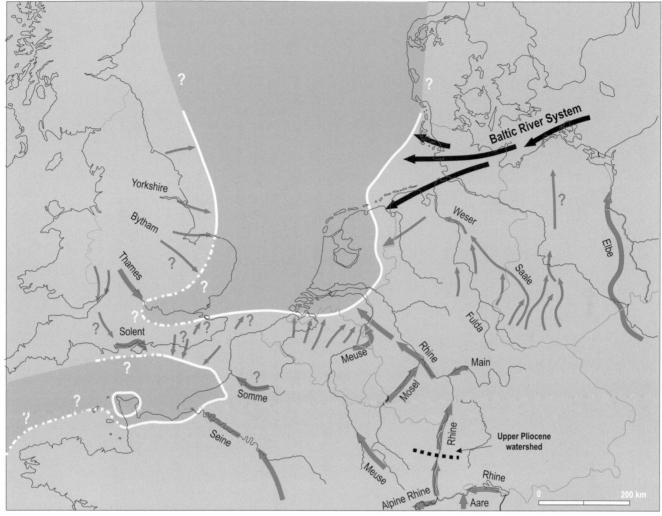

Figure 2.13 Schematic palaeogeographic reconstruction of the drainage system for the Reuverian to the Tiglian (late Pliocene–Early Pleistocene). The Baltic Sea did not yet exist. Weser and Elbe rivers were tributaries of the Baltic river system. The Alpine Rhine initially flowed into the Rhône, but found its connection to the Upper Rhine during the latest Pliocene. Adapted from Gibbard (1988).

After it was determined that the Alps had been glaciated not only three but (at least) four times (Penck & Brückner 1901–09), there were no lack of attempts to prove that a fourth glaciation had also occurred in north Germany. However, neither the separation of the Saalian into two independent glaciations nor the discovery of an allegedly older (Elbe) glaciation could be supported. On the other hand, vegetational and sedimentological studies in the Netherlands have shown that the number of cold and warm events had to be increased. By the 1950s authors such as van der Vlerk & Florschütz (1950) and Zagwijn (1957) had proposed a revised sequence of climatostratigraphic units:

- Weichselian (Tubantian) Cold Stage
- Eemian Warm Stage
- Saalian (Drenthian) Cold Stage

Figure 2.14 Sand clasts in Pliocene so-called 'Kaolin Sands'at Braderup, Isle of Sylt. The sand clasts must have been transported in a frozen state, or else the loose sand would have disintegrated. Photograph by Jürgen Ehlers.

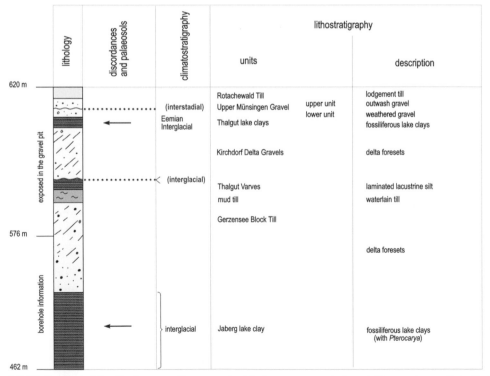

Figure 2.15 Drawing of the Braderup section; cross-bedding measurements in red. The sands are deposited towards the northeast.

- Holsteinian (Needian) Warm Stage
- Elsterian Cold Stage
- Cromerian Warm Stage.

Five earlier stages had to be added:

- Menapian Cold Stage
- Waalian Warm Stage
- Eburonian Cold Stage
- Tiglian Warm Stage
- Pretiglian Cold Stage.

This subdivision later required further amendment. Originally, each of these units was considered to represent either a single warm- or cold-climate period. Later, however, it was found that within some of those periods repeatedly alternating cold (boreal) and warm to temperate climates had occurred. Today they are therefore regarded as complex time divisions, each consisting of several warm and cold intervals. During the cooler periods of the Tiglian Stage, which had originally been classified as a single warm (interglacial) interval, there was even permafrost (Kasse 1988). The Tiglian must therefore be regarded as a climatically complex time-unit (Tiglian Complex). The same is true for the Cromerian Complex Stage which, as currently defined, includes four warm (interglacial) and three cold (glacial) intervals, with at least one of the glacials subdivided by an interstadial (Litt et al. 2007).

The pollen analysis of sediments from the beginnings of the Ice Age is problematic. Almost all pollen samples from the Early Pleistocene units in the Netherlands do not come from organic deposits (peat, gyttja), but from clastic sediments. Such deposits must always be expected to contain a proportion of reworked older palynomorphs, some of which are derived from Tertiary or Mesozoic source rocks. In addition, the chemical and mechanical corrosion of the pollen grains influences the composition of the pollen diagrams, and some fluctuations in the pollen composition may have been caused by changing depositional conditions.

More favourable conditions for the preservation of complete vegetational sequences are present in the Early Pleistocene deposits of northern Germany. In a karst hollow at Lieth on the Elmshorn salt dome, a series of strata with five Early Pleistocene interglacial deposits representing the period from the Tiglian to the Menapian was investigated by Menke (1969, 1975). They are autochthonous peats and muds formed in place which, in contrast to the Dutch profiles, represent true vegetational sequences. The connection to the Pliocene is possible through samples from the Oldenswort borehole (Schleswig-Holstein). More recently, the youngest part of the Early Pleistocene (probably end of the Menapian to early Elsterian stages) has been identified in a karstic depression on the Gorleben salt dome in Lower Saxony (Müller 1992).

The same number of cold and warm periods has been identified in northern Germany as in the Netherlands. The only problem is that, in comparison to the deep-sea stratigraphy, there are still far too few events differentiated on land. How can that be? Should it not be expected from a karst depression, which acted as a sediment trap, that it would contain all climate-relevant historical horizons? Apparently not. The pollen diagrams from the Gorleben interglacials are internally incomplete, a consequence of the variability in the sedimentary environment through time.

Through recent decades several interglacial sites have been identified that are older than the Elsterian Glaciation. Many of these early warm intervals are isolated occurrences with no direct relationship to the rest of the Quaternary stratigraphy. A typical example is the Hunteburg Interglacial. This unit, which was discovered in 1985 in a research borehole 18 km northeast of Osnabrück (Hahne et al. 1994), is stratigraphically difficult to place. The interglacial lake deposits are overlain by Saalian glacial sediments (sands of the Middle Terrace and Saalian meltwater sand) in addition to possibly Elsterian periglacial sediments. The vegetational history of this event indicates that the deposits represent an interglacial within the 'Cromerian Complex' Stage.

Which interglacial could these deposits represent, and can palaeomagnetics help? The Hunteburg Interglacial is reversely magnetized, and therefore belongs to the lowermost part of the Cromerian. Only the Cromerian-interglacial I (termed Waardenburg) is reversely magnetized; Cromerian-interglacial II (Westerhoven) is normally magnetized, the magnetic reversal being the Brunhes–Matuyama. However, the vegetational succession of Hunteburg is incompatible with that from the Netherlands' Waardenburg locality. In Hunteburg *Eucommia* (the 'Chinese rubber tree') is absent, a warmth-loving plant that was widespread in northern central Europe during the Tertiary. Today it is found in southern China. A correlation with the Westerhoven interglacial would appear more appropriate, but the normal magnetization of the latter seems to conflict with this similarity of the vegetational assemblage. Could it be that there is an additional Cromerian interglacial between I and II which has not yet been identified? Or is it the equivalent of the Westerhoven interglacial, and the sediments correspond to the reversely magnetized Lishi event? This question cannot yet be answered (Litt et al. 2007).

2.6.1 ELSTERIAN GLACIATION

The terms Elsterian, Saalian and Weichselian glaciation first appeared (from 1910 onwards) on the Prussian 1: 25,000 scale geological map (Keilhack 1910). In the explanatory notes, however, terms such as first glaciation, second or penultimate glaciation and last glaciation were still used. The terms Elsterian, Saalian and Weichselian were not generally accepted until the 1920s. Woldstedt (1929) used them in his textbook, certainly contributing to their widespread acceptance.

During the Elsterian Stage (Fig. 2.16), northwest Europe was affected by glaciation for the first time with the ice sheet advancing as far as the margin of the Central German Uplands. Although there had also been earlier glaciations, especially during MIS 24–22 and MIS 16 (Donian Glaciation), the ice sheets now advanced for the first time so extensively that the Scandinavian and British ice met in the North Sea. Since the ice blocked the drainage to the north in the North Sea Basin, a huge ice-dammed lake was formed whose waters eventually breached the Weald–Artois threshold and thus created a drainage path south into the Bay of Biscay by about 455,000 years ago. The English Channel, initially a large river valley system, was overwhelmed by the meltwater overflow and this breached the land barrier

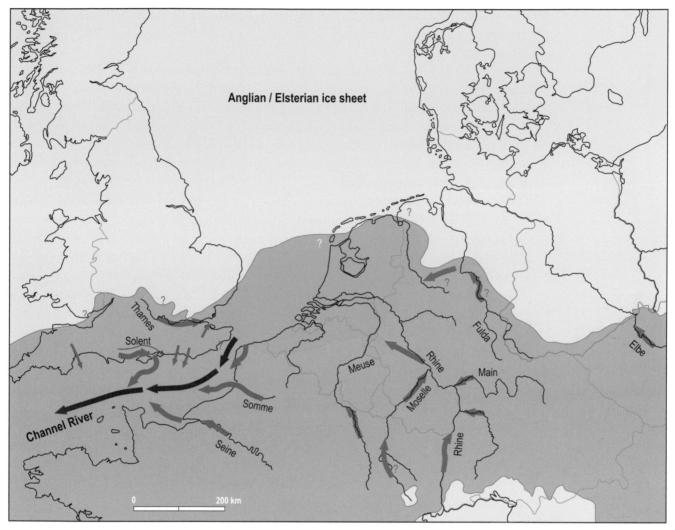

Figure 2.16 The Central European river system during the Elsterian Glaciation and at the opening of the Channel. Adapted from Gibbard (1988).

between southern Britain and northern France. This initial gap was extended in the next warm stage and in subsequent cold stages, until it eventually reached its current width.

The history of the English Channel was not easy to unravel. Investigations in the area of the Channel itself were of little use, because its formation was essentially an erosional event. Research had to be extended further out into the Bay of Biscay, off the Channel mouth. Here Toucanne et al. (2009a, b) evaluated deep-sea core MD01-2448. This 28.8-m-long drill core, which was recovered at 44° 46.790′ N and 11° 16.470′ W at a water depth of 3460 m from the top of Charcot Seamount, occurred in an ideal position to examine the sediment influx from the Channel River and correlate it to the deep-sea marine isotope stratigraphy. The sequence extends back to MIS 35, that is, to the period prior to about 1.2 Ma. A comparison with the globally averaged oxygen-isotope record from the LR04 stack (Lisiecki & Raymo 2005) shows that the core profile is complete.

It was therefore possible to date an important benchmark for the correlation between the continental and marine Quaternary stratigraphy. Here the Elsterian glacial Stage unequivocally corresponds to MIS 12.

The Elsterian Glaciation reshaped the landscape throughout northwest Europe so thoroughly that within the glaciated area virtually all traces of older Quaternary landforms were completely obliterated. The extent of any previous glaciations is unknown. Moreover, the drainage system was changed during the Elsterian Glaciation. The north German and Polish rivers were partly dammed by the advancing ice and diverted to the west or east. At the southern edge of the ice sheet, in the West German Central Uplands, only relatively small traces of dammed-up rivers have been found.

The extent of the Elsterian Glaciation in some areas is still not fully understood. It must be assumed that considerable quantities of Elsterian tills have been eroded in the subsequent cold stages. For example, only patches of Elsterian till are found in Emsland (Lower Saxony). More of the original till cover has been preserved further to the east, such as north of Bremen where Elsterian till is exposed in several outcrops (Wansa 1994).

In southern Lower Saxony and North Rhine–Westphalia, the presence of Elsterian ice is primarily documented by the occurrence of redeposited Scandinavian gravel in the Middle Terrace. Thome (1980) believed there might have been an advance of Elsterian ice into the Münsterländer Bucht. In addition, Klostermann (1985, 1992) regarded an advance of the Elsterian ice to the Lower Rhine as possible. However, there is a lack of evidence for such a far-reaching southwards advance of the Elsterian ice to the west of the Weser River, since no tills of the Elsterian Glaciation have been found in the area (Meyer 1970).

An accurate reconstruction of the ice limit is possible only to the east of the Hartz Mountains. Here the limit of flint distribution coincides with the maximum extent of the Elsterian ice. This flint line is found at an altitude of 300–480 m asl (Wagenbreth 1978). Its height distribution falls significantly towards the west, so that in the Weser Mountains it is at >200 m asl (Kaltwang 1992). There, however, it represents the Saalian Glaciation maximum. Even in western Poland and in parts of the Moravian Gate (Czech Republic), the Elsterian deposits are overlain by sediments of Saalian age; they are only found again at the ground surface east of the Oder River (Marks 2004).

The first Elsterian tills in northwest Germany usually contain relatively high proportions of western Scandinavian boulders (Fig. 2.17). Rhomb Porphyry and other south Norwegian indicators are frequent, together with flint conglomerate from the bottom of the Skagerrak (Meyer 1970, 1983b). The Norwegian rocks are never dominant, their contribution always being less than 10% of the total (crystalline + sedimentary) indicator clasts. Eissmann (1967) has shown that Rhomb Porphyry even occurs in the Elsterian till of the Leipzig area. It is thought that there may have been an early Norwegian/west Swedish ice stream, the eastern boundary of which was located in the Bornholm area, that was pushed aside by a north Swedish/Finnish ice stream during the Elsterian Glaciation. In east Germany and Poland, the 'Baltic' facies of the Elsterian tills is the 'normal' form. This means that the tills contain abundant Palaeozoic limestone and little flint, which is what would be expected due to the position of these regions being much further east.

The southern marginal area of the north European glaciation is dissected by a system of deep channels or tunnel valleys, which are often connected to a network-like pattern.

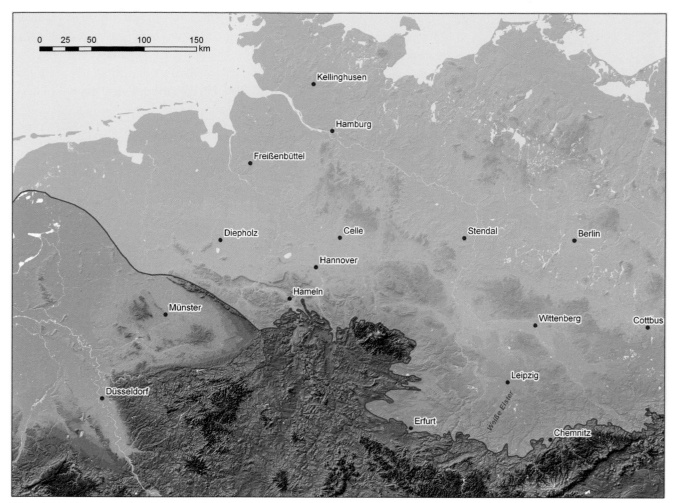

Figure 2.17 The extent of the Elsterian Glaciation in northern Germany.

The most extensive Elsterian channels in Lower Saxony end approximately in the south along a line from Diepholz via Nienburg to Celle. In the west they extend through the Netherlands into the North Sea and into eastern England (Lutz et al. 2009). While the shallow channels, which are generally filled with morainic material, lack the late Elsterian glaciolacustrine clays and also the marine sediments of the subsequent Holsteinian, both are frequently present in the deeper tunnel valleys. This means that the deeper channels were either formed later, or that they remained active as meltwater drainage pathways until the end of the Elsterian.

As the Elsterian ice melted, large ice-dammed lakes formed in front of the ice margin in which silt and clay were deposited. In northwest Germany, these deposits are referred to as 'Lauenburg Clay'. These deposits extend from the Netherlands ('potklei' in the provinces of Friesland, Groningen and Drenthe) as far as Mecklenburg-Vorpommern in Germany. This glaciolacustrine sediment can reach thicknesses of over 150 m. The composition of the Lauenburg Clay reflects the gradual disintegration of the Elsterian ice. While the lower parts are rich in admixed gravel (dropstones) and sand, sorting increases and stratification gradually begins towards the top. If this represents annual layering, the Lauenburg Clay of the Hamburg area was deposited over a period of over 2000 years. The upper part of the Lauenburg Clay is often reddish in colour. Hinsch (1993) attributes this to the fact that the finest sediments of the meltwater of a final, East Baltic-dominated, Elsterian ice advance were deposited here.

2.6.2 HOLSTEINIAN INTERGLACIAL

Towards the end of the Elsterian Glaciation, the sea transgressed into Jutland and northern Germany. Marine late Elsterian deposits can be traced inland as far as Kellinghusen, about 50 km from the present coastline. This early transgression was a result of the glacio-isostatic depression of the land surface. In the Holsteinian interglacial period (Fig. 2.18), a second much more extensive transgression occurred. The Holsteinian climate was warmer than in the two following warm stages, and was effectively rather similar to today's climate.

Figure 2.18 Holsteinian interglacial site of Hamburg-Hummelsbüttel. Photograph by Jürgen Ehlers.

2.6.3 SAALIAN COMPLEX

The climatic deterioration at the end of the Holsteinian was followed by a return to a predominantly cold climate period termed the Saalian Stage. Since several climatic fluctuations occurred during the Saalian – with some of the warm intervals being of interglacial character – the resulting complex stratigraphic unit should be referred to as the Saalian Complex Stage (Litt et al. 2007).

2.6.3.1 Early Saalian

During the Early Saalian, that is, during the period between the end of the Holsteinian Stage interglacial and the first ice advance of the Saalian Complex, periglacial conditions initially prevailed. However, this phase was followed by a renewed intensive warming event. The deposits of this additional warm stage were discovered almost simultaneously by Klaus Erd and Burkhard Menke. Menke presented the entire pollen diagram from this new interglacial in 1964. At Wacken (Schleswig-Holstein) he found a 1-m-thick layer of interglacial deposits separated from the underlying 34-m-thick Holsteinian interglacial sequence by cryoturbated sand. The complete profile of the Wacken interval was published four years later (Menke 1968). Meanwhile, Klaus Erd also reported the results of his studies of a sequence at Pritzwalk–Prignitz. He found a warm period above the Holsteinian-age interglacial deposits, which he called the Dömnitz Interglacial (Erd 1965). These two sequences appear to represent the same climatic event, and are therefore termed the 'Dömnitz Interglacial' (Litt et al. 2007).

With the discovery of the Dömnitz Interglacial, a fundamental principle of Quaternary stratigraphy as then established began to falter. Until then, it was assumed that each glaciation had been followed by only one interglacial period. However, now there was an exception: between the Elsterian and Saalian there were two warm intervals. Or perhaps even more?

Urban et al. (1991) and Urban (1995) discovered another intra-Saalian Interglacial in the Schöningen lignite mine (about 30 km ESE of Braunschweig), which they considered to postdate the Dömnitz Interglacial. Litt et al. (2007) point out, however, that the early phase of the interglacial is missing and consequently a correlation with other sites is difficult. The 'Reinstorf Interglacial' (Urban 1995, 2007) found in Schöningen is likely to be equivalent to the Holsteinian, according to Litt et al. (2007). Within the Saalian Complex there is therefore at least one additional warm event that has been unequivocally established, although there is a strong likelihood that a second also occurred.

2.6.3.2 The Saalian Glaciations

It has been traditionally assumed that northern Germany was affected by two major ice advances during the Saalian glaciations (Fig. 2.19): the Drenthe and the Warthe advances. The concept goes back to early work by Woldstedt (1927), who also coined the term 'Warthe'. The name 'Drenthe' was introduced much later as a consequence of work in the Netherlands (Woldstedt 1954).

Three different tills are found within the Saalian Complex in northern Germany, separated from each other by meltwater deposits. Since the different local stratigraphies are so far contradictory, the following text follows Kabel (1982) who referred to the three ice advances as Older (early), Middle and Younger (late) Saalian glaciations. The Older Saalian ice advance is equated to the Polish Odra Glaciation. This ice advance reached the Netherlands and the edge of the Central German Uplands. In the Netherlands and northwestern Germany the uppermost part of the sequence is a reddish-brown till, the composition of which differs from the rest of the Older Saalian tills. It is characterized by East Baltic-derived clasts and matrix. The lack of flint and the presence of relatively large amounts of dolomite suggest that it was deposited by ice advancing from the NNE–ENE.

It is believed that all three Saalian tills that occur in the Saale–Elbe region were deposited during the Older Saalian Glaciation. Morphologically, two different ice advances can be distinguished. In the first, the so-called Zeitz phase, the ice reached its greatest extent. In the ensuing melting phase, the ice margin receded into the Bitterfeld region. The second advance only reached the Halle–Leipzig region (Leipzig phase).

At the end of the Older Saalian Glaciation the ice melted back to the area of the Baltic Sea. The subsequent ice advance of the Middle Saalian Glaciation began in northwestern Germany with the deposition of thick meltwater sands. In contrast to the Elsterian meltwater deposits which are found mainly in the channels, wide outwash plains were deposited, some of which are several tens of metres thick. In this phase the ice advanced into the area south of the present Elbe valley. Drainage was aligned via the Aller–Weser valley towards the North Sea

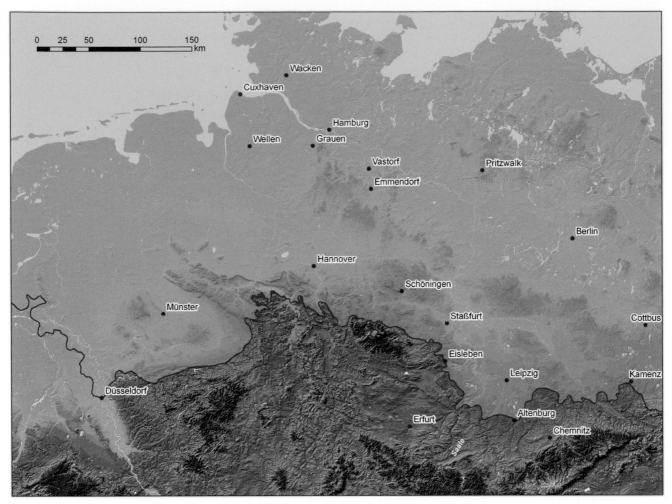

Figure 2.19 The extent of the Saalian Glaciation in northern Germany.

(Meyer 1983c). It is believed that the ice advanced at least as far as the Altenwalder Geest area, a push moraine south of Cuxhaven. In the advance of the Middle Saalian Glaciation, many of the push moraines of the northern Lüneburg Heath were formed.

Correlation with areas further to the east is problematic. The characteristic Middle Saalian till, with its high content of chalk and flint, is not found in Sachsen-Anhalt, Brandenburg or Mecklenburg-Vorpommern. Only a few flint-rich gravel deposits in the Altmark region might be attributable to this advance. It is probable that the composition of the deposits changes further to the east, so that lithostratigraphic correlation is difficult. In Schleswig-Holstein and Hamburg this ice advance has been assigned to the Warthe phase, whereas in Lower Saxony it is regarded as part of the Drenthe phase. The main watershed at the time was between the Warta and Pilica rivers. The Pilica drained in an easterly direction via the Dnieper and Pripyat towards the Black Sea.

After the end of the Middle Saalian Glaciation, the ice probably melted back again to the area of the present Baltic Sea. Large quantities of dead ice remained, which was subsequently covered by meltwater sands derived from the Younger Saalian ice advance. Indicator clast analyses from Brandenburg suggest that the youngest Saalian ice advance in that district represents both the Middle and the Younger Saalian advances in the west. The margin of this ice advance is found in the Fläming, the Lower Lausitz Grenzwall and Muskauer Faltenbogen. It is assumed that the ice pushed forwards across the River Elbe to the Schmiedeberg terminal moraine. The youngest till of the Saalian Glaciation, the 'red Altmark Till', is found widely in eastern Lower Saxony and in the Altmark region. Here it is characterized by an East Baltic clast association. The ice at this time advanced into northern Germany from the NE–ENE. In the Hamburg area the ice advanced directly from the east.

2.6.4 EEMIAN INTERGLACIAL

The term 'Eemian interglacial period' (named after the River Eem in the Netherlands) was coined by Harting (1874). International recognition was achieved over 30 years later through a comprehensive study by Danish geologists. Madsen et al. (1908) were the first to compare marine deposits of the Eemian interglacial in Denmark, northern Germany and the Netherlands, and they adopted Harting's term. Later the name was also transferred to the terrestrial deposits of the last interglacial period following Jessen & Milthers' (1928) detailed investigations.

The climatic evolution of the Eemian interglacial period was similar to the Holocene, but overall the Eemian was warmer. Consequently, it is not surprising that the Eemian global sea level in most areas (but not in north Germany) was higher than that today.

The Dutch Eemian deposits of Amsterdam and Amersfoort have been re-examined in recent years. On the banks of the Elbe near Lauenburg at the so-called *Oberstleutnantweg*, peats of the Eemian Interglacial are exposed at the *Kuhgrund* site (Fig. 2.20). Precise numerical age determinations have now shown that the Eemian Interglacial spanned the period 126–115 ka (Müller 1974a). The Lauenburg peat has been studied by Menke (1992). This site is interesting not so much for its vegetational succession, which does not differ significantly from other Eemian sites, but for the fact that the peat is not covered by younger glacial deposits. This shows that the Weichselian Ice Sheet did not cross the Elbe River.

The vegetation of the Eemian interglacial period is very well known from numerous palaeobiological investigations of organic deposits in Germany (e.g. Litt 1994; see also Chapter 9). It is clear now that pollen profiles with a differing vegetational sequence do not represent any additional warm stages between the Saalian and Weichselian glaciations. They have either been misinterpreted (e.g. Neumark-Nord), stratigraphically incorrectly classified ('Treene Interglacial'; cf. Menke 1985) or found in glaciotectonically disturbed strata ('Uecker Interglacial'; cf. Hermsdorf & Strahl 2006).

Figure 2.20 Eemian peat on the banks of the Elbe at Lauenburg. Photograph by Jürgen Ehlers.

2.6.5 WEICHSELIAN

Early efforts to subdivide the Ice-Age deposits in northern Germany were based solely on geomorphology. The three known glaciations of north Germany were initially unnamed. Keilhack (1909) distinguished only 'terminal moraines of the last glaciation' and 'terminal moraines of the penultimate glaciation'. The age of the moraines was assessed according to the perceived freshness of the landforms. Fläming and Lüneburg Heath (south of the Elbe River) were assigned to the most recent glaciation. Only by more detailed morphological mapping could Gripp (1924) finally determine the approximate Weichselian maximum glacial limit (Figs 2.21, 2.22).

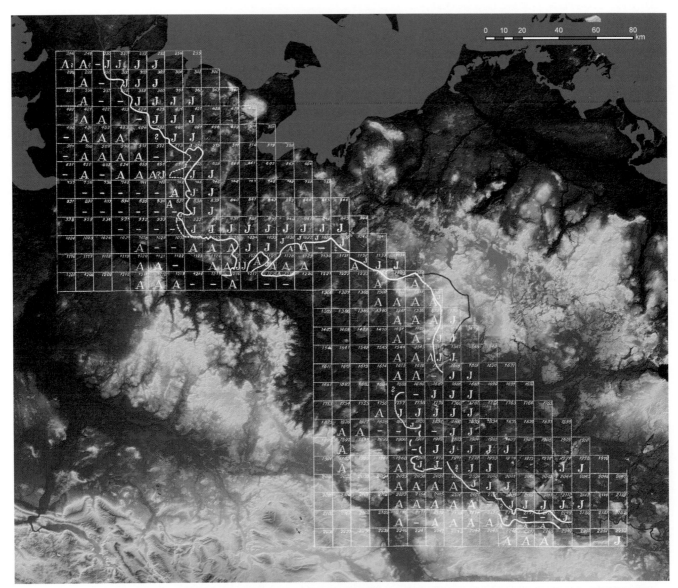

Figure 2.21 The original limit of the Weichselian Glaciation (white line, mapped by Gripp 1924) compared to the present-day Weichselian limit (red line, Ehlers et al. 2011a). A = 'Altmoräne' ('Old Morainic' landscape), J = 'Jungmoräne' ('Young Morainic' landscape).

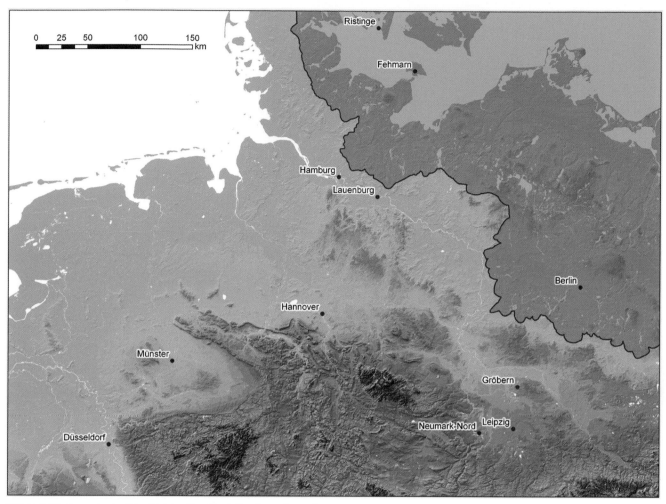

Figure 2.22 The extent of the Weichselian Glaciation in northern Germany.

A year later Woldstedt attempted to determine the limits of the various ice advances in more detail. Unlike previous workers he did not map the terminal moraines as isolated hills, but linked them to continuous ice-marginal positions, including the distribution of the vast outwash plains in his considerations. He distinguished between: Fläming Phase; Jutland Phase (with Brandenburg and Poznan sub-phases; Fig. 2.23); and Pomeranian Phase (Fig. 2.24).

Woldstedt followed Gripp's distinction into older and younger landforms, but left it open whether the two generations of ice margins were separated from each other by an interglacial or interstadial climatic event. In 1935 he published his 'Geological-morphological map of the north German glaciation area' in which he assigned the landforms to three glaciations. Until the advent of radiocarbon dating, this map formed the basis of the north German morphostratigraphy (Lüthgens & Böse 2010).

The maps by Woldstedt and later Liedtke (1975) looked very convincing. Gradually, as knowledge increased through the dating of organic deposits and after a first model-like reconstruction by Boulton & Jones (1979), the first complete map of the isochronous deglaciation of the Scandinavian Ice Sheet was presented in 1981. This chart was based entirely on

Figure 2.23 Weichselian ice margin in Jutland. The Weichselian glacial area is dissected by numerous subglacial meltwater channels (tunnel valleys), while the older morainic areas and outwash plains, beyond the ice limit, have a smoothed relief and relate to the Saalian Glaciation.

radiocarbon dating of appropriate deposits. In the detailed accompanying text, Björn G. Andersen had described and discussed every single radiocarbon-dated site.

This map was based on reliable data, but it was still dependent upon indirect evidence as the glacial sediments could not be dated; instead, the organic strata either over- or underlying the glacigenic deposits were used. The 1980s saw the advent of new techniques that could directly determine the age of the sediments, however, at least if they were of Weichselian age.

The first method that offered new possibilities in this direction was thermoluminescence (TL; Box 2.5). With the help of TL dating, and later also optically stimulated luminescence (OSL; Box 2.5) dating, it was demonstrated that in the Weichselian of Denmark there had been an early advance of the ice sheet which had passed through the Baltic Sea Basin to the area of the Danish islands. This Old Baltic Advance had deposited the 'thin till' (the Ristinge Klint Till) exposed in the famous Ristinge cliff section on Langeland (Houmark-Nielsen 2010).

If this advance passed through the Baltic Sea Basin, then Schleswig-Holstein had to be affected. In 1995 it was shown that there had been an ice advance in Schleswig-Holstein, which was significantly older than the Late Weichselian Glaciation maximum (Marks et al. 1995).

Figure 2.24 Block-strewn Pomeranian Phase end moraine, south of Neubrandenburg. Photograph by Jürgen Ehlers.

BOX 2.5 DEFECTS IN THE CRYSTAL LATTICE: AGE DETERMINATION METHODS

In nature, the internal crystal structure deviates from the ideal lattice of the minerals that is shown in textbooks. Two types of defects are observed: (1) primary damage caused during the mineral formation; and (2) secondary damage that has occurred over time under the influence of alpha, beta or gamma radiation. These defects act as 'traps' for electrons that are emitted by the minerals under radioactive radiation. If the mineral is heated to about 400°C, the electrons are released and the atoms return to their starting point. The release of a large amount of photons in this process leads to a measurable light effect, so-called thermoluminescence (TL).

The TL signal is proportional to the number of the released electrons. It is therefore also proportional to the time during which the mineral was exposed to the radiation. Ideally, there would be a linear relationship between exposure time and thermoluminescence. Unfortunately, this is not the case. During the passage of time, more and more potential electron traps are already filled and the mineral eventually approaches saturation.

In order that thermoluminescence can be used for dating, there must have occurred an event in the past that emptied the electron traps. Since the electrons are released by heat, the electron clock is always set to zero when the material is burned. The method was used first in archaeology in the 1950s to determine the age of pottery or burnt flint.

Glacial deposits are not normally heated. However, if quartz or feldspar crystals are exposed to sunlight for a long time, most of the electron traps are emptied but not all. Since this is a gradual process, it results in a bleaching curve. For successful dating, the sample should have been exposed long enough to sunlight so that most of the electron traps have been emptied. In the laboratory, the natural thermoluminescence of the sample is measured and compared with an artificial TL signal which is generated by exposing the sample to a calibrated radiation source (Wintle 1991).

In 1979, Wintle & Huntley suggested that grains of quartz and feldspar could be used for TL dating to determine the time that had elapsed since the grains were last exposed to sunlight, that is, the time of deposition. This idea was not new; a laboratory in Kiev had published sediment age determinations by TL in Russia since about 1968, but the West was sceptical about this data.

The main problem with TL dating is that the sediment sample must have been exposed long enough to sunlight to erase all the older TL signals, and after that point was never exposed to sunlight until the time of sampling. Loess and aeolian sand meet most of these conditions. The procedure has also been used for dating of meltwater deposits and of tills that should be unsuitable for TL dating. The age limit for potential dates is set by some scientists to about 1 million years; however, the majority of researchers adopt an upper limit for reliable results at an age of around 100,000 years.

A new method of luminescence dating of sediments has evolved from 1985: optically stimulated luminescence (OSL) dating. In OSL dating the same type of defects in the crystal lattice is utilized as in TL dating, but in this case only light-sensitive traps are used. Instead of heat, a laser beam is used to release the trapped light-sensitive electrons (Geyh & Schleicher 1990). Since the total bleaching in sunlight takes just seconds, the method has the advantage that it can also be used for materials that were only exposed to daylight for very short periods of time, for example, meltwater sands (Huntley et al. 1985).

Further progress led to the detection by Hütt et al. (1988) that, when feldspar instead of quartz was stimulated at room temperature in the near-infrared, the luminescence signal yielded a similar result to that of TL. This led to the development of infrared-stimulated luminescence (IRSL).

(continued)

BOX 2.5 DEFECTS IN THE CRYSTAL LATTICE: AGE DETERMINATION METHODS (*CONTINUED*)

The single-grain dating method, developed by Duller (2008), finally increased the accuracy of luminescence dating and is now widely used for dating of sand. OSL is an important tool for the dating of Quaternary deposits, especially from the last 100 ka (Wintle 2008).

While the original luminescence centres become damaged during TL and OSL dating, use of electron spin resonance (ESR) means that microwave energy is released to determine the centres *in situ*. This has the advantage that the sample remains in an unaltered form, available for further study. The period which can be dated by this method spans the entire Pleistocene. The method is applied to limestones, for example, stalactites, travertine, mollusc shells or coral. However, the procedure is not without its problems. One of the basic assumptions is that the centres remain stable over long periods of time, but that is not necessarily the case. Molluscs have been found to be unreliable, and the ESR ages of stalactites and travertine do not always agree with the U–Th ages of the same samples.

TL, OSL, IRSL and ESR have significantly improved the possiblity of dating sediment samples, especially Late Pleistocene deposits. All four methods allow the age determination of materials that otherwise could not be dated.

Preusser came to the same result based on OSL dating of meltwater sediments (Preusser 1999). It turned out that a significant proportion of the Weichselian meltwater sands had to be attributed to an Early Weichselian ice advance (Frechen et al. 2007).

With the help of ^{10}Be dating, (Box 2.6; Fig. 2.25) the deglaciation at the end of the Weichselian has recently been dated in Poland, Belarus and Lithuania at a number of sites (Rinterknecht et al. 2005, 2007, 2008). It must be noted, however, that the ice margin was subdivided into lobes reflecting the advances of various ice streams. The latter most likely did not occur simultaneously, since the various ice streams and lobes had developed their own dynamics (Houmark-Nielsen 2003). Individual segments of the ice-marginal ridges were in fact created at different times (time-transgressive) (Lüthgen & Böse 2010).

The dating of Quaternary events is just one aspect of Quaternary research, however; interpretation must always be based on geological and geomorphological field investigations.

BOX 2.6 COSMIC RAYS AND ROCKS: DATING THE EARTH'S SURFACE

Another method of age determination dates the age of rock surfaces. Radioactive ^{10}Be is generated in the atmosphere by the reaction of nitrogen and oxygen with cosmic radiation. ^{10}Be is bound to aerosol particles and therefore arrives with rain on the Earth's surface as background meteoric ^{10}Be. However, ^{10}Be is also produced at the Earth's surface by secondary cosmic radiation. This secondary radiation occurs as protons lose an electron and form neutrons (and muons) which shower the Earth's surface. This secondary neutron flux can penetrate up to 2 m into rocks and causes a nuclear reaction in the oxygen (the target element for ^{10}Be) present in quartz, which is common in all siliceous rocks. This *in situ* terrestrial ^{10}Be is several orders of magnitude less than the ^{10}Be that is formed in the upper atmosphere. Production rates of ^{10}Be in surface rocks are small (of the order of just a few atoms per year) and such small accumulations of atoms need to be measured on extremely sensitive equipment: an atomic mass spectrometer measures the actual number of atoms in a processed rock sample.

^{10}Be dating is used to determine the age of rock surfaces that contain quartz (siliceous rocks) and ^{10}Be is formed from the target element of oxygen. In siliceous rocks ^{26}Al can also be used because this *in situ* terrestrial cosmogenic nuclide is formed from the target element of silica. Since siliceous rocks contain both silica and oxygen (SiO_2), both ^{10}Be and ^{26}Al can be used to date the same rock surface. Rocks that contain no quartz can be dated using other terrestrial cosmogenic nuclides, such as ^{36}Cl in limestones or basalts where the target elements are calcium or potassium, respectively.

^{10}Be analysis can be used to date surfaces dating from the Late Holocene (several hundreds of years) to the pre-Quaternary (several millions of years). In the time period of the Pleistocene glaciations, it has been primarily applied to erratic boulders or ice-moulded bedrock. In particular, boulders on moraine surfaces have been dated with the hope of measuring directly the age of the ice-marginal position. The applicability of the method can be hampered by several factors, however.

1 The rock surface may have been removed by weathering (the best rock surfaces are those which display striations and therefore zero or minimal erosion).
2 Rock surfaces may be shielded by cliffs or mountains, preventing full bombardment by secondary cosmogenic radiation.
3 Snow cover can attenuate radiation penetration to the rock surface.
4 Boulders may topple or become exhumed, with their surfaces revealed to the atmosphere sometime after their deposition by glaciers.

Consequently, the measured ages are often too young. The scatter of values increases with age, and there is a risk that even the oldest measured value obtained from a >100 ka old moraine in a series of analyses may only represent a minimum age. Other problems include: rock previously exposed on a cliff face can fall onto a glacier and be passively transported supraglacially (on the top of the glacier) or englacially (within the glacier) before being laid to rest as a 'glacial' boulder; or glaciers fail to erode bedrock to depths of >2 m, resulting in the presence of previously acquired terrestrial cosmogenic nuclides. In both these cases the inheritance of terrestrial cosmogenic nuclides that built up during previous exposure mean that calculated ages will be too old (i.e. older than the most recent stable exposure history of that surface). Another significant issue is understanding the production rates of terrestrial cosmogenic nuclides such as ^{10}Be in rocks. It has recently become apparent that production rates are lower than previously thought, meaning that all ages published prior to *c.* 2009 are actually up to 15% older (e.g. Balco et al. 2009; Putnam et al. 2010; Fenton et al. 2011; Briner et al. 2012; Young et al. 2013).

A final challenge is measuring the amount of cosmogenic nuclides in a rock sample. This requires an accelerator mass spectrometer which can count individual atoms and a laboratory dedicated to the preparation and measurement of samples. There are only a few facilities worldwide, especially those that can measure cosmogenic nuclides using atomic mass spectrometry (AMS), although preparation laboratories are more numerous. Examples of AMS laboratories that measure cosmogenic nuclides include: the Scottish Universities Environmental Research Centre (United Kingdom); ETH Zurich (Switzerland); CEREGE (France); University of Cologne (Germany); Helmholtz Centres at Potsdam & Dresden (Germany); PrimeLab (Purdue University, USA); Lawrence Livermore National Laboratory (USA); the Australian Nuclear Science and Technology Organization; and the Australian National University (both Australia).

Despite the challenges associated with the technique, cosmogenic exposure dating has revolutionized glacial geomorphology over the past two decades. Prior to this, glacial landforms were notoriously difficult to date and often relied on radiocarbon dating of sediment cores taken from glacial lakes, which provided minimum ages for the last deglaciation. A review of the potential of cosmogenic exposure dating is provided in Balco (2011).

Figure 2.25 A boulder being sampled for dating on a moraine in the High Atlas, Morocco. Around 500 g of sample from the top 5 cm thickness was taken from the top of the boulders (which are andesites that contain quartz veins and phenocrysts) for ^{10}Be exposure dating. The mass of sample that is required depends on how much quartz is present, the duration of surface exposure and the altitude/latitude of the site. After extensive treatment in the laboratory c. 10–50 g of pure quartz grains were normally required from the High Atlas in order for ^{10}Be to be successfully measured in Late Pleistocene samples. Photograph by Phil Hughes.

2.7 The British Pleistocene Succession

With some minor exceptions, the bulk of the evidence for the pre-glacial Pleistocene is restricted to southern and eastern England, and to East Anglia in particular as the eastern part of this area forms the western margin of the tectonically downwarping North Sea Basin. Although in a marginal setting, sedimentation mostly records sea-level highstand sequences of shallow marine, littoral and sublittoral environments. No direct evidence for glaciation is known from these sequences. In southern Britain, a series of fluvial aggradations comprising predominantly quartz-rich gravels of the Kesgrave Formation were deposited by the ancestral Thames (Hey 1991). These deposits range from the 'Pre-Pastonian' Substage (Tiglian) to the early Anglian and contain ample evidence of cold-climate deposition. Clasts of rhyolitic rocks from the Berwyn Mountains in North Wales (Rose 1994; Rose et al. 1999) are regarded as evidence for repeated contemporaneous glaciation that entered the uppermost part of the Thames catchment.

Pre-Anglian interglacial deposits were first reported from Cromer in East Anglia (see Fig. 2.26 for place names mentioned in this section). Cromerian deposits were subsequently found in various other places throughout Europe, representing at least four warm stages (see

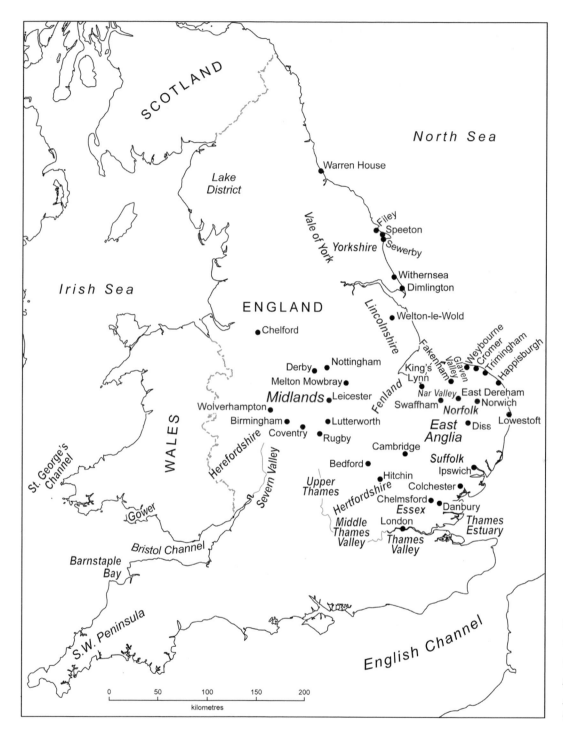

Figure 2.26 Locations mentioned in the description of the British glacial succession.

Source: Gibbard & Clark (2011).

Turner 1996 for a review). The precise stratigraphic position of the individual sites, however, has remained rather unclear. The English Cromerian of West Runton in East Anglia should be correlated either with the Dutch Cromerian II or with a so-far undetected interglacial between the Cromerian II and III. Regarding mammal stratigraphy, the English Cromerian sites are difficult to compare with those in the Netherlands. The stratotype of the Cromerian occurs *sensu stricto* at West Runton; at several other British sites, including Little Oakley and Sugworth, *Mimomys savini* occurs while other *Mimomys* species are missing. In the Netherlands there is no equivalent to this faunal assemblage. However, the English Cromerian *sensu stricto* does compare well with the Thuringian Voigtstedt Interglacial.

The earliest undisputed glacial diamicton known on land in Britain is the Happisburgh (Member) Diamicton, which is generally regarded as being of Anglian age. This marks the arrival of the glacial part of the Pleistocene sequence in Britain. Once again evidence for this glaciation is best developed in eastern and southern central England, although sediments are known to extend as far as the Welsh borderlands region near Hereford and south Wales.

Although it is now generally possible to relate the terrestrial to deep-sea sequences fairly reliably at a coarse scale for the Late Pleistocene and, to a lesser extent, in the late Middle Pleistocene, the situation is progressively and significantly more difficult for earlier periods (Gibbard & West 2000). This is especially so for the early Middle Pleistocene (Turner 1996), the period for which Rose (2009) and Lee et al. (2010) not only identify additional pre-Anglian glacial events indirectly from the occurrence of erratic clasts in the fluvial Kesgrave Formation Thames deposits, noted above, but tentatively assign the identified glaciations to specific marine isotope stages. However, there is no means of reliable 'dating' for the sediments in question.

While there is certainly a longer record of glaciation represented in the British Isles as a whole, the times at which these events occurred must be regarded as undetermined at present. Therefore, as elsewhere in the southern North Sea region, no direct evidence for pre-Anglian (pre-Elsterian: MIS 12) glaciations has yet been found.

2.7.1 ANGLIAN GLACIATIONS

The glaciation in the Anglian Stage was the most extensive of the British Pleistocene (Fig. 2.27). However, since for the greater part of its extent its deposits were overridden by younger glaciations, much of the sedimentary record has been lost. For this reason the Anglian glacial sediments are best preserved beyond the margins of later ice advances, and Eastern England and the adjacent offshore region include extensive sequences of deposits and features formed during this event. In North Norfolk glacial deposits rest on Cromerian Complex Stage deposits and, throughout their distribution area, are overlain directly by Hoxnian Stage and younger interglacial sediments (Preece & Parfitt 2000; Banham et al. 2001; Gibbard et al. 2008; Preece et al. 2009). No evidence of interstadial or higher-rank climatic oscillations has been unequivocally identified intervening within the Anglian sequence; it is therefore generally interpreted as representing a single complex glacial event. However, this was questioned by Moorlock et al. (2000) and a number of subsequent publications. The Anglian is equated with the continental Elsterian Stage. Its equivalance to MIS 12 has recently been definitively demonstrated by both direct means (Toucanne et al. 2009a, b) and numerical dating (Pawley et al. 2008).

The Anglian Glaciation began as the ice advanced from the northeast and overrode the pre-existing sedimentary sequences. This advance deposited a complex suite of interbedded tills and associated meltwater sediments, termed the North Sea Drift Formation. These sediments, best exposed on the northern and northeastern coasts of East Anglia from Weybourne to Lowestoft (Fig. 2.26), can also be traced inland as far as Norwich and south towards Diss (Mathers et al.

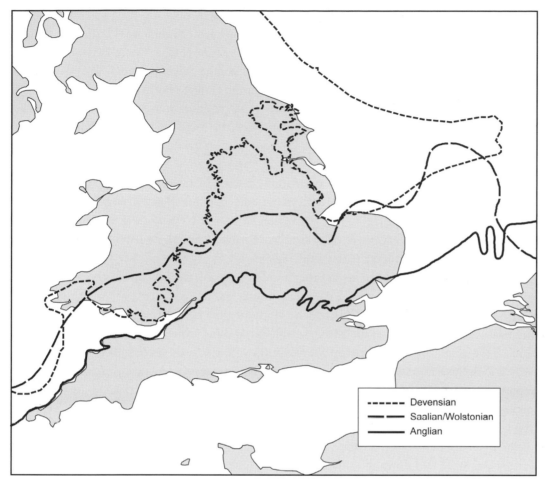

Figure 2.27 The limits of glaciations in the British Isles, representing maximum extent limits. However, the maximum extent lines are known to be diachronous and in some areas these lines are simplifications. For example, for the Devensian limits compare the limits in SW England with those in Figure 2.28 and 2.29.

Source: Gibbard & Clark (2011).

Legend:
- - - - - - Devensian
— — — Saalian/Wolstonian
——— Anglian

1987). Internally this sequence is highly complex, including evidence for a three-fold oscillation of the ice sheet; the intervening phases are represented by deltaic outwash and glaciolacustrine sedimentation (Hart 1987; Hart & Boulton 1991; Lunkka 1994; Gibbard 1995; Gibbard in Clark et al. 2004). For all but the final stillstand phase during which the delta-morainic Cromer Ridge in northeast Norfolk was laid down (Gibbard & van der Vegt 2012), no precise margins are known for these individual advances. The ridge is clearly intimately related to the glacial sequences on which it rests. The sequence includes the glacio-tectonized elements at Trimingham in the east. The sediments are overlain by Hoxnian-age interglacial pond deposits (Preece et al. 2009). OSL dating of the Morston Raised Beach (Late Saalian, MIS 5–7) which also overlies the deposits (Hoare et al. 2009) reinforces this interpretation.

The North Sea Drift Formation sediments contain a suite of igneous and metamorphic erratics of southern Norwegian origin. The rhomb porphyry and larvikite, derived from the Oslofjord area, indicate that the ice sheet originated in southern Norway and crossed the North Sea Basin (Hoare & Connell 2004). Rhomb porphyries and larvikites can be found as far south as Bedford, Hitchin, Ipswich and Cambridge (Ehlers & Gibbard 1991) where they were probably carried by later ice movement. Alternatively, reworking from older pre-existing Scandinavian deposits in the North Sea has been suggested (Lee et al. 2011). However, no such sediments have yet been identified.

During the next phase, ice of British origin advanced through the Vale of York and Lincolnshire into central and western East Anglia. On the basis of lithology (Perrin et al. 1979), erratic content (Baden-Powell 1948), detailed fabric measurements (West & Donner 1956; Ehlers et al. 1987) and fossil assemblages from erratic chalk pebbles (Fish & Whiteman 2001), it can be shown that the ice radiated outwards from the Fenland Basin in a fan-like pattern. It is thought that interaction of this British ice with the Scandinavian ice was responsible for this unusual pattern of ice movement. This ice deposited the so-called Lowestoft Formation till, which has a grey to blue-grey clay matrix (up to 45%), is rich in flint and subrounded chalk clasts and includes erratics that originated on the British landmass.

Advance of the British Lowestoft Formation ice into Essex and Hertfordshire brought it into the region influenced by the River Thames and its tributaries. The substantial series of Thames deposits, termed the Kesgrave Formation, are aligned WSW–ENE to west–east across southern East Anglia. That pre-glacial course of the river was overridden by the ice in all but the western and southernmost parts of the region.

With the continued withdrawal of the Scandinavian ice, the Lincolnshire coast and adjacent offshore area became open for the expanding British ice advancing south-eastwards into Norfolk. Again the Wash and the Breckland Gap directed the ice stream. Ice that flowed over the relatively high ground of the chalk escarpment and into the Fenland Basin led to deposition of the characteristic chalk-rich 'Marly Drift' till facies onto more typical Lowestoft Till near Kings Lynn (Straw 1991). Ehlers et al. (1987) demonstrated that this facies does not represent a separate glaciation but simply locally derived diamicton deposited during a later phase of the Anglian. Chalk-rich till was transported into much of East Anglia, including the Gipping valley in Suffolk, where it overlies the Jurassic clay-rich tills (Lowestoft facies) of the preceding Lowestoft advance (Ehlers et al. 1987). As the ice retreated, vast meltwater formations were laid down as sands and gravels, for example those between East Dereham, Swaffham and Fakenham. Here the gravels rest on the chalk-rich till of the last Anglian ice advance. Similarly, in the area around the Glaven valley in North Norfolk kames, dead-ice topography and an esker are associated with the retreat of the chalk-rich till ice (Sparks & West 1964; Ehlers et al. 1987).

Throughout East Anglia a series of deep, steep-sided valleys have been found cutting through the chalk and associated bedrock. Detailed studies of these 'tunnel valleys', by Woodland (1970), Cox (1985), Cox & Nickless (1972) and van der Vegt et al. (2009), among others, indicate that they are normally filled with glacial sediments, predominantly meltwater sands, gravels or fines. Tills also occur but are less frequent. Closely comparable in form and scale to the *rinnen* of Denmark, northern Germany and Poland, although of shallower depth, they are undoubtedly of glacial origin and probably result from subglacial drainage discharge under high hydrostatic pressure (Ehlers et al. 1984). Their overall depth appears controlled by substrate and ice-lobe characteristics (GRASP 2009).

In northern Norfolk (particularly at Kelling and Salthouse Heaths), substantial proglacial outwash sandur plains, apparently independent of tunnel valleys, developed and mark retreat-phase ice-front stillstand positions (Sparks & West 1964). In the marginal areas, particularly at Corton and along the Cromer Ridge (already noted), the ice front retreated in standing deep water (Bridge & Hopson 1985; Hopson & Bridge 1987; Bridge 1988). A similar phenomenon is found in the Nar valley (Ventris 1986, 1996).

Outwash from an Anglian Ice Sheet in NE Herefordshire deposited the Risbury Formation, which consists of up to 30 m of glaciofluvial and ice-contact glaciodeltaic gravels (Richards 1998). The gravels contain clasts of Devonian and Lower Palaeozoic rocks derived from the west and southwest, suggesting that the glacier originated in central Wales. The Anglian age of these glacial deposits is indicated by their stratigraphical relationships to interglacial sequences, paralleling those found in East Anglia. In the Mathon valley to the west of the Malvern Hills, the formation locally overlies fluvial silts containing a Cromerian Complex-age interglacial fauna and flora (Coope et al. 2002). At the head of the Cradley Brook, it is overlain by silts containing late Anglian–early Hoxnian temperate pollen and molluscs (Barclay et al. 1992).

In general, the pre-Devensian glaciation of southern Wales is referred to as the 'Irish Sea' Glaciation (Pringle & George 1961). This glaciation potentially includes evidence for one or more glacial episodes during which complete glaciation of the province occurred by local ice sheets. The only unequivocal Anglian-age unit is the Llanddewi Formation of SW Gower, which represents the margin of Anglian-age Welsh ice. This glaciation extended across the Bristol Channel as far as the northern coast of the English southwest peninsula, although there is some dispute over precisely how far it reached. However, recent geochronological evidence based on cosmogenic exposure dating from glaciated surfaces on Lundy Island (at the entrance to the Bristol Channel near Barnstaple Bay (Fig. 2.26 and 2.29) indicates that Devensian ice also reached this area. The Devensian ice came not from the direction of Gower but from the west, moving eastwards up the Bristol Channel before retreating between 35,000 and 40,000 years ago (Rolfe et al. 2012).

2.7.2 HOXNIAN INTERGLACIAL STAGE

In Britain, the Hoxnian (Holsteinian) deposits at the classic sites of Hoxne or Marks Tey can be clearly distinguished from earlier as well as younger Ipswichian (i.e. Eemian Interglacial) deposits (West 1980). However, there are also organic deposits from some Middle or Upper Pleistocene sites that do not fit readily into the scheme. They are either atypical equivalents of the known interglacials, or they may represent additional stages (such as the continental Dömnitz).

2.7.3 WOLSTONIAN STAGE GLACIATIONS

The second glaciation in Britain took place during the late Middle Pleistocene, between the Hoxnian (Holsteinian, c. ?MIS 11c; Ashton et al. 2008) and Ipswichian (Eemian, c. MIS 5e) interglacial stages. This glacial episode is less well represented in the Pleistocene record and has to date been little studied and weakly defined (Clark et al. 2004).

Following the classic work of Shotton (1953, 1968, 1976, 1983a, b), the glacial sequence of the English Midlands in the area around Coventry and Birmingham was considered to represent what was termed the Wolstonian Glaciation. This name was also selected for the predominantly cold-climate interval or stage in which the glaciation occurred (Mitchell et al. 1973), the Wolstonian (=Saalian) Stage, broadly equivalent to MIS 11b–6. The sequence comprises a series of glacial sediments identified over a large area in the west and central Midlands (Rice

1968, 1981; Rice & Douglas 1991). Of these, diamictons and associated meltwater sediments provide the evidence for an extensive glaciation of the region as far south as the Cotswold Hills at Morton-in-Marsh in Gloucestershire (Figs 2.27, 2.28).

The glacial sediments of the Midlands region overlie deposits of a pre-existing river system, represented by the Baginton–Lillington Gravel. These deposits are characteristically composed of quartz-rich sediment derived from underlying Triassic bedrock. In contrast to normal long-lived fluvial systems in the region, this SW–NE-aligned system was apparently relatively short-lived since it lacks a terrace-like system. The river which deposited these sediments appears to have formed after the Hoxnian and was overridden by the Wolstonian ice; it can therefore only

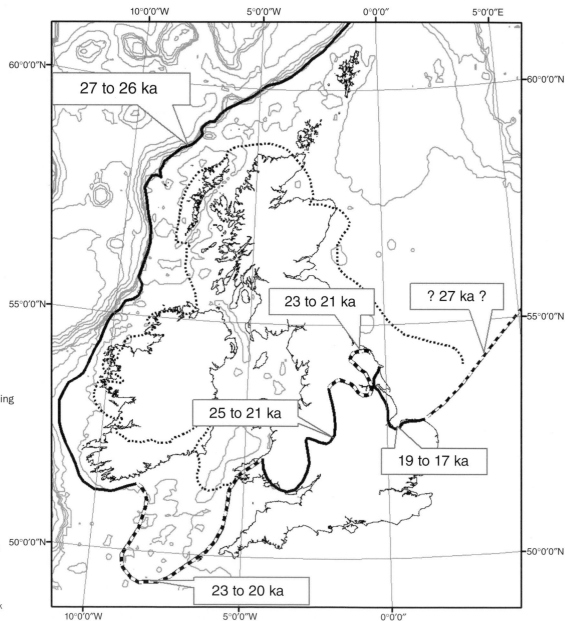

Figure 2.28 The timing of retreat from the maximum phase of the Late Devensian Ice Sheet over the British Isles and Ireland. Some areas saw earlier retreats from more extensive Devensian positions (see Fig. 2.29).

Source: Gibbard & Clark (2011).

have existed for a relatively limited period (c. 140 ka; Shotton 1953, 1983a, b). The reinterpretation of these sediments as the headwaters of a pre-Anglian 'Bytham river', aligned towards East Anglia across the Fenland, found favour for some years, especially when the Wolston Formation sediments were re-assigned to the Anglian Stage (e.g. Rose 1987). However, recent work (Gibbard et al. 2008, 2013) has now confirmed that the Bytham river did not exist in the suggested form and Shotton's interpretation has therefore regained considerable support.

As defined by Shotton (1953), the Midlands Wolston Formation consists of two tills (the Thrussington and Oadby members) with sands and gravels below, between and above. The Thrussington Till is reddish brown and mainly Trias-derived, whereas the Oadby Till contains chalk, flint and Jurassic limestone clasts in a grey matrix derived mainly from Liassic clays. These components suggest deposition by ice from western and eastern sources, respectively. The Thrussington Till was deposited as far east as Melton Mowbray, Rugby and Fenny Compton, and its meltwater formed a large ice-dammed lake between Leicester and Market Bosworth in which up to 25 m of laminated clays and silts (the Bosworth Member) accumulated. The Oadby Till was then deposited over much of the Eastern Midlands as far north as Nottingham and Derby and as far west as Stratford-upon-Avon. It also extends southwards to Moreton-in-Marsh.

In the past there has been controversy over the extent of post-Hoxnian/pre-Ipswichian glaciation in eastern England following the work of Bristow & Cox (1973). More recently, following reappraisals of the glacial evidence in northern East Anglia, Rose and colleagues (Clark et al. 2004) proposed that additional glaciations might have occurred early in the Wolstonian (?MIS 8–10). However, considering the regional stratigraphy (Gibbard et al. 1992; Gibbard in Clark et al. 2004), the validity of this suggestion is questionable.

The advance of the Tottenhill ice lobe into the East Anglian Fenland Basin reached the eastern marginal area and was possibly halted by the rising ground of the chalk hills to the east and south. Here a group of landforms and their underlying deposits represent a series of glaciofluvial delta-fan and related sediments deposited as ice-marginal deltas in a lake at the maximum ice-marginal position (the 'Skertchly Line', Feltwell Formation). This evidence confirms historical descriptions of a glaciation of the Fenland, and demonstrates that sites including Warren Hill, High Lodge, Lakenheath, Feltwell and Shouldham Thorpe are of glacial meltwater origin. On the basis of regional correlation, supported and confirmed by OSL dating (Pawley et al. 2004, 2008; Box 2.6), the glaciation occurred at approximately 160 ka, that is, during the late Wolstonian (=Saalian) Stage.

The recognition of the eastern Fenland ice margin extends the Tottenhill glacial limit south and southwest and indicates that, at its maximum extent, the ice lobe must have occupied the entire Fenland Basin. Independent confirmation of this interpretation derives from the numerical dates and a possible western limit recognized by H.E. Langford (unpublished data, 2007) west of Peterborough, Cambridgeshire and north of Uffington, Lincolnshire. A potential northern equivalent in Lincolnshire is the Welton Till, described from the Welton-le-Wold area by Straw (1991, 2005), which closely resembles red-brown eastern Fenland diamicton in both its stratigraphical position and lithology. However, while it is potentially possible that these observations might represent a different event to that seen in East Anglia, further work is required to clarify the local correlation question.

The evidence therefore indicates that, during the late Wolstonian, a substantial ice-lobe advanced down the eastern side of Britain and entered the Fenland Basin (Fig. 2.26). It then dammed a series of westwards-flowing streams to form shallow glacial lakes that coalesced, culminating in an extensive proglacial lake in immediate contact with the ice front. This lake drained westwards to the North Sea via the River Waveney valley.

2.7.4 IPSWICHIAN

In Britain, the equivalent of the Eemian is the Ipswichian Stage. Interglacial deposits of last interglacial age have been found in numerous places, mainly in river valleys. The type site is at Bobbitshole, south of Ipswich (West 1957). None of the Ipswichian sites in Britain covers the entire duration of the interglacial. Consequently, its vegetational history had to be reconstructed from numerous pieces (Jones & Keen 1993). Relatively long Ipswichian sequences have been found at Beetley, East Anglia (Phillips 1974) and at Wing in the East Midlands (Hall 1978).

2.7.5 EXTENT AND TIMING OF THE DEVENSIAN (WEICHSELIAN) ICE SHEET

During the Devensian Stage, spanning MIS 5d–2, ice sheets waxed and waned over the British Isles reaching a culmination in spatial extent during the Late Devensian Substage. In Figure 2.28 the long-held view of the extent of the Late Devensian Ice Sheet is depicted along with the revised, more extensive version. Instead of a mostly terrestrially constrained and smaller ice sheet ($c.$ 357,000 km^2), a version twice as large is envisioned today, covering extensive areas of current seafloor including the North Sea and continental shelves of Britain and Ireland.

The change in assessment comes from a variety of sources. Ballantyne (2010) reviewed the burgeoning database of cosmogenic exposure ages and demonstrated that the long-argued Late Devensian 'ice-free enclaves' in northeast Scotland and in southernmost Ireland are in fact erroneous, and that the dating evidence is best satisfied by complete terrestrial ice cover here (Fig. 2.28). The assessment reported below on the timing of maximum extent of ice draws directly from this publication and from papers reviewed in Gibbard & Clark (2011).

The most prominent discoveries necessitating the reassessment of maximum ice cover are of extensive moraine systems on the continental shelf surrounding northern Scotland (Bradwell et al. 2008), and which continue anticlockwise around the shelf to southwest Ireland (Ó Cofaigh et al. 2010; Clark et al. 2012). They clearly demonstrate that, at some time, ice extended to the shelf edge (Fig. 2.29) and paused for long enough to build up large moraines.

Scourse et al. (2009) report a 'pronounced increase' in ice-rafted debris (IRD) on the Rosemary Bank and Barra-Donegal Fan at 29 ka which requires that the ice sheet must have grown into the sea. Wilson et al. (2002) have demonstrated that ice must have reached the shelf edge at at least one location by 27 ka because the Barra-Donegal trough mouth fan was being fed with new material. IRD fluxes also occurred at 27 ka further south along the western Irish margin (Peck et al. 2006; Scourse et al. 2009). The simplest interpretation is that ice reached the shelf edge everywhere by 27 ka along the whole boundary from SW Ireland to the Shetlands. This is consistent with the interpretation of Bradwell et al. (2008), who reconstruct ice at the shelf-break from Scotland and across the North Sea to Norway during 30–25 ka, and also follows Sejrup et al. (1994, 2005, 2009).

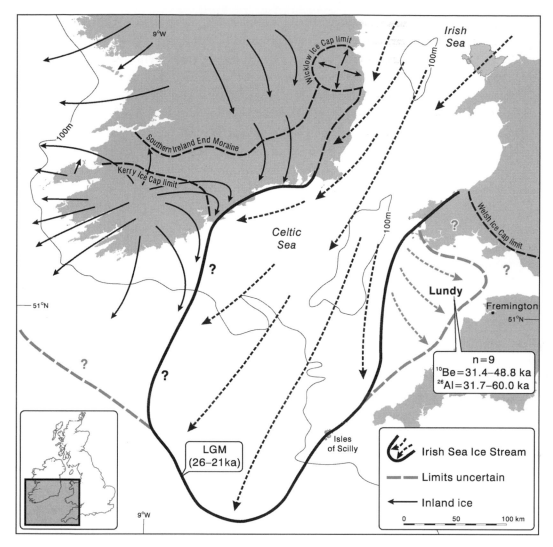

Figure 2.29 Devensian glaciation in the southern Irish Sea region. The Lundy data are from Rolfe et al. (2012). LGM position and ice directional indicators are based on Ó Cofaigh & Evans (2007, their fig. 2).

Source: Rolfe et al. (2012). Reproduced with permission of Elsevier.

Having dealt with the continental shelf we continue anticlockwise around the margin. Advance dates from the Isles of Scilly suggest that these islands were reached after *c.* 25 ka (26.9–24.6 ka; Scourse 2006). This is consistent with dates from the Celtic Sea placing ice advance by a short-lived ice stream after 24.2 ka (Ó Cofaigh & Evans 2007). Ice was at the southern Welsh limit by 23 ka (Phillips et al. 1994; range 25.2–21.2 ka). For the rest of the southern margin of the ice sheet, the picture is more complicated. For example, in the Bristol Channel cosmogenic exposure ages from glaciated surfaces on Lundy island indicate that an ice lobe in this area retreated at 40–35 ka (Rolfe et al. 2012; Fig. 2.29).

The limits recorded by sediment and landform evidence are reviewed by Clark et al. (2004) and Chiverell & Thomas (2010). The youngest date for advance into the Cheshire Plain suggests that ice advanced inland here after 27 ka (Bateman, pers. comm. 2009; range 29–25 ka). However, a woolly mammoth bone dated to 18 ^{14}C ka BP (Rowlands 1971; Bowen 1974) lying below Irish Sea Till suggests that Irish Sea ice did not advance up the Vale of Clwyd, and potentially the Cheshire Plain, until after 21 cal. ka BP. Alternative scenarios include: incursion of ice into the Vale of Clwyd was initially prevented by the presence of Welsh ice; the location existed as an ice-free enclave until 21 cal. ka BP; the bone date is unreliable; or the date reflects an oscillation of the ice margin in this region around 21 cal. ka BP.

Ice advanced down the Vale of York after 23.3 ka (Bateman et al. 2008; range 24.8–21.8 ka) but had retreated to the north by 20.5 ka (21.7–19.3 ka). Dates from Dimlington on Holderness suggest ice did not reach the eastern English coastline until after 22 ka (Penny & Rawson 1969; range 22.5–21.3 ka), and dates from inland Lincolnshire suggest ice did not progress inland until after *c.* 17 ka (Wintle & Catt 1985; range 19.1–14.9 ka). Ice at this position at this time is consistent with a recently published age for a beach deposit (16.6 ka, Bateman et al. 2008; range 17.8–5.4 ka) related to Glacial Lake Humber, the existence of which requires ice damming the Humber Gap. This new date is significantly younger than the previously quoted maximum age for Lake Humber of 26.2 ka (Gaunt 1974, 1976; range 28.1–24.2 ka) and inconsistent with a more recent date for deposition of sands into Lake Humber at *c.* 22 ka (Murton et al. 2009) that is difficult to reconcile with the Dimlington date for ice first reaching the eastern coast. It is suggested that the dates at these sites could reflect oscillations of the ice margin, including sporadic damming of the Humber Gap. In the absence of deglacial dates preceding the 'young' advance dates, it not possible to confirm or disprove this (Gibbard & Clark 2011).

If all of the above dates are accepted there are two possible interpretations: (1) ice did not reach eastern England or the Cheshire Plain until after 17 and 21 ka, respectively; or (2) the dates reflect oscillations of the ice margin within the last glaciation. This implies advance into the Cheshire Plain after 27 ka, followed by retreat to an unknown position north of Wales before 21 ka and a subsequent re-advance south after 21 ka. This could reflect oscillations of the Irish Sea glacier during uncoupling with Welsh ice. In eastern England, the dates could be interpreted as advance after 25 ka (Ventris 1985) followed by retreat to an unknown offshore position, followed by a re-advance at least as far as Dimlington after 22 ka, with ice reaching the Lincolnshire Wolds after 17 ka. The Dimlington dates have been invoked to support a contemporaneous re-advance of the British Ice Sheet with the Tampen Re-advance of the Scandinavian Ice Sheet (Sejrup et al. 1994; Carr 2004).

Of the whole perimeter of the ice sheet, the least-constrained margin is that in the southern North Sea for which virtually no information exists. Given good evidence for ice existence in the northern North Sea, there clearly must have been a southern margin but this has yet to be determined. In Figure 2.28, a line from the eastern England limit is interpolated to the Main Stationary Line in Denmark (Houmark-Nielsen 2004).

2.8 Quaternary History of North America

In North America, the Pleistocene glaciations developed two largely independent ice sheets: the Laurentide Ice Sheet that covered the Canadian Shield and adjoining areas, and the Cordilleran Ice Sheet in the mountains of northwestern North America.

Whereas in Europe the Weichselian glaciers were surrounded by an almost vegetation-free belt >100 km wide, the last cold stage (Wisconsinan) North American ice sheets advanced into a wooded landscape in the Midwest from Illinois to Ohio. Tree stumps below till at numerous sites provide evidence of overridden forests. The same applies to older glaciations in Nebraska and Kansas. Consequently, the question arose whether the Wisconsinan Ice Sheet was ever surrounded by a significant periglacial zone in which permafrost conditions prevailed. This question has to a great extent been answered. In the western Plains of Nebraska, extensive ice-wedge casts have been found up to 50 km south of the Wisconsinan glacial limit (Wayne 1991). In Wisconsin, a broad belt of permafrost features lies outside the glacial limit of the last ice sheet (Black 1965, Clayton et al. 2008) and the same is true of Illinois, where patterned ground can be traced about 40 km south of the Wisconsinan ice maximum (Johnson 1990).

2.8.1 THE GLACIATIONS ARE GETTING OLDER

There has been a recent tendency to replace the traditional stratigraphy based on field investigations by 'dating stratigraphy'. In his overview of the Quaternary, Pillans (2007) rightly points out that a stratigraphy based solely on age determinations brings considerable dangers. An example from the United States illustrates this. From early days in North America, the ^{14}C method was used for dating glacial strata. In Salmon Springs, Washington State there are two strata of glacial tills and meltwater deposits, separated by a 1.5 m thick layer of non-glacigenic sediment, including volcanic ash (tephra), silt and peat. Ideal conditions for an age determination, you might think. The peat was radiocarbon-dated to 71,500±1400 years before present (Stuiver et al. 1978). The two glacial deposits were consequently interpreted as Early and Middle Wisconsin (Weichselian), and numerous other tills were correlated with this key profile.

This fixed point of Pleistocene stratigraphy wavered when Easterbrook et al. (1981) dated the tephra between the two tills using the fission-track method, measuring an age of 840±210 ka. The dating was admittedly very crude but, as the silt immediately above the tephra showed a reverse magnetization, it therefore had to be older than the Brunhes–Matuyama magnetic reversal at 780 ka.

How did this spectacular failure of radiocarbon dating happen? The first mistake was that a dating method, in this case the ^{14}C method, was applied right up to its limits so a slight contamination with younger carbon had resulted in a completely wrong age. The second error lies in the human psyche: the younger age was accepted because it met the scientists' expectations.

The stratigraphic framework of the North American Quaternary has undergone significant changes since then. What was formerly regarded as Nebraskan and Kansan actually represents a number of different glaciations which are mostly much older than previously thought. Further, the so-called Yarmouth Palaeosol, a fossil soil which was used as a time marker for the separation of Illinoian (=Saalian) glacial deposits against older strata, can no longer be correlated with the Holsteinian Stage in Europe. While it was formed during MIS 11–7 in Indiana and Illinois, elsewhere its origin dates back to MIS 13 or 15 (Hallberg 1986).

Duk-Rodkin & Barendregt (2011) provided an overview of the ancient glaciations (Fig. 2.30). It demonstrates that even in the palaeomagnetic Gauss Chron (i.e. during the Pliocene), glaciation occurred in North America. In fact, global glaciations started in the

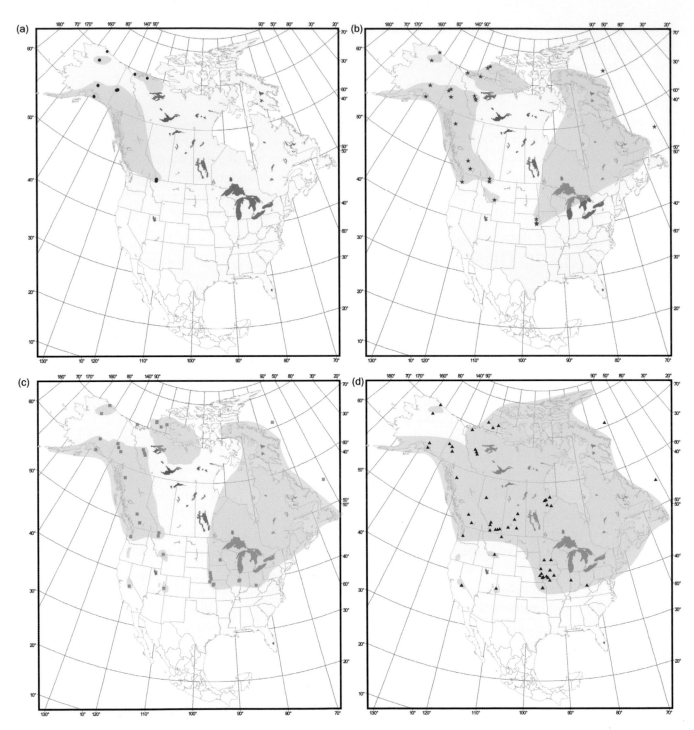

Figure 2.30 Extent of ancient glaciations in North America in the: (a) Gauss Chron; (b) lower Matuyama Chron; (c) upper Matuyama Chron; and (d) Brunhes Chron. Adapted from Duk-Rodkin & Barendregt (2011). Reproduced with permission of Elsevier.

Figure 2.31 The intercalation of glacial deposits with tephra layers allows the age determination of the glaciations in large parts of North America. In some cases the till is interspersed with tephra. The picture shows till with volcanic bombs on the shore of Yellowstone Lake, Wyoming. Photograph by Jürgen Ehlers.

Pliocene (De Schepper et al. 2014). Gao et al. (2012) have shown that this early glaciation was not restricted to the mountains of the northwest and the Mackenzie Delta region, but that the James Bay Lowland was also covered by an ice sheet at about 3.5 Ma.

The small number of dated sites does not yet allow for reliable distribution maps of the early ice sheets. Figure 2.30 can therefore only show the approximate glacial limits. However, it seems clear that extensive glaciers existed in Early Pleistocene not only in the Cordillera but that an early ice sheet also extended from Labrador to Kansas and Nebraska.

The interaction of Quaternary volcanism and glaciation in large parts of North America allowed the building of a well-founded chronostratigraphy (Box 2.7). The volcanism in the Yellowstone area has been active for 17 Ma (i.e. since the Miocene). The last three major eruptions occurred at 2.1 Ma (Huckleberry Ridge), 1.3 Ma (Mesa Falls) and 0.64 Ma (Lava Creek). The caldera of the last eruption is about 40 km long and 25 km wide, and occupies about a quarter of today's Yellowstone National Park area. In that eruption some 1000 km^3 of tephra were hurled into the atmosphere. Deposits of an earlier ice-dammed lake were fragmented and partially baked into a breccia, which is exposed on the shore of Yellowstone Lake (Figs 2.31, 2.32). An overview of the history of the Yellowstone volcano was provided by Morgan et al. (2009).

Figure 2.32 Traces of volcanism at Yellowstone Lake. The volcanic eruption has broken and fritted varved lacustrine clays. Photograph by Jürgen Ehlers.

BOX 2.7 TEPHRA: DATING THE ASHES

The study of tephra layers as a means of dating Quaternary deposits began in the 1930s. The Icelandic volcan-ologist Sigurður Þórarinsson was working together with Lennart von Post, the founder of the pollen analysis, at Stockholm University. While studying Icelandic bogs, he realized that the numerous ash layers of Icelandic volcanic eruptions could be used for age determination. At first his research focused on vegetation history, soil formation and soil erosion. In his PhD thesis (1944), Þórarinsson for the first time defined the terms tephra and tephra chronology. The Greek word tephra (τέφρη) means 'ash', and is a generic term for all components released when a volcano erupts explosively. Tephra can range in grain size from fine dust to large blocks of several cubic metres in size.

Particularly favoured for the application of tephra chronology are regions in the vicinity of volcanoes. In North America, this includes the area around Yellowstone Park in Wyoming. In Germany, numerous tephra lay-ers were formed in the vicinity of the Eifel Maar volcanoes. Today microtephra or cryptotephra studies are being increasingly applied in the northeast Atlantic region for detailed chronostratigraphy of the late glacial period.

Fine volcanic ash can be transported over considerable distances, and is eventually deposited in up to centimetre-thick layers. One problem is that they are usually not continuous, uniform layers, but concen-trated locally and missing in other places. In many cases the layers are mixed with other sediment by bioturbation so that they are usually no clearly delimited lines, but the ash particles may be spread over several decimetres of sediment thickness. The layers that are most visible to the naked eye, such as the Laacher See Tephra (Fig. 2.33) or the Vedde Ash of Iceland, were mapped earliest. Most recently considera-ble attention has been paid to the tephra produced by the massive Toba eruption on Sumatra. The eruption that dates to 74,000 years ago sent vast volumes of ash across the Sea of Bengal that travelled as far as the Indian subcontinent. However, there is still a large area with 'cryptotephra' which are so thinly spread that they cannot be recognized with the naked eye. This is particularly a problem where their con-centration in the host sediment is very low and the fine glass particles are only visible under the microscope. In this way, the Laacher See Tephra was demonstrated to extend to Torino in the south (van den Bogaard & Schmincke 1985), and the Vedde Ash could be traced to northern Germany, southern Sweden and beyond the Baltic Sea to St Petersburg (Alloway et al. 2007).

For the formation of younger tephra layers, there are eye-witness accounts. This is the case for the eruption of Vesuvius in 79 AD, the catastrophic eruption of Mount Tambora on Sumbawa in 1815 (Indonesia) or for Beerenberg eruption on Jan Mayen in 1985 (Fig. 2.34). The first safely dated ash

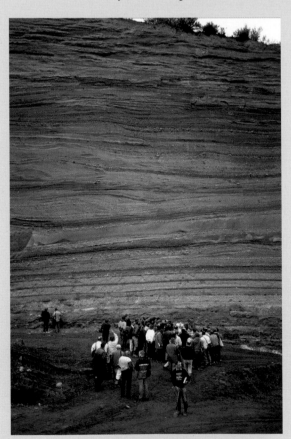

Figure 2.33 Laacher See Tephra in a 30 m high wall at the Wingertsberg quarry in the Eifel. Photograph by Jürgen Ehlers.

Figure 2.34 The 2277 m high Beerenberg volcano on Jan Mayen island is still active. The last eruption occurred in 1985. Photograph by Hinrich Bäsemann (www.polarfoto.de).

layers in Iceland include the historically documented eruptions of Hekla (1693), Katla (1721) and again Hekla (1766). Older tephra layers can either be dated with the fission-track method or by ^{40}Ar/^{39}Ar dating methods, which cover the entire Quaternary in their age range. In addition, indirect dating methods can be applied by dating the surrounding strata using the ^{14}C method or by dendrochronology, TL or OSL. Palaeomagnetic measurements are another important tool for the age determination of tephra.

The great distance to which tephra from a volcanic eruption can spread has been demonstrated recently by the eruption of Eyjafjallajökull in Iceland, which began on 20 March 2010. Relatively little lava was initially produced, but on 14 April a new and more violent eruption occurred in the summit caldera, emitting large amounts of steam and ash into the atmosphere. As a result, air traffic over Europe was temporarily suspended. The satellite image (Fig. 2.35) shows the distribution of the ash cloud on 19 April. The ash did not spread in a single direction, but repeatedly changed its course depending on the wind direction. Whether ash from this eruption will later be found as a datable layer on the ground depends not only on the wind direction, but also on rainfall.

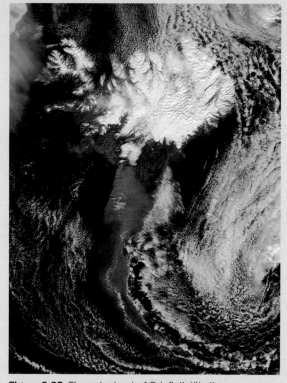

Figure 2.35 The ash cloud of Eyjafjallajökull.

Source: NASA, MODIS, Rapid Response System, Terra, 04/19/2010.

2.8.2 EARLY PLEISTOCENE

The classical subdivision of the North American Quaternary stratigraphy dates to the late nine-teenth century. Chamberlin (1894) distinguished three major glaciations: the Kansan (oldest), the East Iowan (middle) and the East-Wisconsin (youngest). The East-Wisconsin was soon shortened simply to Wisconsin by Chamberlin (1895). One year later Chamberlin (1896) identified the Illinois Glaciation. He was followed by Leverett (1898a–c), who introduced the terms Yarmouth, Sangamon and Peoria for the intervening interglacials. The Aftonian, the term for the pre-Kansan interglacial, had already been introduced earlier by Chamberlin (1895). The original three cold stages were supplemented by Shimek (1909) with the Nebras-kan as the oldest North American glaciation. From this the following stratigraphical sequence evolved: Wisconsinan Glaciation – Sangamonian Interglacial – Illinoian Glaciation – Yarmouth Interglacial – Kansan Glaciation – Aftonian Interglacial – Nebraskan Glaciation.

This subdivision was correlated with the four-fold subdivision of the European Alpine Quaternary stratigraphy and remained valid until recently (cf. Flint 1971).

2.8.3 PRE-ILLINOIAN WARM STAGES

Early and Middle Pleistocene vegetation has attracted relatively little attention in North America. This partially results from a lack of adequate sites and often poor pollen preservation. Smiley et al. (1991) list only 27 sites in the United States from which floral investigations of pre-Wisconsinan Quaternary and late Tertiary sediments have been published. Most of these cover the Sangamonian Interglacial. Dating, however, is often problematic.

2.8.4 PRE-ILLINOIAN GLACIAL STAGES

The pre-Illinoian glaciations reached furthest south in the Central Plains of Nebraska, Iowa, Kansas and Missouri. Correlation of the pre-Illinoian tills is based on radiometric age deter-mination of the Pearlette B, S and O ashes and the Bishop Ash. In western Iowa at least seven major pre-Illinoian till units have been distinguished so far. Clast lithology and heavy-mineral composition allows subdivision into three till groups (A, B and C), and mineralogical criteria and physical stratigraphy are used for further subdivisions. The till units are also separated by some well-developed palaeosols. However, both dating and inter-regional correlation are still poorly developed (Hallberg 1986). In Kansas, all diamictons and interbedded stratified deposits are now referred to as the Independence Formation. The two till members within this formation are correlated with the A2 and A3 tills of Hallberg's classification. They have normal polarity and are of age 0.7–0.6 Ma (Aber 1991).

2.8.5 PRE-ILLINOIAN–ILLINOIAN INTERGLACIAL

Originally the last interglacial before the Illinoian had been defined by the Yarmouth Palaeo-sol, identified in a well section near Yarmouth, Des Moines County, Iowa (Leverett 1898b). Like the Sangamon Palaeosol, it consists in many places of what is referred to as 'accretion gley', that is, a soil sediment of redeposited silt and clay with an admixture of organic material which can be up to 3 m thick (Frye & Willman 1975). However, it has turned out that the so-called 'Yarmouth Soil' represents different times in different regions. For example, in parts of Indiana and Illinois it was formed during deep-sea Isotope Stages 7–11, whereas in other regions it started to form as early as Stage 13 or 15 (Hallberg 1986). The geosols are often composites, reflecting soil formation during more than one stage. For instance, in major parts of the American Midwest from which Illinoian deposits are absent, a combined Yarmouth–Sangamon Palaeosol is found. Micromorphological, textural and mineralogical investigations are used to identify the different processes that have contributed to the formation of this geosol complex.

On the whole, much less is known of the pre-Illinoian–Illinoian warm stage than of its counterpart in Europe. In the Alps, as well as in northern Europe, the two glaciations are separated by at least two fully developed warm stages. Nevertheless, nothing comparable has yet been identified in North America.

2.8.6 ILLINOIAN GLACIATION

Although the occurrence of Illinoian tills has been reported from many states, thorough investigations have mainly been carried out in Illinois where the Illinoian glaciers advanced about 200 km beyond the limits of the subsequent Wisconsinan Glaciation. Willman & Frye (1970) provide a comprehensive review of the ice-marginal positions and till stratigraphy. The Illinoian sedimentary sequence comprises the Glasford Formation, named after Glasford in Peoria County. It comprises tills and outwash deposits of gravel, sand and silt. The Illinoian Stage was originally subdivided into three substages – the Liman, Monican and Jubileean – each of which was represented by a till unit. The substages were separated by ice-free periods, when minor soils were able to develop.

Further investigations have revealed that the Illinoian stratigraphy is more complicated than originally thought (Curry et al. 2011). In addition to the lithological and pedological investigations, much emphasis is put on the application of dating methods (OSL, [10]Be). Detailed investigations of the Illinoian till stratigraphy have also been conducted in Ohio. However, dating of the ice advances is not well constrained, and correlation between different sites is not always possible (Szabo et al. 2011). Moreover, in Pennsylvania the age of all pre-Late Wisconsinan tills still remains uncertain (Braun 2011).

2.8.7 SANGAMONIAN INTERGLACIAL

The Sangamonian Stage was named after a fossil soil that was first found in hand-dug wells in Sangamon County, Illinois. The name was first used by Leverett (1898a), who defined it as 'the weathered zone between the Iowan loess (Early Wisconsinan) and the Illinoian till sheet'. However, Leverett's Sangamon Soil is time-transgressive. It contains several merged soils of Wisconsinan, Sangamonian and possibly late-Illinoian age that may only be differentiated if careful investigations of the pedological parameters and the geomorphological circumstances are undertaken at each site (Follmer 1978, 1982).

The Sangamonian of North America, the last interglacial before the Holocene, is therefore defined (by some) differently from the Eemian Stage of Europe. This must be considered wherever stratigraphical comparisons are attempted. While the Eemian is restricted to MIS 5e, the Sangamonian, according to the Geological Survey of Canada and the Illinois Geological Survey, represents the entire MIS 5. This means that it spans the period 130–75 ka and includes the Brørup and Odderade interstadials of Europe (e.g. Fulton 1989, Curry & Follmer 1992). Others, for instance the US Geological Survey and most participants of IGCP projects, follow the European concept (e.g. Richmond & Fullerton 1986; Barnett 1992). The latter position is supported by the fact that the Laurentide Ice Sheet began to grow during MIS 5d, and the Appalachian ice cap during MIS 5b. Moreover, on Baffin Island the maximal ice advance of the last glaciation occurred as early as MIS 5d (Miller et al. 1992).

A Sangamonian-age buried swamp in Washington DC yields a last interglacial pollen assemblage with *Quercus* and *Carya*, suggesting climatic conditions similar to those in the Holocene (Knox 1962). Thirty pre-Wisconsinan interglacial sites have been identified in the Atlantic Provinces of Canada (Mott & Grant 1985), seven of which are clearly correlated with the Sangamonian *sensu stricto* (MIS 5e). One of these (Woody Cove, Newfoundland) spans the whole interglacial (Brookes et al. 1982). The climatic optimum was reached early in the stage, as compared with the Holocene. During all of MIS 5 there seems to have been non-glacial conditions in this region. The climatic optimum of the Sangamonian was significantly warmer, particularly during the summers, and more continental (Mott 1990).

One of the key sections for the study of the Sangamonian is in the Don Valley Brickyards in central Toronto in Canada. The last interglacial Don Beds (now referred to as the Don Formation) were first described by Coleman (1894). Besides the deciduous trees which now grow in the area, they contain a number of plant taxa that presently occur much further south. They include *Chamaecyparis thyoides*, *Gleditsia*, *Fraxinus quadrangulata*, *Maclura pomifera*, *Quercus stellata*, *Quercus muhlenbergii* and *Robinia pseudoacacia*. The pollen assemblage suggests that the average climate during the Sangamon was about 3°C warmer than today (Karrow 1990).

Vertebrate remains of last interglacial age have been recovered from 20 sites in Canada, from Nova Scotia to the Old Crow River in Yukon. The fauna includes a number of species which are now extinct, such as *Megalonyx* sp. (ground sloth), *Castoroides ohioensis* (giant beaver), *Mammut americanum* (mastodon), *Mammuthus primigenius* (mammoth), horses, *Camelops hesternus* (western camel) and *Bison latifrons* (giant bison) (Harington 1990).

The Sangamon Geosol from the American Midwest, after which the last interglacial in North America was named, is not very well suited as a reference. Its formation started in the Illinoian and it continued well into the Wisconsinan. MIS 5e, which is normally referred to when the climate or coastlines of the last interglacial are discussed, covers a much shorter interval. Consequently, the Sangamonian *sensu lato* should be abandoned (Otvos 2014).

2.8.8 WISCONSINAN GLACIATION

2.8.8.1 Early Wisconsinan

As long as the Sangamonian Stage of North America is regarded as representing either MIS 5e only or all of MIS 5, the length of the following stage (the Wisconsinan) will vary accordingly. If the Sangamonian is regarded as including all of MIS 5, the Wisconsinan is shorter than the European Weichselian and does not start before about 80,000. After the Sangamonian *sensu stricto*, a drier climate resulted in the midwestern region (e.g. in the Pittsburg Basin, central Illinois) in a change from deciduous forest to prairie conditions (Grüger 1972, Teed 2000). In more humid areas, such as Scarborough, Ontario, open woodland existed close to the ice margin (Dreimanis et al. 1989). In Beringia, however, glaciations started very early. In Chukotka as well as in parts of Alaska valley glaciers extended rapidly after MIS 5e (Elias & Brigham-Grette 2013).

The Laurentide Ice Sheet began to form in northeastern Labrador probably as early as MIS 5d (Vincent & Prest 1987). Simultaneously, major ice advances occurred in Greenland (Funder et al. 1991, 1994; Israelson et al. 1994). However, it was not before the Early Wisconsinan Substage that a major ice sheet formed in continental North America.

There has been much discussion concerning whether or not Early Wisconsinan glaciers advanced into the United States. The Whitewater and Fairhaven tills in southwestern Ohio and southeastern Indiana were formerly regarded as remnants of such glaciations (e.g. Fullerton 1986). However, recent investigations have revealed that the interstadial deposits separating the strata in question from undoubtedly classical Wisconsinan till may be considerably older than assumed.

In New England along the coast as well as inland, remnants of a till sheet older than the Late Wisconsinan Glaciation have been preserved in numerous places. For instance, the lower till at Sankaty Head in Nantucket and the Montauk Till Member on Long Island were deposited during a glaciation that was at least as extensive as the last glaciation; these tills are probably pre-Wisconsinan in age, however. At Sankaty Head the till is overlain by marine deposits, from which a detrital coral yielded a U–Th age of 133±7 ka. The fauna in these deposits indicates that they were deposited in seawater warmer than present. The underlying glacigenic strata therefore probably date from the Illinoian or are even older (Oldale & Colman 1992).

This interpretation corresponds with the results of investigations in Maine, where Weddle (1992) showed that the area had undergone only one Late Wisconsinan Glaciation during the last cold stage. This event is represented by a sequence of ice-marginal deposits, involving several till units and changing ice-movement directions. On the other hand, in maritime Canada four tills from different ice centres have been found overlying the Sangamonian Interglacial beds. It is however unlikely that they are all Late Wisconsinan in age, and this might also apply to adjacent Maine. Nova Scotia was crossed by Early Wisconsinan ice of the Caledopia Phase (Stea et al. 2011), and early Middle Wisconsinan tills are present around Hudson Bay. In other parts of Canada including the Prairies, several till sheets postdating interglacial deposits have been recognized; the dating is once more uncertain however.

2.8.8.2 *Middle Wisconsinan*

The Middle Wisconsinan includes a lengthy interstadial complex from about 65 to 25 ka, from which no major ice advances have been recorded (Dredge & Thorleifson 1987). The age of tills which were formerly attributed to Middle Wisconsinan ice advances, such as the Titusville Till, has been questioned by more recent research (see review in Clark & Lea 1992). Pollen spectra in northern Illinois during the Middle Wisconsinan, for the period 47–24 ka, are still dominated by *Pinus* and *Picea* with some *Betula* and *Salix*, which indicates forest or open woodland (Heusser & King 1988). In the Toronto area, the Thorncliffe

Formation was laid down during the Middle Wisconsinan Substage. This has been dated by radiocarbon and thermoluminescence techniques to between >50 ka and 28 ka (Hicock & Dreimanis 1992).

2.8.8.3 Late Wisconsinan

In the Midwest, southeast of the Great Lakes, the Late Wisconsinan begins with deposition of the Peoria Loess. This loess was deposited under the influence of the advancing Late Wisconsinan Ice Sheet. The loess units are time-transgressive, aeolian deposition having begun about 25 ka in northern Illinois and about 23 ka further to the south (Curry & Follmer 1992). Peoria Loess is probably the thickest Last Glacial loess in the world. It is more than 48 m thick at Bignell Hill in central Nebraska (Bettis III et al. 2003). The primary loess sources were the valleys of the Mississippi and Missouri rivers. Here, vegetation changed from steppe to periglacial conditions. At Wedron, for instance, macrofossils of arctic and subarctic plants were recovered from proglacial lacustrine sediment, including *Dryas integrifolia* (arctic avens), *Vaccinium uliginosum* (arctic blueberry), *Selaginella selaginoides* and *Betula glandulosa* (Garry et al. 1990). Loess sedimentation at Wedron ended at 22 ka when the site was overridden by the ice sheet (Curry & Follmer 1992).

Little is known about the build-up phase of the Late Wisconsinan Laurentide Ice Sheet. On the basis of the sea-level record in Atlantic Canada, Quinlan & Beaumont (1982) have suggested that a major ice dome must have formed during an early glaciation phase in the central Labrador highlands. This coincides with the observations of Klassen (1983) and Klassen & Bolduc (1984) that striae directions indicate a pre-Late Wisconsinan ice dispersal centre between the Churchill and St Lawrence rivers. From this centre, ice seems to have flowed radially in all directions. In the southwest the Harricana Interlobate Moraine is interpreted as having formed at the contact between the Labrador and Hudson ice. It consists largely of glaciofluvial material (Vincent 1989).

In the western part of North America, the Late Wisconsinan Laurentide Ice Sheet was the most extensive Pleistocene ice sheet. It was less extensive than earlier glaciations in the south and in the western Arctic. It had several flow centres, so that extent and timing of glaciation and deglaciation were different across the continent. Along the southern ice margin from Montana to Connecticut, an early glacial maximum was reached around 22–20 ka. In other places, for instance in Washington, Iowa and Pennsylvania, a later ice advance during 16.5–14 ka reached furthest south. However, further to the north the maximal ice sheet advances became increasingly younger. In Newfoundland the Wisconsinan maximum did not occur before 10 ka (i.e. the Younger Dryas), and on Baffin Island the glaciers reached their north-easternmost positions around 8500 (Prest 1983). In the eastern Arctic, Holocene ice extended further than any preceding glaciation.

The southern margin of the Laurentide Ice Sheet was largely subdivided into major and minor lobes, in contrast to the situation in Europe where the Weichselian Ice Sheet at its maximum possessed a rather straight margin. In the west in Montana, the shape of the St Mary, Milk River, Shelby, Havre and Missouri valley lobes was largely controlled by the pre-glacial topography (Fullerton & Colton 1986). North and South Dakota were covered by the extensive James Lobe, and further to the east the Des Moines Lobe advanced via Minnesota into Iowa (Hallberg & Kemnis 1986). Between the Des Moines Lobe and the Green Bay Lobe of Wisconsin, a complex pattern of minor sub-lobes developed in the surroundings of the ENE–WSW-trending Superior Lobe (Matsch & Schneider 1986). In the east the Green Bay Lobe, Lake Michigan Lobe and Huron Lobe followed the forms of the lake basins. Temporarily the Lake Michigan Lobe and Huron Lobe were separated by the Saginaw Sub-lobe (Eschman & Mickelson 1986). Further to the east the Ontario-Erie Lobe abutted a number of sub-lobes. It was only in New England that the southern ice margin of the Laurentide Ice Sheet did not have a lobate form (Mickelson et al. 1983).

The Wisconsinan Glaciation shaped major parts of the North American landscape, especially that of the Great Lakes region. Belts of end moraines can be traced around the southern margins of the lakes. Some of these end moraines differ greatly from the push moraines of northwestern Europe. Internally, some consist almost entirely of till (Wickham et al. 1988). For this reason, Mickelson et al. (1983) consequently define an end moraine as 'a ridge composed predominantly or entirely of till and formed at the ice margin during the last episode of till deposition.' These ridges often show a very subdued topography. In Europe, similar end moraines are found for instance in Estonia at the Pandivere ice-marginal position. Push moraines seem to be rare; however, large-scale glaciotectonic deformations are exposed at Ludington on the eastern shore of Lake Michigan, for example (Larson et al. 2003).

As in Europe, the end moraines also provided the first means by which the Wisconsinan Glaciation could be subdivided. They formed the basis of the morphostratigraphical schemes by Leverett (1929) and Leighton (1960). Later it was found that some of those landforms are inherited features, and that parts of many moraines in the Great Lakes region do not mark halts of the last ice sheet (Mickelson et al. 1983). Morphostratigraphy was replaced by a chronostratigraphical approach by Frye & Willman (1960), later visualized in time-distance diagrams (Frye et al. 1965). They subdivided the Late Wisconsinan into two principal phases: the Woodfordian and the Valderan. Intensive geomorphological and stratigraphical investigations in the 1970's, however, did show that the scheme required revision (Mickelson & Evenson 1975).

The major ice advances and retreat phases of the Late Wisconsinan occurred within a very short period of time. This implies that ice movement was extremely fast, at rates comparable to the flow of the recent ice streams of Antarctica. Reconstructions of ice thickness in the lobate areas suggest that the ice sheet must have been very thin (Clark 1992). Indeed, Clark concludes that very low driving stresses existed for the ice lobes, enabling them to advance by sliding and/or subglacial sediment deformation at rapid rates.

In New England the retreat of the Laurentide Ice Sheet was accompanied by a marine transgression into coastal Massachusetts, New Hampshire and both coastal and central Maine. Rapid ice-marginal recession in the Gulf of Maine region was probably caused by the development of a calving bay, while the margin of the grounded ice sheet in southern New England retreated much more slowly. After 13 ka, the remaining ice cap in New England was separated from the Laurentide Ice Sheet by rapid ice retreat along the St Lawrence valley and subsequent transgression of the Goldthwait and Champlain Sea (Mickelson et al. 1983).

Final disintegration of the stagnant ice mass resulted in accumulation of ice-contact stratified deposits and a lack of end moraines in northern central New England. Another region with widespread ice-decay features in the marginal area of the Laurentide Ice Sheet is the area of the James and Des Moines lobes in parts of Minnesota and South and North Dakota. Here the landscape of the lowland areas is characterized by a thin, hummocky cover of supraglacial till, while the till cover is thicker and the landforms are higher in the uplands (Mickelson et al. 1983).

In the mountains of western North America, the Cordilleran Ice Sheet formed through coalescence of large piedmont glaciers in the Rocky Mountains and Cascade Ranges in Canada and flowed south into the Unites States. It extended south to about 47°30′ latitude north. The extent and shape of the Cordilleran Ice Sheet was largely controlled by topography. Major ice lobes developed in the west. The Juan de Fuca Lobe and the Puget Lobe were constrained by the Cascade Range and the Olympic Mountains and, east of the Cascade Range, the large Okanogan Lobe and several smaller lobes occupied north–south-trending valleys.

In western Washington, the Puget Lowland was covered by the Late Wisconsinan Cordilleran ice. The deposits are well dated. The last major glaciation, the Late Wisconsinan Fraser Glaciation, can be subdivided into three advances, the oldest of which occurred at about 21–19 ^{14}C ka BP. After a short period of ice melt it re-advanced to its maximal position, which has been radiocarbon dated to 15–14 ^{14}C ka BP in Washington (Easterbrook 1986, 1992).

The southern margin of the Cordilleran Ice Sheet lay in Washington, Idaho and Montana. The easternmost lobe of the Cordilleran Ice Sheet, the Flathead Lobe, reached just south of Lake Flathead, Montana, where it coalesced with Alpine glaciers from the Rocky Mountains (Richmond 1986). Alpine mountain glaciers reached further south. In Washington in the northern part of the Columbia Basin, the ice sheet blocked drainage of the Columbia River, leading to the formation of the large ice-dammed lakes Columbia and Missoula.

Further to the south in the Great Basin, numerous pluvial lakes formed at periods during the Quaternary. Around 120 lakes existed in the Great Basin during the Late Wisconsinan, the largest of which were lakes Lahontan and Bonneville. Lake Lahontan covered a maximum area of 21,000 km^2, which is almost the size of the present Lake Erie. Lake Bonneville, originally described by Gilbert (1890), reached a maximum size of 51,700 km^2, slightly smaller than the modern Lake Michigan. The former reached its last maximum at about 17–15 ka. The lake sediments represent major parts of the Quaternary, but dating and correlation of the individual units is difficult. However, it is clear that during the late Illinoian the lake levels had risen almost as high as during the Late Wisconsinan; shore deposits of the respective stratigraphical unit, the Alpine Formation, have been identified in various places (Currey 1990; Morrison 1991).

During the Late Wisconsinan, Lake Bonneville rose to the level of the Red Rock Pass, allowing overflow to the Snake River. Catastrophic drainage down the Portneuf and Snake rivers followed, with a maximum discharge of about 935,000 m^3 s^{-1} (Morrison 1991). The hydraulic conditions of the flood have been reconstructed from a detailed survey of the geomorphological and sedimentary features (O'Connor 1993).

Reconstructions of the Late Wisconsinan glaciations of North America have shown ice-free corridors reaching hundreds of kilometres north and south (Fulton 1984; Dyke & Prest 1987; Bobrowsky & Rutter 1992). However, at the western edge of the Laurentide Ice Sheet in Alberta, Canada, this concept has been challenged. It has recently been postulated that in this region the Late Wisconsinan Ice Sheet advanced much further than its predecessors. Liverman et al. (1988) found that in the Grande Prairie region of western Alberta the only till from an eastern source was the Late Wisconsinan Glaciation. According to Young et al. (1994), the only Quaternary deposit underlying the Late Wisconsinan till sheet, the Saskatchewan Gravels and Sands, were deposited during 42,910–21,300 years ago. If these dates are correct, then the Wisconsinan Laurentide Ice Sheet may have been the only one to have coalesced with the Cordilleran Ice Sheet, and that confluence extended very far to the south. This in turn would mean that the ice-free corridor postulated for this area did not exist during the last glaciation.

Since the main drainage divide in North America is near the US–Canadian border, the Laurentide ice was forced to flow uphill on its way south. This effect was enhanced by the fact that, during the Late Wisconsinan, the central parts of the ice sheet were already isostatically depressed and most of the areas north of the Great Lakes were at or below sea level. The extent of maximum isostatic depression is not known. In the west, the present land surface of the formerly glaciated area slopes from more than 1000 m asl near the Rocky Mountains in Montana and Alberta to below sea level in Hudson Bay. From directional indicators it is known that in Alberta the ice flowed westwards, straight uphill (Mickelson et al. 1983).

A major sector of the Laurentide Ice Sheet drained through Hudson Strait towards the North Atlantic. Extensive field investigations have shown that the Hudson Strait was occupied by a large ice stream, with a catchment area extending far beyond Hudson Bay, including major parts of Keewatin. Erratics from these areas have been found on the islands in western Hudson Strait (Laymon 1992). The Heinrich Layers of ice-rafted detritus in the North Atlantic (Heinrich 1988) have been correlated with major meltwater outbursts from the Laurentide Ice Sheet (Andrews & Tedesco 1992; Andrews et al. 1994). According to Clark (1994) the ice sheet had become unstable; under the thick ice cover the glacier sole became warmer, resulting in a deformable bed that had triggered the outbursts.

2.8.9 HOLOCENE

In contrast to the situation in Europe, at the beginning of the Holocene major parts of North America were still covered by the Laurentide Ice Sheet.

Morphologically, and with regards to its position in the centre of the glaciated area, Hudson Bay is the North American equivalent of the Baltic Sea. The oldest tills in North America already contain clasts from the Canadian Shield, so that Precambrian rocks must have been exposed at that time. The Shield in general seems to have experienced net glacial erosion of only a few tens of metres. Greater erosion occurred near the outer zones of the ice sheet than close to the centre (Dyke et al. 1989). This is in common with northern Europe, where Tertiary weathered bedrock has been preserved at the surface in Finland, for example.

Traditionally, it was assumed that Hudson Bay had been covered by the Laurentide Ice Sheet throughout the Wisconsinan (Denton & Hughes 1981). However, marine evidence suggests that no related ice shelf existed across Baffin Bay and the northern Labrador Sea (Aksu 1985). Additionally, the Missinaibi Beds of the last interglacial in the Hudson Bay Lowlands are overlain by a till sequence, the individual units of which are separated by thin and discontinuous sand seams. Dredge & Cowan (1989) interpret these seams as minor subglacial meltwater features, whereas Andrews et al. (1983) and Shilts (1984) believe they indicate that the bay might have been ice free periodically during the last cold stage. In the latter case, the core of the Laurentide Ice Sheet would have been less stable than had been thought, but would be susceptible to rapid 'draw-down' if the fast-flowing ice stream through Hudson Strait could drain into the open sea (Dyke et al. 1989).

The instability of the Laurentide Ice Sheet is also reflected in an early Holocene abrupt ice-stream advance which may have occurred at about 9.9–9.6 ka at the mouth of Hudson Strait. This Cold Cove Advance resulted in increased iceberg release from the calving ice front over 200 km long in open, *c.* 500 m deep water. It may have been the cause of a brief cooling period noted in several high-resolution climate records of the North Atlantic for the period immediately following the Younger Dryas (Kaufman et al. 1993).

The last advances of active ice in the Hudson Bay area are the so-called Cochrane Re-advances, named after the Cochrane district south of James Bay and discussed by Antevs (1925, 1928). This author considered them to be equivalents of the European Younger Dryas end moraines. However, it became clear later that they were much younger. A review of the geological and geomorphological evidence has shown that they represent minor surges (50–75 km) of an unstable ice margin into ice-dammed lakes that fringed the entire southern margin of the ice sheet. Varve counts indicate that the individual surges lasted only about 25 years. The lake sediments have been dated to about 8300 years. It is thought that the lakes led to a thinning of the ice sheet and a flattening of the glacier profile, contributing to the rapid deglaciation of Hudson Bay (Dredge & Cowan 1989).

At this phase Hudson Bay ice was calving rapidly on its northern side into Hudson Strait and Boothia Strait (west of Southampton Island) and the southern margin, which was bordered by lakes, also disintegrated rapidly. Eventually, buoyant forces of the invading sea lifted the entire remaining ice sheet, causing catastrophic drainage of the southern ice-dammed lakes. This drainage is recorded in 'drainage horizons', consisting of rounded pebbles of varved clay overlain by pebbly sands (Dredge & Cowan 1989).

At this stage of deglaciation, about 8 ka, the Tyrrell Sea invaded the isostatically depressed Hudson Bay lowlands and flooded a vast area, reaching as much as 300 km beyond the present coastline. At about the same time, the Baltic Sea in Europe finally connected to the ocean in its Litorina Phase. The highest shoreline of Hudson Bay is found at 180 m asl at the southern end at James Bay and in an area north of the Nelson River which was formerly covered by the Keewatin ice. Isostatic rebound has subsequently led to gradual regression to the present sea level, forming flights of elevated beach ridges. This process continues; the remaining uplift has been estimated at about 150–300 m (Walcott 1970).

2.9 The Course of the Ice Ages: A Global View

A global perspective on the extent and timings of Quaternary glaciations is provided in the introduction to Ehlers et al. (2011a, b). The key findings, by the same authors as this book, are described in the following sections.

2.9.1 PLIO-PLEISTOCENE GLACIATIONS

Evidence of glaciation is widespread throughout the Quaternary and indeed the Neogene in the Northern Hemisphere (de Schepper et al. 2014). The longest sequences are restricted to Alaska and the adjacent Northwest Territories of Canada which, together with Greenland and the Rockies, preserve evidence of glaciation from the Neogene to the present. In northern

Canada and Alaska, the oldest till and accompanying ice-rafted detritus in marine settings dates from the early Miocene, with regionally widespread glaciation occurring in the Pliocene and regularly throughout the Pleistocene (Haug et al. 2005; Duk-Rodkin & Barendregt 2011). In adjacent British Columbia a comparable sequence is found, particularly in the north (Clague & Ward 2011). Similarly, in Greenland and Iceland glaciation began in the Miocene, occurring regularly through the Pliocene and onwards to the present-day in the mountains (Geirsdóttir 2011). Likewise, in Norway's adjacent offshore and the neighbouring Barents Sea, glaciation is recorded from the early Miocene, early Pliocene and Plio-Pleistocene (Knies et al. 2009; Mangerud et al. 2011; Vorren et al. 2011). In the Rockies of the USA a much shorter glacial sequence occurs, although a Plio-Pleistocene-aged till is known from California (Gillespie & Clark 2011). In Europe, glaciation before the Middle Pleistocene is generally only represented by ice-rafted material outside the mountain regions (e.g. in the Netherlands, lowland Germany, European Russia and Britain; Ehlers et al. 2011c; Gibbard & Clark 2011; Laban & van der Meer 2011; Velichko et al. 2011).

In the Southern Hemisphere glaciation is much longer established, as noted above. Here the ice had already formed by the late Eocene–early Oligocene in East Antarctica (Miller et al. 1987; Ingólfsson 2004; Tripati et al. 2005) and built-up in a step-like pattern through the Neogene. The present polar conditions were already established by the Early Pleistocene after 2.5 Ma (Ingólfsson 2004). A similar history is known from the Piedmont areas of Argentina and Chile, where substantial ice caps were established by 14 Ma (Heusser 2003; Rabassa 2008). Till deposits interbedded with basalt flows indicate the occurrence of glaciation even before the Pliocene–Pleistocene boundary (c. 2.6 Ma) and widespread lowland glaciation became established at 2.05–1.86 Ma (?c. MIS 68–78), followed by the 'Great Patagonian Glaciation' that took place at 1.15–1.00 Ma (c. MIS 30–34; Coronato & Rabassa 2011; Martínez et al. 2011). Further north, there is little documented evidence of tropical Andean glaciers from the Plio-Pleistocene. It is estimated that the tropical Andes have attained most of their present elevation only from c. 6 Ma. However, the earliest glaciation recorded in the Bolivian Andes dates from at least 3.25 Ma (La Frenierre et al. 2011). Furthermore, in Colombia the first glaciations are dated to near the Gauss–Matuyama magnetic reversal at 2.6 Ma (Helmens 2011). The earliest records in Australasia are found in New Zealand from the Plio-Pleistocene (2.6 Ma; MIS 98–104; Barrell 2011).

2.9.2 EARLY AND MIDDLE PLEISTOCENE GLACIATIONS

The 'glacial' Pleistocene effectively begins with extensive glaciation of lowland areas, particularly around the North Atlantic region, and the intensification of global cold period (glacial) climates in general. It coincides with the 'Middle Pleistocene transition' (1.2–0.8 Ma) when the 100 ka Milankovitch cycles became dominant and caused the cold periods to become sufficiently cold for long enough to allow the development of continental-scale ice sheets.

The till sheets of the major glaciations of the 'glacial Pleistocene' are found throughout Europe. In northern Europe, till sheets characterize large areas of the lowlands and are also found at the floors of the adjacent seas (Gibbard & Clark 2011; Johansson et al. 2011; Kalm et al. 2011; Karabanov & Matveyev 2011; Lee et al. 2011). New investigations have revealed the large degree to which the North Sea floor was shaped by repeated glaciation

(Graham et al. 2011). Further south in central and southern Europe, till is restricted to the mountains and piedmonts (Calvet et al. 2011; Giraudi 2011; Nývlt et al. 2011; Urdea et al. 2011; van Husen 2011; Woodward & Hughes 2011). In northern Europe, widespread lowland glaciation began in the early Middle Pleistocene shortly after the Brunhes–Matuyama palaeomagnetic reversal (780 ka). The phases represented include the Weichselian (Valdaian, MIS 4–2), Saalian (Dniepr and Moscovian, MIS 6, 8 and 10), Elsterian (Okan, MIS 12) and the Donian (Narevian, Sanian, MIS 16). More limited glaciation may also have occurred in the circum-Baltic region during the latest Early Pleistocene (MIS 20 and 22). The evidence for Early Pleistocene glaciations in this region is restricted to Latvia (Zelčs et al. 2011), Poland (Marks 2011) and possibly Lithuania (Guobytė & Satkūnas 2011), although current research in central Jylland, Denmark, may also reveal evidence for pre-Cromerian glaciation in this area (Houmark-Nielsen 2011). Curiously, evidence for early Middle Pleistocene glaciation is absent from the North Atlantic and Norway, although it is certainly present in Denmark, the Baltic region and European Russia. In the Italian Dolomites, glaciation became established in MIS 22 (Muttoni et al. 2003). Comparable evidence is also found from north of the Alps in Switzerland and southern Germany (Fiebig et al. 2011). Further to the west in the Pyrenees, the oldest glaciation identified is of late Cromerian age (MIS 16 or 14; Calvet 2004). Widespread lowland glaciation is first seen in North America in MIS 22 or 20 (Barendregt & Duk-Rodkin 2011; Duk-Rodkin & Barendregt 2011). From this point onwards, major ice sheets covered large regions of the continent during the Middle Pleistocene pre-Illinoian events MIS 16, 12, 8 and 6 (Illinoian *s.s.*) and the Late Pleistocene MIS 4–2 (Wisconsinan). In Mexico, the oldest moraines on volcanoes have been dated at 205–175 ka and probably correspond to an advance early in MIS 6 (Vázquez-Selem & Heine 2011). Evidence from east Greenland suggests that the southern dome of the ice sheet may almost have disappeared during the Eemian Stage interglacial (*c.* MIS 5e). However, dating of the basal ice in the Dye 3 ice core has shown that ice was present over this locality from at least MIS 11 (Funder et al. 2011).

According to Chinese investigations glaciation of Tibet and Tianshu is not recorded before the Middle Pleistocene, of which the MIS 12 glaciation was the most extensive. Four discrete Pleistocene glaciations have been identified on the Qinghai–Tibetan Plateau and the bordering mountains. These four main Pleistocene glaciations are correlated to MIS 18–16, 12, 6 and 4–2. This apparently delayed glaciation of the Himalayan chain might reflect late uplift of high Asia. The Kunlun Glaciation (MIS 18–16) was the most far reaching. Subsequent glaciations have been successively less extensive, probably caused by increasingly arid climates resulting from the progressive Quaternary uplift of the plateau (Shangzhe et al. 2011). On the contrary, Kuhle (2011) maintains that the last glaciation was most extensive when he envisages an ice sheet covered Tibet. This is increasingly a minority view however, especially the concept that the Qinghai–Tibetan Plateau was almost entirely covered by an ice sheet. Many workers rigorously maintain that Pleistocene glaciations over the Qinghai–Tibetan Plateau were much more restricted (including Lehmkuhl & Owen 2005 Shangzhe et al. 2011). Indeed, it is now generally well established that a large ice sheet did not cover the Tibetan Plateau, at least not during the Late Pleistocene. The situation during earlier cold stages of the Pleistocene is still open to speculation, and further research is needed to ascertain the maximum extent and timing of Pleistocene glaciations over large parts of Tibet and the neighbouring Qinghai province.

As in Europe and North America, glaciation increased in intensity throughout the South American Andes from 800 ka to the present day. In the northern Andes there is some evidence of glacial deposition prior to the Late Pleistocene glaciations in Venezuela and Colombia (Helmens 2011; Kalm & Mahaney 2011). However, in the southern Andes Late Pleistocene glaciations were less extensive than during the Early Pleistocene events (Coronato & Rabassa 2011; Harrison & Glasser 2011; Martínez et al. 2011). Both easternmost Tierra del Fuego and the Falkland Islands are thought to have remained largely unglaciated during the Pleistocene (Clapperton 1993), though there are traces of glacial scouring (Ehlers & Gibbard 2008). In Australasia, following a 1 Ma break the glacial record continues in MIS 12, followed by MIS 6, 4 and 3. In Tasmania, the earliest Pleistocene ice advances are thought to have been of age c. 1 Ma, but may be older. Middle Pleistocene ice advances occurred during MIS 10, 8 and 6 (Colhoun & Barrows 2011). The presence of glaciations in this area during MIS 8, revealed by cosmogenic exposure dating, is interesting because a record of glaciation during this interval is lacking in many other parts of the world.

The succession of glaciations in MIS 16, 12, 6 and 4–2 is striking in that it is repeatedly found in numerous areas of the world; the absence of records of glaciations during MIS 18, 10 or 8 probably reflect the fact that later glaciations were more extensive. However, it is possible that glaciations during MIS 18, 10 and 8 were, in fact, more widespread than currently realized, simply because many dating techniques provide minimum ages for glacial deposits and landforms. This is true for U-series dating of cemented moraines (Woodward & Hughes 2011) and also when applying cosmogenic nuclide analyses to date 'ancient' surfaces (e.g. Colhoun & Barrows 2011). Lack of precision when dating glaciations of the Middle Pleistocene and earlier in terrestrial settings means that some uncertainty will remain, despite the development of better dating techniques.

2.9.3 LATE PLEISTOCENE GLACIATIONS

During the last glacial cycle of the Late Pleistocene, the extent of the glaciation of the Southern Hemisphere differed very little from that of the Pleistocene glacial maximum. Glaciers in Antarctica still reached the shelf edge, and on New Zealand, Tasmania and in South America the glacier tongues were only slightly smaller than during earlier events (Barrell 2011; Colhoun & Barrows 2011; Coronato & Rabassa 2011). On mainland Australia, local mountain glaciation occurred (Colhoun & Barrows 2011). In many parts of the Northern Hemisphere, glacial ice reached an extent very similar to the Quaternary glacial maximum. In North America, the differences are also very small. Again, most parts of Canada were ice covered, including the shelf areas (Stea et al. 2011). It is the same in Greenland (Funder et al. 2011) and Iceland (Geirsdóttir 2011).

In Europe and on mid-latitude mountains around the world, however, the situation was different. New evidence suggests that the North Sea was not fully glaciated during the Weichselian glacial maximum, but slightly earlier during MIS 4 (Graham et al. 2011). A similar situation has been invoked for the Bristol Channel area of the British Isles, where cosmogenic exposure ages show ice was retreating during MIS 3 (Rolfe et al. 2012). Further north over the Barents Sea, glaciation during MIS 4 was more extensive than the later Weichselian glaciations. During the Late Weichselian an ice sheet covered the Barents Sea and extended well into the Kara Sea, but hardly touched the Russian mainland and did not reach onto the Severnaya Zemlya islands (Vorren et al. 2011). In glaciated mountain areas outside the major ice sheets, such as in Italy and Greece, the maximum Middle Pleistocene glaciations were sometimes markedly greater in extent than the local last glacial maxima (Giraudi 2011; Woodward & Hughes 2011). This is attributed to a change in equilibrium-line altitude (ELA), which has a much bigger impact on glacier size in areas characterized only by mountain glaciation than in areas where ice covered the lowlands during multiple glaciations. However, this was not the case everywhere. For example, in Romania and Turkey only Late Pleistocene glaciations have been found (Sarıkaya et al. 2011; Urdea et al. 2011). Mountains often display contrasting geochronologies between areas and sometimes even within the same mountain range (Hughes & Woodward 2008). The former situation is likely to reflect regional differences in moisture supply, while the latter situation, whereby different geochronologies are presented for the same mountain area, are more likely to reflect problems with the reliability or interpretation of dating techniques or results (Ehlers et al. 2011b).

As in parts of Europe, the maximum ice advance of the last glaciation occurred in MIS 4 (and in some places through to MIS 3) in the mid-latitude mountains of Japan (Sawagaki & Aoki 2011) and Taiwan (Böse & Hebenstreit 2011). In New Zealand, there is now convincing evidence of extensive glaciation in MIS 3 with significant yet smaller glacier advances during MIS 2. Putnam et al. (2013) presented the results of 73 [10]Be ages from moraines at Lake Ohau on South Island, New Zealand. Based on this data they identified successively smaller glacier advances at 138.6±10.6 ka, 32.52±0.97 ka, 27.4±1.3 ka, 22.51±0.66 ka and 18.22±0.5 ka. The MIS 2 glaciers were therefore smaller than the MIS 3 advance. There is also evidence of the largest Late Pleistocene glaciations earlier than MIS 3. For example, in the Cascade Plateau area of SW New Zealand, Sutherland et al. (2007) argued on the basis of cosmogenic exposure ages that the most extensive glaciation of the last glacial cycle occurred early, at *c.* 79 ka during MIS 5a. Later glacier advances also occurred at *c.* 58 ka and 22–19 ka. In NW Nelson, South Island, New Zealand, [10]Be exposure ages suggest that glaciers expanded to the most extensive positions of the last glacial cycle during MIS 4–3 (Thackray 2009). Similarly, McCarthy et al. (2008) used cosmogenic and luminescence ages to argue that MIS 4 cirque glaciers were similar in size or slightly more extensive than MIS 2 glaciers. Moreover, Barrell (2011, fig. 75.2) identified several areas where MIS 4 glaciers were more extensive than during MIS 2. It is therefore clear that global glaciations during the last glacial cycle, whether characterized by lowland ice sheets or mountain glaciers, at high, middle or low latitudes, were asynchronous (Hughes et al. 2013). While many glaciers did advance at the global Last Glacial Maximum (Clark et al. 2009), many areas saw bigger glaciations earlier in the last glacial cycle. The reason for this is likely to be due to global variations in changing moisture supply with the global LGM being not only very cold, but also very dry (with the LGM defined by global dust peaks at 27.5–23.3 ka; Hughes & Gibbard 2014). The latter phenomenon caused the retreat of glaciers in many areas, despite cold air temperatures at the global LGM.

Shallow supraglacial meltwater stream disappears in a moulin, Kverkjökull, Iceland. Photograph by Jürgen Ehlers.

Chapter 3
Ice and Water

Ice and water have shaped the landscapes of the formerly glaciated areas. This is true for Iceland, as well as for the Alps. Even in the ice shield of Antarctica, where little water is visible on the surface, water at the bottom of the ice contributes to the shaping of the subglacial landforms.

3.1 The Origin of Glaciers

The total area of glacier ice on the Earth covers an estimated 14.9 million km^2, which represents 10% of the land surface. 12.6 million km^2 are located in Antarctica and 1.7 million km^2 in Greenland; the remaining 4% is distributed in the form of numerous small mountain glaciers and ice caps over the rest of the world (Table 3.1). Melting of the Antarctic ice would result in a global sea-level rise of 57 m, whereas melting of the Greenland Ice Sheet would raise sea levels by only 7 m (IPCC 2007). During the glacial maxima, the total volume of glacier ice was at least 2.5 times as large as today (Denton & Hughes 1981). As a result of a warmer climate since the early 1970s, the combination of glacier melting and ocean thermal expansion caused about 75% of the observed global mean sea-level rise. Globally, glaciers have in recent decades lost 301 ± 135 Gt a^{-1} (gigatonnes per year) and contributed the equivalent of 0.83 ± 0.37 mm a^{-1} to global sea levels over the period 2005–2009 (IPCC 2013).

Glaciers only form where the winter snowfall does not completely melt during the following summer, so that the snow of the next winter adds to the total volume. The amount of snow that is required in order to survive the following summer depends on the temperatures of that summer; the higher the temperature, the greater the amount of melting. If summer

The Ice Age, First Edition. Jürgen Ehlers, Philip D. Hughes and Philip L. Gibbard.
© 2016 John Wiley & Sons, Ltd. Published 2016 by John Wiley & Sons, Ltd.
Companion website: www.wiley.com/go/ehlers/iceage

TABLE 3.1 AREA AND EXTENT OF SNOW AND ICE ON EARTH AND POTENTIAL RISE IN SEA LEVEL (SEA-LEVEL EQUIVALENT, SLE) AT ITS MELTING.

Cryospheric component	Area (10^6 km^2)	Ice volume (10^6 km^3)	Potential sea-level rise (m)
Snow on land	1.9–45.2	0.0005–0.005	0.001–0.01
Sea ice	19–27	0.019–0.025	0
Glaciers and ice caps	0.51–0.54	0.05–0.13	0.15–0.37
Ice shelves	1.5	0.7	0
Ice sheets	14.0	27.6	63.9
Greenland	1.7	2.9	7.3
Antarctica	12.3	24.7	56.6
Seasonally frozen ground	5.9–48.1	0.006–0.065	0
Permafrost	22.8	0.011–0.037	0.03–0.10

Source: IPCC (2007). Reproduced with permission.

temperatures are low then a modest amount of snowfall can survive the summer; if summer temperatures are high, then a lot of snow is required to sustain a glacier.

Summer temperatures vary with latitude and altitude. With increasing distance from the poles low temperatures are restricted to higher altitudes above sea level, so that glaciers near the equator are confined to the highest mountain regions. The only three glaciated mountains of Mexico (at *c*. 19° N) are all higher than 5000 m. The limit of glaciation in the subtropics (*c*. 38–23° N) is even higher, due to the dominance of high-pressure systems and low precipitation. For example, on the edge of the Atacama Desert in South America, mountains remain unglaciated up to over 6000 m altitude. Similarly, no glaciers are present in North Africa next to the Sahara Desert, despite the mountains reaching over 4000 m in the High Atlas. In the Kunlun Shan mountains range in central China glaciers only occur on mountains that reach over 5000 m (Fig. 3.1), and in the interior of Tibet only on peaks that exceed 5800–6000 m.

Figure 3.1 Glacier in the Kunlun Shan, China. Photograph by Phil Hughes.

Seasonality can also be important. In maritime regions where there is only a small difference between summer and winter temperatures, winters can be too warm for significant snowfall year-on-year despite relatively low mean annual temperatures. The mountainous western parts of the British Isles and Ireland are a good example. Conversely, in continental areas very high seasonality with hot summers and cold winters causes a lot of summer melting. This eliminates any snowfall that occurs in the winter, which can be small in amount because of dry conditions inherent in continental areas. Finally, tropical glaciers can experience both snow accumulation and melting all year round since temperatures do not change significantly over the year. Consequently, such glaciers can experience a daily balance of snow accumulation and melting. Glaciers are therefore controlled not only by winter precipitation (snowfall) and summer temperatures (melting), but also by annual and daily temperature ranges (Hughes & Braithwaite 2008).

As noted earlier, a sufficient amount of precipitation is another prerequisite for the formation of glaciers. High evaporation rates and low rainfall in cold-climate environments may lead to the formation of cold periglacial deserts, but not to the formation of glaciers. A west coast location at higher latitudes generally provides extremely favourable conditions for the formation of glaciers (e.g. Norway). The westerly winds bring high rainfall and its high-latitude location means that winters are cold enough for snow. The decisive factor in this case is the seasonal distribution of precipitation. The summer precipitation falls as rain and often contributes little to the mass balance of the glaciers. Most of it does not freeze but drains together with the glacial meltwater, running off rapidly on the surface or in tunnels in the glaciers (Sugden & John 1976).

High precipitation is required for the *formation* of large ice sheets; on the other hand, low precipitation is sufficient (at low summer temperatures) for their *preservation*. The inner areas of major ice sheets are among the most arid areas of the world. In the northern part of the Greenland Ice Sheet only 150 mm of annual precipitation occurs; this is only in the form of snow however, so that it completely balances the small amounts of melting that occur on the outlet glaciers at this latitude. The largest losses of mass in places such as Greenland occur not as a result of summer melting but as a result of ice calving into water where the glaciers reach the sea. This can cause rapid transfer of snow and ice from the centre of the ice sheet to the sea and produce ice streams. However, this mode of glacier behaviour is not sustainable for the arid ice sheets of Greenland and Antarctica and is usually restricted to a few localities. It is possible that rising sea levels could increase the rate of calving and result in the ice sheet collapse. This has been suggested as a possible scenario for the West Antarctic Ice Sheet, which has an ice volume equivalent to *c.* 5 m of global sea level (Pollard & DeConto 2009).

Glaciers are more usually formed in areas that have a favourable relief, such as lee-side positions in the mountains. Most of today's glaciers in Scandinavia are on the north-eastern slopes, which are the coldest positions. In addition, the highest snow accumulations are found in the lee of mountains (Østrem et al. 1973). This situation is common across the Northern Hemisphere where glacial cirques have a strong aspect preference to the northeast (Evans 1977). Windblown and avalanching snow is important and can dramatically increase the amount of snow accumulation, enabling glaciers to form at altitudes well below those at which they would form by direct precipitation alone. Large accumulation of snow by both wind and avalanche is very common for small glaciers at low altitudes at higher latitudes, such as the Polar Urals where Dolgushin (1961) called such phenomena 'Polar Ural type' glaciers. However, such glaciers are not restricted to the high latitudes. For example, glaciers in Montenegro and Albania (at 40–45° N) are some of the lowest-altitude glaciers (1980–2420 m) at this latitude (42.5° N)

(a)

(b)

Figure 3.2 (a) Snow accumulation in the cirque containing the Debeli namet glacier, Montenegro. This picture was taken in May 2006. (b) The Debeli namet glacier in September 2006. This glacier survives because total snow accumulation, including that by avalanching and windblown snow, is twice that which falls by direct precipitation alone (Hughes 2008). Photographs by Phil Hughes.

in the world (Hughes 2008, 2009a; Fig. 3.2). Around 4–5 m of annual snow accumulation (water equivalent; this equates to a powder snow depth of up to 50 m) is necessary to sustain these glaciers, located less than 100 km from the tourist beaches of the Adriatic Sea (Fig. 3.2). Such high levels of accumulation are likely to be facilitated by large inputs of snow from wind-blown snow and avalanching snow in addition to direct precipitation; drifts exceeding depths of 5 m are common in Montenegro even at modest altitudes (Fig. 3.3).

A variety of combined factors are involved when considering the question of the origin and spreading of glaciers. This becomes especially clear when it is borne in mind that, within large ice sheets, some ice-free oases can occur such as the dry valleys in Victoria Land (Fig. 3.4). An even larger example from the Quaternary glacial areas is the Driftless Area in Illinois and Wisconsin (USA).

Following Ahlmann (1935), three types of glaciers can be distinguished: temperate, subpolar and polar glaciers. The base of a temperate glacier is at the pressure melting point throughout, and ice and snow are melting on its surface (at least outside the polar regions) in the so-called ablation zone all year round. In contrast, the polar glaciers see very limited superficial melting, and their base is frozen to the bedrock. The subpolar type of Ahlmann represents a transition in form between the two extremes. Such transitional forms exist in many cases, where glaciers may exhibit temperate base parts in areas while other parts are frozen to the rock surface; these are called polythermal glaciers (Liestøl 2000). Previously it was assumed that Alpine glaciers all belong to the temperate type; today we know that at least parts of some very high-altitude Alpine glaciers may also have negative basal temperatures. On the Monte Rosa Plateau basal ice temperatures of −2 to −3°C can be found. The classic tripartite division is too coarse adequately to characterize all glaciers. Most glaciers do not fall exactly into one of the categories, but can be divided into different zones with different characteristics (Winkler 2009a).

Figure 3.3 Windblown snow drifts in the Durmitor massif, Montenegro. May 2006. Photographs by Phil Hughes.

Figure 3.4 Taylor Valley, a dry valley in Antarctica, with the Commonwealth Glacier in the background. Photograph by Hans-Christoph Höfle.

In the upper area of a glacier, snow accumulation is predominant. This part is called the accumulation zone. In the lower part of the glacier, the ablation zone, the melting of ice is dominant. The boundary between the accumulation and ablation zones forms the equilibrium line of the glacier. In the present glaciers of the Alps, the accumulation area is about twice the size of the ablation zone (Hoinkes 1970) and the ratio of the accumulation area versus ablation area is commonly about 0.5–0.8 (Benn & Evans 2010). The equilibrium line is defined as the line where the net balance at the end of a balance year is nil. It usually has a complicated history and cannot be constructed from a map. Since the equilibrium line altitude (ELA) varies from year to year, it is usually necessary to average the measurements over time. The height of the equilibrium line or ELA is used to compare the general climate conditions of different glaciated areas. On glaciers where no measurements exist and for all reconstructed palaeoglaciers the ELA has to be calculated. Several different formulae exist; note that different calculation methods can lead to different results (Nesje & Dahl 2000).

Freshly fallen snow on the glacier is transformed into ice, which initially results in old snow (Fig. 3.5). The change is relatively expeditious (usually over a few days) since the temperature of the crystals is usually near the melting point, meaning that the molecules can move relatively freely. The movement of the molecules follows the principle that the surface of the crystals is reduced and the free energy is minimized. In this way, snowflakes with their intricate forms are slowly transformed into spherical shapes. In addition, larger crystals tend to grow at the expense of smaller crystals, since in this way the free energy is also reduced. Simultaneously with these processes a general settling will take place, in the course of which the interstices between the particles become smaller. Bridges are formed at the contact points between grains.

The old snow of the previous year is referred to as 'firn'. The lower limit of firn cover at the glacier surface is known as the firn line (Hoinkes 1970). Molecular diffusion leads to recrystallization and further compaction of the firn. When the density reaches c. 0.8 g cm^{-3}, the pores between the crystals close and trapped air remains in the form of gas bubbles. The firn has turned into ice (Winkler 2009a).

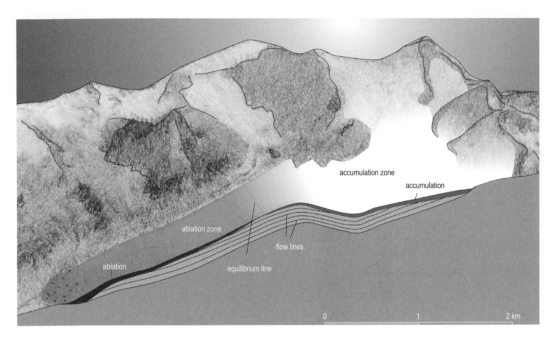

ablation zone
accumulation zone
accumulation
flow lines
ablation
equilibrium line
0   1   2 km

Figure 3.5 Longitudinal section through a valley glacier. Newly formed ice from the accumulation area of the glacier flows downvalley. The ice formed in the upper part of the glacier ice moves closest to the glacier sole. The formation of ice continuously decreases towards the equilibrium line. In the ablation zone melt processes steadily increase further downhill.

3.2 Recent Glaciers: Small and Large

3.2.1 HOW DOES A GLACIER MOVE?

When the ice reaches a thickness of about 30 m, it begins to move. However, the precise threshold thickness for ice movement is dependent on the surface slope of the ice as well as the conditions at the base of the glacier (see below). The glacier movement is made up of two components: internal deformation and basal flow.

Wherever the base of the glacier is at the pressure melting point, in addition to internal deformation basal sliding occurs. This is most effective when the glacier slides on a water film (Weertman 1964). Water-filled cavities at the glacier base therefore play a major role in determining the flow velocity. The water-filled cavities expand with increasing hydrostatic pressure, leading to a reduction in friction and an increased flow velocity (Lliboutry 1968). This explains the observation that glaciers flow faster in summer than in winter, at least in the ablation zone (Sugden & John 1976).

The horizontal flow behaviour of a glacier is not completely uniform, but is determined by the mass balance and the shape of the land surface. Nye (1952) highlighted the need to distinguish two different types of glacier flow: compressive flow and extending flow. Compressive flow leads to a reduction of glacier movement if increased ablation reduces the ice load, or if the glacier sole is concave. On the other hand, extending flow occurs where the ice thickness increases and where there is a convex glacier sole. Compressive flow in a glacier results in upward trajectories; extending flow results in downward trajectories.

The flow velocity of glaciers can be relatively constant (usually a few metres to tens of metres per year), but sometimes significant periodic fluctuations occur. In such surges, a special type of rapid glacier advance, the flow rate of the ice can increase by 10–100 times the normal value. The rapidly advancing ice can be divided into three zones: (1) a wave of thickening ice under compressive flow conditions; (2) a zone of high velocity with a very broken ice surface downstream of the wave crest; and (3) a zone of ice expansion and decreasing

TABLE 3.2 FLOW RATE OF GLACIAL SURGES.

Ice stream	Velocity
Rutford Ice Stream (Antarctica)	>400 m a^{-1}
Whillans Ice Stream (Ice Stream B)	827 m a^{-1}
Jakobshavns Glacier (West Greenland)	6–12 km a^{-1}

Adapted from Clarke (1987) and other sources.

speed (Sugden & John 1976). A surge wave sometimes does not reach the edge of the glacier; where it does however, it can lead to catastrophic glacier advances. Brúarjökull in Iceland in 1963–64 advanced up to 8 km, reaching velocities of up to 5 m hour^{-1} (Thorarinsson 1969). At the end of such a surge, the advanced ice often stagnates until a sufficient ice thickness has been reached for a further advance.

According to studies of North American glaciers, surges occur at a frequency of one per 15–100 years (Meier & Post 1969). The exact cause of the surges is unknown, but it is clear that they are each accompanied by an above-average supply of meltwater (Sugden & John 1976).

Large ice streams exhibit some significant, highly variable flow rates as listed in Table 3.2. The catchment area of such an ice stream can be a thousand times greater than that of a typical rapidly advancing mountain glacier (Figs 3.6–3.8). Hughes (1992) pointed out that about 90% of the ice flow from the Greenland and Antarctic ice sheets into the sea is accomplished by such ice streams. Similar conditions are assumed to have existed during periods of collapse of the Pleistocene ice sheets.

Figure 3.6 Mountain glaciations will leave a number of prominent landforms that are clearly visible in satellite images. These include numerous cirques, many with cirque lakes at the upper end of the valleys, U-shaped valleys and tongue basins, many of which are filled with lakes. This example shows the Wind River Mountains, Wyoming.

Source: US Geological Survey, Landsat 5 TM Path 37, Row 30, 9/8/1999.

Figure 3.7 Tongue basin with adjoining end moraines. Wind River Mountains, Wyoming. Photograph by Jürgen Ehlers.

Figure 3.8 Glacial erosion transforms the V-shaped valleys created by rivers into U-shaped glacial valleys. Example from northern Sweden, north of Tärnaby. Photograph by Jürgen Ehlers.

3.2.2 DEVELOPMENT OF AN ICE STREAM NETWORK

For the development of Alpine glaciers, the local land surface is of crucial importance; glaciers cannot form on steep rock walls. On the other hand, gently inclined slopes can gradually develop firn depressions in which corrie glaciers form, from which (under a positive mass balance) valley glaciers eventually evolve. On the south side of the Alps the snow line is around 200 m higher compared to the north side of the Alps due to the higher temperatures (von Klebelsberg 1948–49). The different elevations of the northern and southern foothills of the Alps also play a role (about 500 m in the north versus *c.* 100 m in the south). Accordingly, the Quaternary glaciers on the northern edge of the Alps penetrated far into the foothills, whereas in Italy they did not reach beyond the foot of the mountains.

During the peak of the last glacial the Alpine snow line was about 1200 m lower than today. The accumulation area of glaciers had therefore been greatly expanded. The Alpine valleys were traversed by mighty rivers of ice. An ice stream network was formed which was surmounted by only a few mountain peaks. At Bolzano, the distance between the peaks flanking each side of the Adige River valley was *c.* 40 km (von Klebelsberg 1948–49).

Ehlers et al. (2011a) provide an overview of the maximum extent of the ice sheets during the last glaciation. The Alps were not equally affected by glaciations throughout the area (Fig. 3.9). In the west and in the central Alps only higher peaks rose above the ice stream network, the surface of which was at about 2500 m above sea level and higher. To the east, the area of large-scale ice cover was restricted to the large glaciers in the Enns and Mur valleys and a lobe in the Drava River valley. Towards the east, peak elevations were lower but an increasing continentality is felt with a noticeable lack of rainfall. Consequently, only smaller local glaciers and ice caps could form.

(a)

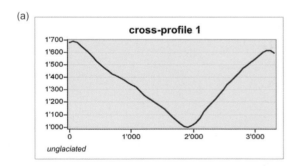
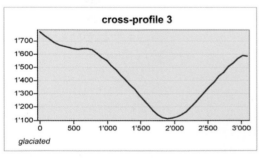
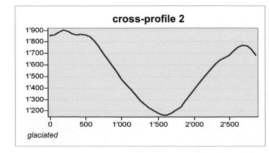
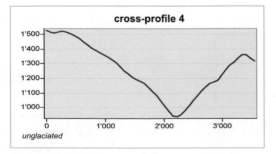

Figure 3.9 The formerly glaciated valleys of the Alps can be distinguished from unglaciated valleys from the digital terrain model. (a) Cross-sections through four Alpine valleys

(b)

(c)

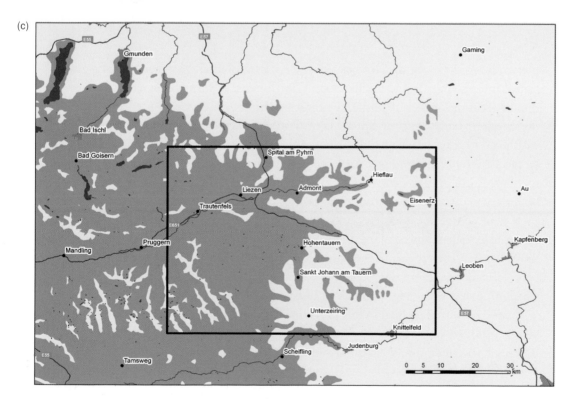

Figure 3.9 *(Continued)* (b) location of the cross-sections relative to the Würmian ice margin; and (c) location map.

With increasing ice thickness, the narrow Alpine valleys could no longer adequately drain the massive ice of the central Alps. This eventually resulted in flow over the Alpine passes, resulting in transfluence of central Alpine ice to the northern edge of the Alps. Consequently, glacier tongues which were much larger than would have been expected from the relatively small catchment areas developed in the valleys of the Ammer, Loisach and Isar (van Husen 1987). In the Iller and Lech valleys, where this overflow did not occur, only relatively small glaciers have formed.

3.2.3 DEVELOPMENT OF ICE SHEETS

A glacier in a mountain range that is evolving gradually from a corrie glacier to a valley glacier initially flows downvalley, following the natural slope. If the ice supply is large enough, the individual valley glaciers will unite at the foot of the mountains to form a piedmont glacier. When the ice accumulation in the central valleys becomes so large such that the valley can no longer manage the collected runoff from the tributary valleys, ice will cover the adjacent hills and the glaciers of adjacent valleys will unite to form an ice sheet. If the ice mass continues to rise, an icecap of many hundreds of kilometres in diameter may result.

The great Pleistocene ice sheets were not uniform bodies of ice, but were usually controlled by several more or less independent ice centres whose dynamics differed significantly from each other. As early as the end of the last century, Tyrrell (1898) took the view that the former glaciation of North America did not consist of a single ice sheet but was based on two separate centres, one in Keewatin and one in Labrador. Later he added a third centre in Patricia (south

BOX 3.1 LIMIT OF GLACIAL POLISH

Determining the extent of past glaciers and ice sheets has always been a major goal of glacial morphology. While the horizontal distribution of former glaciers is defined by the distribution of their deposits, the limit of glacial polish can be used to determine the vertical extent. The limit represents the boundary between an upper zone of strongly weathered rock (especially by frost shattering), and a lower zone which is characterized by glacial erosion and minor postglacial weathering. Where a clear boundary between weathered rock and glacier-polished rock surface is seen, the explanation is simple: the boundary marks the upper limit of glacial erosion during the Last Glacial Maximum.

It is of course possible that peaks above the line were covered with cold ice, while temperate glaciers flowed along the valleys. Moreover, it is possible that the grinding limit does not correspond to the Last Glacial Maximum but represents either an older or a younger glacier advance. Finally, it is also possible that the lower limit of frost weathering may suggest an upper limit of glacial polish.

Exposure dating highlights this problem. Briner et al. (2003) have shown that glacier-transported clasts overlying weathered tors in eastern Baffin Island showed an age of 17–11 ka, while the tor itself dates back to more than 60 ka. A similar result was obtained by Marquette et al. (2004) for the mountain regions of Labrador and NW Quebec. The measured $^{26}Al/^{10}Be$ exposure ages for tors and boulders show that intricate rock structures may have survived the last and probably also earlier glaciations under a protective blanket of cold ice without morphological alteration.

of Hudson Bay). Today it is believed that there were two separate ice centres in Keewatin and Nouveau Quebec/Labrador. An additional centre was found further to the north, in the area of the Foxe Basin (Ives & Andrews 1963).

It is clear that the glaciation of the Rocky Mountains stood only in a relatively loose contact with the Laurentide Ice Sheet. Both were vast expanses of ice separated from each other by an ice-free corridor. However, there are differing views about its extension. Some authors restrict the length of the ice-free corridor to a few hundred kilometres; others suggest it was more than a thousand kilometres long both to the north and the south of the contact zone. The closing of this gap (Dyke 2004) only occurred some 15 ka ago.

The northern European glaciation area also had several ice centres. Individual ice sheets formed on the British Isles and in the Timan–Urals region.

There are still many uncertainties about the extent of glaciation in locations such as Siberia or in the highlands of Tibet, and even the extent of the northern European Ice Sheet has been greatly revised in recent decades. When Aseev (1968) tried to reconstruct the Weichselian Scandinavian Ice Sheet, he considered a maximum altitude of the ice of 2500 m above sea level. In southern Norway however, a series of mountain peaks were not covered by ice during the last ice age. Using the upper glacial limits at those peaks, new ice sheet profiles could be constructed (Nesje & Sejrup 1988). It turned out that the ice had been much lower than assumed a few years earlier. For other areas, however, no concrete data on ice thickness are available. In contrast to Aseev, today we know that the ice in the north and east had a much greater extent. It has been proven that the Bering Sea was glaciated, and that in the east the Scandinavian Ice Sheet was in contact with the glaciation of the Timan–Urals area.

In the area of the Nordic glaciation of Europe, there were few peaks that towered over the ice sheet. Consequently, the ice could expand relatively freely in all directions during the glacial maximum, controlled only by the mass balance of the glacier. The North European Ice Sheet reached its greatest expansion east of the mountains. Precipitation was brought by the westerly winds and fell largely as snow, which was deposited in the lee position. Due to the steep slope of the continental margin in the north and the Norwegian Trough in the south, no extensive glaciation could form west of the Scandinavian Mountains. The result was an ice sheet, the apex region of which was strongly shifted to the northwest.

The great ice sheets of the Pleistocene have spread very rapidly. This is best demonstrated for the glaciation of the last cold stage. Ehlers (2011) noted that, during the last cold stage, the front of the Scandinavian Ice Sheet advanced by as much as $100–150$ m a^{-1} into northern Germany. In North America, Goldthwait (1959) demonstrated that the Laurentide ice spread in Ohio at a rate of $17–119$ m a^{-1} by radiocarbon dating of tree trunks overridden by the ice. In Germany, Junge (1998) showed that the ice of the older Elsterian Glaciation in Saxony pushed forward at an average rate of 400 m a^{-1}, demonstrating that even large ice sheets can advance amazingly rapidly. In contrast to the rapid advance of Alpine glaciers due to the shorter distances and the particular nature of the glaciation (ice stream network), the rapid development of the North European and North American ice sheets are difficult to explain using traditional concepts; other possible explanations must be sought. Two possibilities have been discussed in the international literature: the *deformable bed* (Box 3.2) according to Boulton & Jones (1979); and *instantaneous glacierization* (Box 3.3) by Ives et al. (1975).

BOX 3.2 DEFORMABLE BED

Each glacier whose ice exceeds a thickness of about 60 m moves through internal deformation. Since the amount of deformation increases with increasing distance from the glacier sole, the glacier moves more slowly at depth than at greater heights. In a cold glacier that is frozen to the bedrock, internal deformation is the only movement; at the base of the glacier, this movement is equal to zero. With a larger ice thickness or, if the bottom of the glacier is at the pressure melting point, a thin film of water forms on which the glacier slides downhill. The result of this movement at the bottom of the glacier is the same as at the surface. The basal sliding of glaciers with a warm base can account for up to 90% of total glacier movement.

From glacier studies in Iceland, Boulton & Jones (1979) have shown that there is a third type of glacier movement (Fig. 3.10). Not only is the ice itself moving, but the underlying unconsolidated substratum is incorporated into the glacial movement. According to Boulton & Jones, this mechanism might have contributed significantly to the high rate of flow of Pleistocene glaciers in northern Europe. A layer of rearranged material can often be found at the base of tills, the grain size distribution of which is similar to the underlying deposits and the structure of which is reminiscent of the overlying till. The original strata are considerably sheared and stretched. Grube (1979) referred to this sediment type as 'sohlmoräne' (sole till). The sole till usually ranges from a few centimetres to up to about 2 m in thickness (Ehlers & Stephan 1983). The primary layering is mostly undisturbed. Occasionally horizontal displacements along bedding planes can be observed, but these are comparatively rare.

If the mechanism envisaged by Boulton & Jones was effective on a larger scale, it must have been confined to the upper metres of the layers. Where two tills of different age are superimposed, the upper part of the older deposit has been partly incorporated in the younger glacial movement and the long axes of its clasts have been rearranged in the direction of the new ice advance. Tills lying on top of one another often differ significantly in their material composition, although they are in themselves largely homogeneous sediment bodies. Although large amounts of deformed material are incorporated into the new till during the ice advance, the thickness of the visibly deformed zone does not exceed a few metres in most tills.

In many cases in which a pronounced sohlmoräne is lacking, the deforming bed is likely to have been limited to the rock debris that the respective glacier itself brought along (Alley et al. 1986, 1987). While it was traditionally assumed that maximum ice thickness in northwest Europe exceeded 2500 m (Aseev 1968), under the assumption of subglacial deformation Boulton et al. (1985) arrived at a maximum ice thickness of only *c.* 2000 m. Around Copenhagen, where Aseev assumed 1000 m of ice, Boulton et al. calculated only *c.* 500 m. However, it must be borne in mind that the conditions at the bottom of the ice sheet have not been constant during the entire glaciation and that repeated changes between a frozen bed and basal deformation may have taken place (Lambeck et al. 2010).

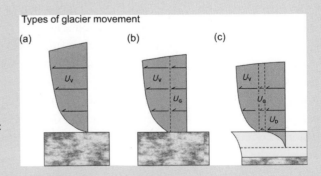

Figure 3.10 Different types of glacier movement: (a) internal deformation U_V; (b) basal sliding U_G; and (c) subglacial deformation U_D.

BOX 3.3 INSTANTANEOUS GLACIERIZATION

It has traditionally been assumed that the Pleistocene continental ice sheets developed from local mountain glaciations. In the foothills, the individual glaciers would gradually merge into a piedmont glacier, from which an ice shield eventually evolved. This model for the formation of ice sheets may seem plausible for the Scandinavian region, but in the area of the Laurentide Ice Sheet in North America the only suitable high ground is found on the extreme eastern edge of the ice sheet. A closer look reveals that this is not a real mountain range, but the upper edge of the Canadian Shield which rises abruptly from the sea and slopes very gently inland. Traces of an incipient glaciation from there could not be found. Ives et al. (1975) therefore assumed that the Laurentide ice had formed spontaneously (i.e. instantaneous glaciation). They assume that snow accumulation on a large area resulted in the formation of large ice sheets over the course of a few centuries. This mechanism is certainly thought likely for the Laurentian Ice Sheet in North America (Clark & Lea 1992).

If the rapid development of the North European Ice Sheet could be explained in this way, it would mean that, at least during the initial phase of glaciations, no continuous transport of Scandinavian rock material to northern Germany was possible. The result would be the widespread formation of local tills. Such local tills would (with the exception of the oldest ice) be similar in composition to the next-oldest glacial deposits, from which they had been derived (Ehlers 1990).

3.3 Dynamics of Ice Sheets

The flow pattern of a glacier or ice sheet is not static, but adapts in the course of time to the mass balance and changes in the temperature regime of the glacier. Clear expressions of this are shifts in the position of the ice divide or changing directions of ice advances, known for example from Denmark, which was invaded by Norwegian, Swedish and Baltic ice (Houmark-Nielsen 2011). To reconstruct these changes of glacier dynamics, an examination of the glacial deposits and landforms is required (Fig. 3.11).

When overridden by a glacier, pre-existing landforms are converted in the direction of ice movement. This is most clearly visible in the formerly glaciated areas of the far north, where landforms are clearly visible in aerial photographs or satellite images due to the lack of vegetation. The glacigenic overprint is also clearly depicted in a digital terrain model (Fig. 3.11). The significance of the large landforms is lower than that of small forms. Large surface forms of hard rock cannot be created during a single glaciation; their present morphology results from repeated, unidirectional movement of the ice of several glaciations (Figs 3.12, 3.13). Consequently, the large landforms in Scandinavia show nothing but the general ice flow direction from the high altitudes of the Scandinavian mountains down into the lower areas (Lundqvist 1990). The formation of the southern Swedish lake fans falls into this category; the corresponding flow pattern was repeated over and again for certain periods, probably during all glaciations.

(a)

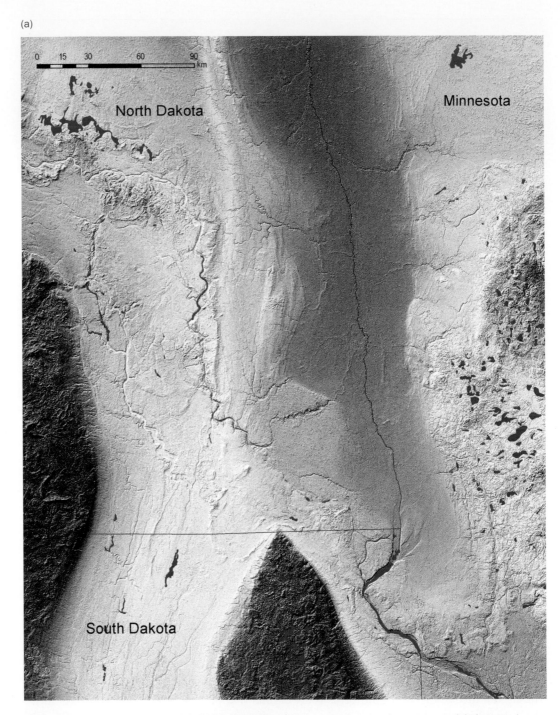

Figure 3.11 Digital elevation models highlight landforms that cannot be seen in maps or aerial photographs; at the border between North and South Dakota and Minnesota the terrain surface is overprinted by glacial ice advances: (a) SRTM (Shuttle Radar Topography Mission) elevation model

(b)

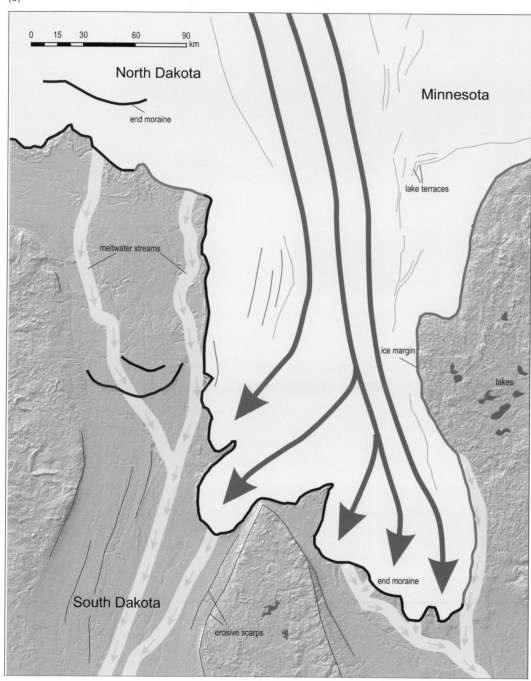

Figure 3.11 *(Continued)* (b) interpretation.

Figure 3.12 Glacially sculptured rocky headland in a Norwegian fjord. Photograph by Jürgen Ehlers.

Figure 3.13 Ice-moulded bedrock on the summits of the Rhinog mountains, North Wales. This evidence illustrates that ice over-ran this mountain range from an ice centre to the east (from right to left). However, in nearby mountains less than 2 m of bedrock has been removed, resulting in cosmogenic nuclide inheritence from previous surface exposures (e.g. in the nearby Aran Mountains, Wales; Glasser et al. 2012). It is therefore likely that multiple glaciations are required to remove larger depths of bedrock across wide areas. Photograph by Phil Hughes.

The surface area overridden by the glacier shows in many cases larger or smaller, more or less streamlined landforms whose long axes strike in the direction of ice movement; the small-scale forms are referred to as flutes, and the large forms as drumlins (Box 3.4). They consist mainly of morainic material and are usually several hundred metres long and somewhat less than half as wide. In extreme cases, forms occur of up to 3 km in length. The term drumlin comes from Ireland (from the Gaelic word for ridge = druim), where extensive drumlin swarms of several hundred square kilometres piqued the interest of geologists. They include forms of well-pronounced streamlined ridges with a steep front, and a gently sloping distal surface as well as steep symmetrical forms. Some contain a rocky core, while others consist entirely of sediments (McCabe 1991). The rock-core drumlins have a close relationship to the glacially shaped knobs known as roches moutonnées (Fig. 3.14). Drumlins usually occur in groups of well over one hundred individual separate forms, referred to as drumlin swarms (Figs 3.15, 3.16), and are commonly found in Ireland, Wisconsin and southern Germany.

There have been many studies on the formation of drumlins. Theories for the origin of drumlins include those which attribute drumlins to subglacial deposition, deformation or erosion by ice (Boulton 1987; Hart et al. 1997) or catastrophic subglacial meltwater floods (Shaw 2002). Neither theory is fully accepted. The meltwater hypothesis in particular has generated heated debate (e.g. Benn & Evans 2006; Shaw 2010). It is possible, and indeed probable, that drumlins are formed by a range of different processes and no single theory can explain every drumlin. One significant obstacle to our understanding of drumlin formation is the inability to observe processes at the base of modern, existing ice sheets, with all theories being based on observations of drumlins formed by previous ice sheets. However, some insight into the actual processes responsible for drumlins

Figure 3.14 Roche moutonnée in the forefield of Nigardsbreen glacier, Norway. Photograph by Jürgen Ehlers.

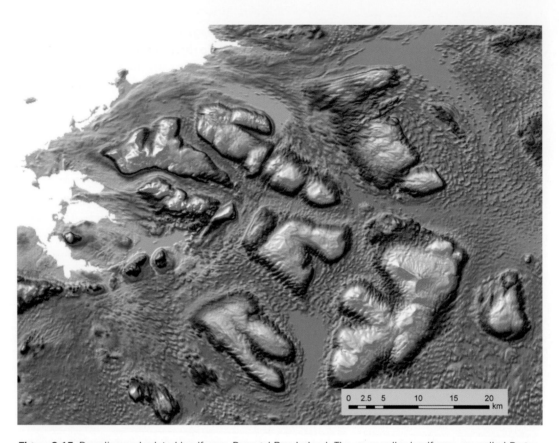

Figure 3.15 Drumlins and related landforms, Donegal Bay, Ireland. The cross-valley landforms are called *Rogen Moraine*.

Figure 3.16 Wisconsin drumlin swarm in the satellite image.

Source: NASA Aster satellite image of 28 October 2009

comes from the Antarctic, where the formation and rearrangement of drumlin-like structures has been observed at the base of rapidly flowing ice streams (Smith et al. 2007; Larter et al. 2009).

In addition to drumlins and crag-and-tail features (rock cores with a train of till), which strike largely parallel to ice movement directions, distinct minor forms referred to as flutes are also found. Comprising streamlined morainic matter, they are mostly less than 2 m high and have a width of up to 50 m. Such forms are barely recognizable from the ground, but show up clearly in the digital terrain model (Fig. 3.11).

Glaciofluvial deposits may also provide clues to the former ice movement direction, although they are of lesser significance than the morainic landforms. For instance, the elongated ridges of meltwater sediments deposited in tunnels under the ice (eskers) are generally nearly parallel to the glacier flow direction, while sandur-like meltwater deposits usually have a steep proximal edge corresponding to the former ice margin.

BOX 3.4 GLACIAL STRIAE

In general, elongated glacial landforms and glacial striae (striations) point in the same direction. While major landforms can be observed directly from aerial photographs or satellite imagery, the inclusion of glacier striae (Fig. 3.17) requires extensive site work. The time-consuming and laborious collection of measurements is offset by the fact that

Figure 3.17 Glacial striae and iron precipitates in the lee of obstacles in the Nigardsbreen glacier forefield, Norway. Photograph by Jürgen Ehlers.

(continued)

BOX 3.4 GLACIAL STRIAE *(CONTINUED)*

often several stages of ice movement can be reconstructed. Whilst in an evaluation of major landforms generally no more than two generations can be distinguished (e.g. small drumlins or flutes which are superimposed on larger, older forms), three or more generations of glacial striae can often be distinguished. A problem arises from the fact that, as a result of being overridden by the ice, the glacial striae have only one clearly defined property: their direction. A sequence of different ice movement directions is reflected by younger glacial striae crossing older systems.

The finding that glacial striae can be used to reconstruct ice movement direction is not new; Kleman (1990) has already provided an overview of the current state of knowledge and the problems of striae interpretation. The quality of the glacial striae depends to a large extent on the type of hard rock on which they are found. However, it can be assumed that glacial striae originally formed on almost all types of rock. The bases of temperate glaciers always contain debris, so striations occur wherever the ice moves over solid rock. In contrast, for cold glaciers it is usually assumed that the ice moves by internal deformation and that no sliding over the bedrock takes place. Nevertheless, Shreve (1984) considered that glacial abrasion may also take place in cold glaciers; the scarring effect is certainly lower than for temperate glaciers, however.

The conservation of glacial striae differs from place to place. In Scandinavia (including the archipelago of eastern Sweden and southern Finland) for example, it can be assumed the entire glacial striae inventory of the last glaciation is only preserved in areas with thin till cover that were recently isostatically raised above sea level. Kleman (1990) pointed out that most of the measurements derived from glacial striations come from freshly excavated road cuts, where the striae were found under 0.4–2 m of diamicton. A till cover of about 0.5 m thickness is sufficient to provide protection against weathering. The striae that are found under such till cover are usually older than the surviving scratches on exposed bedrock. Where there are tills of different ages, there are also striae of very different ages. In areas where there is a widespread till cover, it is possible that the most recent striae (which represent the last ice movement direction) are weathered beyond recognition and only older striae can be identified.

The distribution pattern of glacial striae in Scandinavia is not a reflection of actual flow lines, as the striae did not form simultaneously. Even if we consider only the youngest sets of striae, a time-transgressive sequence of directional signs is obtained. This is particularly true near the ice centre, where slight shifts in the ice divide or changes in the configuration of the ice sheet result in significant changes in the direction of ice movement.

In addition to glacial striae, other minor forms of glacial erosion may also be used to reconstruct the direction of the ice motion. They include sickle marks as well as the parabolic cracks described, amongst others, by Chamberlin (1888) and Schwarzbach (1978). Wintges & Heuberger (1982) and Wintges (1984) have discussed these small surface forms in detail.

Landforms that offer evidence of the former glacier movement are not limited to the central parts of the former ice sheets, but are also found in the outer areas. The parallel ridges and troughs of the Stader Geest, Syker Geest and Oldenburg/East Frisian Geest in northern Germany (Fig. 4.34) and the Drentse plateau in the Netherlands reflect the directions of former ice movement. A glance at the map shows that these forms cannot all be the same age. The direction of the valleys in the Syker Geest deviates by about 70° from the direction of the valleys in adjacent areas. Ehlers (1990) and Höfle (1991) assume that the NNW-SSE-orientated structures are older. Höfle (1991) suggests that the older forms have been preserved under stagnant ice.

3.4 Meltwater

The hydrology of recent glaciers has been described in detail by Röthlisberger & Lang (1987); the following discussion is based primarily on this source. The flow behaviour of glacial rivers is controlled mainly by heat conduction and energy balance at the Earth's surface. It differs fundamentally from the drainage behaviour of other rivers. Precipitation falling as snow generally has a negative effect on the meltwater runoff, since radiation is reduced during the precipitation event and the high albedo of freshly fallen snow subsequently lowers the temperature at the glacier surface.

The outflow of glacial meltwater is subject to a daily and annual cycle. Both are primarily controlled by solar radiation and the associated fluctuations in air temperature. The daily maximum discharge follows a short time after the radiation maximum. Daily fluctuations are naturally most extreme during the summer months. Immediately after the end of the summer ablation season, meltwater runoff decreases exponentially. The renewed start of the runoff occurs in spring a short delay after the beginning of the melting period (as the snow cover of the glacier and the subglacial drainage system have a significant retention potential); 90% of the total runoff occurs within a few weeks.

Above the altitude of 3500–4000 m in the Alps, close to 100% of the precipitation falls as snow. Much of the precipitation falling in the catchment area of a glacier is therefore initially stored. At the Aletsch Glacier in the Swiss Alps, for example, the meltwater drainage is essentially limited to the months of May–October. Considerations of the water balance are normally based on a period of a hydrological year, which lasts from early October until the end of September the following year. As of late May, mass balance is dominated by glacier melting over accumulation. Melt predominates throughout during unfavourable years (e.g. 1975–76), and the glacier has a negative mass balance. In contrast, in years with high rainfall it has a positive mass balance. In the long term, a sequence of years with positive mass balance results in the advance of the glacier.

In addition to the regular fluctuations of the runoff, non-periodic fluctuations and extreme flood events can occur. There are three processes that result in greatly increased flow: extreme melting rates; heavy rain; and sudden eruptions of meltwater from the discharge of ice-dammed lakes or water dammed within the glacier. Extreme melting rates occur in the Alps and in the Scandinavian glaciers, especially in connection with summer high-pressure weather conditions. In each case most of the snow melts from the lower part of the glacier tongue, which has a low albedo and therefore warms up more easily. Heavy precipitation falling as rain up to great heights leads to an increased runoff, especially if it occurs in combination with high melt rates (e.g. summer thunderstorms in the late afternoon or evening). Sudden eruptions of dammed meltwater can be triggered by changes within the unstable subglacial meltwater drainage system.

The majority of meltwater is formed at the glacier surface. Superficial melting rates fluctuate in the area of the Alpine glaciers according to altitude, and can vary between 0.1 and 10 m a^{-1} water column. On the other hand, the basal melting rate due to friction and geothermal influences is on the order of 0.01 m a^{-1}. Very little water is created solely through the glacier flow. In addition to meltwater and the rainwater that falls on the glacier, valley glaciers also experience a lateral flow of surface water and groundwater from the valley slopes, contributing to the total runoff.

The type of drainage depends to a large extent on the shape of the glacier surface. Snow and firn react similarly to an unsaturated pore aquifer. The meltwater seeps into this layer, until it encounters a layer of low permeability (usually the glacial ice). The water moves like common groundwater, forming a water table, and the runoff is controlled by gravity according to Darcy's law. Unlike ordinary pore aquifers, the permeability of the snow changes over time due to the conversion to firn and ice. The dewatering of the aquifer occurs in three directions: discharge at the firn line into channels at the glacier surface; runoff into crevasses; or seepage through the ice.

In the ablation zone of a glacier, meltwater runoff can be observed at the glacier surface. The water usually disappears into the glacier after a short distance. Larger meltwater streams that flow right to the edge or the front of the glacier are the exception. Most of the water disappears into moulins or crevasses, but seepage through the ice is not to be underestimated. In the transition region between the accumulation zone and ablation zone on the glacier surface, a permanently frozen thin layer is found. The thickness and extent of this layer are controlled by climatic factors. In winter and early spring this layer may extend over the entire ablation zone if it is not covered by wet snow, as may be the case with glaciers in humid maritime areas (Röthlisberger & Lang 1987).

In temperate glaciers, part of the infiltration takes place through the gaps between the ice crystals. Air bubbles in the cavities and the deformation and recrystallization of the ice restrict this type of drainage. An essential part of the downward drainage from the glacier surface to its base is therefore accomplished through moulins, where the water initially drops in free fall into the depths. Moulins are usually found near the upper edge of a zone of crevasses, and it can be assumed that cracks in the ice form the starting points for their formation. The water does not drop right down to the glacier sole, but vertical shafts alternate with gently sloping or horizontal sections. Entrained sediment is deposited in basins that form at the bottom of the moulins. These sediment accumulations finally reach the glacier surface in the ablation zone of the glacier, where they occur as isolated dirt cones (Fig. 3.18; Röthlisberger & Lang 1987).

Meltwater drainage at the glacier base is in tunnels under the ice (Figs 3.19, 3.20). These are either cavities in the ice (Röthlisberger, or R channels) or in the rock (Nye, or N channels), or are a combination of both. There is a strong tendency to fill the voids, particularly for large glaciers with significant movement at the glacier sole, causing meltwater drainage under pressure (Röthlisberger & Lang 1987). Under these conditions there will be increased

Figure 3.18 Dirt cone on the surface of Kverkjökull, Iceland. Photograph by Jürgen Ehlers.

Figure 3.19 Entrance to an abandoned meltwater tunnel under Nigardsbreen glacier, Norway. Photograph by Jürgen Ehlers.

Figure 3.20 Abandoned meltwater tunnel at the base of Nigardsbreen, Norway. Photograph by Jürgen Ehlers.

subglacial meltwater erosion. A subglacial tunnel valley formed this way in Iceland in 1996 was described by Russell et al. (2007).

A special case of meltwater runoff is the drainage of glaciers in volcanic areas where catastrophic meltwater outbursts occur termed jökulhlaups (Box 3.5; Figs 3.21, 3.22), mainly known from the area of Iceland's glaciers (Björnsson 2009). The largest historically documented jökulhlaup (from the Katla) was said to have had an outflow of 100,000–300,000 m^3 s^{-1}. The drainage of ice-dammed glacial lakes can also result in strong outburst floods of meltwater; outflows of up to 3000 m^3 s^{-1} were measured in Iceland (Björnsson 1992).

Figure 3.21 Volcanic eruption under Vatnajökull, Iceland, October 1996. Photograph by Magnús Tumi Guðmundsson.

Figure 3.22 Jökulhlaup in Iceland: icebergs floating in the swirling meltwater. Photograph by Magnús Tumi Guðmundsson.

BOX 3.5 VOLCANIC ERUPTION UNDER THE ICE

When a volcano erupted in March 2010 in south Iceland, the biggest concern was that the adjacent ice caps could be affected; the crack from which lava flowed into the environment was situated right between the Myrdalsjökull and Eyjafjallajökull. This concern proved unfounded however, and the glaciers were not affected.

Volcanic eruptions under the ice are not uncommon in Iceland. The last major eruption occurred in October 1996 under the Vatnajökull ice cap, where about 3 km^3 of ice were melted. The meltwater was first stored under the ice in the Grímsvötn caldera, but on 4 November 1996 at 9:30 the ice dam broke. On 5 November at 07:20 the meltwater reached the ice margin, and the first outbreak occurred at the eastern edge of Skeiðarájökull. During the day two additional outlets formed further to the west. The two main rivers Gígjukvísl and Skeiðará, through which the main discharge ran off to the sea, recorded a peak drainage of 33,000 and 23,000 m$_3$ s^{-1}, respectively. Fifteen hours after the start, the outflow reached a volume of 50,000 m^3 s^{-1} and was at that time the second-most powerful river in the world. The flood covered an area of 750 km^2 and roads and bridges were destroyed (Fig. 3.23).

(continued)

BOX 3.5 VOLCANIC ERUPTION UNDER THE ICE (CONTINUED)

Figure 3.23 The end of the jökulhlaup. Bridges have been destroyed and the ring road is impassable. Photograph by Þröstur Þorsteinsson.

Even without volcanic activity, water can accumulate at the bottom of large ice sheets. For instance, it is well known that there are subglacial lakes under the ice of the Antarctic. With the help of the ICESat satellite (ice, cloud and land elevation satellite) it was possible to create an overview of these lakes for the first time. In the period from 2003 to 2008, 124 lakes were recorded. These lakes fill up with water through periods of months or years, and can sometimes drain suddenly. They are most widespread in the coastal areas of Antarctica and among the major Antarctic ice streams. With the help of high-precision altitude measurements, it can be shown that these lakes take part in the glacial water cycle. Sudden water outflow from these lakes may trigger rapid glacier advances (Fricker & Scambos 2009; Smith et al. 2009).

Till profile with intercaleted, washed-out large sand lens; Brodtener Ufer cliff at Travemünde, Schleswig-Holstein. Photograph by Jürgen Ehlers.

Chapter 4
Till and Moraines: The Traces of Glaciers

4.1 Till

4.1.1 WHAT IS A TILL?

The deposits of glaciers are usually made up of a poorly sorted mixture of clay, silt, sand, gravel and stones, traditionally referred to as morainic deposits. The term 'moraine' is used for both a landform (i.e. a terminal moraine) and a sediment (i.e. ground moraine). However, glacial sediment should be referred to as 'till' (Piotrowski 1992; Fig. 4.1) with 'moraine' reserved for its surface morphological expression. The consolidated deposits of earlier geological eras have been referred to as tillites for a long time (Fig. 4.2). The word comes from the Scottish *till*, which originally meant nothing more than a stony, clay soil. As Scotland was a centre of early Quaternary research, this term has been universally applied to glacial deposits. Geikie (1863) originally used till both as a descriptive term and a genetic concept. Today the word till is only used as a genetic term. It refers to a deposit formed in a certain way, not its composition. The non-genetic term for poorly sorted sediments is diamicton. Diamictons come in all shapes and sizes and can be formed by a range of unrelated processes, including as a result of gravitational mass flows (including landslides); tsunamis; volcanic eruptions; and glacial processes. However, the properties of diamictons (such as clast shape, roundness, fabric and textural features) as well as their geomorphological and or/stratigraphical context can help to determine an origin and justify a genetic label, such as till.

The Ice Age, First Edition. Jürgen Ehlers, Philip D. Hughes and Philip L. Gibbard.
© 2016 John Wiley & Sons, Ltd. Published 2016 by John Wiley & Sons, Ltd.
Companion website: www.wiley.com/go/ehlers/iceage

Figure 4.1 Till on Iceland. The dark colour of the matrix is caused by the reworking of basaltic rocks. Photograph by Jürgen Ehlers.

Figure 4.2 Tillite (lithified till) on Iceland. The glacial striae have been caused by the latest glaciations. Photograph by Jürgen Ehlers.

There are many different types of till, as listed in Table 4.1. Their classification has been the subject of investigation of the INQUA Commission on Genesis and Lithology of Quaternary Deposits. The final report of this commission was presented at the XII INQUA Congress in Ottawa (Dreimanis 1988). Detailed descriptions of the various types of till are also covered in glacial textbooks such as *Glaciers and Glaciations* (Benn & Evans 2010) *Glacial Geology*

TABLE 4.1 SUMMARY OF THE DIFFERENT TYPES OF TILLS.

Type of till	Clast shape	Clast size	Clast fabric	Clast packing	Clast lithology	Overall structure
Lodgement till	Rounded edges, spherical form, striated and faceted; classic evidence of active transport and subglacial abrasion	Imodal or multimodal	Strong fabric aligned closely with ice flow direction	Dense and well consolidated	Dominated by local rock types	Massive and structureless, sheared or brecciated clasts, boulder pavement or clusters may be present
Subglacial meltout till	Rounded, spherical form, striated and faceted; these characteristics are less pronounced than for lodgement till	Bimodal or multimodal; sorting associated with dewatering and sediment flow may be present	Strong in the direction of ice flow, but with greater range of orientations than lodgement	Well consolidated but less marked than for lodgement till	Variety of clast types, local and exotic, depending on pre-meltout history	Usually massive structureless sediments, but can contain folds and flow structures; stratification can be present
Deformation till	Dominated by the inherited clast shapes from the sediment being deformed	Diverse range	Strong fabric in direction of shear	Densely packed and consolidated	Diverse range	Fold, thrust and fault structures may be present if shear homogenization is low; in cases of high stresses then the sediment can be completely homogenized
Supraglacial meltout till	Angular clasts that have undergone passive transport; a minority of clasts exhibit evidence of active transport and abrasion (rounding, striae, etc.)	Typically coarse and unimodal; some sorting where meltout occurs	Fabric is unrelated to ice flow; weak fabric and spatially variable	Poorly consolidated with low bulk density	Very variable; depends on catchment size	Stratification can occur, but often structureless
Flow till	Wide ranging, but dominated by angular and non-spherical without striae	Usually coarse and unimodal but occasionally well-sorted	Variable; some flow units can exhibit a strong fabric associated with gravitational movement	Poorly consolidated with low bulk density	Variable	Flow structures, crude sorting; sand and silt layers in cases where meltwater is involved.
Sublimation till	Clasts typical of active transport including striae, faceting, rounding and spherical	Bimodal or multimodal	Strong in the direction of ice flow	Low bulk density, loose and friable	Variety of clast types	Usually stratified and may preserve englacial fold structures

Adapted from Bennett & Glasser (2009, table 8.1).

(Bennett & Glasser 2009). In this chapter, we provide a brief overview of some of the classic till types with reference to case studies and discuss clast transport in glacial systems and the geochemistry of tills. We then discuss the different types of moraines and finish with a case study of the Weichselian dynamics of the Scandinavian Ice Sheet, the largest ice sheet in Europe of the last ice age.

After the advent of the glacial theory in the late nineteenth century, till was initially perceived in northern Europe as superficial morainic debris of the Scandinavian Ice Sheet. This idea was only refuted after Nansen's crossing of Greenland. Nowhere had the Danes found any mountains that towered over the Greenlandic Ice Sheet (nunataks), and consequently the ice sheet had no superficial moraines. Morainic material only came to the surface of the ice at the outermost ice margin (Mohn & Nansen 1893).

Since the deposition of till takes place below the glacier, this process cannot be observed directly. Information on the nature of the deposition can only be derived from the structural properties of the deposited morainic matter. Basically there are two processes. In the formation of lodgement till the sediment is deposited grain by grain from the actively moving glacier at the glacier sole. Meltout till, on the other hand, is the sediment of stagnant glacier ice, deposited by basal melting. The entrained sediment gradually melts out and sinks to the ground. As a rule, both types of till occur in the Quaternary glaciation areas.

During the formation of lodgement till (Fig. 4.3), larger grains stay in motion longer than smaller grains. A small grain of sand will be deposited as soon as it comes into contact with the underlying till that has been already deposited. When a larger rock fragment comes in contact with the underlying till, it continues to move as most of it is still enclosed in the ice. The increased friction leads to an increasing slowing down of the transport. As the ice moves on, the clast is tilted forward so that platy stones are finally arranged in tile-like storage (Ehlers 1990) which is why the longitudinal axes of clasts in lodgement till tend to dip upglacier. Depending on the ice thickness and on the speed of the glacier, lodgement till is either deposited everywhere or only on the upglacier side of obstacles.

In glaciers whose base is at the pressure melting point, basal meltout till is deposited. It can often be observed in outcrops that almost undisturbed lenses of meltwater sands are included in the basal portions of the till, whose internal structure is virtually undisturbed by glacial thrust and shear. As the presence of such sand lenses is incompatible with the genesis of lodgement till, the till in question must be meltout till.

The meltout of morainic matter occurs not only at the glacier sole, but also on the ice surface. In this case we refer to it as supraglacial meltout till. In northern Europe, till sheets are rarely overlain by such a supraglacial meltout till. Firstly, this is because transport of rock debris in the northern European ice sheet was largely restricted to the basal portions of the ice, so there was little starting material for the formation of supraglacial meltout till to begin with. Secondly, in many cases the top layers of the morainic deposits have been truncated by later ice advances, or altered by periglacial processes and soil formation, to such a degree that any initially present supraglacial meltout till can no longer be identified today as special till type. One area in which supraglacial meltout till (and flow till) is widely found is the ice-decay landscape in the peripheral regions of the Baltic Sea. In those areas the lodgement till from the active advance phase of the ice is limited to a few millimetres to centimetres at the base of the unit in extreme cases, and the mass of the

Figure 4.3 Lodgement till (top) and meltout till (bottom), Langeland, Denmark. The penguin in the bottom image is 12.5 cm high. When were these tills deposited? The geographer Albrecht Penck originally had the view that a layer of glacial debris several metres in thickness might be moved beneath the ice. Based on his investigations in Greenland, Von Drygalski (1897) came to the conclusion that the morainic material had been transported in the basal portions of the ice. This view prevailed, but today we know that some sediment transport also takes place below the ice. Photographs by Jürgen Ehlers.

glacial debris consists of subglacial and partly supraglacial meltout till (superficial moraine; Stephan & Ehlers 1983).

In their investigations in the St Lawrence Valley, Canada, MacClintock & Dreimanis (1964) found that the upper part of till sheets from earlier ice advances is reworked by younger ice advances in such a way that the microstructure adjusts to the direction of the new ice advance. The reorientation may extend to a depth of 10 m.

In many cases, older sediments underlying the tills are not completely undisturbed, but have been incorporated into the ice movement. If the pressures are sufficiently high then underlying tills are deformed to such as degree that any structural evidence of deformation, such as folds, are obliterated and the sediments become homogenized (Hart & Boulton 1991). A layer of rearranged sediment a metre or so thick is often found at the base of the tills in northern Germany, the particle size distribution of which is more similar to the underlying strata than to the overlying till. This 'sohlmoräne' (Grube 1979) belongs to the deformed base according to Alley et al. (1987). It has been proposed to refer to such layers as deformation till. As these sediments do not show significant deformations or have the composition of a true diamicton, it would be more accurate to refer to 'deformed subglacial sediments' (Stephan & Ehlers 1983).

The distribution of till is not equal in all formerly glaciated areas. In the Swiss Mittelland region (Fig. 4.4), till sheets of several tens of metres thickness are found. In most other areas in the Alps, however, significantly less till is present than in the Nordic glaciated areas. This deficiency is more pronounced in the central areas of the glaciation compared to the margins. In particular, till is preserved in positions where the advancing glaciers met rock obstacles as the compressive flow meant no crevasses could be formed. This meant that the morainic material

Figure 4.4 Earth pillars at Euseigne, Switzerland. Photograph by Jürgen Ehlers.

was not only deposited but was also protected from meltwater erosion. On the lee side of obstacles, however, under the influence of extending flow, unrestricted meltwater erosion is found (Schlüchter 1980b).

All tills that were deposited under the ice display evidence of over-consolidation, due to compaction by glacier load and the associated squeezing-out of the water content. As a result, they react like a bulky rock. They are very sensitive to rewetting after previous dehydration, which can lead to the formation of steep slopes and, in extreme cases, to the formation of the famous earth pillars (e.g. near Bolzano in South Tyrol, Pont-Haute south of Grenoble or in Euseigne, Valais; van Husen 1981).

In addition to the above-mentioned types of till there are two other types of morainic deposition: flow till and subaquatic till. In contrast to the morainic deposits, these deposits are secondary tills because they have not been directly deposited by a glacier. Instead, they consist of redeposited morainic material (Dreimanis 1988).

Flow till, which has been redeposited by mass flow at the ice front, was described by Boulton (1968). When Gripp (1929) first observed this type of sedimentation on Spitsbergen, he referred to the resulting sediments as mud flows and not as tills. Although it may seem practical to count these sediments as till (in samples from boreholes they are indistinguishable from 'real' till), genetically they are the result of mass movements (sediment flows) and not glacial deposits. Supraglacial flow tills are not common in areas where there have been repeated glaciations. Most flow tills which are formed on the snout of an advancing glacier are immediately reworked during subsequent overriding. Flow tills only have a good chance of preservation if deposited in greater thickness and in a protected location (e.g. small depressions). For instance, Wansa (1991) described flow tills from the edge of Elsterian channels in an opencast lignite mine at Gräfenhainichen, just north of Leipzig in Germany.

Subaquatic (waterlain) till is morainic material which has been deposited from the glacier front or glacier sole into bodies of standing or running water (Dreimanis 1988). Where glaciomarine deposits are missing, the occurrence of subaquatic till is limited to former major and minor ice-dammed lakes. Genetically, this type of diamicton is more of a lacustrine sediment than a glacial deposit. Corresponding sediments in the Alpine region are known of, for example from the Lake Garda area.

The discovery that many tills previously classified as terrestrial deposits were actually deposited under water was made in the 1970s. Since the end of the nineteenth century, the sedimentary sequence of the Scarborough Bluffs on the north shore of Lake Ontario, near Toronto, Canada (Fig. 4.5) was always interpreted as a sequence of morainic deposits with intercalated outwash and lacustrine sediments (Karrow 1969). Today, it is assumed that these layers are part of a subglacial channel filling.

Among the largest ice-dammed lakes that formed during the Quaternary are those which formed in North America in front of the Laurentide Ice Sheet. In Europe, large lakes also formed in front of the Elsterian ice sheet and formed incompletely infilled channels that can be traced across northern Germany (Fig. 4.6). The base of the lake sediments often consists of a diamicton (subaquatic till), grading upwards into lacustrine deposits (known as the Lauenburg Clay).

Figure 4.5 Scarborough Bluffs, Ontario. Morainic deposits (left) and glaciolacustrine silts (right) deposited in a subglacial meltwater channel. Photographs by Adriaan Janszen.

Figure 4.6
Glaciolacustrine clays of
a Late Weichselian ice-
dammed lake exposed
in the construction site
for the Travemünde ferry
terminal (near Lübeck,
Germany). Photograph
by Jürgen Ehlers.

4.1.2 TILL: A MIXED BAG?

In the British Isles the lowlands tills are dominated by fine-grained matrix with occasional boulders. This characteristic led to the term boulder-clay being applied to describe tills in the nineteenth century in the British Isles. Particle-size composition of till matrix is particularly useful in determining till genesis, as certain till types have matrix particle-size distributions which fall within characteristic envelopes on a ternary diagram (Sladen & Wrigley 1983). For example, subglacial lodgement till has a greater concentration of fines in relation to supraglacial, meltout or flow tills arising from clast crushing beneath the former glacier (Boulton et al. 1974).

The particle-size distribution of till is to a considerable extent determined by the composition of the local substratum. In continental Europe this applies to both the northern glaciations (Rappol 1983) as well as the Alpine region (e.g. Cammeraat & Rappol 1987). The tills in northern Germany generally have a clay content of 10–20%; the only exceptions are the sandy Elsterian till in the Elbe–Weser triangle (Höfle 1980; Wansa 1994) and some extremely clayey local tills.

The particle-size distribution can be used as a criterion for distinguishing tills of different ages. The Older Saalian till of Lower Saxony is often highly sandy (Fig. 4.7a), while the Middle Saalian till is often rather clay-rich (Fig. 4.7b). The Younger Saalian till (at least in the Hamburg area) takes an intermediate position. The use of particle-size distribution for stratigraphic correlations is limited, however (Stephan 1987). In addition to a widespread 'typical' lithofacies, every major ice advance deposited one or more other facies. Their respective composition is partly due to differences in the local underground and partly due to differences in glacier dynamics. Within the Older Saalian till in northern Germany, for example, a red facies is found which is also more clay-rich. A highly sandy facies often occurs at the base of the Middle Saalian till in the Hamburg area, which is similar in grain-size distribution to the Older Saalian till. In addition to the more sandy 'normal' facies in the Younger Saalian till, a red, clay-rich till is found (Vastorf type; Fig. 4.7c).

The tills of mountain glaciations have a significantly higher proportion of coarse boulders than comparable deposits associated with glaciations that reached the lowlands (Fig. 4.8). In the southern foothills of the Alps in Switzerland and in Austria, the gravel portion of the tills is usually 20–50% (e.g. van Husen 1977; Schlüchter 1981; Grottenthaler 1989) while in the lowland north German tills it is mostly well below 5%. In the British Isles local mountain glaciations produced large blocky boulder moraines during the last phase of mountain glaciation. The large size of moraines of this late-glacial phase is often inconsistent with the small size of the former glaciers and the short intervals available for cirque glacier activity. For example, the Younger Dryas (the last phase of significant glaciation in the British Isles) lasted only 1200 years during 12.9–11.7 ka, yet appears to have produced large moraine assemblages relative to the size of the glaciers (e.g. Hughes 2002). It is likely that rock falls associated with both periglacial activity and pressure release occurred as the last ice sheet retreated, exposing rock surfaces, causing rocks to fracture as the weight of ice was unloaded and removed (Ballantyne 2002). Subsequent re-advances of glaciers during periods such as the Younger Dryas then had ample debris material available for moraine building, explaining their large size.

(a)

(b)

(c)

Figure 4.7 Till types from North Germany: (a) sandy Elsterian till from Wellen, Lower Saxony; (b) clay-rich Middle Saalian till from Grauen, Lower Saxony; and (c) red Younger Saalian till (Vastorf Till) from Emmendorf, Lower Saxony. Photographs by Jürgen Ehlers.

Figure 4.8 Alpine till with striated clast. Photograph by Jürgen Ehlers.

4.1.3 CLAST TRANSPORT

Even before the advent of glacial theory in the late nineteenth century, the boulders of northern Europe were regarded as witnesses to former southwards-directed transport from the mountains of Scandinavia and the British uplands. All the coarser constituents (cobbles and boulders; see Table 4.2 for grain-size definitions) in the northern European lowland tills (such as north Germany and East Anglia in England; Box 4.1; Fig. 4.9) are far-travelled. However, studies in Scandinavia have shown that the smaller pebbles and finer matrix of till material often has a local signature. In general, there will be a frequent change between erosion and redeposition with the result that the composition of the tills always corresponds to a large extent to the local rock substrate. Lindén (1975) examined this relationship in a 48-km-long north–south transect west of Uppsala (Sweden; Fig. 4.10). The changes of the subsurface rocks are immediately reflected in the clast inventory of the tills. This is true for both the larger pebble components (20–60 mm) and for the finer pebble components (5.6–20 mm).

Lowland tills often contain a very high proportion of local material. The striking clasts of Scandinavian origin in East Anglia, England make only a small proportion of the overall till. The far-travelled components are surrounded by a matrix of completely different composition. In the tills of northern Germany, the fraction above 2 mm diameter comprises no more than an average of 3% of the till. The remaining 97% are sand, silt and clay, mostly derived from local source materials such as quartz sands and clays from older Tertiary or Cretaceous substrate rocks. The transport of debris on the glacier, known as supraglacial transport (superficial moraine; Fig. 4.12), can play a large role in the case of mountain glaciers. Lateral moraines, close to steep valley sides, can sometimes be dominated by angular supraglacial

TABLE 4.2 WENTWORTH CLASSIFICATION OF GRAIN SIZE. THIS SCALE MODIFIES AN EARLIER VERSION BY JOHAN UDDEN, AND THE PHI (LOGARITHMIC) SCALE WAS DEVELOPED LATER BY WILLIAM KRUMBEIN.

Millimetres	Wentworth grade	Phi (Φ) scale
>256	Boulder	−8
>64	Cobble	−6
>4	Pebble	−2
>2	Granule	−1
>1	Very coarse sand	0
>1/2	Coarse sand	1
>1/4	Medium sand	2
>1/8	Fine sand	3
>1/16	Very fine sand	4
>1/32	Coarse silt	5
>1/64	Medium silt	6
>1/128	Fine silt	7
>1/256	Very fine silt	8
<1/256	Clay	>8

BOX 4.1 WHERE DOES THE 'OLD SWEDE' ERRATIC COME FROM?

During dredging operations to deepen the Elbe navigation channel in September 1999, a large stone was found at a depth of 15 m; early estimates suggested an erratic of about 140 tonnes (Fig. 4.11). It was decided that the largest boulder ever found in Hamburg should definitely be salvaged. A first attempt with a floating crane on 18 September failed, however; when lifting the colossus out of the Elbe the ropes tore off and the boulder sank back into the depths. However, the second attempt on 23 October succeeded. The stone has now been securely placed on the banks of the Elbe at Övelgönne. The boulder actually weighs 217 tonnes and has a circumference of 19.7 m. Length and width are 7.9 m and 5.2 m, and its height is *c.* 4.5 m.

Where does 'Old Swede' come from? For an accurate determination of its composition a sample of the rock was needed. As the giant boulder was to be placed under protection, the usual geologists' method of taking a big hammer to cut off a piece was not expedient. Instead, a core had to be drilled (which is now stored in the Geological Survey). The hole in 'Old Swede' was subsequently plugged with another granite core; the scar on the east side of the boulder is still visible upon close inspection.

The mineral composition of 'Old Swede' shows that it is a grey, non-porphyritic granite from eastern Småland (Sweden). Its composition is similar to granite deposits near Vilkensved in Växjö. The glacier must have approximately followed the course of today's Baltic Sea basin when it transported the rock to Hamburg.

'Old Swede' is the single largest boulder in Hamburg, but there are even greater erratics elsewhere. The largest boulder in northern Germany is the 'Buskam' located at Göhren off the coast of Rügen. It lies so far from the beach that it's true size was difficult to measure. Its perimeter is given as 40 m in the older literature and its volume was estimated at 600 m³. A modern survey, in 2004 has, however, shown that it is much smaller than previously thought: its perimeter is only 27.5 m and its volume *c.* 206 m³. Even after having shrunk to nearly a third of its original size, it still remains the largest German erratic.

material (e.g. Boulton 1978; Lukas et al. 2012). However, for large ice sheets the supraglacial material is often restricted to the marginal areas where melting of the overlying ice reveals englacial debris to be revealed or exhumed at the ice surface (Dreimanis 1990).

In the Nordic glaciations the transport of debris is usually restricted to the lowest parts of the ice (Fig. 4.12). The debris may go through repeated cycles of deposition and reworking, resulting in a mixture of all rock types which are overridden by the ice on its way to the deposition zone. The far-travelled components will become progressively more diluted in

Figure 4.9 Fine-grained matrix-supported till containing erratic cobble clasts at Thursaston, Wirral peninsula, NW England. These deposits were formed at the base of the Irish Sea Ice Sheet during the last glaciation. Photograph by Phil Hughes.

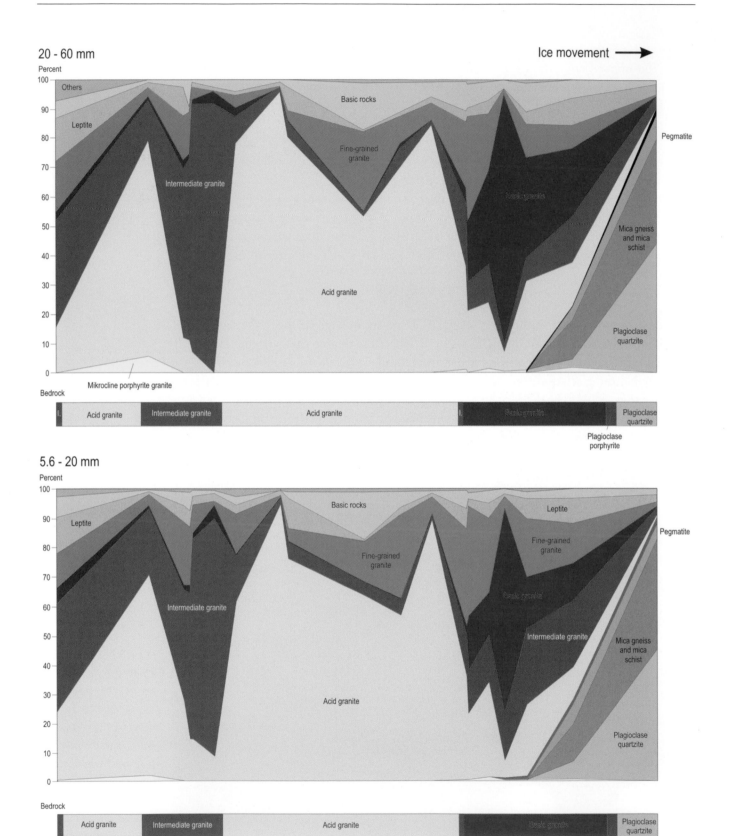

Figure 4.10 Till clast composition in relation to changing bedrock conditions in Sweden.

Source: Lindén (1975).

(a)

(b)

Figure 4.11 (a) Giant erratic 'Alter Schwede' (Old Swede) at the River Elbe near Hamburg and (b) the scar where the core sample was drilled. Photographs by Jürgen Ehlers.

(a)   (b)

Figure 4.12 Supraglacial and subglacial transport of morainic material by the glacier: (a) rock debris on the surface of Tsijiore Nouve glacier, Switzerland. Photograph by Jürgen Ehlers. (b) Banded, debris-rich ice at the base of Taylor Glacier, Antarctica. Photograph by John Shaw.

this way, and the local material enriched. Local material will also be incorporated, where the 'bedrock' consists of an older till. For this reason, tills are usually composed of a mixture of very different materials from which a significant portion has been reworked by more than one glaciation.

Most of the debris transported in the glacier ice moves directly above the glacier sole. This debris-rich layer can be several metres thick. The debris concentration in the basal zone is on average about 25% by volume, but varies in layers from fractions of a percent up to about 90% (Lawson 1979). The main part of the glacial abrasion and crushing of the rock takes place in this zone (Dreimanis 1990). The basal zone is overlain by largely rock-free ice, in which only some isolated particles or rock fragments are present. Bands of debris-rich deeper ice are raised up into this zone by compressive flow. Since further breaking of the rock is unlikely and the wear is minimal, relatively pure material can be safely transported in this zone over large distances (Dreimanis 1976). This is probably the way in which completely unmixed layers of till, including the so-called 'red till', have been transported from Scandinavia to northern Germany.

In contrast to most other tills, the 'red till' of the late ice advances in northern Germany consists of relatively pure far-travelled material (Meyer 1983a). This is evident not only in the matrix but also in fine gravel analyses (Ehlers 1990). In the fine gravel fraction the local components are represented mainly by quartz. In a broader sense, flint and chalk may also be counted among the local attributions; both are rarely present in the 'red tills'. To explain the origin of a till without admixture from the local subsurface, it must be assumed that it was transported in the higher parts of the ice.

Despite all the mixing, the boulder content of a certain till usually has a specific composition which, with some variations, is characteristic of the till. This applies to the fine gravel fraction as well as to the coarse gravel. The only difference is that some crystalline and sedimentary rock types can be exactly determined in the coarser fraction, while the fine gravel analysis is limited to the analysis of rock groups. The specific sediment composition allows correlations between adjacent boreholes and outcrops. Under favourable conditions, correlations over long distances (up to hundreds of kilometres) are possible. However, a change in clast composition perpendicular to the ice movement direction should be taken into account. The specific composition is only valid for a particular lithofacies, not for the total till deposits of a certain time period such as the 'Drenthe' or 'Warthe' phase. It can therefore only be used for chronostratigraphic interpretations when supported by further information. The glacigenic deposits represent a succession of different facies types. The differences in composition of these deposits are due to changes in glacier dynamics and in the ice movement direction.

Indicator analyses can be used to find ore deposits hidden under thick till (Eriksson 1983). This is primarily the case for transport over small and medium distances. In contrast, where long-distance sediment transport plays a role, the mixing of material of different age, with multiple rearrangements and repeatedly varying transport directions, can considerably complicate the evaluation in relation to the ice dynamics. Drake (1983) reviewed the nature of what he termed 'ore plumes in till'. These are boulder trains, ore trails and elements fans which are produced in tills as ice overrides ore bodies. These features have three-dimensional (3D) shapes resembling smoke plumes drifting from the ore source. The maximum ore concentration may be some distance from the ore source (often 1–10 km), and only extends back to the bedrock source at depth. Knowledge of glacial dynamics is therefore required in order to successfully utilize ore plumes in till to locate ore-bearing rocks (see Box 4.2).

For the ice advance from Scandinavia to northern Germany, there were basically three distinct directions available.

1 The northwestern route led from Norway and central Sweden over the Kattegat and the Danish islands into Schleswig-Holstein. Along this transport path, southern and central Swedish rocks and debris from the Oslo area were transported. Tills from this flow direction contain relatively little chalk and flint.
2 The middle path led from the Swedish peninsula between Bornholm and Rügen and then through the Western Baltic. Tills from this ice movement direction contain rocks from southern and central Sweden and Bornholm and are usually rich in chalk and flint, derived from the Cretaceous bedrock areas of the western Baltic Sea floor.
3 The eastern route followed the eastern edge of the Baltic Sea Basin. Mainly rocks from the Åland Islands were transported this way, together with rocks from the bottom of the Baltic Sea and from the Baltic Republics.

BOX 4.2 'THERE IS PLENTY OF GOLD SO I AM TOLD ...'

The gold rush in California was over within a few years, but the hope for quick wealth remained. The quest for the coveted metal moved into ever less-accessible areas of the Far West. Gold was found in 1860 in the Clearwater River and a little later also in the Salmon River in Idaho. Subsequently, the area was first affiliated to the United States as a 'territory' and in 1890 as a new federal state (Idaho). The gold-seekers of the Wild West did not bother to look for the corresponding outcrops; they just washed the precious metal from the gravels of the rivers where it was enriched because of its high specific weight. They found very little.

In the early 1930s, several miners in Idaho joined forces to form a company and looked for someone who was willing to dig for them for gold at the Yankee Fork. Finally they managed to interest the Silas Mason Co. from Shreveport, Louisiana for the project. After examination of rock samples, they hoped to reclaim gold at a value of about $16 million. They formed a subsidiary company that was to carry out the dredging.

The Bucyrus-Erie Company of South Milwaukee was awarded the contract to build the excavator in 1939. Among other machinery, the company had produced the steam shovels that were used in the construction of the Panama Canal. The dredge was completed in 1940, then disassembled into its component parts and taken by train to Mackay, Idaho. From there it was transported the last 130 km by truck to the Yankee Fork and then reassembled.

The dredge was 34 m long, 16 m wide and 19 m high, and weighed 988 tonnes (Fig. 4.13). It was powered by two seven-cylinder Ingersoll-Rand diesel engines that

Figure 4.13 The Yankee Fork gold dredge, Idaho. Photograph by Jürgen Ehlers.

powered the dredging operation. A total of 72 buckets, each of volume 226 litres, began to dig into the gold-bearing gravels. The Yankee Fork was a stream in which no ship of this magnitude could operate. The excavator had to excavate its own pool and consequently dig its own way. The dredged material was washed in the ship, then the waste was dumped again on the backside. From 1940 until August 1952, the excavator dug through the gravel bed of the Yankee Fork. Alas, no major commercial success was achieved. The company changed hands several times and the work was eventually stopped when the last boundary of the claims had been reached. There the excavator was abandoned, where it still remains today. Throughout its operation a total of nearly 5000 m^3 gravel were processed, ruining the valley over a distance of 9 km. In the process, gold and silver was recovered with a total value of US$ 1,037,322. The whole enterprise had cost US$ 1,076,100.

Source: website of the US Forest Service, Salmon Challis National Forest and parksandrecreation. idaho.gov.

Within the ice stream network of the Alpine glaciations, the paths of the glaciers were largely predetermined by the relief. There was usually only some marginal contact between the individual glaciers. However, the catchment areas of the glaciers have often changed during the glaciations. Under certain conditions, during its peak ice flowed across the mountain passes into neighbouring valleys; this process is reflected in the composition of the corresponding sediments (Doppler 1980; Dreesbach 1985). By indicator counts, the areas of influence of different ice lobes can also be delimited. In the overlapping range of Loisach, Ammer and Lech glaciers, the deposits of the Loisach glacier can be distinguished due to their higher proportion of crystalline rocks as compared to the other glacial deposits (Piehler 1974).

4.1.4 CLAST ORIENTATION

The stones are not randomly distributed within the tills, but have a certain orientation. Richter (1932) demonstrated that clast orientation forms in the active ice by glacial movement in such a way that the long axes are arranged parallel to ice movement. This orientation is preserved when the debris is eventually deposited as till. Consequently, an investigation of the till fabric allows the ice movement direction to be reconstructed. The use of clast fabric for determining former ice flow directions was first demonstrated in Caithness in Scotland by Miller (1850), followed by other studies in the nineteenth century such as those by Jamieson (1865) and Bell (1888). Richter (1932) applied the technique to clasts in tills of northern Germany and was the first to investigate a significant number of sites. The results demonstrate how the ice moved inland to build the classic Pomeranian moraines of this region. Holmes (1941) developed the technique further and established the classic technique as a fundamental tool in glacial sedimentology (West & Donner 1956).

About 50–100 measurements are required in order to obtain a clear orientation maximum. However, measurements do not provide a clear result in every case: a secondary, so-called B-maximum perpendicular to the main peak may occur. In most cases this is much weaker, but it can become the orientation maximum in thrust zones and areas with compressive flow.

The orientation of the long axis depends on clast shape; stones with a more pronounced long axis are better orientated than rounded stones (Krüger 1970). The grain size fraction of the investigated clasts also plays a role. Fine gravel and sand grains are usually much less orientated than coarser material. The reason for this is that larger debris 'float' more or less in the fine-grained matrix, while the smaller particles often come into contact with like-sized grains, thereby hampering the orientation. The spatial orientation of fine particles may also be determined in thin-section or by radiography (Ehlers 1990). The measurement of finer particles allows for greater differentiation within the investigated till units. On the other hand, more measurements are needed to obtain a secure representation of the structure of the entire till layer.

Clast orientation measurements can also be used to investigate the genesis of a diamicton. While the vast majority of clasts lie almost horizontally in lodgement till, with long axes dipping slightly in the direction from which the glaciers came, the 'drop stones' in a glaciola-custrine sediment that have rained out from floating icebergs often stand vertically. In a cliff on southern Langeland (Denmark; Fig. 4.14) for example, fabrics of the deposits of an ice-dammed lake and the till of the last ice advance are easily distinguished. However, clast fabric as a tool of interpreting till genesis is not always clear-cut; Bennett et al. (1999) highlighted the problems of relying on till fabrics to determine till types.

Figure 4.14 Clast orientation measurements in till and underlying glaciolacustrine deposits on the Isle of Langeland (Denmark). In the lake sediments many stones are upright, and there is no clear orientation of the long axes. Photograph by Jürgen Ehlers.

In addition to till *macro*fabric (the measurement of visible clast orientations), till *micro*fabric can help determine the processes leading to its formation. Till microfabric analysis looks at the microscopic structure using thin-sections, and Ostry & Deane (1963) demonstrated its use for determining the direction of movement of the ice which deposited the till. It has since been widely used to differentiate between different processes leading to till formation and identifying different till types (Fig. 4.15), especially deformation tills (e.g. van der Meer 1993; Menzies et al. 2006). Micromorphology is especially useful where macrofabric analysis is not possible, such as in borehole samples (e.g. Carr et al. 2008).

Macro- and microfabric orientation measurements are most useful in areas where there is doubt about the direction in which the ice has moved. This is especially the case where the large ice sheets of northern Europe and North America are involved. In the Alpine region and in upland areas, the directions of glacier advances are often controlled by the general morphological situation.

The composition of the glacial deposits provides essential information about the path that the ice took. The most comprehensive information can be obtained by determination of the coarse components: the clasts. Where no exposures are available, investigations have to resort to finer fractions such as fine gravel. The study of indicator clasts, described in the following section, makes a decisive contribution to the reconstruction of glacier dynamics. Note, however, that although indicator clast assemblages provide evidence of former ice movement, they are not age specific.

(a)

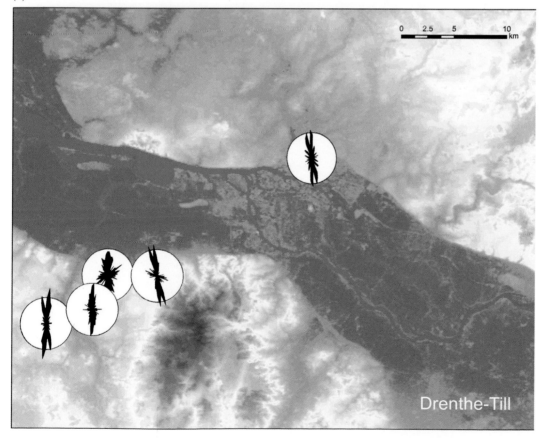

Figure 4.15 In some cases the deposits of different ice advances can be distinguished based on their clast orientation. Till fabric measurements from three tills from the Hamburg region in Germany: (a) Older Saalian advance

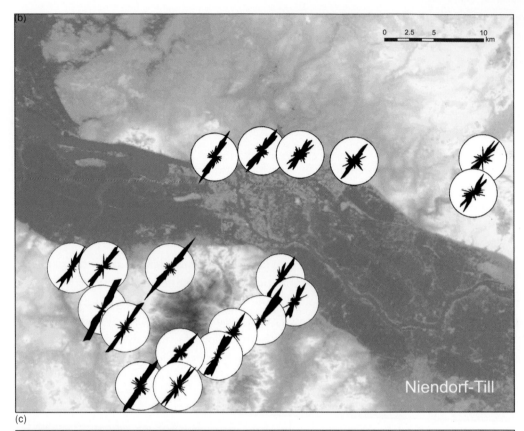

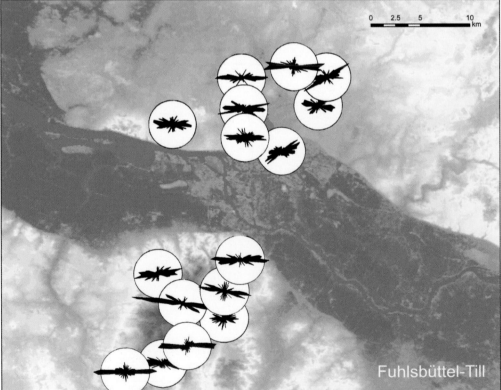

Figure 4.15 (Continued) (b) Middle Saalian advance; and (c) Younger Saalian advance.

4.1.5 INDICATOR CLASTS

Indicator clasts provide information on the movement directions of Pleistocene glaciers. Indicator clasts are rocks derived from a narrowly defined area of origin, and are clearly identifiable. The evaluation is mainly restricted to crystalline rocks. Some workers go so far as to only use crystalline rocks for their counts.

Erratic boulders were noted by Victorian scholars in the British Isles, and their lithologies were identified and linked back to potential source areas. In 1873 Mackintosh published a paper on the erratics of northwest England. Distinctive lithologies were found far from their upland sources, often from Scotland and the English Lake District, now resting in the lowlands. Mackintosh noted that erratic clasts were often used as building materials and in the city of 'Chester, for a very long time past, it has been customary to pave the streets with small erratic boulders and pebbles, and to place large boulders against or at the corners of the walls of houses'. In Manchester, numerous erratic boulders were removed during excavations in the nineteenth century expansion of the city. One example is a large (3 ×2 × 1 m) volcanic boulder from the English Lake District, some 150 km distant to the northwest, now residing in the Old Quadrangle of the University of Manchester (Fig. 4.16). Erratics noted by Mackintosh (1873) included the boulders formed in Shap Granite. This distinctive pink granite (Fig. 4.17) can be traced to a small area near the village of Shap in the eastern English Lake District (Harkness 1870).

As in Britain, the North German Pleistocene stratigraphy has been shaped to a large extent by indicator stratigraphical investigations. Gerd Lüttig developed a method in which the latitude and longitude of the source areas of identifiable indicator clasts are noted. The average of those values provides a 'theoretical indicator centre' (*Theoretisches Geschiebezentrum*, TGZ) for the sample, which in many cases allows the stratigraphic classification of tills (Lüttig 1958). Other methods include the reconstruction of the route that the ice has taken (Smed 2002) or distinguishing certain areas of origin (Zandstra 1988). The results of such investigations are only reproducible when the type and number of rocks are given as well as the index numbers.

Indicator clasts have been described extensively in the literature: Hesemann (1975) describes 170 indicator clasts; Smed (2002) mapped and described 157 types of rocks that can be used as indicator clasts; and Zandstra (1988) distinguished 209 types of indicator clasts (Fig. 4.18). To identify those rocks, a comprehensive understanding of the geology of Scandinavia is needed as well as a basic knowledge of geology and mineralogy.

Some of the most important indicator clasts can be easily distinguished. The mere frequency or absence of those stones provides a rough indication of the source regions of that particular till. For example, from the east the ice has brought the Åland quartz porphyry from the area around the Åland Islands. In the dense, often reddish-brown, matrix phenocrysts of pink K-feldspar are often found, *c.* 1 cm large, and plagioclase is missing. Particularly striking and characteristic are the large, round, usually dark grey-coloured quartz grains; these have been fractured by the crystallization and almost always contain radial or irregularly distributed fine veins of feldspar (Fig. 4.18d).

The second typical crystalline rock of eastern origin is the Rapakivi (which means 'rotten stone') granite (Fig. 4.18c), found on the Åland Islands and in mainland Finland. This type of rock is characterized by 3-cm-large K-feldspar ovoids, often surrounded by a ring of whitish or

(a)

(b)

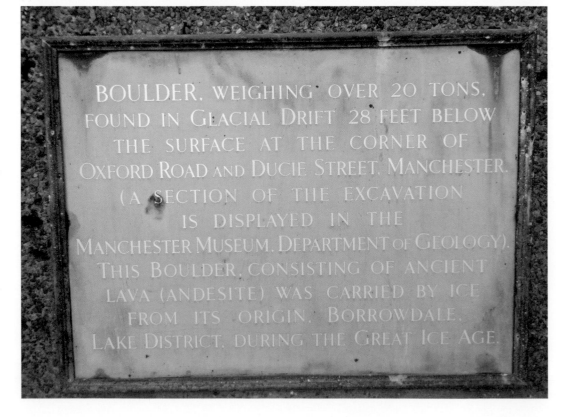

Figure 4.16 (a) Volcanic erratic boulder in the old Quadrangle at the University of Manchester. This boulder has a source over a hundred kilometres away in the English Lake District. (b) Text on a plinth adjacent to the erratic. Photographs by Phil Hughes.

Figure 4.17 Shap Granite erratics have spread widely from their original area. In this example, a granite boulder rests on limestone bedrock in the NW Pennines, England. Photograph by Phil Hughes.

grey plagioclase. On fields or in gravel pits, the distinctive feldspar 'eyes' with their weathered and dark circles are easily recognized.

The source areas of some rock types can only be determined from the distribution of their erratics. The Red Baltic Porphyry (Fig. 4.18b) has a dense, strikingly red matrix (often brick red), in which small reddish-brown or dark-grey phenocrysts of quartz can be seen. Small phenocrysts of feldspar also occur. Also characteristic are inclusions of a dark-green, fine-grained

(a)   (b)

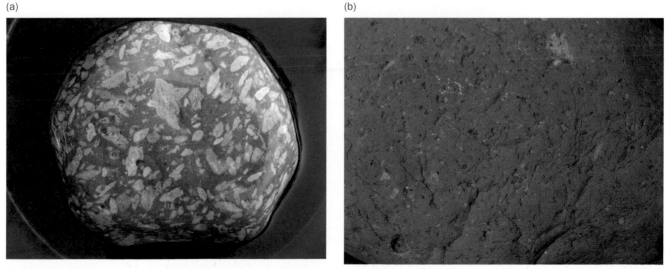

Figure 4.18 Crystalline Scandinavian indicator clasts: (a) Rhomb Porphyry; (b) Red Baltic Porphyry;

(c)

(d)

(e)

(f)

(g)

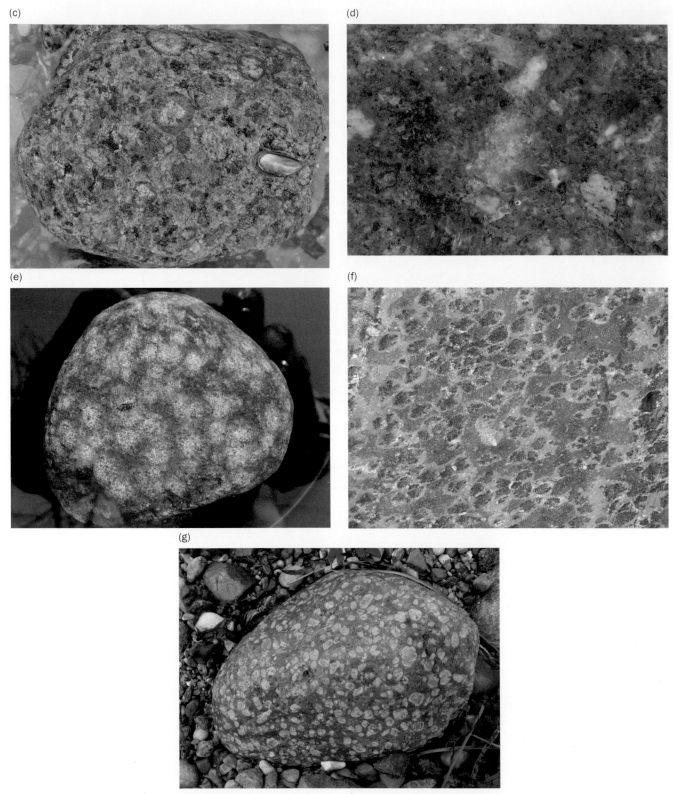

Figure 4.18 (*Continued*) (c) Åland Rapakivi; (d) typical quartz grain in an Åland quartz porphyry; (e) Kinne Diabase; (f) Västervik Spotted Rock; and (g) Påskallavik Porphyry. Photographs by Jürgen Ehlers.

foreign rock (xenolith). The Red Baltic Porphyry outcrops on the Baltic Sea floor to the southeast of the Åland Islands.

Brown Baltic porphyry originates from the seabed southwest of Åland. Its matrix is grey-brown to reddish-brown (coffee brown) and individual grains are visible with a magnifying glass. The stone has many inclusions; most notable are the pink- to reddish-brown potassium feldspars, but almost-white plagioclase and many grey quartz crystals and dark augite crystals also occur, some of which are altered to greenish chlorite.

One of the most striking indicator clasts is the Påskallavik Porphyry from the southeast Swedish Baltic coast. Numerous phenocrysts of feldspar are found in the grey-pink, brown or blackish matrix, the corners of which are usually rounded (Fig. 4.18g). The outsides of the phenocrysts are often brighter than the inner parts; they are Kalifeldspar crystals with plagioclase seams.

Granites from Småland in many cases have a special feature by which they are easily identifiable: the quartz crystals are not grey but of a slightly bluish to blue colour. Another rock from Småland which is easily identifiable is the Västervik spotted rock (Fig. 4.18f). The red component is composed mainly of alkali feldspar (microcline), plagioclase and some quartz. The dark spots are cordierite crystals, which are interspersed with muscovite, feldspar and quartz. The dark colour comes from tiny enclosed scales of biotite.

Easily identifiable rocks in central Sweden include the Dala Porphyries from Dalarna. The most common and easily identifiable type of Dala Porphyry is the Bredvad Porphyry. It is a bright red rock, the surface of which is mostly weathered to bright pink, containing numerous individual small crystals of kalifeldspar. In contrast to the Red Baltic Porphyry, the quartz is missing. Of the bright-green plagioclase grains, usually only the many four-sided holes are left over by the weathering.

The Kinne Diabase (Fig. 4.18e) is an unmistakable indicator, which originates from the area between lakes Vänern and Vättern. Its surface is mottled by weathering in a conspicuous manner. It is an important indicator clast for ice advances that have not followed the Baltic Sea depression, such as the 'Northeast Ice' of the Weichselian Glaciation maximum in Denmark.

The source areas of indicator clasts are not evenly distributed across Scandinavia, which has to be taken into account. In studies of the different tills on the Danish Isle of Langeland the main difference between the 'thick till' of the northeast ice and the other two Weichselian tills seemed to be that the thick till contained much less indicator clasts than the other tills. This picture changes if garnet amphibolites are also counted as indicator clasts. The garnet amphibolite from southwest Sweden (Småland and Halland-West) is actually a basic migmatite, containing hornblende, quartz and biotite and conspicuous patches of reddish-violet garnet.

The westernmost easily recognizable indicator clasts come from the Oslo area; most notable are the Rhomb Porphyries. They are mostly grey- to brown-coloured volcanic rocks, the matrix of which generally contains a large number of light-coloured feldspar crystals. Many of them look lozenge-shaped, such as the eponymous rhombs, but twinned crystals and star-shaped forms also occur. Also from the Oslo area is the Rhomb Porphyry Conglomerate.

Oslo rock types are rarely found in Germany. There are certain tills where they occur more frequently than in others (e.g. the Elsterian tills of northwest Germany; Meyer 1970), but the bulk of indicator clasts are from other areas of origin. The spread of the Oslo erratics ranges across the North Sea well into East Anglia, Yorkshire and the Scottish east coast. A large block of a Norwegian indicator has even been found on the Shetland Islands.

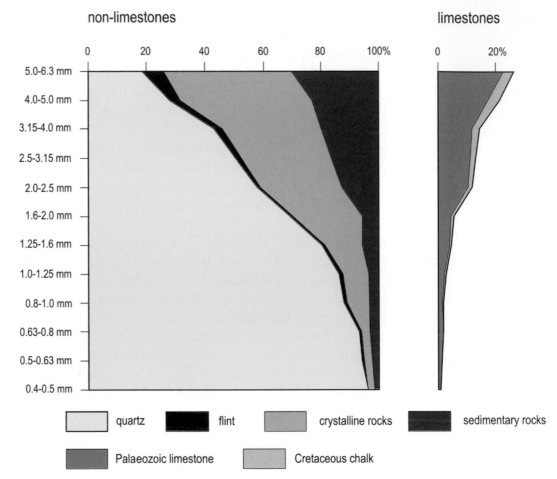

non-limestones   limestones

Figure 4.19

Composition of different grain-size fractions of Elsterian till from a depth of 34–35 m from the Dradenau KB 42 core drilling in Hamburg-Waltershof. The finer the material, the higher the quartz content.

quartz   flint   crystalline rocks   sedimentary rocks

Palaeozoic limestone   Cretaceous chalk

4.1.6 FINE GRAVEL

The petrographic composition of the Quaternary sediments is heavily dependent on grain size (Fig. 4.19). In the composition of a typical Elsterian till sample (high quartz content, relatively low carbonate content) from Hamburg, it can be seen that the fractions <2 mm are clearly dominated by quartz. In indicator statistical investigations, the possible accuracy of the result is also determined by the sample size. A larger sample allows for a more precise determination of the percentages of the various components. However, the accuracy will not increase linearly. Statistical studies have shown that a meaningful sample size is *c.* 300 grains.

When counting the 3–5 mm fraction using the Dutch method (Zandstra 1983), some 30 types of rock are distinguished. Most of these rock types are represented in such small quantities that they must be combined with others to allow statistically meaningful analysis. Limestones and non-limestones should be distinguished, and the group of non-limestones subdivided into quartz, flint, crystalline and sedimentary rocks.

Fine gravel analysis (Fig. 4.20) does not provide any additional information to that which could be determined by indicator counts. The great advantage of the method, how-ever, is that core samples can be analysed due to the smaller sample size. While the area of

(a)

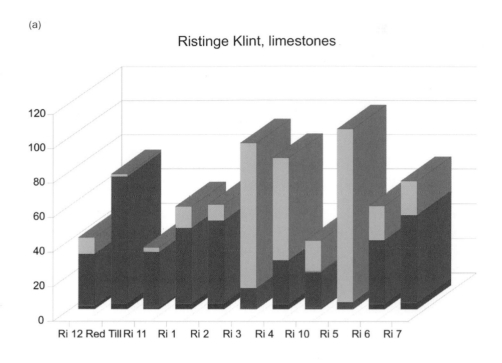

(b)

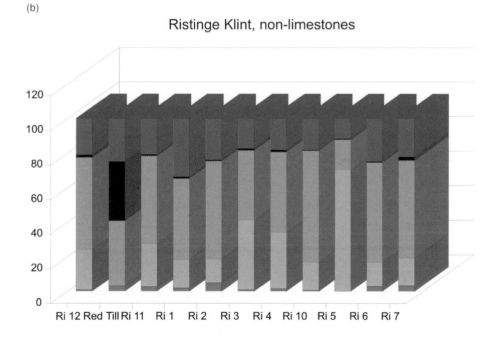

Figure 4.20 Fine gravel analyses from till samples from Ristinge Klint, Langeland, Denmark: (a) limestones; and (b) non-limestones. The three tills can be clearly distinguished. (The raft of reddish till found in the western part of the section is strikingly different from the rest of the samples.)

origin of the individual components cannot be determined as accurately as in indicator clast analysis, fine gravel analysis also allows a rough breakdown of sediments into different areas of origin.

Fine gravel analysis requires less sample material than an indicator count but it is still a laborious process that takes hours, including the preparation of the samples. There have been various attempts in northern Germany to use other analysis methods, such as an evaluation of the heavy minerals or the clay minerals and the geochemical composition of the glacial deposits. Convincing results have not been achieved in any case, and the high cost of the analyses has meant that no new attempts have been made in this direction. Hans-Jürgen Stephan (1998) concluded that, to determine parameters such as particle size distribution, carbonate content and fine gravel composition are the best means to provide a stratigraphic classification of the various tills. Their usefulness of such methods has been proven by many examples.

4.1.7 QEMSCAN: COMPLETE ANALYSIS

With the help of QEMSCAN analysis (an automated mineralogical analysis method based on electron beam technology from the FEI Company) it is possible to investigate sediment samples in a single pass in terms of grain size, grain shape, carbonates, heavy and light minerals and clay minerals. For North German Quaternary deposits, this method was first used on Elsterian channel fill sediments from the Hamburg area. In March 2010, Ehlers and Andrea Moscariello sent till samples from 20 drill holes in the Hamburg area for QEMSCAN investigation to Canada. These samples were taken from four different tills, and included:

- five samples of Niendorf till (Middle Saalian Glaciation);
- four samples of red Drenthe till (Older Saalian Glaciation);
- eight samples of 'normal' Drenthe till (Older Saalian Glaciation); and
- three samples of Elsterian till from a core drilling.

The results, depicted in Figure 4.21, were as follows. As expected, the proportion of quartz in the red Drenthe till is the lowest. The low quartz content is well known from the fine-gravel counts from the Hamburg area. Surprisingly, however, the quartz content of the Elsterian tills does not differ from that of the Drenthe till; higher values would have been expected on the basis of fine gravel composition.

There are significant differences in the mica content. The Niendorf till differs from the other three by a low mica content. The red Drenthe till contains significantly more biotite than the other tills. The Elsterian till contains extremely high amounts of muscovite and extremely low levels of biotite.

For calcite and dolomite, the expected peaks can be seen in the samples of the red Drenthe till. Odd outliers result from the fact that in some samples large single grains of calcite and dolomite were recorded. These can be easily identified in the corresponding planar representations of the studied area (Fig. 4.22).

Pyrite is expected to be the most frequent mineral in the Elsterian till. The high values are derived from the reworking of Miocene mica clay which, in many parts of Hamburg, is directly underlying the Elsterian deposits. The high concentration of gypsum and anhydrite in the Elsterian till may be due to the fact that the examined samples were taken in the lee of the Langenfelde salt dome, the caprock of which the glaciers had partially eroded.

The heavy minerals also show clear differences between the tills, where the composition of the Elsterian tills in particular differs from the other samples.

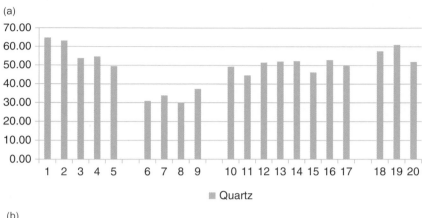

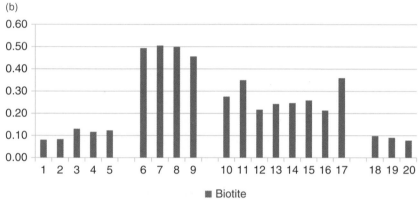

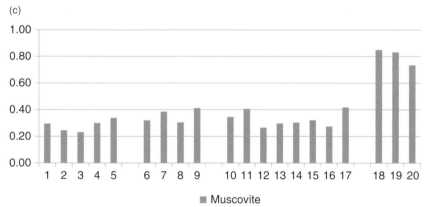

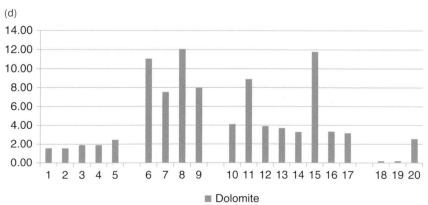

Figure 4.21 QEMSCAN analyses of till samples from Hamburg, Germany: (a) quartz; (b) biotite; (c) muscovite; (d) dolomite

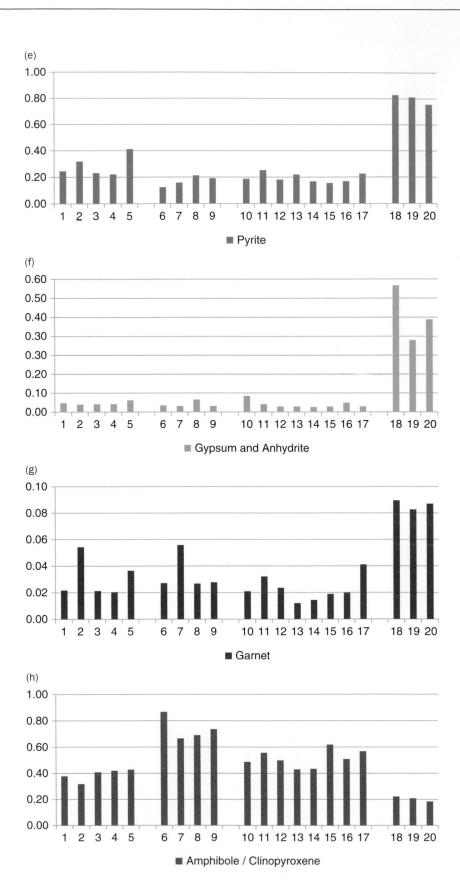

Figure 4.21 (*Continued*)
(e) pyrite; (f) gypsum and
anhydrite; (g) garnet;
(h) amphibole and
clinopyroxene

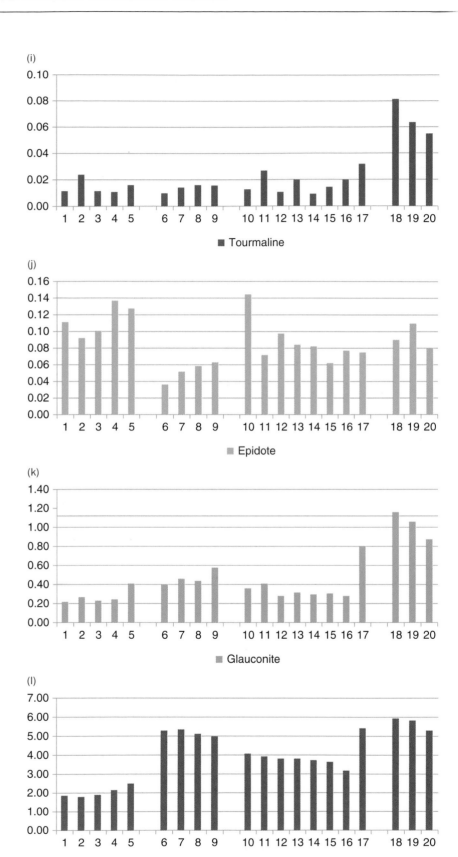

Figure 4.21 (*Continued*)
(i) tourmaline; (j) epidote;
(k) glauconite; (l) illite

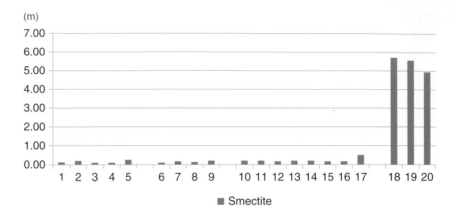

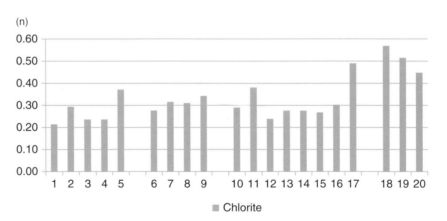

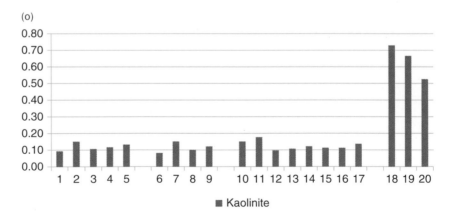

Figure 4.21 (*Continued*)
(m) smectite; (n) chlorite;
and (o) kaolinite.

The clay mineral composition is also distinctly different. The Niendorf till contains significantly less illite than the other samples; on the other hand, red Drenthe till and Elsterian till contain very high values. The Elsterian till contains extremely high values of smectite and kaolinite. The clay mineral determination by QEMSCAN, however, is regarded as relatively uncertain.

These results suggest that it might be possible with the help of QEMSCAN analysis to distinguish the various tills in the Hamburg area. Such an assignment would be

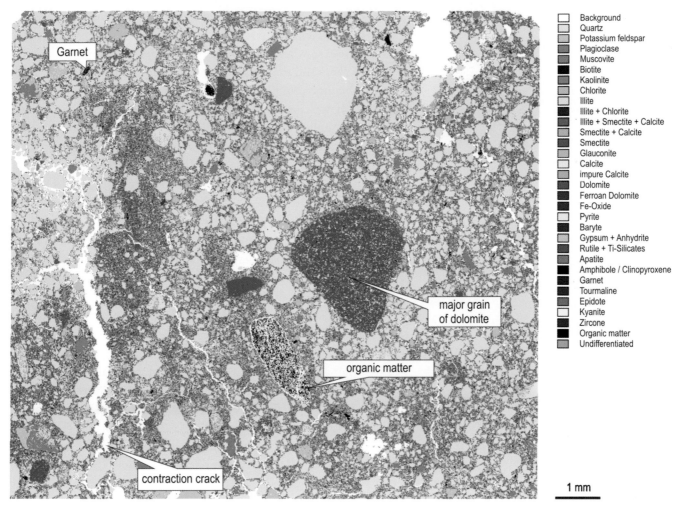

Figure 4.22 QEMSCAN analysis general view image used to detect irregularities. In this sample several large grains of dolomite are visible, which account for the irregularity in Figure 4.21d (sample 11).

Source: SGS. Reproduced with permission.

very helpful for reliable groundwater modelling in the near-surface area because, despite the large borehole density in the city of Hamburg (over 230,000 geologically recorded drilling records), the stratigraphic correlation in many cases is difficult or impossible. Further analysis is required before it can be known over which distances the differences can be traced.

4.1.8 SPECIAL CASE MICROFOSSILS: THE PRE-EEMIAN OF LANGELAND IN DENMARK

Redeposited microfossils are present in the matrix of most tills. The pollen content of the tills is essentially due to the reworking of older sediments. The most strongly represented units are normally those that were deposited in the last period before the ice advance. The Late Weichselian tills of Finland therefore primarily contain pollen from Weichselian interstadial

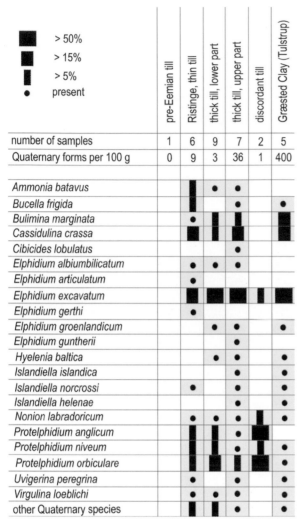

	pre-Eemian till	Ristinge, thin till	thick till, lower part	thick till, upper part	discordant till	Græsted Clay (Tulstrup)
number of samples	1	6	9	7	2	5
Quaternary forms per 100 g	0	9	3	36	1	400
Ammonia batavus		▮	•	•		
Bucella frigida		▮		•	▮	•
Bulimina marginata		•	▮	▮		▮
Cassidulina crassa		▮		▮		
Cibicides lobulatus				•		
Elphidium albiumbilicatum		•	•	•		
Elphidium articulatum		•				
Elphidium excavatum		▮	▮	▮	▮	▮
Elphidium gerthi		•				
Elphidium groenlandicum			•	•		•
Elphidium guntherii				•		
Hyelenia baltica			•	•		•
Islandiella islandica				•		•
Islandiella norcrossi		•		•		•
Islandiella helenae				•		•
Nonion labradoricum		•	•	•	▮	•
Protelphidium anglicum		▮	▮	•	▮	
Protelphidium niveum		▮		•	▮	•
Protelphidium orbiculare		▮	▮	▮		•
Uvigerina peregrina		•		•		•
Virgulina loeblichi		•	•	•		•
other Quaternary species		▮	▮	•		•

Legend: ▮ > 50% ▪ > 15% ▪ > 5% • present

Figure 4.23 Redeposited Quaternary foraminifera in the tills from Ristinge Klint, Denmark. For comparison: foraminifera from the marine Eemian Græsted Clay of Tulstrup. Adapted from Sjørring et al. (1982). Reproduced with permission of Dansk Geologisk Forening.

deposits which are especially characterized by high levels of birch (*Betula*), non-tree pollen and many spores (20–40%). However, where Eemian sediments have been reworked, pollen of *Pinus* and *Alnus* also occur in addition to *Betula* pollen, while the number of non-tree pollen is comparatively low.

It was possible to distinguish reworked Eemian and Holstein pollen assemblages in tills in Estonia. Consequently, it was possible to use the palynological studies of tills to determine the age of the deposits. The Estonian literature is quoted in Dreimanis et al. (1989). Corresponding studies in Ontario, Canada (Dreimanis et al.1989) have produced less favourable results, however. Of the 13 samples analysed, only two have shown significant proportions of reworked interglacial pollen; the remaining samples were characterized by high *Pinus* values, attributed to Early Weichselian or Early Saalian reworking of cold phase or interstadial deposits.

Palynological investigations of tills have also been carried out in Denmark (see overview by Petersen 1983). These observations have shown that a substantial portion of the till matrix originated from the reworking of local material.

The sometimes very high content of marine microfossils in Denmark offers the possibility of foraminifera analysis for the Quaternary stratigraphical classification of tills (Fig. 4.23). A fundamental work in this regard is the article by Petersen & Konradi (1974). The application of such investigations for Quaternary stratigraphical questions can be found in Sjørring et al. (1982).

4.2 Moraines

4.2.1 END MORAINESS

If not deposited as till, the debris which a glacier takes up on its way will be transported to the ice margin. If the ice margin remains stationary for a long time, the melting out of the debris accumulates and creates a depositional end moraine (Fig. 4.24). With a receding glacier, this process can lead to the formation of a series of annual moraines. Depending on its position with respect to the glacier, the debris that accumulates on the edge of the glacier is called either a lateral or a terminal moraine. The term *moraine* dates back to Agassiz (1841) and comes from the area of the Alpine glaciations. Early glacial geological research in northern Germany largely concentrated on the search for phenomena known from the area of mountain glaciations. Consequently, Gottsche (1897b) mapped the terminal moraines of Schleswig-Holstein in northern Germany as accumulations of coarse rock debris. Today we know that a significant portion

Figure 4.24 End moraines of a former ice cap outlet glacier in the Orjen Massif, Montenegro. Moraines are glacial deposits expressed by their surface form. Photograph by Phil Hughes.

of the block accumulations he found are in fact the result of meltwater activity, that is, they are sediments that were deposited directly at the ice margin.

The clearest end moraines are often those found in cirques and valleys of mountain regions (Fig. 4.24), as the end moraines are visible in conjunction with other morphological features such as cirques (Figs 4.25–4.27). However, not all glaciers leave a clear terminal moraine. Sometimes glacial limits in mountain cirques and valleys are marked by 'drift' or boulder limits (Gray & Coxon 1991). Sometimes the evidence of localized glacier occupation is quite subtle and the significance of boulder limits is questionable. For example, in Glen Sannox on the Isle of Arran, Scotland (Fig. 4.28), bouldery mounds can be traced in the lower parts of the valley below 100 m asl. This led Gemmell (1973) to argue for a large valley glacier during the Loch Lomond Stadial (Younger Dryas), the last phase of significant glaciation in Scotland. However, moraines are also present further upvalley and Ballantyne (2007) stated that these represent the Loch Lomond Stadial glacier limits and that there is no convincing evidence for the downvalley limit of an earlier glacier advance in this glen.

4.2.2 PUSH MORAINES

Not everything on the ice edge has been deposited as undisturbed sediments. Glacier advances often lead to disturbances of the older layers and to the formation of push moraines. The oldest descriptions of Alpine push moraines date from the sixteenth–seventeenth centuries. Johnstrup (1874) was first to recognize that corresponding deformations also played a role

Figure 4.25 Small end moraines in a cirque below Prutaš in the Durmitor Massif, Montenegro. Photograph by Phil Hughes.

Figure 4.26 Arcuate end moraines at the head of Keskadale in the English Lake District viewed from the west near Newlands Hause (see Fig. 4.27). Photograph by Phil Hughes.

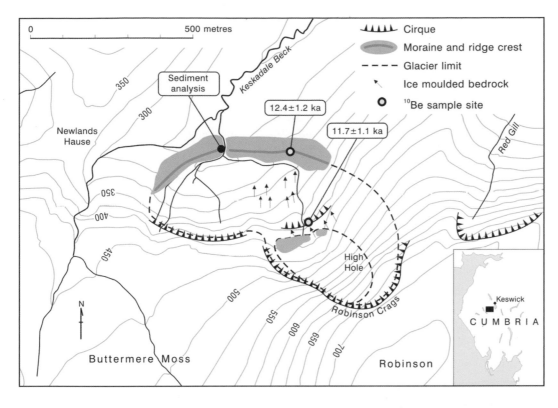

Figure 4.27
Geomorphological map showing the features at the head of Keskadale (Fig. 4.26).

Source: Hughes et al. (2012).

Figure 4.28 Glacial boulders in lower Glen Sannox, Isle of Arran, Scotland. Photograph by Phil Hughes.

in the area of the former Nordic ice sheets. He found that the tectonic disturbances of the chalk on the Danish Isle of Sjælland did not continue at depth. From this, he concluded that the disturbances must have been produced by glacial pressure. However, it was only after studies on recent glacier margins in Svalbard (Gripp & Todtmann 1926; Gripp 1938) that the great importance of thrust end moraines for the shaping of the glacial landscape in northern Germany was recognized (Box 4.3). Thrust moraines have been recognized in many different settings; for example, stacked and tilted diamicton units on both Mount Smolikas

BOX 4.3 THE SEARCH FOR THE END MORAINES

At the end of the nineteenth century it was clear that north Germany had been affected by at least two glaciations, the glaciers of which must have left behind end moraines. But where were they? They had to be found to draw precise maps of the former glaciations. In his 'Geological map of the duchies of Schleswig and Holstein', Forchhammer (1847) mapped the boundary between till and glacial sand. Was that the limit of the latest glaciation?

Christian Carl Gottsche (Fig. 4.29), whose habilitation in 1880 was on the subject of 'The sedimentary erratics of the province of Schleswig-Holstein', obtained leave of absence

Figure 4.29 Christian Carl Gottsche.

Source: Zeitschrift der Deutschen Geologischen Gesellschaft (1909).

from the *Hohe Oberschulbehörde* and set out in the summer of 1892 to find the missing end moraines. The project was financed by a 'generous travel scholarship' from the Hamburg Geographical Society. The study took Gottsche three years and was completed in 1896 when he revisited the area and made some revisions.

His map of the terminal moraines of Schleswig-Holstein at a scale of 1:750,000 is now obsolete, however. Gottsche had taken the accumulations of large quantities of boulders as a sign of the outermost ice edge. Between Haderslev and Blumenthal that position is approximately correct but, further to the east, the ice of the last glaciation advanced much further to the south than Gottsche had thought. The investigation never lost its value however because of the accurate descriptions. In addition, as Gottsche readily admits, he was lucky in that 'my excellent friend, the pharmacist Frucht-Braunschweig, had the great kindness, to accompany me for 14 days with his photographic apparatus.' To this happy circumstance we owe images of the 'almost unlimited block fields' (Fig. 4.30) excavated in the Kaiser-Wilhelm-Kanal (Kiel Canal) construction site and the richness of stones at various other sites which are no longer accessible.

STEINGEWINNUNG AUS BLOCKPACKUNG: STENTEN, W. DES WALDES.

Figure 4.30 Stone extraction from a block accumulation at Stenting (Schleswig-Holstein, west of the forest).

Source: Gottsche (1898).

and Mount Tymphi in Greece is best explained by proglacial thrusting, possibly in front of a surging glacier in response to changes in the dynamic mode of a polythermal glacier (Hughes et al. 2006a). Englacial thrusting in polythermal glaciers is also thought to explain hummocky moraine in some areas of Scotland (Bennett et al. 1999); see Section 4.2.3.

The big push moraines at the margins are commonly associated with corresponding excavation zones. In the German literature they are often referred to as 'tongue basins' (*zungenbecken*); in the English literature, however, moraines and depressions are referred to as 'hill-hole pairs'. The former basins have often been subsequently filled with sediment and are no longer visible on the land surface, particularly in the areas of older glaciations.

For the formation of push moraines, two different possibilities are discussed: thrust of frozen ground and thrust of unfrozen ground.

If the base of the glacier is not at the pressure melting point, the glacial ice freezes firmly to the ground. As the ice continues to be in motion, there is a shear pressure which can lead to the shearing off of the material. In deeply frozen ground, some layers are more conducive to sliding than others. Since clays freeze later than sand, they are sheared off preferentially. Accordingly, clays are often found at the base of the individual thrust slices (Fig. 4.31).

Figure 4.31 Ice-pushed chalk on Møn, Denmark.
Photograph by Jürgen Ehlers.

Permafrost is not absolutely necessary for the shearing off of sediment at the ice margin. A rapid ice advance favours shearing, since it produces a high pore-water pressure. When a water-saturated clay is overridden by a rapidly advancing glacier, the sudden load means that the weight is wholly or partly carried by the pore water, so that the shearing occurs easily. In fine-grained sediments, the pore pressure cannot be dissipated fast enough (van der Wateren 1985).

Not all terminal moraines were formed during late oscillations of the ice margin in the withdrawal phase of the glaciation. Older push moraines, which are overrun by a renewed ice advance, may retain their surface shape. It is now believed that the Dammer Berge hills in northern Germany were formed during the advance phase of the Older Saalian Glaciation (van der Wateren 1987). If the ice had first pushed forward to its outermost margin at the edge of the uplands, and if the push moraine had been formed by a late oscillation of the ice during the melting phase, then the Older Saalian till must have been included in the push moraine. However, the pushed layers contain no till. In fact, they are in some places overlain by morainic deposits (Meyer 1980). The whole situation therefore suggests that the Dammer Berge hills were overridden subsequent to their formation. Those terminal moraines of the so-called Rehburg phase are therefore unique. So far, the melting glaciers of Iceland, Greenland and the Alps have not revealed older, overrun moraines of this magnitude anywhere in their retreat.

BOX 4.4 RISTINGE KLINT: CROSS-SECTION THROUGH A PUSH MORAINE

At Ristinge Klint on Langeland (Denmark), a relatively complete sequence of Late Pleistocene deposits is exposed. The sequence begins with a shell-bearing marine clay from the Eemian Interglacial. This is overlain by a fossil-free clay (the 'shiny clay' of Madsen et al. 1908), underlain by a pre-Eemian till at at least one point (Sjørring et al. 1982).

Above the Eemian clay an approximately 5 m thick layer of fine sand follows, the so-called 'white sand'. Ice-wedge pseudomorphs have been found repeatedly in this part of the sequence; Friis & Larsen (1975) described them from the overlying sand. We found examples in the Eemian clay that seem to have originated at the contact between clay and overlying sand. This suggests that there was a break between the marine deposits and the subsequent Late Weichselian ice advances. During this period of non-deposition, the area was dry land as the formation of ice wedges requires terrestrial conditions.

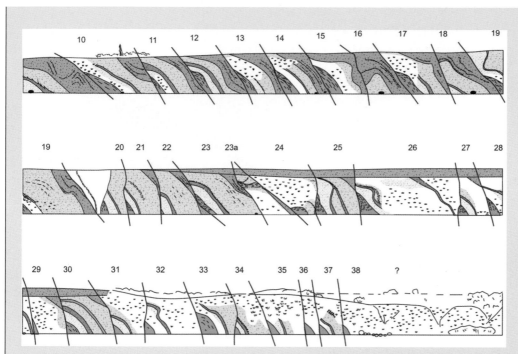

Figure 4.32 Geological profile of Ristinge Klint, Langeland, Denmark

The 'white sand' is overlain by the so-called Ristinge Klint Till (Fig. 4.32), deposited by an Old Baltic ice advance at the beginning of the Weichselian Glaciation (Houmark-Nielsen 2007). The Ristinge Klint Till is poor in clasts, and its thickness is <1 m everywhere. It was broken by ice thrusts into several small fragments, some of which are offset against each other like steps.

The Ristinge Klint Till is superimposed by the 'yellow sand' of Madsen et al. (1908). The sand is predominantly fine grained, but also includes local gravelly layers. A layer of gravel (very rich in flint) is often found, especially at the surface of the Ristinge Klint Till. The lithological composition is similar to that of the overlying East Jylland Till. There are no periglacial forms in these sands, interpreted as glaciofluvial outwash of the ice advance that deposited the next till.

The 'yellow sand' is overlain by the clast-rich East Jylland Till, which contains a lot of chalk and flint. This layer reaches a thickness of up to 5 m (at the western end of the cliff), and contains up to four well-developed stone layers. At the western end of the cliff, where only East Jylland Till is exposed, several layers of stones allow the reconstruction of large folds (Sjørring 1983).

Fabric measurements near the western end of Ristinge Klint (Fig. 4.33) in the seemingly undisturbed till gave no clear result. The lack of orientation may result either from deposition into water or from glaciotectonics.

Stone layers such as those found in the cliff at Ristinge have not been found elsewhere on Langeland. However, they are present in the cliffs on Ærø where the East Jylland Till also occurs.

(continued)

BOX 4.4 RISTINGE KLINT: CROSS-SECTION THROUGH A PUSH MORAINE *(CONTINUED)*

Figure 4.33 The western part of Ristinge Klint (top) and the raft of reddish till (below). Photographs by Jürgen Ehlers.

The thrust slices are superimposed with the remains of a third, discordant diamicton in two places. This is the Bælthav Till from the young Baltic ice advance, which has pushed and overridden the older deposits.

The presence of the two sand layers between the tills may have played a crucial role in the formation of a push moraine at Ristinge. The sand strata apparently wedge out towards the west end of the cliff where the thrust scales merge into large folds, which only occasionally include Eemian clay.

Investigations into the internal composition of the Uelsen and Dammer Berge end moraines have shown that they are not composed of thrust slices, but that nearly horizontally overlapping thrust nappes dominate.

Similar to the formation of high mountains, it is not easy to explain how the relatively thin nappes moved intact over long distances. This kind of deformation is only possible if the friction is assumed to have been considerably reduced. For the formation of the nappes a plane of décollement had to develop, on which the sediment packages could move with minimal resistance. Only the fine-grained sediments (clays and silts) are suitable for this, found at the base of many thrust zones. In this context, it should be noted that the terminal moraines of the Rehburg phase are located a short distance north of the northern edge of the Central German Uplands. Here Tertiary and older clays move to near the surface, into the realm of glaciotectonic stress (van der Wateren 1987, 1992).

The presence of suitable sediments in the subsurface is indeed a prerequisite for the formation of push moraines, but it is not sufficient to trigger the compression process. In Hamburg for instance, Miocene clays form almost the entire base of the Quaternary sediments without any noticeable glaciotectonic deformations. In addition, a rapid glacier advance is required in which a high pore-water pressure builds up to promote the shearing of the sediment (van der Wateren 1992).

While usually Tertiary or Cretaceous clays provide the planes of decollement in the thrust zones of the Saalian Glaciation, the sliding planes within the Weichselian thrust zones are sometimes provided by older till. In the Halkhoved moraine in Jutland, meltwater sands and till from the Saalian Glaciation have both been thrust by the Weichselian glaciers. Frequently, however, marine or glaciolacustrine clays are involved in the thrust zones, such as in the case of Lønstrup Klint (Pedersen 2005, 2006) or in Ristinge Klint on Langeland (Kristensen et al. 2000).

Push moraines also occur in the foothills of the Alps. For example, Schindler et al. (1978) describe slight glaciotectonic deformations from the Rhine glacier area. However, although frequently encountered in the marginal areas of the Nordic glaciations, substantial (up to several tens of metres deep thrust zones) deformations appear to be missing from the inner-Alpine region. Instead, there are widespread lateral moraines and depositional end moraines (*ufermoränen*), delineating the former ice-marginal position. A portion of the ufermoränen consists of elongate alluvial sediment cones, which is especially true of the moraines at higher topographic positions (Röthlisberger 1976). Thrust material can be involved. However, inclined sedimentary layers are not a safe indication of ice thrust, and it cannot be ruled out that these are the effects of subsidence over thawing stagnant ice (particularly when seen in smaller exposures). With any renewed glacier advance, the various types of marginal moraine can be covered by till. The outer shape of the deposits generally precludes the drawing of direct conclusions on the internal composition and origin (Schlüchter 1980a).

4.2.3 HUMMOCKY MORAINES

The term 'hummocky moraine' is used to describe irregular morainic topography within former glacier limits, and has been applied to a wide range of moraine types in different parts of the world (e.g. Hoppe 1952; Gravenor & Kupsch 1959; Sissons 1967; Aario 1977; Sharp 1985; Çiner et al. 1999).

Sissons (1967, 1979a, b) regarded hummocky moraine in Scotland as evidence of *in situ* glacier stagnation, a view which influenced many of the British glacial geomorphological studies in the 1970s and early 1980s. This interpretation has become less popular, however, and workers such as Eyles (1983), Bennett & Glasser (1991), Bennett & Boulton (1993) and Bennett (1994) argued that hummocky moraine topography can be formed during active glacier retreat. For example, in the Cairngorm Mountains, Scotland, Bennett & Glasser (1991) reinterpreted hummocky moraine mapped by Sissons (1979b) as closely spaced recessional moraines. It is also possible that, in some cases, hummocky moraine is the product of the decay of detached ice blocks from an actively retreating glacier (Eyles 1983). In addition to these theories of formation by ice stagnation and active retreat, Hodgson (1982) concluded that hummocky moraine is formed by the subglacial deformation of coarse debris including older till deposits. A genetic link between certain occurrences of hummocky moraine and subglacial fluting has also been considered by other workers (e.g. Donner & West 1955; Peacock 1967; Gray & Brooks 1972). Another mode of formation was proposed by Bennett et al. (1998), who suggested that some types of hummocky moraine are formed via englacial thrusting in polythermal glaciers.

Despite the problems of genetic definition, the term 'hummocky moraine' is very useful for describing the overall appearance of many areas of moraines in formerly glaciated areas. This point is clearly evident in the work of Çiner et al. (1999, 2015) in the Taurus Mountains of Turkey and Hughes et al. (2006b) in the Pindus Mountains of Greece. As such, the morphological term 'hummocky moraine' is still used, with moraine genesis considered individually at different sites based on available sediment exposures and overall landform assemblage.

4.2.4 GLACIER DYNAMICS OF THE WEICHSELIAN GLACIATION

The dynamics of large ice sheets can only be reconstructed if it is possible to determine the sequence of individual ice advances. While evaluation of glacial striae provided a rough picture of the last Weichselian ice movement (in Scandinavia especially), equivalent information from the marginal zones of the ice sheet took much longer to obtain. The reconstruction of regional glacial advance directions of the ice sheet in northern Germany began with Richter's (1936) map of the fabric measurements in Weichselian till in Pomerania. The situation was less favourable in the old morainic landscape, and Woldstedt (1938) had to settle for a reconstruction of ice movement directions entirely based on morphological evidence. The latter includes the strike direction of the parallel valleys and ridges of the East Frisian-Oldenburg Geest (Fig. 4.34).

Ehlers & Stephan (1983) put together the fabric measurements scattered in numerous individual publications in order to obtain a picture of the glacier movement in northern Germany. New studies by Speetzen & Zandstra (2009) from the Weser-Ems region are now available. For Denmark, one of the most extensive documentations of the stratigraphy and ice movement directions has been put together (Houmark-Nielsen 1987, 2004, 2007). Denmark is a key area for understanding the ice dynamics because Norwegian, Swedish and Baltic ice advances are clearly distinguishable; the cliffs of the Baltic Sea coast also provide a good insight into the Quaternary stratigraphy.

The example of the Weichselian Glaciation demonstrates the typical sequence of north European glaciation. Two major ice advances occurred in the early phase of glaciation, which both reached Denmark. A later foray of Baltic ice reached the Danish islands from the south during marine oxygen isotope stage (MIS) 4, at *c.* 64 ka. This so-called Old Baltic ice advance deposited the Ristinge Klint Till on Langeland (Ristinge Klint). The ice followed the depression of the Baltic Sea.

Also during the Weichselian glacial maximum, Denmark (North Jutland) was first reached by a Norwegian ice advance (Andersen & Pedersen 1998). This ice advance reached as far as the Limfjord. A Norwegian ice advance far into Denmark was only possible at the beginning of glaciation, when the ice divide was located far enough in the west. After the ice melted into the Kattegat the first major ice advance from Norway arrived, equivalent to the Brandenburg advance in north Germany. A second major ice advance quickly followed; in Denmark it reached the 'Main Stationary Line' in Jutland and in Germany it created the end moraines of the Frankfurt Phase. The Baltic Sea Basin played no role in controlling the ice movement during the Weichselian glacial maximum; the ice spread from the centre towards the edges of the ice sheet. Such a situation is only conceivable if at that stage no deformable bed was available.

(a)

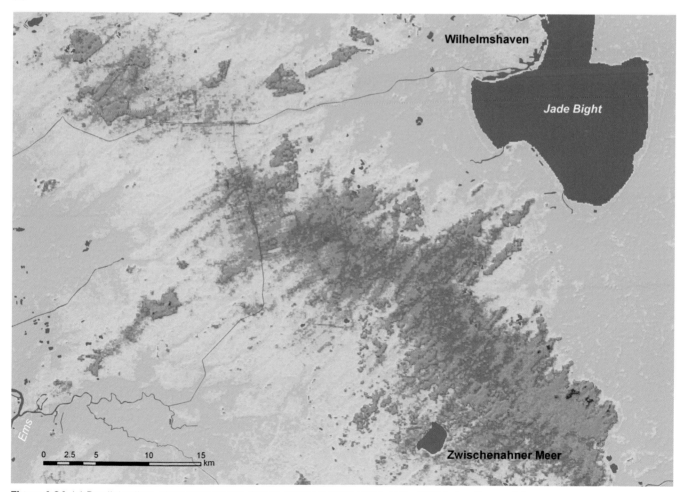

Figure 4.34 (a) Parallel valleys of the East Frisian–Oldenburg Geest made visible in the SRTM elevation model

(b)

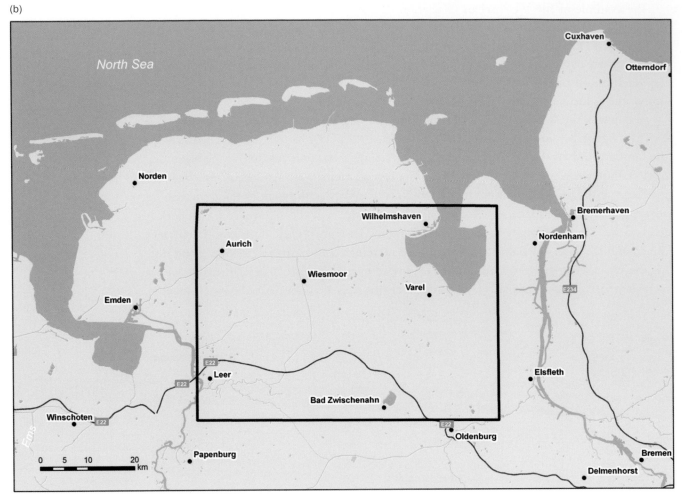

Figure 4.34 *(Continued)* (b) location map.

During the subsequent Pomeranian Phase of the Weichselian Glaciation, the ice movement was again to a very large extent determined through the shape of the Baltic Sea bed (Stephan et al. 1983; Ehlers et al. 2004). A Baltic ice stream advanced from the southeast and south into the area of the Danish islands. Unlike the situation during the Brandenburg Stage, however, the ice divide was situated further to the east, resulting in a slightly different ice movement direction from the early glaciation phase.

The ice extent depicted in Figure 4.35 does not represent the maximum extent of the Weichselian ice, but the extent sometime after the peak of the glaciation. A correspondingly smaller extent of the ice sheet is shown.

(a)

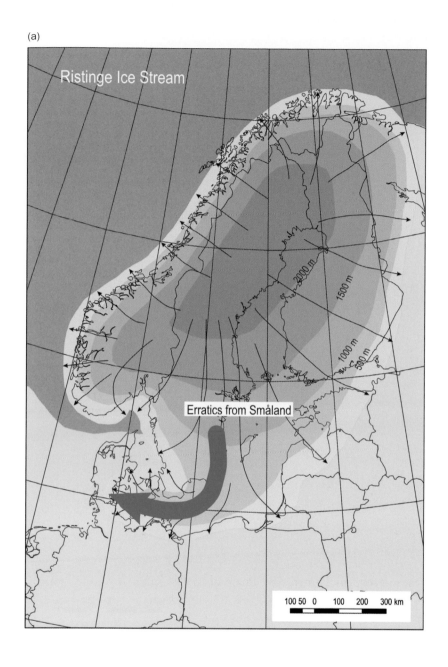

Ristinge Ice Stream

2000 m

1500 m

1000 m

500 m

Erratics from Småland

100 50 0   100   200   300 km

Figure 4.35 Glacier dynamics of the Scandinavian Ice Sheet during the Weichselian. (a) An early ice advance, the Ristinge Ice Stream, reached Denmark.

(b)

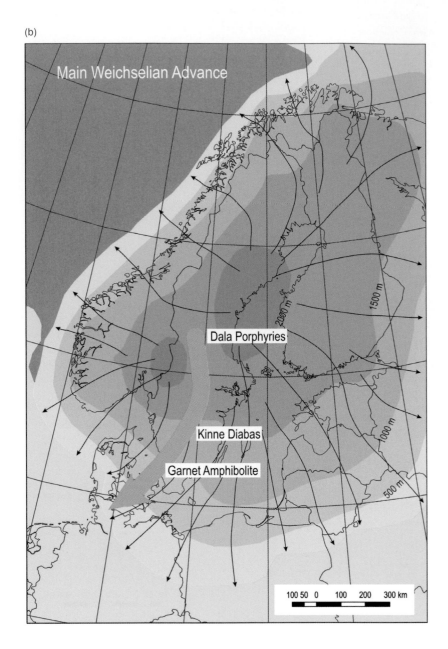

Figure 4.35 (*Continued*) (b) In the subsequent Main Weichselian Ice Advance, the Baltic Sea Basin played no role in controlling the ice movement.

(c)

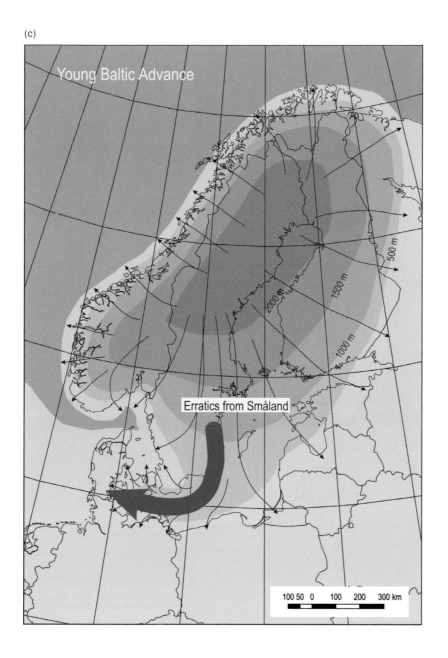

Figure 4.35 (*Continued*) (c) During the Young Baltic Advance, ice movement was again controlled by the Baltic Sea Basin.

(d)

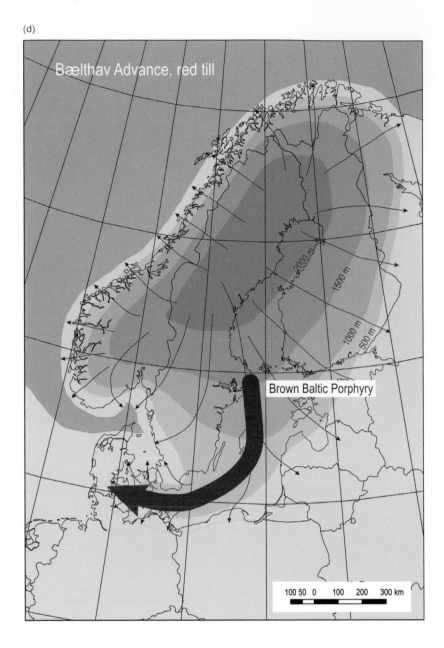

Figure 4.35 (*Continued*) (d) The last ice advance, the Bælthav Advance, brought rock material from more easterly *source* areas to Denmark.

The rapid ice movement at the beginning and end of the glaciation exploited the deformability of water-saturated morainic material at the base of the ice sheet. This helps to explain both the rapid ice build-up and the rapid ice decay because, under these conditions, a much lower ice thickness can be assumed than in the traditional glaciation model. We must assume that during the Saalian and Elsterian glaciations a similar rotation of ice movement directions also took place.

Giant pothole in the forefield of Briksdalsbreen glacier, Norway. Photograph by Jürgen Ehlers.

Chapter 5

Meltwater: From Moulins to the Urstromtal

Among the first indications that the European Central Mountains might once have borne glaciers are the giant potholes. Potholes are formed where meltwater penetrates to the sole of the glacier in a moulin and, in a swirl of water and debris, generates deep holes in the rock surface. Nevertheless, caution is required with the interpretation: similar forms can also arise in cases of normal river drainage or even on a rocky coast. Where appropriate gravel and water are available, the process of pothole formation may be set in motion.

5.1 Fjords, Channels and Eskers

5.1.1 FJORDS

Fjords are characteristic landforms of formerly glaciated areas. They were created by glacial erosion of former river valleys and, unlike normal glacial troughs, are now inundated by the sea. They are found for example in Greenland, Alaska (Fig. 5.1), Chile and New Zealand; the fjord coast of Norway is particularly strongly incised. The Sognefjord (Fig. 5.2) is 204 km long and, at the lowest point, its base lies about 1500 m below the water table. It is partially infilled by young sediments of up to 200 m thickness. The Sognefjord is the longest and deepest fjord in the world. As the adjacent mountains rise 1200–1800 m above the sea, this makes a total depth of c. 3000 m. At the mouth of the fjord, a rock sill is found at c. 150–200 m depth. It has been calculated that the excavation of Sognefjord alone would require the erosion and removal

The Ice Age, First Edition. Jürgen Ehlers, Philip D. Hughes and Philip L. Gibbard.
© 2016 John Wiley & Sons, Ltd. Published 2016 by John Wiley & Sons, Ltd.
Companion website: www.wiley.com/go/ehlers/iceage

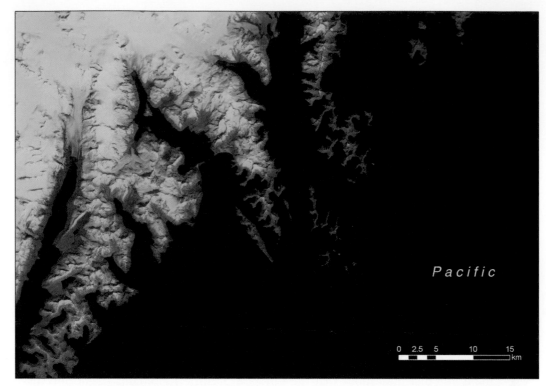

Figure 5.1 Fjords in Alaska in the satellite image. In the low-lying parts, former cirques can be seen which became 'drowned' in the postglacial sea-level rise. The narrow ridges between those cirques are called arêtes.

Source: US Geological Survey Landsat ETM satellite image 2003.

Figure 5.2 The Sognefjord is 204 km long and, at the lowest point, the bedrock lies about 1500 m below the water table. Photograph by Jürgen Ehlers.

of 2000 km^3 of rock. If this were to be evenly distributed over a country the size of Germany, it would result in a sediment layer of about 5 m thickness (Andersen & Borns 1994).

The fjords usually follow tectonically pre-drawn lines. Their formation is not solely a consequence of glacial excavation. Subglacial meltwater erosion must also have played a significant role. A characteristic of the glacially shaped valleys is that they consist of several basins subdivided by rock thresholds. Where the thresholds now lie above sea level, lakes have formed. The depressions did not form in a single ice advance, but in many successive glaciations. It is believed that the Sognefjord is the result of more than a million years of repeated glacial and meltwater erosion (Andersen & Borns 1994; Ehlers 2009).

5.1.2 CHANNELS

When studying meltwater activity, it is useful to distinguish between the drainage beneath the ice (subglacial area) and the runoff beyond the ice margin (proglacial area). While meltwater accumulation prevails in the proglacial area, the subglacial drainage is characterized by a preponderance of marked erosion. The southern margins of the north European glaciations are therefore dissected by systems of channels of up to several hundred metres depth, which predominantly extend radially from the centre of the ice towards the former ice margin. The older channels are usually completely filled with sediments and levelled to the relief of the adjoining ground. Their courses on land can only be reconstructed with the help of drillings. Only the most recent channels that were formed during the last glaciation are still clearly visible on the ground surface. These forms are often referred to in the literature as 'tunnel valleys' because they were formed by meltwater that flowed in tunnels beneath the ice. The infilling of tunnel valleys produces very thick accumulations of sediment.

Some of the best-developed tunnel valley systems in Europe occur in Germany. In the Hamburg area, Gottsche (1897a) noticed that very thick accumulations of glacial deposits had formed. The large borehole density in the Hamburg area enabled further investigations. Koch (1924) suggested that the deep 'basins' (as he called them) were caused by a combination of erosion by glacial ice and subglacial meltwater. He drew a first map of the 'prediluvial land surface under and around Hamburg'. This represented some of the earliest work on tunnel valleys, a concept which spread to other countries, such as the British Isles, only decades later.

Modern geological investigation of the channels began in East Germany. Hannemann (1964), Cepek (1967) and von Bülow (1967) described the deep Quaternary channels. First detailed investigations of the glacial channel systems were conducted by Eissmann (1967, 1975). He not only had a large number of boreholes available for such studies in the area around Leipzig, but the channels could also be followed some miles in the outcrops of the lignite mines. A comprehensive cartographic representation of the Elsterian channels found in North Germany was presented by Stackebrandt (2009).

The Elsterian glacial channels in the marginal areas of the Nordic ice sheets have been under detailed investigation in recent years. Further east in Poland, Belarus, Latvia, Lithuania and Estonia corresponding channel systems are known. The channels in Belarus and Poland are very similar to the forms found in northern Germany, so it is assumed that they have formed in the same way. The partial strong overdeepening, the interconnection of the channels and reticular formation argue against an origin as 'normal' river valleys (Kuster &

Meyer 1979). Seismic surveys in the North Sea have clearly shown that they cannot be fluvial landforms (Huuse & Lykke-Andersen 2000; Praeg 2003; Kristensen et al. 2007, 2008).

The depth of the channels varies. The basis of the Elsterian Reeßelner Rinne channel in North Germany is given as 434 m below sea level. Incisions to a depth of more than 400 m below sea level have also been found in the southern North Sea. The significantly smaller depth of the channels in Estonia for example (with a maximum depth of about 60 m below sea level; Noormets & Floden 2002) is due to the fact that the ground there consists of hard rock (Palaeozoic limestones, dolomites and sandstones), while in northern Germany the floor is formed of Quaternary deposits of easily erodible unconsolidated rock.

The intensity of channel formation has not been the same everywhere and in every glaciation. For example, in northern Germany:

- only small channels (up to 100 m deep) were formed during the Weichselian Glaciation (Fig. 5.3), ending at the proximal ends of outwash plains;
- there were no significant channels formed during the Saalian Glaciation; instead, there were extensive sandur planes; and
- channels of up to *c.* 400 m deep were formed during the Elsterian Glaciation.

However, these differences do not apply everywhere. Seismic surveys in the North Sea (Figs 5.4, 5.5) have shown that channels were formed there during all three glaciations (Carr 2004). The oldest channel system is most strongly pronounced in the southern North Sea. In

Figure 5.3 Lake Schmalsee, a subglacial meltwater erosion channel north of the Weichselian ice margin near Mölln, Schleswig-Holstein. Photograph by Jürgen Ehlers.

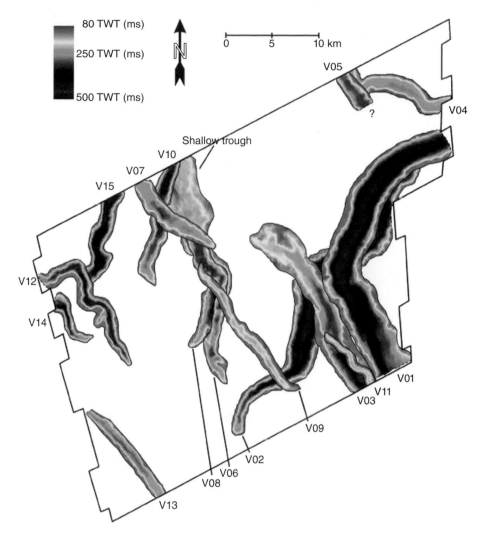

80 TWT (ms)

250 TWT (ms)

500 TWT (ms)

N

0   5   10 km

V05

?

V04

Shallow trough

V10

V07

V15

V12

V14

V01

V11

V03

V09

V02

V06

V08

V13

Figure 5.4 Fifteen generations of Pleistocene subglacial channels at the bottom of the North Sea. The channels are completely filled with sediment and only visible in the 3D seismic data. In the section shown there are two underground salt domes (SD), and the channels run around the salt domes.

Source: Kristensen et al. (2007). Reproduced with permission of John Wiley & Sons.

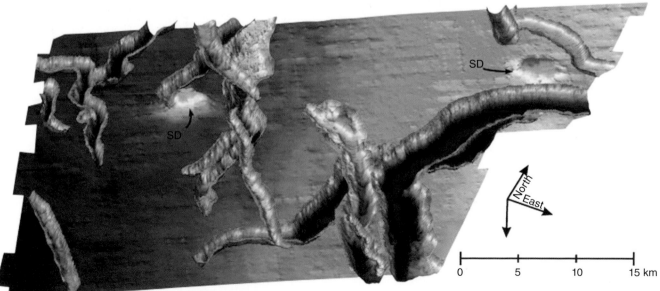

SD

SD

North

East

0   5   10   15 km

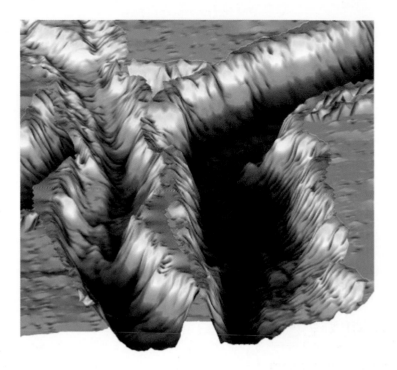

80 TWT (ms)

250 TWT (ms)

500 TWT (ms)

North
East

0   1   2   3 km

Figure 5.5 Detailed view of two glacial meltwater channels on the North Sea floor as depicted in 3D seismic data. The irregular base typical for subglacial channels is clearly visible in both channels.

Source: Kristensen et al. (2007). Reproduced with permission of John Wiley & Sons.

the central North Sea, the most deeply incised forms are Weichselian in age demonstrating that the experience of northern Germany cannot be readily transferred to other areas.

The causes of the channel formation have long been controversially discussed. Ussing (1903) correctly interpreted the troughs of the Weichselian Glaciation as tunnel valleys that were formed by meltwater erosion beneath the ice. He pointed out that the channels all ended at the former ice margin and that they were adjacent to relatively high-lying outwash plains. He assumed that the meltwater flowed under conditions of high pressure in tunnels beneath the ice and, when it eventually exited the glacier, its velocity decreased and the sediment load was deposited.

Paul Woldstedt concluded that glaciers and meltwater might have interacted with the channel formation (Woldstedt 1926). A key argument for the glacigenic formation of channels is their similarity to the overdeepened valleys in the Alps and Alpine foothills or to the Norwegian fjords, where these forms can be largely explained by glacier erosion. Some of the alleged tunnel valleys are also very wide (e.g. Bay of Kiel and Eckernförde Bay in Germany) and cannot be purely a result of meltwater erosion, resembling tongue basins rather than real meltwater channels. However, the proportion of the respec-

tive processes in the formation of those landforms is viewed differently by different authors. In his investigations in shallow channels in Hamburg, Bruns (1989) demonstrated that glacial ice was involved in their formation. In essence, however, they are dominantly formed by subglacial meltwater erosion (Piotrowski 1997).

It is difficult to reconstruct the exact shape of the channels using boreholes. While in reconstructions of Elsterian channels in north Germany the features usually follow a regular U-profile, evaluation of North Sea seismic data shows that this is the exception. Only about one-third of the deep channels have a simple U-shaped profile, while the remainder demonstrate substantially more complicated cross-sections (Kristensen et al. 2008). In a few cases this is an artefact caused by the same channel being cut several times by the seismic line (in the case of curved channels). Detailed analysis of seismic profiles only 1 km apart demonstrated that the channels are often composed of a system of parallel channel parts. The same is true for many Weichselian channels of the mainland (see Galon et al. 1983). This suggests that the channels are unlikely to have been formed by a single event, but rather by a sequence of several similar events. The shapes of the channels, with their local overdeepenings and irregularities, clearly disproves an extraglacial origin as river valleys. Subglacial meltwater erosion in most cases is likely to have played the dominant role in forming the channels.

In the exploration of the buried channels in continental areas, drilling is not always necessary. Small features, such as the channels discovered by Mathers & Zalasiewicz (1986) at the edge of the Elsterian Glaciation in East Anglia (Great Britain), have been explored using simple resistance measurements (Figs 5.6, 5.7). Electromagnetic measurements from a helicopter (HEM) have been used in north Germany. The University of Aarhus in Denmark applied a transient-electromagnetic system (SkyTEM) that, under favourable conditions, allows the reconstruction of the course of Quaternary channels to a depth of 200 m (BURVAL Working Group 2009; Jørgensen & Sandersen 2009).

The erosive effect of the meltwater is often underestimated. It is usually assumed that glaciers abrade the underlying rock during their advance. The pre-Quaternary ground of Norfolk in England consists mostly of chalk, which is mined in numerous pits. Above the chalk there is a thin, often patchy blanket of glacial Elsterian till. The high amount of chalk contained in the till shows that glaciers have incorporated large quantities of local material.

However, a closer look reveals that between till and bedrock there is often a thin layer of meltwater deposits. The drawn sections and photographs from Hillington illustrate the situation (Fig. 5.8), where all kinds of transitions between pure chalk and chalk boulders can be observed in a matrix of meltwater sediments. The sand is also found in cracks and crevices of a few decimetres depth into the bedrock (Ehlers 1990).

Figure 5.6 Resistance measurements can be used to explore the subsurface geology and to detect buried channels. Here the EM-31 device of the Canadian Geonics company is seen in operation. Photograph by Steve Mathers.

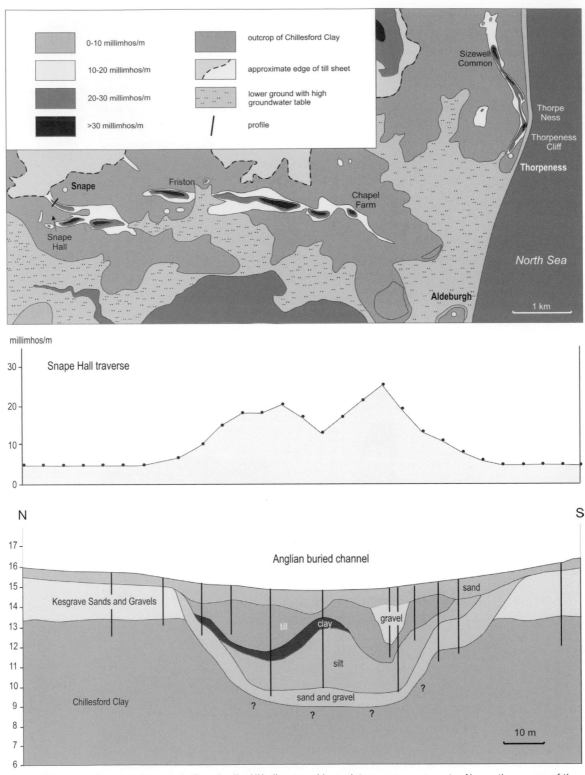

Figure 5.7 Small Elsterian channels in East Anglia, UK, discovered by resistance measurements. Above: the course of the channels; bottom: resistivity profile and cross-section based on boreholes at Snape Hall, Suffolk.

Source: Mathers & Zalasiewicz (1986), British Geological Survey © NERC All rights reserved IPR/125-15CT.

NW   SE

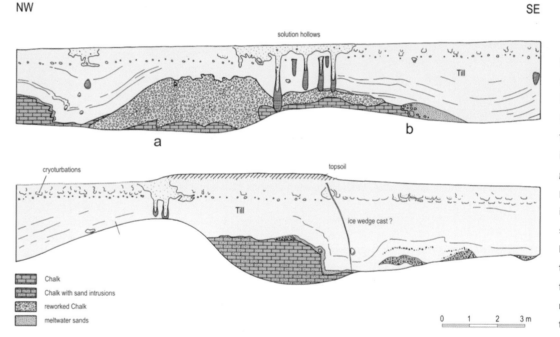

Figure 5.8 Traces of meltwater erosion at Hillington, Norfolk, UK. The upper part of the chalk has been incorporated in the glacier movement, and plucking has resulted in the formation of steps. Meltwater sand has been washed into the clefts of the highly fractured chalk. Ice movement was from left to right.

In northern Germany, a substantial portion of glacial erosion is due to meltwater activity. This is evident from the fact that several 100-m-deep Quaternary channels are filled mainly by meltwater deposits. In those sediments, the proportion of local material (i.e. quartz) is always much higher than in the tills. This is clearly visible in the results of fine gravel analyses.

When examining quartz grains from Norwegian till under an electron microscope (Fig. 5.9), it is immediately obvious that these are freshly broken, have sharp edges and are unweathered. On the other hand, the sand fraction of a north German till very often reveals rounded edges and grain surfaces showing signs of chemical weathering dating from well before the Ice Age (Fig. 5.9). These quartz grains are derived from reworking of Tertiary lignite sands. At least one-third of the quartz grains of the north German till sample comes from erosion of local material.

Figure 5.9 Sand grains under the electron microscope. (1–6) Till from the Oslo Fjord area, Norway and (7–13) till from Hamburg. The sand grains from Norway show fresh fractures; the grains from Hamburg traces of strong weathering. Photographs by Jürgen Ehlers.

This proportion is much higher in meltwater sands. Ohse (1983) found that the quartz to feldspar ratio was 12:1 in Saalian meltwater sands, while in fresh granite it was about 1:2. In fresh till in the Hamburg area however, the ratio is 6:1. The very high percentage of quartz in the glacial sands reflects the reworking and incorporation of Tertiary sands.

5.1.3 ESKERS

Subglacial meltwater not only erodes deep channels, but also piles up embankment-like ridges known as eskers (Fig. 5.10). These often consist of very coarse sand and gravel of varying grain sizes, as well as pebbles and boulders. The pebbles are often well-rounded. Eskers are long, railway-embankment-like walls of varying width and height, up to tens of kilometres long. In some cases they form branches, known as an esker network. Usually they run along the deepest parts of valleys or channels, but they can also cross mountain ranges since the direction of the subglacial drainage is determined through the slope of the ice surface (Flint 1971). Eskers can form in active ice, but can only be preserved when the surrounding ice is no longer active; their distribution therefore reflects the direction of the last ice movement. The alignment of eskers therefore corresponds to the general strike of the most recent glacial striae.

(a)

(b)

Figure 5.10 (a) Eskers at Folldal, Norway. (b) Exposure in an esker at Folldal, Norway. The coarse, well-rounded pebbles of these eskers, the 'Rolling Stones', provided the starting point for the discussion of the great stone flood. Photographs by Jürgen Ehlers.

Another type of esker are the so-called 'engorged eskers' first described by Mannerfelt (1945). These form during an advanced stage of ice decay, when meltwater is transported in feeder tunnels laterally downslope to the central drainage tunnel. They often run slightly diagonally because the tunnels tend to be drawn out in the direction of ice movement.

While large eskers in Canada, Sweden and Finland can be followed over tens of kilometres, only relatively minor esker trains can be found in northern Germany. The latter comprise very small features, such as the two eskers in Ahrensburg and in the Stellmoor tunnel valley in Hamburg (Grube 1968) or the Gellendin esker in Mecklenburg (Schulz 1998).

Only a few eskers are known from the Alpine area. Troll (1924) describes some from the marginal zone of the Inn-Chiemsee glacier. Other examples are controversial; Hantke (1978) shows cross-sections of eskers from the Glatt valley (north of Zurich), but points out a number of wrongly interpreted forms from other parts of Switzerland. Van Husen describes smaller eskers from the Enns (1968) and from the Traun valleys in Austria (1977).

5.2 Outwash Plains and Gravel Terraces

The vast accumulations of meltwater sands and gravel at the margin of recent glaciers have long attracted the attention of geologists. After a trip to Iceland, Torell (1858) concluded that the heath in northern Germany had been formed during the Great Ice Age in the same way as today's outwash plains at the edge of Vatnajökull (Figs 5.11, 5.12). It took 17 years for this interpretation to be finally accepted in Germany. The term 'sandur' was introduced to the German literature by Keilhack (1883). In most cases today, the Germanized word '*sander*' is used instead of the Icelandic *sandur* (plural *sandar*). In the English-speaking world, the term 'outwash' is more frequently used (in conjunction with 'plain', 'fan' or 'surface', etc.) and follows a long tradition which began in North America. Salisbury (1902) used the term 'outwash plain' extensively in the 800+ page study of the glacial geology of New Jersey, and provided detailed definitions of its geomorphological context. The term outwash plain was then

Figure 5.11 Skeiðará-Sandur, outwash plain at the southern edge of Vatnajökull in Iceland, with the typical braided drainage system. Photograph by Jürgen Ehlers.

applied to Pleistocene sand and gravels, and spread to the British Isles (e.g. Charlesworth 1927, 1929). In South Wales, Charlesworth (1929, p. 348) recognized that outwash sands and gravels were associated with glacier retreat and noted that this "led to the dissolution of the piedmont ice into its feeding glaciers and the deposition of recessional moraines, such as those which, with their outwash sands and gravels, were laid down about Caerphilly, after the withdrawal of the Rhymney Glacier". The processes occurring on recent outwash plains have been studied by Krigström (1962) in Iceland and Church (1972) on Baffin Island.

An essential part of the sediment movement on the recent outwash plains takes place during extreme flood events that occur after heavy rainfall in summer and in the course of jökulhlaups. When draining a small ice-dammed lake on Baffin Island, for example, the meltwater flow increased from less than 5 m^3 s^{-1} to 200 m^3 s^{-1} within 24 hours, and then dropped again within 2 hours down to 20 m^3 s^{-1}. This flow behaviour is considered typical for such meltwater outbursts. It is attributed to the fact that an ice tunnel grows gradually larger by enhanced melting of the ice walls, so that it reaches the largest diameter when the dammed-up water supply is exhausted (Church 1972).

Figure 5.12 Sandur sediments at the edge of Kverkjökull glacier in Iceland. Photograph by Jürgen Ehlers.

Krigström (1962) has divided the Icelandic sandar into three major zones in which different processes prevail. In the proximal zone, a few narrow, deep meltwater streams flow in relatively stable, slightly incised beds. Further downstream in the middle zone, the streams split into numerous river arms which move relatively quickly laterally. This is the realm of the classical outwash plain, where the rivers are wide and shallow. In many sandar which discharge into lakes or the sea, a third, distal zone is found which combines the meltwater flows into a sheet-like layer of torrential runoff with a delta deposited at the end. The water here is mostly shallow, but there are also deeper troughs.

The longitudinal profile of an outwash plain is usually slightly concave and the cross-section can be very irregular; height differences of several metres can be observed. Strong accumulation in major floods, especially around the main river channel, eventually results in frequent radical changes to the channel course. The recent sedimentation of outwash occurs in the form of channel fill, deposition of sediment sheets and levees. In outwash on Baffin Island and Iceland, only a rudimentary stratification is often observed. Where layers are visible, horizontal stratification prevails. The long axes of elongate clasts are predominantly orientated parallel to flow direction. In contrast to recent outwash in Iceland or northern Canada, the glacial outwash deposits in northern Germany (and North America) are mostly cross-bedded.

Weiss (1958) and Hölting (1958) have shown that in outwash sediments the grain size decreases with increasing distance from the ice edge, while the sorting simultaneously increases. The particle-size distribution of meltwater sands often shows more than one maximum. This is probably due to the interaction of two modes of transportation: traction and saltation. In traction the sediment is dragged along the bottom of the water body, while in saltation the

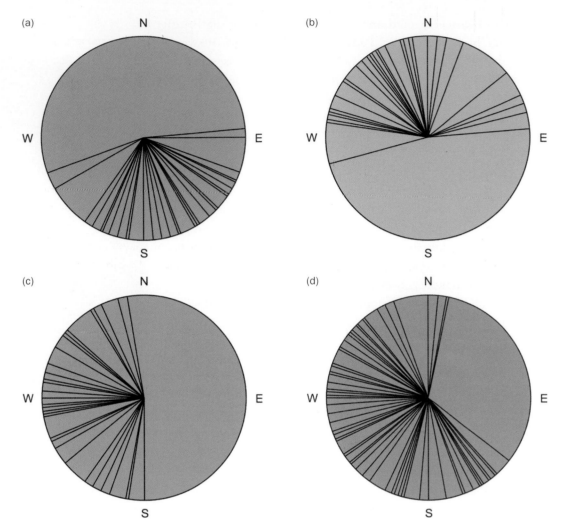

Figure 5.13 Flow directions (a) on the Skeiðará-Sandur, Iceland, and (b–d) in the Harburger Berge hills near Hamburg. The wide range of drainage directions is typical of a braided river system.

grains move by sudden jumps. Evidence of the presence of ground frost during outwash formation occur, but rarely. Syngenetic ice wedges or occasionally (more frequent) sand clasts are found within the sedimentary series.

While the drainage of the Weichselian ice margin can usually be reconstructed relatively easily based on the morphological setting, this does not apply to the older glaciations where only an analysis of the sediments can shed light on the former drainage patterns. The meltwater sands of northern Germany are mostly slightly gravelly, mixed-grained sands. The cross-bedding dips in the direction in which the water has flowed (accretion on the leeward side). It is therefore sufficient to measure the direction of dip in a number of such cross-bedding bodies to reconstruct the direction of water flow. With their strong sediment content and changing water levels, meltwater rivers are not closely tied to narrow drainage channels. In a braided river system like those on Iceland, a wide variation of runoff directions naturally occurs (Fig. 5.13). The directions within a single outwash complex scatter characteristically over a range of more than 180°. Nevertheless, the mean flow direction can be reconstructed by means of a large number of measurements (experience has shown that around 50 measurements are required).

Cross-bedding measurements are regularly used in the palaeogeographic reconstruction of the glacial river systems in northern Germany. Illies used this method first to reconstruct the Pleistocene history of the Lower Elbe river region (Illies 1952a). The results of his measurements show that the meltwater streams of the Saalian were not directed towards the recent Elbe river valley.

With the limited number of available exposures, it is often difficult to make a reliable stratigraphical assignment of the meltwater sands; the first classification of the sand body often has a provisional character. New exposures will later complete the picture, sometimes requiring a reassessment of earlier results. Each additional series of measurements contributes to a better understanding of the drainage history. An example is the study of the Saalian drainage in the Hamburg area; the drainage of the Middle Saalian Glaciation initially gave a confusing picture. Comparing the measurements of Grube (1967) with those of Ehlers (1978), the drainage on both sides of the Elbe River is directed away from the Elbe. The elements of such a drainage system could not have been in operation simultaneously. The deposits north and south of the Elbe are found at a different level; the sands north of the Elbe are located at a level of 0–20 m, while the sands found south of the Elbe lie at altitudes of >25 m. The drainage of the Middle Saalian Glaciation was interpreted by Ehlers (1990) as follows (Fig. 5.14). When the ice from the northeast advanced into the Hamburg area, the drainage of the ice margin was first directed through the valley northwest of Hamburg towards the Pinnau River. When this drainage path was blocked by the advancing ice, the meltwater had to find a new path to the southwest towards the Weser-Aller *urstromtal*. That was when the higher-lying sands south of the Elbe were deposited.

The age of outwash is not always easy to determine, even if the deposits are located at the current ground surface. In the early period of Quaternary research in northern Germany, sandur plains were generally regarded as related to the nearest terminal moraine; today it is known that this is not always the case. In a number of cases it could be shown that sandurs were subsequently overridden by the ice. This is the case for the thick meltwater deposits on the western edge of the Harburger Berge hills (Fig. 5.13), but also for some large outwash plains of the Lüneburg Heath. Even where no overlying till is detected, the age assignment is not always clear.

Regarding the stratigraphic classification of glaciofluvial deposits, the same petrographic methods which are used in till stratigraphy can be applied. However, the composition of the meltwater sands and associated tills is not completely consistent. The meltwater sands contain a significantly higher proportion of local materials. In the Alpine area, petrographic studies can be used to distinguish between different gravel bodies from different source areas (e.g. crystalline versus limestone).

The outwash accumulations at the margins of the Nordic glaciations correspond to the gravel trains of the Alpine Glaciation. In the area of Alpine Glaciation, the thickest meltwater gravels are always found below the corresponding till. Corresponding gravel beds of the last ice age, which are overlain by Würmian till, have been described from the Inn, Enns and Gail valleys. These meltwater gravels continue beyond the respective glaciation areas in the gravel terraces of the proglacial area. The tributaries from non-glaciated catchments only had enough debris to contribute significantly to the accumulation process towards the end of the accumulation phase. Studies on the Enns Valley and in the Vienna region show that, in each case, only the upper parts of the terrace gravels bear a significant admixture of local material (van Husen 2004).

(a)

(b)

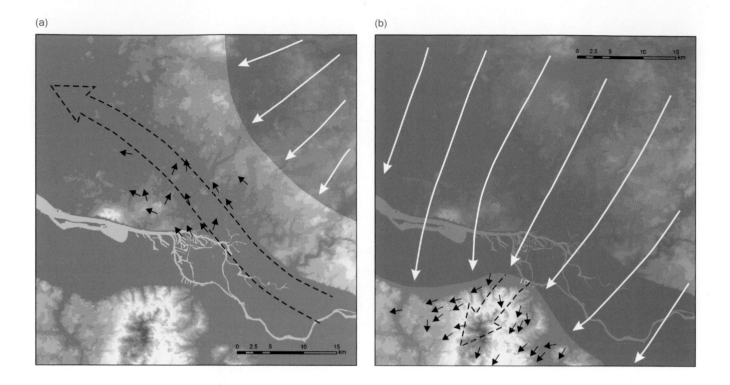

(c)

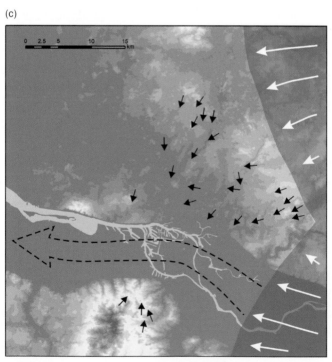

Figure 5.14 Drainage directions of the Saalian meltwater streams in the Hamburg area. (a) Middle Saalian Glaciation (older phase): The ice is coming from the NE and drainage to the NW is still open. (b) Middle Saalian Glaciation (younger phase): the ice margin has advanced and the runoff is now directed to the SW towards the Weser. (c) Younger Saalian Gaciation: the ice is coming from the east and the drainage through the Elbe Valley is free.

While the meltwater sands in northern Germany are completely unconsolidated (apart from local cementation by iron or carbonate), calcareous gravel in the Alpine foothills of the so-called Nagelfluh conglomerates are often consolidated. The degree of cementation is higher in older deposits, but the increase is not linear. While Würmian gravels are largely unconsolidated, Nagelfluh is found in many layers of Rissian gravel. Older gravels (especially when covered by till or on valley slope) are often cemented to concrete-like Nagelfluh (Schreiner 1997).

5.3 Ice-dammed Lakes

The ice of the major glaciations blocked the drainage of many rivers, so that ice-dammed lakes were formed in front of the ice sheets during both the advance and retreat phases. At the end of the last glaciations in North America, Lake Agassiz covered parts of North Dakota, Minnesota, Manitoba, Saskatchewan and Ontario (Fig. 5.15). It reached a size of 440,000 km^2 and was larger than any lake existing today (the Caspian Sea has an area of 386,400 km^2). The deposits of this relatively shallow lake were scarred by drifting icebergs; these scratches are clearly visible on aerial photographs and satellite imagery (Fig. 5.15).

Sandy-gravelly deltas accumulated where meltwater streams entered the lakes; clay and silt were however deposited in the basin interior. The delta foresets dip at an angle of approximately 30° into the lake, covered by thin, subhorizontal topsets (see Box 2.3). The surface of the delta lies at about the former water level.

The still-water sediments in the basin interior often have a fine rhythmic layering and are referred to rhythmites. Annual layers are known as varves (Box 5.1). The bright summer layers are mainly composed of inorganic material and the dark winter layers are characterized by higher organic content. In contrast to the gradual transition from summer to winter layers, the transition from the winter to summer layer is characterized by a sharp boundary. This is caused by the abrupt change in sedimentation conditions when the snow melts and the ice cover breaks open. The 'advance varved clays' of the Leipzig area are found at the base of each till unit. However, they are extremely thin and each comprise only a few years, indicating rapid ice advances.

Jung (1998) examined the varved clays ('bändertone') of the Leipzig area in Germany. For example, the Dehlitz-Leipzig varved clay at the base of the first Elsterian ice advance into the Leipzig lowlands only reaches a thickness of c. 2 m and includes a maximum of 88 varves. It must be assumed that some of the varves were eroded by the glacier. The advance rate of the glacier must have been high, but probably <110–120 m a^{-1} (Grahmann 1925). The ice lake where this varved clay was deposited covered an area of about 750 km^2 (Eissmann 1975). Even when the ice of the Saalian Glaciation blocked the drainage of the Weser and Leine rivers, the local lakes apparently existed for only a few decades (Gassert 1975). Recent studies of the ice-dammed lakes on the edge of the uplands were conducted by Winsemann et al. (2003, 2007). There are only relatively sketchy ideas on the extent and possible overflow of ice-dammed lakes from Lower Saxony and North Rhine-Westphalia, however. A drainage between ice edge and upland margins to the west might be the most likely possibility (Klostermann 1992).

(a)

(b)

Figure 5.15 Deposits of Lake Agassiz, an ice-dammed lake in North America. (a) False-colour satellite image; the stripes are giant striae produced by icebergs.

Source: US Geological Survey, LandSat 7 satellite image of 11 June 2002. (b) Location map.

BOX 5.1 ANNUAL LAYERS IN LAKE SEDIMENTS (VARVES)

A varve is the deposition of a year in a lake. It consists of a light-coloured, coarse summer layer and a dark, fine-grained winter layer. The latter is formed when ice covers the lake in winter, allowing the fine organic particulate matter to be deposited. The Swedish geologist Gerard de Geer is considered the founder of varve chronology, and he provided a complete varve chronology for Sweden as early as 1940. The timescale has since been revised twice, and the age of the so-called 'zero varve' has increased by 118 years. By measuring the annual layers at the mouth of the Ångermanälv, it was possible to extend the timescale up to the present day. The stratification is a result of seasonal fluctuations in flow behaviour. The entire Swedish varve chronology today includes 10,429 annual varves; the margin of error is estimated to be +35/−205 years (Strömberg 1989). Varves can form in clastic sediments, but also in biogenic sediments (muds) or evaporites.

For the establishment of a varve chronology, a number of measurements in outcrops or core drillings is required. The significance of the results depends on the frequency of characteristic layer sequences. Correlations over a distance of more than 10 km are generally problematic. The method of sampling and analysis are described in Strömberg (1983). Annual layers can only be preserved where the sedimentary layers are not disturbed by activities of a soil fauna. With their water temperatures of around zero degrees, ice-dammed lakes provide favourable preservation conditions. The best varves form in fresh water; the flocculation of clay particles in salt water leads to a blurring of the annual layers, which makes measurements impossible in extreme cases. The annual varves can, under favourable conditions, be subdivided by fine so-called 'daily varves', the individual runoff events reflected in the annual cycle. Where thick daily varves occur, they may easily be confused with the annual layering. Varve measurements are best carried out in outcrops. Outcrops are scarce however and in most cases drilling is required. Layers can be repeated in the cores due to landslides. Such errors can only be detected by using multiple boreholes a short distance from each other.

Extensive ice-dammed lakes formed in North America at the end of the last glaciation. In the area of the present-day Great Lakes, Lake Agassiz reached a maximum size of 440,000 km^2 (Teller 1985). Similar dimensions reached the Baltic Ice Lake at the end of the Weichselian Glaciation. The early successes of varve measurements in Sweden, Finland, North America and Argentina encouraged Earth scientists to attempt risky long-distance correlations. When they failed, the whole method fell into disrepute. When radiocarbon dating evolved after 1949, it was initially thought to be an easier and better way of dating. Today however we know that ^{14}C dating is not only less accurate, but also fails in certain periods due to different atmospheric carbon concentrations (for example in the Younger Dryas); these 'blind spots' can be overcome

(continued)

BOX 5.1 ANNUAL LAYERS IN LAKE SEDIMENTS (VARVES) *(CONTINUED)*

by studying the annual layers in lake deposits, such as the Eifel maars (Zolitschka et al. 2000; Figs 5.16, 5.17).

Varve measurements have also been carried out in the north German lowlands. There are not only abundant meltwater sands, but also lake deposits which formed after the melting of the Weichselian ice. Rhythmites are found in the bottom sediments of the Berlin lakes, for example (Pachur & Röper 1987). In a similar position Cimiotti (1983) could count about 1000 varves in the late glacial varved clays at Oldesloe (Schleswig-Holstein). Due to the large distances between sites these results cannot be correlated, and a standard varve chronology for north Germany is therefore still missing.

Interesting results have come from studies of the annual layers of diatomites from the last and penultimate interglacial periods in the Lüneburg Heath area. By counting the summer and winter layers, the approximate duration of the interglacials could be determined.

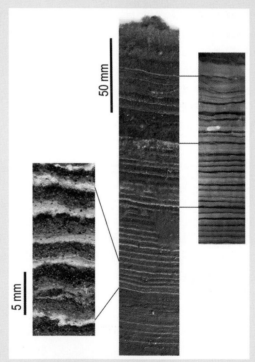

Figure 5.16 Seasonally layered lake sediments from Lake Sacrow near Potsdam, Germany. Centre: photograph of the carbonate-organic varves of a frozen core. The light layers represent summer calcite precipitate. Right: radiograph of the same frozen core. The calcite precipitates are the clearly visible dark layers. Left: micrograph of varves from Lake Sacrow under polarized light, revealing the internal structure of the annual layers. In most cases three sublayers are recognizable: most notably is the white summer calcite layer. It is overlain by the grey detritus layer of autumn/winter. Below the calcite layer (and not always well developed) is a relatively dark spring layer of diatoms. Photographs by Bernd Zolitschka.

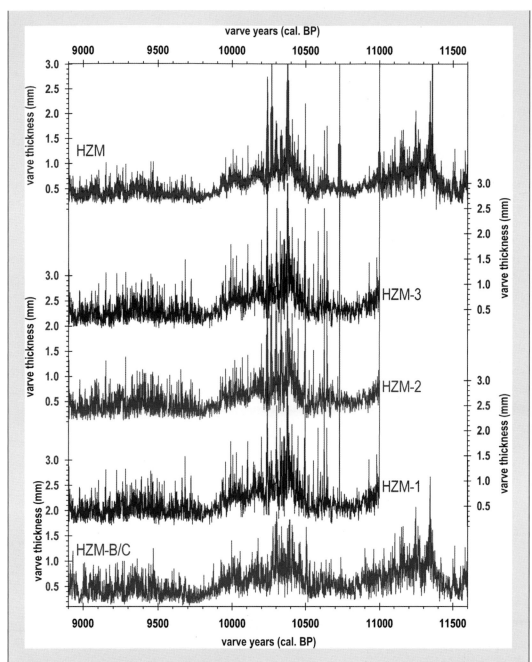

Figure 5.17 Varve thickness measurements of four parallel sedimentary sequences from Holzmaar lake, Westeifel volcanic field. Counts and thickness measurements by B. Zolitschka (HTM-B/C) and B. Rein (HTM-1, HTM 2, HTM-3) were performed on thin sections of the early Holocene varve sequence. The average of the four determinations is shown in the upper graph (red) together with a moving average over seven points (blue). These curves show the Holzmaar system's response to the early Holocene Preboreal (11,400–11,000 cal BP) and Boreal (10,500–10,200 cal BP) climatic oscillations. Both climate fluctuations resulted in thicker varves, caused by increased minerogenic sediment entry.

Source: Bernd Zolitschka.

Many more ice-dammed lakes existed in the decay phase of the Nordic ice caps. In the vast system of lakes in northern Germany at the end of the Elsterian Glaciation, basin sediments some hundreds of metres thick were deposited (Fig. 5.18). In Sweden, the late-glacial varved clays of the Weichselian Glaciation allowed a complete reconstruction of the chronological sequence since the area became free of ice. The calculated glacial limits run perpendicular to the orientation of the most recent glacial striae. Typical retreat rates are on the order of 200–300 m a^{-1} (maximum 1000 m a^{-1} and minimum 50 m a^{-1}; Fig. 5.19).

Due to the varying speed and glacial progress in the Alps, a wide variety of glacial damming of valleys and the formation of large ice-dammed lakes occurred. These were most common in the eastern Alps. In general, a sequence of coarse, local material at the base is overlain by fine-grained sediments. In some cases, particularly thick '*bänderschluffe*' (lacustrine silts) were deposited (van Husen 1977, 1980). Even during the ice decay phases, ice-dammed lakes formed at different points in the Alpine valleys. The high sediment supply resulted in these basins usually filling up very quickly. Where the water level remained constant over time, some large deltas formed. A striking example is the terrace at St Jakob in southern Carinthia (Penck & Brückner 1901–09; van Husen 1981).

Rhythmites were also deposited in the lakes at the Alpine margin. Kelts (1978) found a distinct annual lamination in Lake Zurich in which bright summer layers were caused by the chemical sedimentation in the summer months, while the deposition of organic detritus in the winter months resulted in dark layers. However, not all rhythmic layers are classified as varves. For example, Lake Walen has a rhythmic layering which is not due to the seasonal changes in deposition; the sedimentation in Lake Walen changed fundamentally in 1811 when the Linth river was forced to flow into the lake, doubling the catchment area of the lake. This event is easily identifiable in the sedimentary sequence. Studies by Lambert & Hsü (1979) have shown that an average of two 'varves' per year was deposited from that date. These are due to turbidity

Figure 5.18 Grimsmoen, a delta of a Weichselian ice-dammed lake in Folldal, Norway. Photograph by Jürgen Ehlers.

Figure 5.19 Deposits of ice-dammed lakes: varves overlying Saalian till in Neumark-Nord opencast mine (left); varved silt in Sweden (right). Photographs by Jürgen Ehlers.

currents associated with flood events of the River Linth that occur twice a year on average. The sediments of Lake Geneva also have varves; detailed studies were published by Moscariello (1996). The Laacher See tephra was found in the lake sediments.

5.4 Kames: Deposits at the Ice Margin

Kames are meltwater sediments deposited in the ice or against the ice margin. In shaping the kame landforms, the ice played a passive role (Fig. 5.20). Within a glacier, sediment can be concentrated at various points. Where meltwater falls down a moulin for example, debris is accumulated at the glacier sole. Possible starting points for sediment accumulation are also lakes and ponds, upward directed tunnels or debris-rich bands of ice which are crossed by a meltwater course. Above all, during the melting of the glaciers and ice sheets large amounts of sediment and water are released and redistributed. The resulting landforms vary in size and shape; they can be conical, flattened, elongated or completely irregularly shaped features that occur singly or in groups (Gray 1991).

The relief of the island of Langeland (Denmark) is characterized by a large number of prominent hills which range in height from a few metres to 37 m, have a round or oval outline and a diameter of *c.* 50–300 m. These are known in Danish as *hatbakker* (hat-shaped hills; Fig. 5.21). Smed (1962) mapped about 1200 of these features on Langeland alone. He points out that they consist mainly of laminated material which is often dislocated, sometimes

Figure 5.20 Moulin kame on Langeland, Denmark. The gravel hill was accumulated in a glacier mill during the Weichselian Glaciation. Photograph by Jürgen Ehlers.

Figure 5.21 Two examples of *hatbakker*, the so-called 'hat-shaped hills' on Langeland. These are kames which were created in the decay phase of the Weichselian ice age. Photograph by Jürgen Ehlers.

dramatically. The layers may be vertical, and the strike direction is parallel to the orientation of the whole hill. Similar *hatbakker* have been observed elsewhere in Denmark, but not at this concentration. In the literature, the *hatbakker* are regarded as dislocated kames. It is thought that after their formation they were disturbed by glacial ice. Milthers (1959) assumed that they had been pushed by ice from an ESE direction.

In order to determine the origin of those mounds in the field, Jürgen Ehlers and Grahame Larson have mapped the available outcrops. Several of the *hatbakker* are seen in cross-section in the cliffs on Langeland (Fig. 5.22). The exposures were mapped and all the other cliffs on Langeland were inspected.

(a)   (c)

(b)   Bagenkopsbjerg

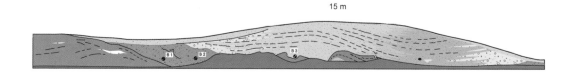

Dovns Klint

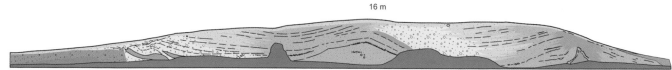

Figure 5.22 (a) Cliff sections through *hatbakker* on the Danish island of Langeland; (b) cross-sections through the Dovns Klint and Bagenkopsbjerg *hatbakker*; and (c) photograph from Bagenkopsbjerg. (a, c) Photographs by Jürgen Ehlers.

There is one real push moraine on Langeland exposed in the cliff section at Ristinge Klint (see Box 4.4). However, the *hatbakker* differ fundamentally from the moraine. While predominantly fine-grained sand and clay layers alternate with tills and marine deposits in the thrust moraine, in the cores of the *hatbakker* almost exclusively coarse-grained sand and gravel are found. The rest of the cliff outcrops on Langeland show mostly undisturbed till, overlying glaciolacustrine silt in places.

The exposures studied in the southern part of Langeland have shown that the *hatbakker* are real kames. The only disturbances that have been found resulted either from loading or sagging during dead-ice decay. It appears that the kames were formed during the decay phase of the Weichselian in glacier mills (moulins) and crevasses; they were probably not overridden by active ice. The sediments represent basin deposits that have been poured in from a westerly direction. Many of the south Langeland *hatbakker* lie on a till plateau, which rises a few metres above its surroundings. The cliff outcrops show that this plateau is composed of till from the Young Baltic ice advance, superimposed on glaciolacustrine silt.

In the northern part of Langeland, instead of *hatbakker*, plateaux are found in various places that rise a few metres above their surroundings. In these forms, there are no exposures. Their proximity to the moulin kames and their similarity to comparable landforms in kames landscapes of Latvia and North America (Wisconsin) suggest that they are plateau kames. The internal structure of the plateaux are similar to the *hatbakker* kames; the only difference is that they occupy larger areas and are predominantly composed of lacustrine deposits.

With regard to their form and their internal construction, the *hatbakker* match moulin kames as they have been described from other locations. These landforms typically occur in the interstices between two ice lobes. In the case of Langeland, the area of kame formation was between the stagnant Baltic ice and the new ice stream which progressed northwards during the Bælthav Advance between the islands of Langeland and Lolland.

While *hatbakker* are irregularly distributed in the southernmost part of Langeland, further north they are arranged in lines which run mostly parallel to the coast in a NNE–SSW direction.

Kame terraces are bodies of meltwater sediment that were deposited between the glacier and adjacent valley slopes. The term is accredited to Salisbury (1894, 1902) who wrote extensively about glacial sediments and landforms in North America. Most kame terraces were formed by glaciofluvial processes; glaciolacustrine forms also occur. Kame terraces occur primarily at the margin of valley glaciers and are less frequent at the margins of major ice sheets. Very good examples of valley kame terraces are found at Loch Etive (Fig. 5.23) and Loch Etteridge in Scotland (Gray 1991). In the Alps, kames are known primarily from the morainic amphitheatres of the Southern Alps. In the northern foothills of the Alps they are found in the Iller glacier area (Habbe 1986a, b) and in the Traun glacier area (van Husen 1977). The terraced gravel plains of the western Rhine glacier area are also basically kame terraces.

In northern Germany kames are certainly more widespread than has been previously assumed (Figs 5.24). The largest documented case (11 km long, several hundred metres wide) is located south of Kiel, between Einfeld and Blumenthal (Stephan et al. 1983). A spectacular kame is Kryižu kalnas (Hill of the Crosses) at Šiauliai in Lithuania (Fig. 5.25).

A thick kame terrace is located on the southern edge of the Teutoburg Forest. This is a sand body *c.* 300–400 m wide, up to 15 m thick and *c.* 15 km long, the stratification dipping slightly to the south. The feature ends at the Münsterländer Bucht with a steep edge.

(a)

(b)

Figure 5.23 Kame terrace at Loch Etive, Scotland: (a) kame terrace surface with dead-ice hollow; and (b) the kame terrace deposits. Photographs by Jürgen Ehlers.

Figure 5.24 Kame at Groß Zecher, Schaalsee Lake, Schleswig-Holstein. Photograph by Jürgen Ehlers.

Figure 5.25 Kryižu kalnas (Hill of the Crosses) at Šiauliai in Lithuania is also a kame. Photograph by Jürgen Ehlers.

Where glaciers blocked drainage, ice-dammed lakes were formed. This was the case in many places, not only in mountain areas but wherever the ice advanced into rising terrain. All the rivers were dammed in northern Germany when the ice sheets advanced towards the margin of the uplands. In the Appalachians in North America for example, whole series of ice-dammed lakes were formed. An even larger ice-dammed lake was produced when the Cordilleran Ice Sheet in Washington advanced into the northern part of the Columbia Basin and blocked the drainage of the Columbia River, creating ice-dammed lakes Columbia and Missoula. In the Late Weichselian there were at least six periods when the tremendous water masses were released, leading to the formation of the Channeled Scablands (Fig. 5.26), a melt-water-eroded landscape of approximately 150 × 200 km (Smith 1993).

When American geologist J. Harland Bretz discovered the traces of this Missoula Flood, he fared no better than other explorers before him: he was not believed. Since he expected this

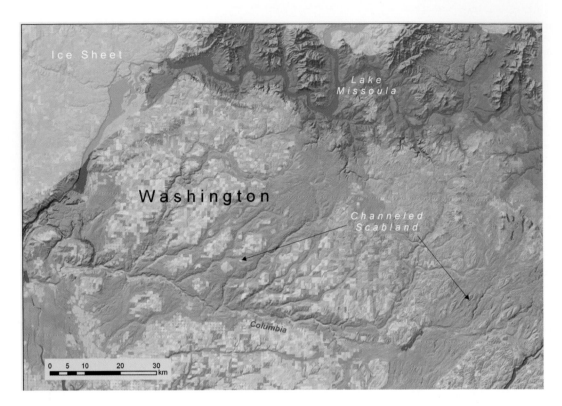

Washington

Ice Sheet

Lake Missoula

Channeled Scabland

Columbia

0 5 10   20   30
                     km

Figure 5.26 The traces of the Missoula Flood are clearly evident in the Channeled Scablands in Washington State.

reaction, in the first essay in which he described the landforms of the Channeled Scablands (Bretz 1923), he carefully avoided any reference to the catastrophe that must have formed them. Later he became more specific, but still only earned disapproval. It appeared far too fantastic that such huge floods could ever have occurred. It was not until decades later that other geologists were gradually convinced, providing a plausible explanation for the disaster: it had been caused by the drainage of an ice-dammed lake (Lake Missoula).

Catastrophic floods occurred during the Quaternary in several places (Baker 2008). The giant boulder in Idaho (Fig. 5.27), observed by Ed Evenson on a field trip in 1979, also dates from the outbreak of a large ice-dammed lake (Glacial Lake East Fork) which the Wild Horse Glacier had dammed. It is not the only block that was relocated by that flood, but the largest. The volume of the ice-dammed lake is estimated at 1.3 km^3, and the outflow reached probably a maximum of about 30,000 m^3 s^{-1} (Barton 2008).

The 'Altai Flood' affected the upper reaches of the Yenisei river. In the western Altai (Sajan), glaciers blocked the drainage of the intramontane Kuray and Chuya basins. This resulted in the formation of giant ice-dammed lakes, the outbreak of which caused the largest known glacial mega-flood (Baker et al. 1993; Grosswald & Rudoy 1996; Rudoy 2002). In the Darkha-dyn Khotgor basin, the largest in the upper reaches of the Yenisei, there was a palaeolake with a capacity of 373 km^3, or about one-sixth of the size of Lake Missoula (Komatsu et al. 2008). Gravel dunes downstream testify to the power of the water released (Fig. 5.28).

According to Shaw (2010), the creation of streamlined landforms in major parts of Canada goes back to an even larger flood. The majority of scientists are sceptical, however. Instead, they

Figure 5.27 Erratic block beyond the limits of glaciation in Idaho, the result of a meltwater flood from the outbreak of an ice-dammed lake. Photograph by Jürgen Ehlers.

Figure 5.28 Gravel dunes in the Todza Basin, Altai, which are traces of a catastrophic outflow of an ice-dammed lake. Note the car for scale. Photograph by Keenan Lee.

prefer to interpret those features in Canada's Prairie provinces as being formed by fast-flowing ice streams (Evans 2010).

5.5 Urstromtäler

Heinrich Girard, Professor of Mineralogy at the University of Halle, first pointed out that north Germany was crossed by a series of wide valleys that can be traced over long distances, but which today lack any major river (Girard 1855). He distinguished three such valleys. Berendt (1879) saw those features, which he referred to as 'main valleys' in the context of the Pleistocene ice sheets. In addition to the three described by Girard, Berendt added another more southerly valley which could be traced from Wroclaw to Hannover. He realized that the valleys could not have formed simultaneously, but must have been active one after the other. Keilhack (1898, 1899, 1904) highlighted the close relationship between the main valleys and the associated ice margins. He coined the term 'urstromtal' for those features (1898).

An *urstromtal* is a glacial meltwater stream which runs more or less parallel to the former ice margin. It originated at the current (European) main watershed and, in its time, it drained an entire sector of the continental ice sheet. In the English literature either the term 'glacial valley' or 'ice-marginal valley' is used; in the Slavic literature the word 'pradolina' is used. The occurrence of such glacial valleys is limited to the margins of the northern European ice sheets. In the Alpine foreland, an *urstromtal*-like situation only occurred in the western Rhine river valley. In North America, the main southwards drainage towards the Mississippi remained intact throughout the glaciations. Only the north- and northwest-ward draining rivers in northern Germany and Poland were blocked by the advancing ice and diverted either to the west or to the southeast.

Four major westwards-directed *urstromtäler* were recognized in Germany towards the end of the last century: Breslau(=Wroclaw)–Magdeburg–Bremen; Glogau–Baruth; Warsaw–Berlin; and Thorn(=Toruń)–Eberswalde (Figs 5.29, 5.30).

While the Wroclaw–Magdeburg glacial valley eventually drained via the River Weser, the other three glacial valleys drained through the Lower Elbe. Keilhack (1899) assumed that each of the glacial valleys had to be correlated with one of the main end moraines.

The Lusatian section of the Wroclaw–Magdeburg–Bremen glacial valley (Keilhack 1913) is definitely older than the Weichselian: in the Nochten area, Eemian deposits overlap the Lower Talsandfolge in that valley (Cepek 1995). The overlying Upper Talsandfolge (with embedded Early Weichselian interstadials) is not a glacial valley formation, but a normal river deposit (Wolf & Alexowsky 1994).

Other glacial valleys have occasionally been postulated, such as the Pomeranian *urstromtal* of Keilhack (1899) or the Reda–Łeba *urstromtal* and the Putnica *urstromtal* of Augustowski (1965). Liedtke (1981) has rejected those terms because of the small size of the respective drainage systems. In addition to the westwards-directed features however, the Pilica–Pripyat *urstromtal* should be mentioned because of its drainage to the east towards the Dnieper.

Despite the short duration of their active phase, the ice-marginal valleys were not formed at once but in several steps. Consequently, the eastern parts of the valleys have some well-developed terraces. The glacial valleys today no longer have a consistent gradient; major river

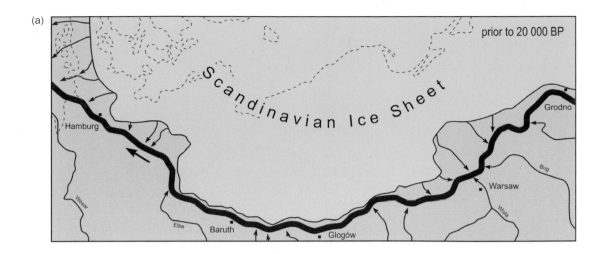

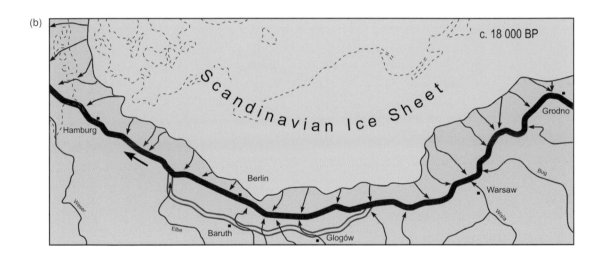

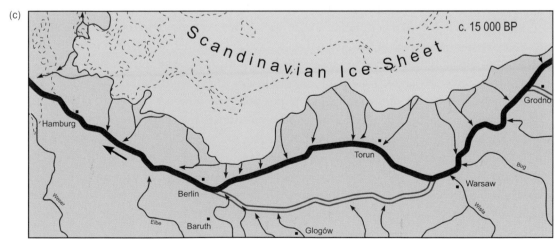

Figure 5.29 Development of the *urstromtal* drainage in the Weichselian Glaciation. (a) Glogau–Baruth *urstromtal*; (b) Warsaw–Berlin *urstromtal*; (c) drainage shifting to the north.

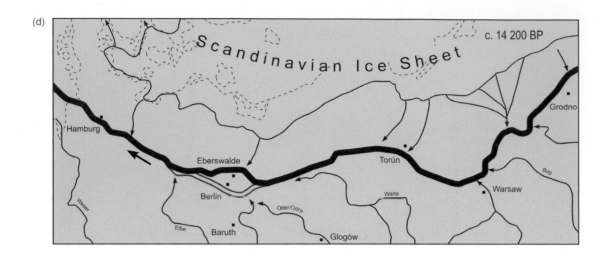

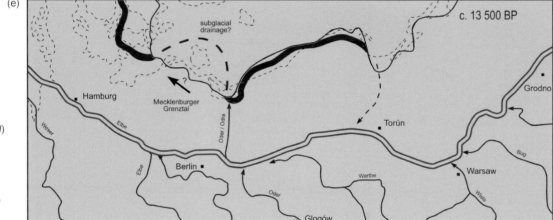

Figure 5.29 (Continued)
(d) Thorn–Eberswalde
urstromtal; and (e)
relocation of the
drainage to the edge of
the present Baltic Sea.

crossings (such as the Oder or the Vistula) have caused gradient reversals. For instance, the Oder east of Eberswalde lies 34–36 m below the level of the *urstromtal* and at Müllrose (south of Frankfurt/Oder) 21–23 m. The Vistula at Fordon lies 30 m deeper than the *urstromtal* valley floor. Detailed studies of the sediment composition have however shown that the glacial valleys have undoubtedly been used by one continuous stream; the gentle slopes (in the case of Thorn–Eberswalde glacial valley, about 1:13,000–1:16,000) were sufficient. The Lower Volga, for instance, has a gradient of only 1:34,000 (Liedtke 1981).

The *urstromtal* valleys are relatively young features. Similar drainage systems must have existed during the pre-Weichselian glaciations, but none have so far been reconstructed. The landforms have been reshaped by subsequent glaciations and additionally blurred by periglacial overprint. Illies (1952b) thought fine sands in the Bremen area were the remains of an Elsterian age *urstrom*. Those sands are, however, widespread distal meltwater deposits, which cannot be interpreted as an *urstrom*. Further, the Hunte-Leda depression is not an *urstromtal*

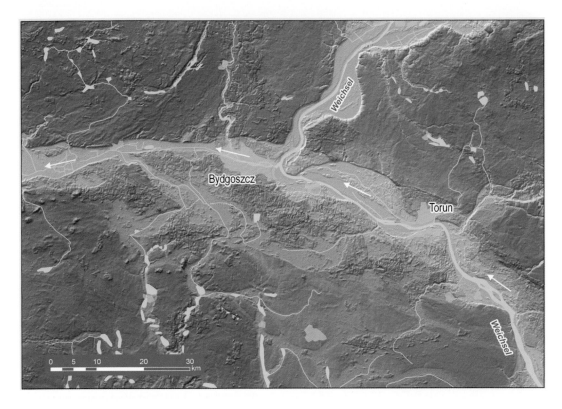

Figure 5.30 Thorn–Eberswalde *urstromtal* (white arrows) branching off from the Vistula (Weichsel) river valley near Bydgoszcz in Poland. The ragged relief on the edge of the glacial valley is caused by dunes.

of the Older Saalian Glaciation (Meyer 1983b). Liedtke (1981) suggested the short duration of the maximum ice advances might have prevented *urstromtal* formation. Varve counts suggest that during the Elsterian Glaciation and the Older Saalian Glaciation maximum ice advances lasted only a few hundred years, during which the meltwater might simply have been dammed up in the valleys of the uplands.

Even the well-known drainage route parallel to the ice margin at the northern edge of the uplands can no longer be detected morphologically. Corresponding sediments suggest, however, that such a drainage formerly existed. From sedimentological studies, we know that the central German rivers (Elbe, Saale and Weser) flowed westwards through the Netherlands towards the North Sea for the first time in the Menapian (Zandstra 1983). In the Saale glacial period, the relationship with the advance of the continental ice sheet is clear. It has been demonstrated that the Middle Terrace of the River Weser (a sediment body of >30 m thickness and several kilometres width) can be traced along the northern edge of the Teutoburg Forest in a westerly direction (e.g. Hinze 1982).

The oldest recorded glacial valley in NW Germany, the Aller-Weser valley, was active during the Middle Saalian Glaciation. To the east it can be traced to the southern edge of the Letzlingen Heath area (north of Haldensleben). Whether a continuous Magdeburg–Breslau–Bremen valley could have formed during the Younger Saalian Glaciation has been questioned by Meyer (1983b). After Liedtke (1981), however, such a valley can be safely traced east beyond Breslau. It is only the further connection to the Warta to the east which is unclear. Liedtke (1981) points out, however, that the outflow of the Warta must have been to the west, because the passes leading to the east are invariably higher. During the Middle Saalian Glaciation the main watershed was between the Warta and Pilica rivers. The Pilica drained to the east via the Pripyat into the Dnieper (Pilica–Pripyat *urstromtal*; Rózycki 1965).

During the Weichselian Glaciation the Lower Elbe valley between the Havel River mouth and the North Sea was constantly available as an ice-marginal drainage path that never had to change its course. Further east, however, four to five different main drainage routes are found, which one after another were occupied by the Weichselian *urstrom*. The main watershed was located further to the east during the Weichselian Glaciation than during the Saalian Glaciation. The beginning of the Glogau–Baruth *urstromtal*, the oldest of the Weichselian glacial valleys, is found in the area of Minsk. Further east, the drainage went into the tributaries of the Dnieper River. The easternmost segments of the glacial valley through which the meltwaters of the Brandenburg Phase flowed are found at an altitude of *c.* 190 m asl (Liedtke 1981).

The Frankfurt ice margin is connected with the Warsaw–Berlin glacial valley. This and the other Weichselian glacial valleys are much shorter than the Glogau–Baruth *urstromtal*, but their headwaters were still in the area between Vilnius (Lithuania) and Molodechno (Belarus). Liedtke thought that during the Pomeranian phase a Thorn–Berlin *urstromtal* first took the drainage. While the drainage path via Netze, Warta and Oderbruch was already open in the east, because of the relatively far southerly position of the ice margin further to the west the drainage was directed through the Buckow gate and into the Berlin glacial valley. During the Weichselian ice decay, drainage went initially via the Thorn–Eberswalde glacial valley that was in function until after the formation of Rosenthal end moraines. According to Liedtke (1981), during the formation of the Velgast end moraines the ice had already melted so far back that the westerly drainage could follow the Mecklenburger Grenztal valley into the Baltic Sea and then through the Belt into the North Sea. This latest *urstrom*, the Netze–Randow *urstrom* of Liedtke, lost its function when the mouth of the Vistula was finally free of ice.

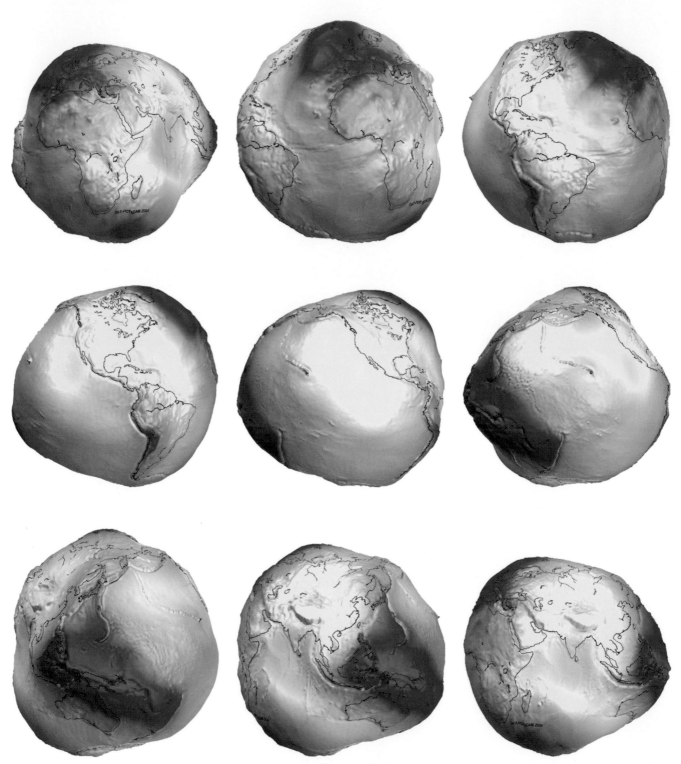

The geoid, the actual shape of the Earth, height exaggerated 15,000 times. The Earth is round, but it is not a ball. With the help of satellite measurements we have a very clear idea of what the shape of its surface actually looks like. (a–i) Globes modelled by GFZ Potsdam illustrate the deviations from the ideal shape.

Source: GFZ German Research Centre. Reproduced with permission.

Chapter 6
Maps: Where Are We?

The traces of the Pleistocene glaciations have been charted for more than a hundred years by geologists and printed on maps. If the scale and coordinate system are given the results are reproducible, a basic requirement of scientific work. However, reproducibility is not everything. Different scales and map projections complicate any comparison. Any differences at any given point can only be gauged when placing one layer of information in the correct position over the other; this is only possible with a digital geographic information system (GIS). Only in this way can the multitude of maps available be organized and compared (Box 6.1).

BOX 6.1 A MAJOR TASK

At the INQUA Congress in Berlin in 1995 at a meeting of the Commission on Glaciation it was decided to set up a working group *Extent and Chronology of Quaternary Glaciations* headed by Jan Mangerud and Peter Clark. A new assessment of the global limits of glaciation seemed overdue. The last such attempt (*The Last Great Ice Sheets* by Denton & Hughes 1981) was almost 25 years old and no longer reflected the current state of knowledge.

Jürgen Ehlers was in Berlin but not at the crucial meeting, so this *Work Group 5* was initiated not only without his participation but also without his knowledge. One year later, however, he was asked if he would be willing, perhaps together with Phil Gibbard, to take over the leadership of the working group. Why not? The last volume of their '*Glacial Deposits*' books had appeared in 1995. This new project was obviously a very big undertaking, but they had just demonstrated that they were able to handle large projects successfully. They discussed the matter and agreed.

(continued)

The Ice Age, First Edition. Jürgen Ehlers, Philip D. Hughes and Philip L. Gibbard.
© 2016 John Wiley & Sons, Ltd. Published 2016 by John Wiley & Sons, Ltd.
Companion website: www.wiley.com/go/ehlers/iceage

BOX 6.1 A MAJOR TASK *(CONTINUED)*

They had no idea what difficulties they would face. They were not simply creating postcard-sized pictures for any textbook, but a map with features which could actually be checked in the field. At least a scale of 1:1,000,000 was required, and it was suggested the already available International Map of the World (also known as World Map) at a scale of 1:1,000,000 be used as a base map.

Phil Gibbard and Jürgen Ehlers met in England, and it seemed most appropriate to use the extensive map collection of the Department of Geography in Cambridge. As they stood in the room housing the map collection and saw the multiple map cabinets with drawer after drawer of the International Map of the World 1:1,000,000, they realized the world was bigger than they had initially thought.

In addition, they found that although the collection included all published maps, the collection was not complete. The International Map of the World (Fig. 6.1) would eventually have to include 2500 map sheets

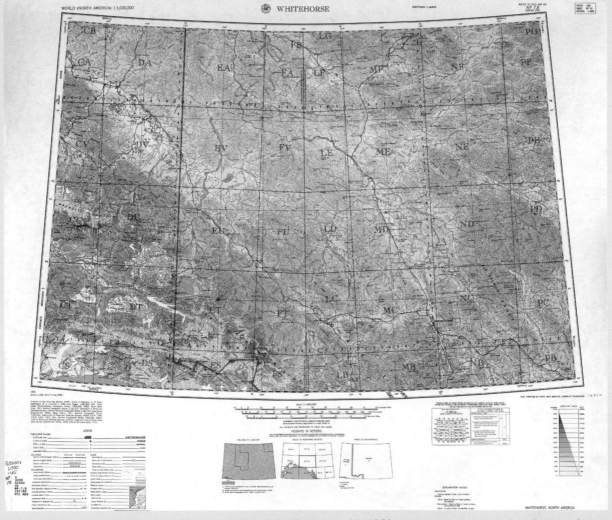

Figure 6.1 Whitehorse Sheet of the International Map of the World printed in 1960, depicting the border region of Alaska–Canada. White spots at the southwestern edge of the map still bear the inscription: *UNSURVEYED*.

Source: Perry-Castañeda Library Map of Collection, University of Texas, Austin. Reproduced with permission.

to cover the whole Earth. Even if the printed maps were limited to the land areas, there were still about 1000 sheets. Overall, however, only about 800 maps had been completed. Although this covered the greater part of the land surface of the Earth and could certainly be useful, the digitization of this stock as a base map for the glacial limits would have exceeded the technical capabilities of the processors as each individual sheet would have to be scanned and georeferenced.

Plan A had not worked; they needed a base map that already existed in digital form. They decided on the Digital Chart of the World (DCW, Fig. 6.2). Since digital maps were not available for all participating colleagues, they had no choice other than plot the maps on paper (DIN A0 size) and post them. The glacial limits and other features of interest were then entered by hand, sent back to Hamburg and digitized manually by Ehlers. The mapping was completed by the INQUA Congress in Reno in 2003, and the three books with CDs containing the map files appeared one year later (Ehlers & Gibbard 2004a–c). An expanded and revised edition was published in 2011 (Ehlers et al. 2011a).

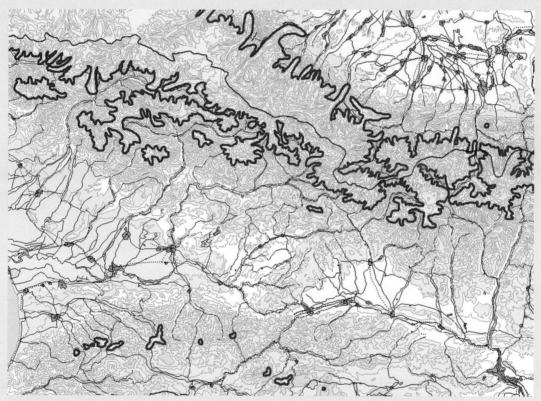

Figure 6.2 Limits of the Weichselian Glaciation in the Caucasus, presented on the DCW base map.

6.1 Digital Maps

The first free vector map of the world is the Digital Chart of the World (DCW) from 1992. This map was originally created by Environmental Systems Research Institute (ESRI) for the US Defense Mapping Agency (DMA). The database essentially consists of the Operational Navigation Charts (ONC) of the DMA. To make the amount of information manageable the Earth's surface was divided into 2094 tiles, most of which have a size of 5 × 5°. The subdivision is not ideal however because most maps and satellite images today use the Universal Transverse Mercator system (UTM), which uses strips of 6° and not 5° width. The original data format (Vector Product Format, VPF) was later converted into ARC/INFO (now ArcInfo). The data were initially sold by ESRI, but today they can be downloaded free of charge and are free of copyright.

The quality of the DCW is determined by the demands of the military, among others. The ONCs are flight navigation maps, where the map content is focused on features that a pilot can recognize at medium or low altitude (500–2000 feet); some of the data layer specifications therefore contain objects that may appear strange to the impartial observer. For example, in the CLPOINT layer for Germany (Cultural Landmark point features), nothing but lighthouses are found. More serious are the deficiencies in the altitudinal representation. The height data of the DCW are in feet, and the elevations are shown in different shades. In some parts of the world, elevation data are missing altogether. The biggest gaps are in South America and Africa, and in some cases there are complete tiles without elevations.

The horizontal accuracy of the map is given as 1600–7300 feet and the vertical accuracy as 160–2100 feet. Those numbers represent the inaccuracies of the ONC, plus the digitization error. The gross height errors do not arise from the fact that somewhere a mountain might be missing, but mainly result from positional errors of the individual objects.

Road signs are not among the features that the low-flying pilot can distinguish, so little care was taken with place names. Not all major cities have a name, and not all places have the right name. For example, the German city of Bochum (379,000 inhabitants) is not listed by name, but the town of Gummersbach (52,000 inhabitants) 60 km to the south is shown. Since the map is old, it still refers to St Petersburg as Leningrad and Chemnitz as Karl-Marx-Stadt. The age of the base maps on which the DCW is based ranges from the mid-1960s to the early 1990s.

Today, higher-resolution digital topographic maps are available. However, VMAP1 (Vector Map) and VMAP2 data are only partially (VMAP1) or not at all (VMAP2) available to the public (Fig. 6.3).

The quality of the elevation data was considerably improved soon after by the terrain model GTOPO30 (Global Topographic map with horizontal grid spacing of 30 arc-seconds). This global digital elevation model was created in 1998 by the USGS and can be downloaded for free from numerous websites. Based on various sources, GTOPO30 has a horizontal resolution of *c.* 1 km and provides an excellent base for overview maps. For most of the world, it is based on Digital Terrain Elevation Data (DTED). While the terrain model appears flawless in global overviews, some artefacts are evident under closer inspection. One of these consists of diagonal stripes, especially visible in Africa and the Middle East. More disturbing in the formerly glaciated areas are noticeable 1 × 1° blocks, inherited from the production of the DTED model, which are most visible in areas with small height differences. Although these defects can be masked by the choice of altitude, they are extremely disruptive for certain analyses (e.g. in the reconstruction of ice-dammed lakes in northern Russia and Siberia; Fig. 6.4).

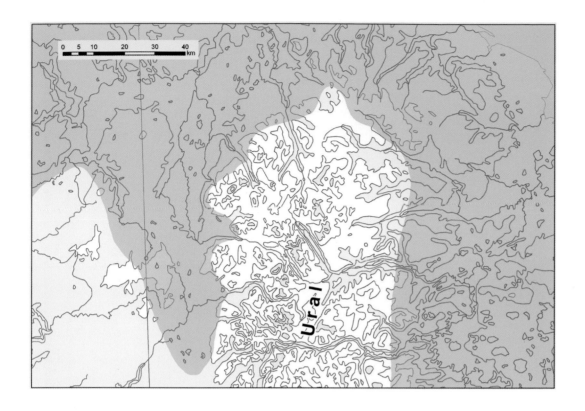

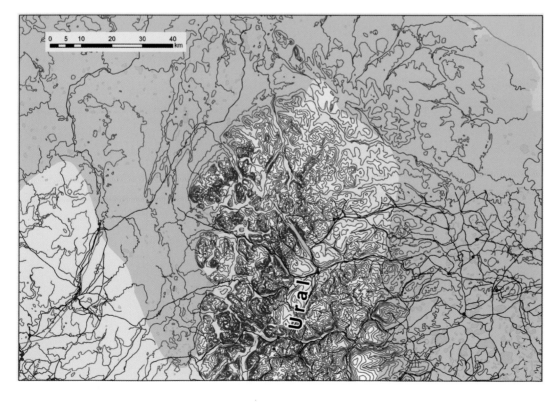

Figure 6.3 Comparison of the DCW (top) with the VMAP1 + GTOPO30 (bottom). The extent of the Early Weichselian Ice Sheet (MIS 5b) in the northern Urals is shown in blue.

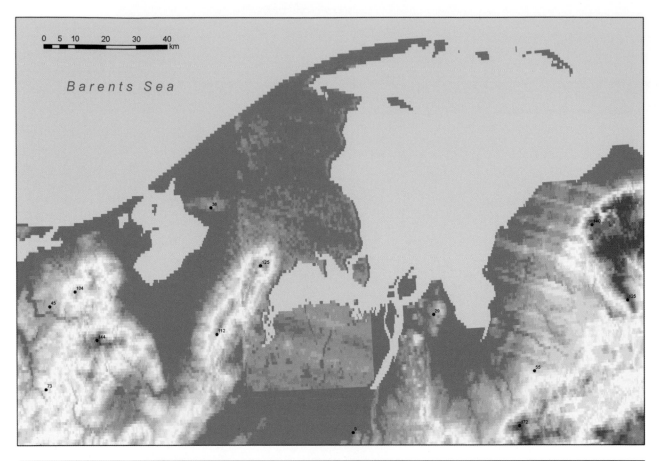

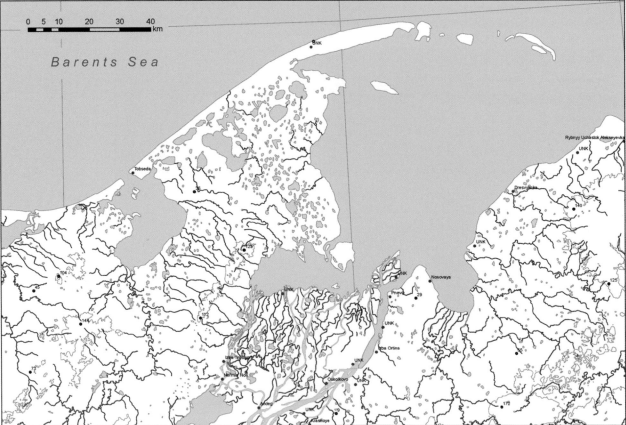

Figure 6.4 Top: Errors in the GTOPO terrain model, an example from Siberia. The altitudes were chosen so that the diagonal stripes and blocks appear most clearly. Bottom: The same area shown in VMAP1.

Once the relief of the land surface was taken care of, underwater topography was next considered. The ETOPO (surface relief of Earth integrated with land topography and ocean bathymetry) terrain model combines the height information of GTOPO data with the coarser information available for water depth (http://www.ngdc.noaa.gov/mgg/fliers/01mgg04.html).

Uniform, higher-resolution (90 m) digital terrain data were then provided by the Shuttle Radar Topography Mission (SRTM) in February 2000, covering all land areas of the Earth between 60° N and 58° S. (The inclination of the orbit of the satellite precludes the collection of data from the polar regions.) Data with a resolution of 30 m are freely available for North American (but not for the rest of the world). The first published data (version 1) included faulty pixels without height information and other measurement errors, and the water surfaces were not shown as level. Improved-quality data were issued in 2005 (Version 2), later revised in 2009 (Version 2.1).

Although the radar altimeter provides very accurate height data, its main disadvantage is that it does not always measure the elevation of the actual land surface as objects such as a house, a tree or a cow may be encountered. Consequently, the boundaries of forests are clearly visible on SRTM images. The other disadvantage of no data from subpolar and polar regions was resolved by turning to other sources. One possibility was to use the data recorded by the Advanced Spaceborne Thermal Emission and Reflection (ASTER) radiometer onboard the Terra satellite from 1999. ASTER produces not only normal satellite images, but in the near infrared spectrum it records both a vertical image and then the same area looking backwards at 27.6°. The ASTER images are therefore suitable for stereoscopic evaluation, with some restrictions. In contrast to the radar data, the ASTER images are influenced by clouds and haze meaning that a large number of repeat measurements may be required for useful information. Theoretically, the resolution of 30 m is better than that of the freely available SRTM datasets. Comparison with SRTM elevation data of the same area shows that the real resolution is lower, however. The ASTER data have been freely available since June 2009 (Fig. 6.5).

It is also possible to download 'real' topographic maps from various websites, but not always free of charge (exceptions are mainly sites in the US and Canada). The available data include georeferenced topographic maps down to a scale of 1:24,000, and their download in most cases is free.

A website providing access to maps produced by the former Soviet Union, all in Russian Cyrillic text, was launched in Belarus in 2003 (Fig. 6.6). The Russian topographic map set at 1:500,000 is almost completely available, and larger scales are available for many areas. For example, the entire Alpine region is covered at a scale of 1:50,000. 'Poehali' was the name of the original project ('Let's go!'), the words of Yuri Gagarin, the first human to fly into space. A total of 59,321 maps is now available from this source (http://mapstor.com/).

The individual sheets of the Russian maps use a grid that is based on degrees latitude and longitude, and they also give the UTM coordinates. However, they cannot be easily fitted to a globally valid GIS. The Russians use a different datum. The maps are not based on the World Geodetic System (WGS84) or European Terrestrial Reference System (ETRS89) but on Pulkovo (the St Petersburg observatory), resulting in deviations in position that cannot be ignored at a scale of 1:200,000. The Russian 6° meridian stripes differ slightly from the 'western' UTM stripes, so that inclusion in a GIS requires a tedious conversion of the maps.

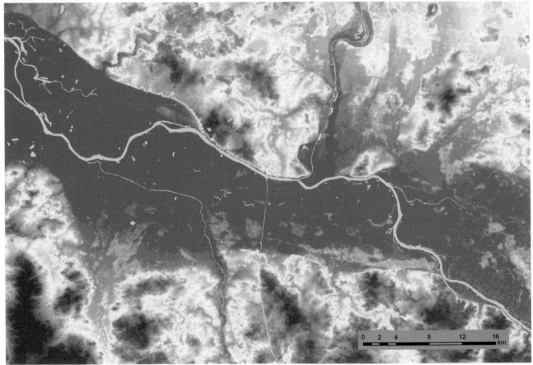

Figure 6.5 Comparison of the ASTER terrain model (top) with the SRTM terrain model (below) for the Elbe river valley near Lauenburg, Germany.

The map in Figure 6.6 shows the northern edge of the Lake Traun area. The Würmian terminal moraine, which can be traced from about Пинсдорф via Клейнрейт to Гшвандт, is clearly visible. Some of the map sheets bear slight traces of use, but most are new. Basic knowledge of Cyrillic script is helpful.

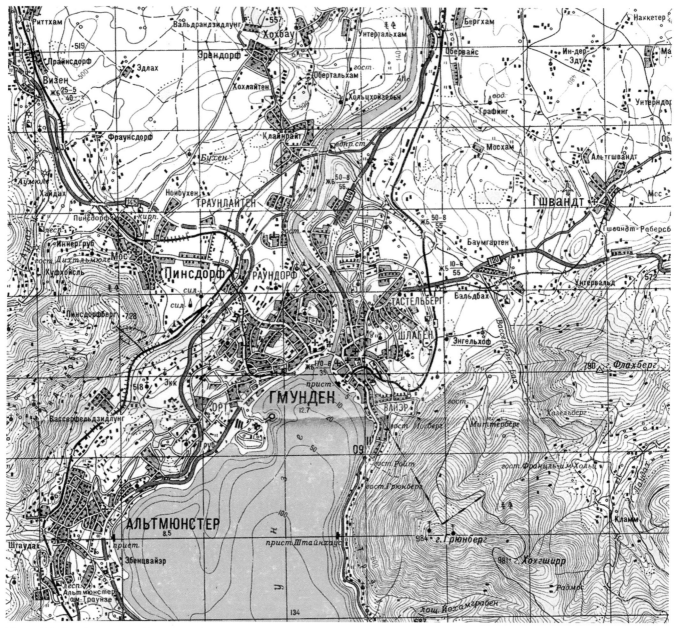

Figure 6.6 Russian topographic map 1:50,000, sheet Gmunden.

Source: Mapstor Poehali.

It should be mentioned that other vector data are also freely available for many parts of the world. A good source is Open Street Map (http://www.openstreetmap.org/). As its name suggests, this is primarily a road map with a few additional information layers (such as lakes, forests, etc.), but without guarantees regarding completeness and accuracy (Ramm & Topf 2008). Google Maps (https://www.google.com/maps) now covers almost the entire globe (with the exception of the polar regions) and provides accurate coverage of both street planning and topography at 20 m contour resolution. Topographic maps within Google Maps can be viewed interchangeably with satellite imagery.

6.2 Satellite Images: Basic Data for Ice-Age Research

Maps will always suffer the problem of aging, and the specified date of publication is not usually the same as the date of data acquisition and processing. Even where the processing status is indicated, there is no guarantee of accuracy. This is especially true for information regarding the land surface. The various editions of the topographic maps of the Isle of Sylt in north Germany at 1:25,000 give the impression that the migrating dunes of the Listland area suddenly ceased moving sometime around 1930. In reality, it was only the cartographers who ceased moving; no changes were recorded for a few decades (Ehlers 2008).

Aerial photographs have been available since the late 1920s, and show – with certain restrictions – the actual situation at the time of recording (Albertz 2009). The use of Earth observation satellites began in 1959 with the launch of the first spy satellite by the United States. The project involved many failures; taking the pictures was technically difficult because the images were recorded on negative film, and it was only after recovery of the films that it could be determined whether the pictures showed anything useful or not. The data recorded by the early spy satellites were released in 1996 and are freely available from NASA (Fig. 6.7; Box 6.2). The

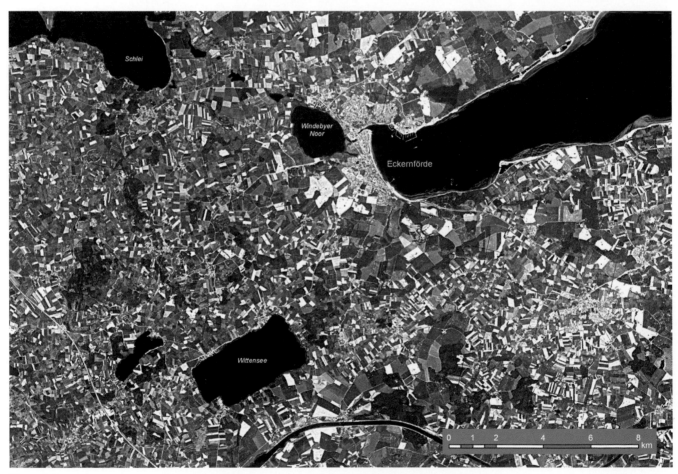

Figure 6.7 Image acquired by an American KH-9 Hexagon spy satellite from the Eckernförde area in 1975.

Source: US Geological Survey image from 6 November 1975, released 2002.

BOX 6.2 FREEDOM OF INFORMATION ACT

The US Freedom of Information Act (FOIA) was approved by President Lyndon B. Johnson on 6 September 1966 and went into effect the following year. This law allows the full or partial disclosure of previously unreleased information and documents of the Government of the USA. The Act defines records that are compulsorily subject to disclosure, outlines the process of disclosure and defines nine exceptions. The ninth exception is 'geological and geophysical information and data, including maps, concerning wells'.

So nothing has changed for geologists? Not quite. More important than the wording of individual sentences is the spirit of this law which states that the findings, which the state acquires by means financed by the citizen, should also belong to the citizen. Satellite images, which were always available but forbiddingly expensive, are now offered free of charge. Even older photographs from spy satellites have become accessible. Maps and aerial photographs in the US are, in almost all cases, also available for free. The download and publication of the American satellite images are also free of charge. This is not the case in countries such as the United Kingdom and Germany where, despite having their own equivalents of the Freedom of Information Acts, maps produced by government agencies are not freely available. However, maps published by the Ordnance Survey (the national mapping agency in the UK) are covered by Crown Copyright; this means that any maps published longer than 50 years ago are freely available to the public.

quality of the early spy satellite imagery was relatively poor, but the images of a later generation (2002 release) have a resolution of at least 0.7–1 m.

Since the launch of Earth Resources Technology Satellite-1 (later named LandSat-1) in 1972, anyone can obtain satellite images. The LandSat images can be downloaded today from the Internet (http://glovis.usgs.gov/). The satellite records data in various spectral bands, and the user puts together the most suitable combination of bands for his or her purpose. For LandSat-7 Enhanced Thematic Mapper (ETM) images, the following specifications apply:

- Band 1: blue visible light of wavelength 0.45–0.515 μm, 30 m ground resolution, shows haze in the air, suspended solids and algae in the water;
- Band 2: green visible light of wavelength 0.525–0.605 μm, 30 m ground resolution;
- Band 3: red visible light of wavelength 0.63–0.690 μm, 30 m ground resolution;
- Band 4: near infrared of wavelength 0.75–0.90 μm, 30 m ground resolution, strongly reflected by chlorophyll;
- Band 5: mid-infrared of wavelength 1.55–1.75 μm, 30 m ground resolution, shows the difference between clouds (dark) and snow (light); reflection increases with increasing iron content of the soil;
- Band 6: long-wavelength infrared/thermal infrared of wavelength 10.4–12 μm; can help to distinguish rocks due to their different thermal (heat) infrared radiating properties; divided into Band 61 (low gain) and Band 62 (high gain);

(a)   (b)

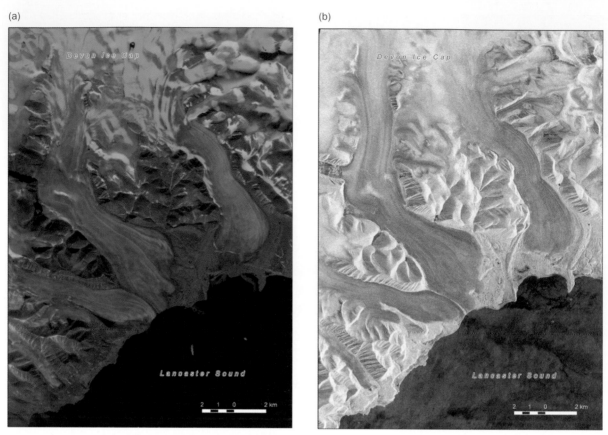

Figure 6.8 Satellite image of part of Devon Island, Canadian Arctic. (a) Landsat-7 ETM image from 2001. Source: US Geological Survey, Landsat 7 ETM, Path 36, Row 7, 20/07/2001. (b) RadarSat-2 image of 2009.

Source: RADARSAT-2 Data and Products © MacDonald, Dettwiler and Associates Ltd. (2010), Radarsat 2 Scene PDS_0103877 on 09/16/2009.

- Band 7 (Fig. 6.8a): mid-infrared of wavelength 2.09–2.35 μm, 30 m ground resolution, shows the moisture content of the surface; and
- Band 8: panchromatic of wavelength 0.52–0.90 μm, 15 m ground resolution; this higher resolution can be used in combination with other bands to obtain imagery up to a scale of *c*. 1:75,000.

Three bands are often combined to so-called 'composites' (RGB colours). If bands 1–3 are used (visible light) and band 1 is set to blue, band 2 to green and band 3 to red, a colour image is created that looks similar to a normal colour photograph. This colour combination is particularly well suited to map coastal morphology. If bands 2–4 are used vegetation appears in various shades of red; on the other hand, water is almost black. The combination of bands 2, 4 and 7 or 3, 4 and 7 allows the moisture content of the land surface to be observed. These combinations are suitable for detecting vegetation under stress, as well as different soil types.

Channels 4, 5 and 7 were used to create the image of Kerguelen from 11 January 2001 (Fig. 6.9). This combination allows glaciers and snow to appear in bright blue, clearly different from the white clouds at the northern and eastern margins of the image. The Îles Kerguelen are a group of about 300 small and one large island, Grand Terre, which is shown in the centre

Figure 6.9 Satellite image of Îles Kerguelen, Landsat 7, channels 4, 5 and 7. A rare recording from the southern summer, showing the islands almost cloudless.

Source: US Geological Survey, Landsat 7 ETM, Path 139, Row 94, 27.11.2001.

of the image. The islands were heavily glaciated during the Ice Age, but probably not completely covered by ice (Hall 2004).

Not all mapping problems can be satisfactorily solved using satellite imagery which is limited to visible or near-infrared light, as many parts of the Earth are under a cloud cover most of the year. For some parts of the Earth, only one or two cloud-free images have been recorded since the launch of the LandSat-3 satellite. Attempts to map the Kerguelen Islands with high-resolution Ikonos satellite imagery resulted in a large number of images showing nothing but the surface of the dense cloud cover.

With radar satellites however, images of excellent quality can be acquired regardless of cloud cover and time of day. When natural disasters strike rapid assessment of the damage is a prerequisite for targeted assistance, utilizing the (not freely available) radar imagery of Envisat. Radar imagery is also useful for recording glacier fluctuations in subpolar areas, such as that recorded by the Canadian RadarSat-2 satellite with its high resolution and three-dimensional view of the terrain (Fig. 6.8b).

Some high-resolution satellite images are – in combination with aerial photos – also shown in Google Earth. The high accuracy of the images with a resolution better than 1:5,000 is impressive. The positional accuracy of the photos is better than 2 m for many regions, with individual trees and bushes visible in some areas. Google Earth has greater versatility in viewing than Google Maps and also allows polar regions to be viewed; the latter is restricted in Google Maps because of the projection. Google Earth is fast becoming the favoured tool of researchers because of its ease of use, versatility and accuracy. Images can also be published as long as certain criteria are adhered to, such as maintaining the Google Icon and not manipulating the imagery.

6.3 Projections and Ellipsoids

Since the Earth is round but a map is flat, the map can never provide a perfect image of the Earth's surface. Near-perfection can only be obtained by setting priorities. Are correct angles needed? Then a conformal Mercator projection should be best. Should the area be correct? A suitable projection might then be the Albers conic projection. Are distances more important? If so, consider a geographic projection. Unfortunately, not every author states on which projection his or her map is based. In Denton & Hughes (1981) for example, provide no details of which projection they used.

What shape of the Earth do we want to assume? The shape of the Earth can be represented in different ways. Either it is considered as a geoid, that is, as an idealized surface of the sea level (without the natural fluctuations of 1–5 m) or as a customized, rotational symmetric ellipsoid.

The cartographers of past centuries, whose job it was to represent their own country as perfectly as possible, have normally adopted a picture of the Earth that came closest to the respective local conditions: that was not the geoid. In Germany, maps are traditionally based on the Bessel ellipsoid of 1841 which adapts well to the mean curvature of the Earth all over Europe and South Asia.

If a geodetical survey is based on this ellipsoid (i.e. when all geodetic measurements are to be projected on it), then the vertical deflections should remain as small as possible. The ellipsoid is therefore positioned in such a manner that the central area of the measurement network resembles the mean curvature of the globe as closely as possible. If two neighbouring states use the same reference ellipsoid, they may still position it slightly different and the resulting grid will be different. The two coordinate systems are similar, but the location of places will vary by a few hundred metres.

The geodetic datum of a map series indicates the zero point and the reference surface, referred to by all location data (coordinates) as well as height and depth values. The geodetic datum is fixed by a reference ellipsoid, which varies from one country to another. For example, in the United Kingdom the mapping grid of the Ordnance Survey is based on the OSGB36 datum, which is fixed to the Airy ellipsoid. In Germany, the geodetic datum and fundamental point is fixed by the Bessel ellipsoid. Internationally, WGS84 is most often used as a reference ellipsoid. The European ETRS89, however, is based on the GRS80. They differ only minimally, for example in the parameters for the flattening of the ellipsoid. For practical applications and geoscientific maps, the difference is negligible. Satellite imagery uses the UTM system.

Photograph of Severnaya Zemlya from the Zeppelin polar expedition in 1931. The nine-lensed camera allowed extreme wide-angle shots.

Chapter 7
Extent of the Glaciers

7.1 Exploring the Arctic by Airship

26th July 1931. The Russian icebreaker *Malygin* had reported bad weather from Franz Josef Land and the Norwegian research vessel *Quest*, which was on its way from Spitsbergen to Franz Josef Land, fought against storm-force winds and ice. It was unfavourable conditions for an Arctic expedition by airship, but LZ 127 *Graf Zeppelin* was on its way north. Its Captain, Dr Hugo Eckener, decided not to change course but to cross the tail of the cyclone.

Returning home was not an option. The airship had to prove its abilities, as it was experiencing strong competition with the improved technical capabilities of aeroplanes. Umberto Nobile's disastrous North Pole expedition (his airship had crashed in 1928 on the Arctic ice) was not forgotten. The airship offered considerable advantages as compared to aircraft in 1931: it could operate at altitudes from near zero to over 1500 m; it could fly as slowly as desired, and even remain standing on the spot if the scientists required; it could remain in the air non-stop for many days; and it could carry a large payload. The polar expedition was attended by 15 scientists and journalists, in addition to the 31 person crew of the airship.

Everything went well. The flight took them from Leningrad to Franz Josef Land (meeting with the icebreaker *Malygin*), then continued via Severnaya Zemlya and the Taimyr Peninsula, via Novaya Zemlya back to Leningrad. In just four days, the expedition members had performed a large range of meteorological and geophysical measurements and surveyed parts of the Arctic islands for the first time with the help of aerial photographs. As well as a Zeiss camera, the nine-lensed panoramic camera of *Photogrammetrie GmbH* (Munich) proved invaluable. All participants of the expedition were full of praise for the technical possibilities. However, it would be the last Arctic expedition by *Zeppelin*.

The Ice Age, First Edition. Jürgen Ehlers, Philip D. Hughes and Philip L. Gibbard.
© 2016 John Wiley & Sons, Ltd. Published 2016 by John Wiley & Sons, Ltd.
Companion website: www.wiley.com/go/ehlers/iceage

The major benefits were offset by the significant costs. Aero Arctic, the International Research Association for the Study of the Arctic with aircraft (chairman Hugo Eckener since the death of Frithjof Nansen), covered a significant portion of the costs of the polar expedition but these were not sufficient. The German *Reichspost* had to help; special stamps were issued and the expedition was financed to a considerable extent by specially delivered air mail. The meeting with the ice-breaker *Malygin* was primarily arranged to exchange bags of letters and postcards. Approximately 50,000 items of mail were transported this way. The Ullstein publishing house in Berlin had bought the exclusive rights to cover the expedition, which brought in more much-needed funds.

Severnaya Zemlya, the 'North Country', had only been discovered in 1913 and at the time of the *Zeppelin* expedition it was only very incompletely known. The Executive Committee of the Communist Party of the Soviet Union (CPSU) in 1926 had abolished the old Czarist names of the archipelago (Nicholas II Land) and the individual islands (St Alexandra, St Olga, St Mary, St Tatiana, St Anna) and had replaced the female saints by more modern terms. The largest islands were now called October Revolution, Bolshevik, Komsomolets and Pioneer Island. Apart from that, little had been done. After the *Zeppelin* expedition, part of the islands could be properly mapped.

At that time, Novaya Zemlya ('New Land') had only been partly explored. It consists of two large islands separated by the narrow Matotschkin Straits. The North Island is the fourth-largest island in Europe and (in contrast to the South Island) largely covered by glaciers. For the first time large parts of the islands could be inspected at a glance from the *Zeppelin* and numerous aerial photographs were taken. The achievements of the expedition were published in a supplement to the prestigious geographical journal *Petermanns Geographische Mitteilungen*.

Today it is possible to map these difficult-to-reach areas with the help of satellite images; the entire Severnaya Zemlya archipelago is covered in a single image (Figs 7.1, 7.2). A comparison of the ice cover of the north island of Novaya Zemlya with the mapping of 1931 reveals that the position of ice margins have remained unchanged over the last 78 years. Because of increased winter precipitation, the glaciers of Novaya Zemlya are so far unaffected by global warming (Zeeberg & Forman 2001; Zeeberg 2002).

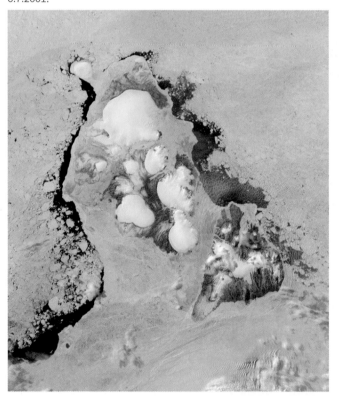

Figure 7.1 Terra satellite image of Severnaya Zemlya. Large parts of the uninhabited islands are covered by ice sheets.

Source: NASA MODIS Rapid Response System, Terra, 6.7.2001.

7.2 Glaciers in the Barents Sea

The fact that the glaciers of the Arctic had a greater extent in the past compared to today was already known to the participants of the 1931 expedition. Professor Samoilovich wrote: 'On this part of our aeronautic journey [Taimyr] everywhere traces of ancient glaciation could be seen. All the hills had a smooth, rounded character, which gave them quite often the appearance of glacier-moulded rock humps. Along the entire route the ground was strewn with erratic blocks' (Berson et al. 1933).

The islands on the edge of the Barents Sea are still glaciated, and there was never any doubt that the glaciers during the Ice Age were more extensive than today. Elevated shorelines on Svalbard testify that the archipelago had subsided under the weight of the ice and uplifted again after the ice melted. The raised beaches seemed to be very old, however. Early radiocarbon dating had yielded an age of over 40,000 years, so it was assumed that during the Weichselian glaciers on Svalbard were not much larger than today.

Two Norwegian scientists, Jan Mangerud from the University of Bergen and Otto Salvigsen from the Norwegian Polarinstitutt, presented the results of more detailed investigations of the area in 1984. They proved that these terraces had been subsequently overrun by glaciers. This meant that the glaciers of the Weichselian Glaciation had not only been more extensive than previously thought but a continuous ice sheet had existed, the centre of which was located somewhere to the east of Spitsbergen.

Figure 7.2 Landsat 7 satellite image of part of the northern island of Novaya Zemlya on 24 June 2002.

Source: US Geological Survey, Landsat 7 ETM, Path 187, Row 7, 24/06/2002.

Russian geographer Mikhail Grosswald had previously suggested that a giant glacier had once covered the whole area of the Barents Sea. A Swedish polar expedition in 1966, in which Grosswald took part, seemed to confirm this view. Since the results were published in English, that idea was more widely accessible and quickly found new followers. When George Denton and Terence Hughes published *The Last Great Ice Sheets*, their synthesis of the extent of the last glaciation (1981), they discussed this possibility. Many of Grosswald's Russian colleagues remained sceptical, however. How could this ice sheet have come about? Was it not more likely that Arctic glaciers from the islands and from the northern Urals had advanced far into the Barents Sea? Grosswald went one step further, however. He assumed that not only the Barents Sea but also the adjacent Kara Sea to the east had been completely glaciated.

It was possible to pursue this issue further after the end of the Cold War. From 1996 to 2002, Jan Mangerud and colleagues from many other countries joined the project *Quaternary Environment of the Eurasian North* (QUEEN) funded by the European Science Foundation, in which detailed field studies were conducted in northern Russia. It was very quickly established that the Barents Sea had in fact been glaciated. However, no significant glaciers had extended from the northern Urals. The Urals are a relatively low mountain range and, while they reach 1894 m at Mount Narodnaya, the northernmost Urals rarely exceed altitudes of 1000 m.

Most of the areas that were covered by the northern European ice sheet are now under water. However, if a large ice sheet had covered both the Barents Sea and the Kara Sea at that time, then it must have left its mark on the islands between the two seas on Novaya Zemlya. There is in fact a series of raised beaches, the highest of which are now over 140 m above sea level. Unfortunately, as a result of Russian nuclear bomb testing, Novaya Zemlya is a place where on-site investigations cannot be carried out without some health risks. Jan Mangerud knew that his Norwegian colleague, geologist Olaf Holtedahl, had led an expedition to Novaya Zemlya in 1921. From the old records it was known that one of the participants, the Quaternary geologist O. T. Grønlie, had not only mapped numerous marine terraces in detail, but had also taken samples of mollusc shells which could possibly be dated. But where were the samples? Mangerud looked in the natural history museums of Oslo and Tromsø, but in vain. The material seemed to have been lost during the Second World War.

However, in December 2003 there was a surprise: the presumed lost samples reappeared in the Museum of the University of Tromsø. Dating of the mollusc shells with the ^{14}C method showed that the archipelago had been free of ice 35–27 ^{14}C ka BP. This meant that the great ice advance, which had led to the isostatic depression of Novaya Zemlya and to the subsequent uplift of the marine terraces, must have occurred during the Early Weichselian Glaciation (MIS 4 and 3; Mangerud et al. 2008).

Much became clear: the Kara Sea had also been glaciated. However, the ice advances from those glaciation centres in the direction of the mainland were much older than Grosswald had thought. During the Weichselian glacial maximum at *c.* 20 ka, the Barents Sea and the Kara Sea ice sheets had not extended onto the Russian mainland. The relatively fresh-looking terminal moraines in the far north are older; they were formed by ice advances at *c.* 40 and 70 ka. The Barents–Kara Ice Sheet reached its maximum extent as early as 90–80 ka (Svendsen et al. 2004), well before the Last Glacial Maximum which occurred at *c.* 23–19 ka (see Section 7.3). Only the northern edge of the Taimyr Peninsula was touched by a Late Weichselian ice advance out of the Kara Sea after 20 ka. On the other hand, mammoth tusks found on Severnaya Zemlya were dated to 25–19 ka, showing that the islands were not completely ice covered during the Weichselian Pleniglacial.

How could the shelf seas ever become glaciated? Elverhøi et al. (2002) gave a simple explanation: during a major glaciation, large amounts of water are bound in the ice sheets and consequently the sea level drops. The shallow shelf sea becomes covered with sea ice, and this floating ice cover gradually comes into contact with the seabed. The ice is fixed, and with additional supply of snow it develops into a glacier. In some ways this represents a special form of 'instantaneous glacierization'. This process may have been facilitated by the initial glaciation of the Norwegian mainland and the islands. Those areas were pressed downwards under the ice load, while the bottom of the Barents Sea was slightly raised by isostatic adjustment.

Based on the available data, we know that at the end of the Weichselian the ice sheet disintegrated very quickly. Rapid decay started from the floating ice margins by calving into the rising sea.

7.3 Isostasy and Eustasy

Charles Lyell selected the Serapis Temple of Pozzuoli near Naples as a frontispiece for his magnum opus, the *Principles of Geology* (Fig. 7.3). In fact, these pillars were extremely well suited to illustrate the main concern of the author: the Earth was not shaped by violent catastrophes, but by slow-acting processes that continue to this very day. The columns (they are not in fact a

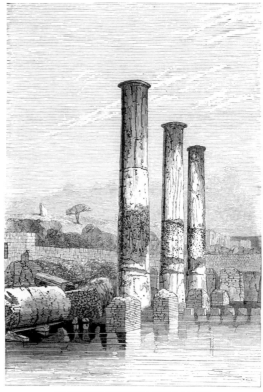

VIEW OF THE TEMPLE OF SERAPIS AT PUZZUOLI IN 1836.

Figure 7.3 The Temple of Serapis in Pozzuoli 1836, frontispiece to Charles Lyell's *Principles of Geology*.

Source: Lyell (1830–33).

Figure 7.4 The Serapis Temple in 1995. Photograph by Jürgen Ehlers.

temple, but part of an ancient marketplace) were of course originally built on dry land. When Lyell visited the site (Fig. 7.4), they stood with their feet in the water. Much more significant, however, is the fact that traces of paddock activity were easily identified at a height of *c.* 4 m, demonstrating that the columns had not long ago stood in much deeper water.

The Mediterranean is an area in which many traces of earlier highstands of the sea can be found. First reports of elevated beaches in Algeria, southern Italy and Sicily had already been published at the end of the nineteenth century. Depéret (1918) identified five different terrace levels. At first he correlated the marine terraces with the river terraces of the Alps. He believed that the four sea-level highstands were equivalent to the four known glaciations. It was not until ten years later, when the principle of eustatic sea-level fluctuations had been developed, that it was recognized that the sea-level highstands did not correlate with glaciations but rather with warm stages.

Initially, many scientists assumed that the Mediterranean sea-level highstands might be detected worldwide; gradually doubts emerged, however. The Sicily Terrace was so high that even if all the polar ice melted it would still be much higher than the resulting sea level. Zeuner (1945) attempted to explain this discrepancy by the fact that the level of the oceans had steadily fallen since the Tertiary. However, recent studies and critical analyses of the existing data could not confirm this assumption. Today we know that the marine terraces of the

Mediterranean have been displaced by young tectonics, so that regional correlations are difficult. Scientists must therefore resort to reconstructing regionally valid sea-level curves. For that, the exact dating of the marine terraces and other preserved water level marks are of crucial importance.

Broecker (1965) was the first to attempt U–Th dating of the fossil coral reefs of Eniwetok Atoll (Pacific) as well as raised beaches from Florida and the Bahamas (Broecker & Thurber 1965). He found two sea-level highstands at 80 and 120 ka before present, which roughly coincided with the assumed last interglacial period. Could two highstands have occurred during one interglacial? Further studies on the coastal terraces of Barbados showed that not two but three highstands were present, dated to *c.* 82, 105 and 125 ka. Additional data from elevated coral reefs in New Guinea confirmed these results (Bloom et al. 1974).

Corals (Box 7.1) are ideal for Quaternary research. They grow just below the surface of the sea and form hard reefs which may, even after the death of the animals, survive for millennia. Corals require water that is always above 18°C and live under water. The upper limit of their

BOX 7.1 DANGEROUS CORAL REEFS

On 16 March 1889 His Majesty Kaiser Wilhelm's gunboat *Adler* (Fig. 7.5) which had anchored in the port of Apia (Samoa) was caught in a typhoon and thrown onto the coral reefs; the *Adler* was one of four warships, two German and two American, to have fallen victim to that storm. What were so many warships doing in distant Samoa?

Figure 7.5 SMS gunboat *Adler* on the coral reef in Apia, Samoa, March 1889.

Source: Jürgen Ehlers, Disposed Geographical Society Hamburg.

The young colonial power of Germany claimed that Samoa was 'under German protection'. The other Pacific colonial powers (the USA and Britain) did not agree, however. Chancellor Otto von Bismarck decided to 'emphasize' Germany's interests by the deployment of three warships. American President Cleveland answered by also sending a squadron to Samoa. In March 1889 six warships were therefore anchored in the harbour of Apia: the American steam frigate *Trenton* accompanied by the sloop *Vandalia* and the gunboat *Nipsic*; and the German corvette *Olga* and the gunboats *Adler* and *Eber*. On 15 March the British frigate *Calliope* also arrived.

The barometer fell and it was obvious that a huge typhoon was approaching. As none of the would-be colonial powers wished to give up first, all the six German and American ships remained in the harbour awaiting the deadly hurricane in the proximity of the dangerous coral reefs: an unwise decision. The result was that two American and two German vessels were thrown on the reefs and destroyed. A total of 93 German and 52 American sailors died in this demonstration of military steadfastness. Only the steamship *Calliope* managed to escape unharmed.

habitat is the low water level and their lower limit is determined by the waning light with increasing depth, which the corals need for photosynthesis. Coral can form atolls, but they can also grow along the coasts in the form of so-called fringing reefs. Samoa, for example, consists of a series of volcanic islands, each surrounded by a fringing reef.

Coral reefs are to some extent able to adapt to sea-level changes. If the sea level rises slowly, the reef grows upwards. If however the sea level drops, the corals fall dry and die. The reef remains as a marine limestone terrace. On an emerging coast, in areas such as the Huon Peninsula of New Guinea, there are series of coral terraces. The terraces can be used to to date changes in sea level by using, for example, the uranium–thorium (U–Th) method (Woodroffe 2007). The aragonite skeletons of corals contain several ppm uranium and also negligible proportions of palladium and thorium, providing theoretically ideal conditions for dating. Unfortunately, the reality is not quite so straightforward.

Older fossil corals contain a higher percentage of ^{234}U, as would be explained by the normal intake from sea water. In addition, those corals have higher proportions of ^{230}Th. Consequently, they yield ages which are too high. The reason for this discrepancy is unknown; only corals which meet the requirements of a closed system are therefore suitable for dating. In this case, the ratio of ^{234}U to ^{238}U should correspond to the value that would be expected in sea water.

Ideally, dates should be checked using other methods. For instance, in their studies on Barbados Schellmann & Radtke (2004) compared electron spin resonance (ESR) ages and uranium–thorium ages; by combining these dating techniques they come to reliable ages for the coral terraces (Figs 7.6, 7.7).

The formation of large ice sheets has a strong influence on the global water balance. Significant amounts of water are held back on the continents in the form of large ice sheets during the cold stages, resulting in a sea-level drop of up to 150 m. With the melting of the ice sheet, sea level rose again. These changes are called eustatic fluctuations of the sea level. During the

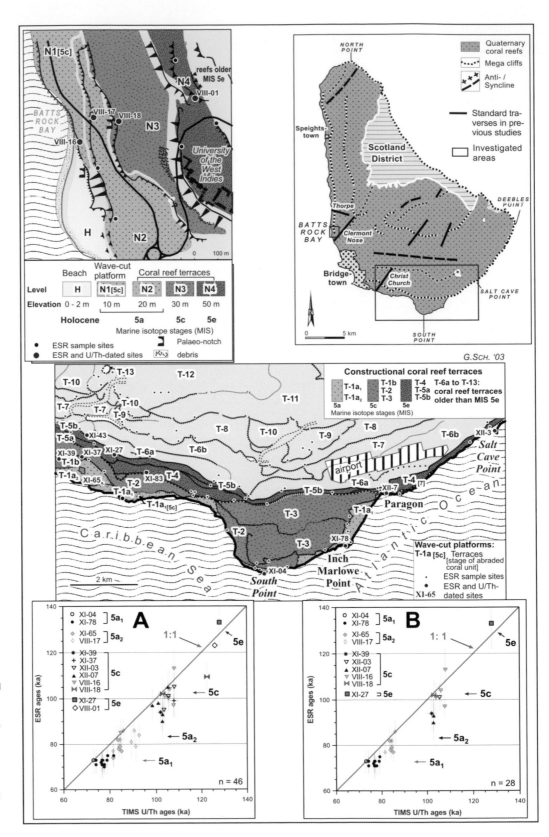

Figure 7.6 Coral terraces from the west and south coast of Barbados dated with ESR and thermal ionization mass spectrometry (TIMS) U–Th: (a) all U–Th data and (b) only U–Th data with $^{234}U/^{238}U$-default values between 141 and 157° are included.

Source: Schellmann & Radtke (2004). Reproduced with permission of G. Schellmann.

Figure 7.7 View of the First High Cliff, near Rendezvous Hill, southwest coast of Barbados, showing the T-5a-terrace and the lower lagoon area of the T-2 terrace.

Source: Schellmann & Radtke (2004). Reproduced with permission of G. Schellmann.

last glacial cycle the lowest global eustatic sea stand occurred during 30–20 ka. Based on U–Th ages from corals, Thompson & Goldstein (2006) suggest that the lowest sea levels occurred during 24.6–23.1 ka based on a number of different studies from around the world. Based on the sea-level records from Barbados, Peltier & Fairbanks (2006) estimate that the global sea-level lowstand started slightly earlier at 26 ka.

When continental plates are covered by massive ice sheets, the ice places a considerable burden on the continental blocks. The rigid Earth's crust is on average 30 km thick and 'floats' on the mantle. This consists of three distinct layers that differ in their deformation behaviour. The top layer is *c*. 200 km thick and formed of relatively hard lithosphere. This is super-imposed on the 500-km-thick soft asthenosphere, and below that lies the 2200-km-thick mesosphere which is relatively hard again. Under the weight of the glacial ice caps the Earth's crust was pressed downwards, and when the ice sheet melted it began to rise again. These movements are called isostatic adjustments.

The extent of isostatic crustal movements in the formerly glaciated areas is difficult to assess. In Scandinavia and North America, the postglacial land uplift has been reconstructed by evaluating raised shorelines. However, this method only provides information for the period after the onset of postglacial transgression and not on the glacial maximum. Svendsen & Mangerud (1987) assumed that central Scandinavia was depressed by *c*. 450 m at 10.3 ka before present. Recent modelling suggests that the maximum depression during the Weichselian cold stage exceeded 600 m (Lambeck et al. 2010).

The Earth cannot be compressed like a sponge. When parts of the Earth's crust are depressed under the weight of the ice, the adjacent areas must be lifted up. The 'forebulge' results from horizontal displacements in the asthenosphere; evidence that such extra-marginal isostatic movements had indeed taken place was obtained from studies in western Norway.

Svendsen & Mangerud (1987) showed the postglacial lowering of marginal Sunnmøre to be *c.* 20 m, which is interpreted as compensation for the post-Weichselian uplift in the glaciation centre. Recent work has shown that 8000 years ago at the German North Sea coast a forebulge with a height of *c.* 7.5 m remained (Vink et al. 2007), which disappeared by *c.* 4800 years ago.

The postglacial uplift of formerly glaciated areas is still continuing. The Late Weichselian Glaciation centre in Scotland is still rising by up to 1.7 mm a^{-1} (Gehrels 2010) and the centre of the Scandinavian ice sheet is rising by 9.3 mm a^{-1} (Donner 1995). From the distribution of postglacial land uplift, it can be inferred that virtually the entire area of the ice sheet was affected by isostatic subsidence and subsequent uplift. Accordingly, more widespread subsidence and uplift areas must be assumed for the more extensive Elsterian and Saalian glaciations in Europe.

The isostatic depression of the glaciated areas resulted in late glacial marine incursions in many cases, which in turn affected the flow behaviour of the melting ice sheets. Where the ice margin started to float, large icebergs were detached under tidal influence. The resulting mass loss had to be offset by more rapid subsequent ice flow. This led to lowering of the glacier surface and an expansion of the catchment area. Strong ice loss led to the formation of a bay in the ice edge (calving bay). The fast-flowing ice streams resulting from rapid ice retreat by calving may therefore have triggered a chain reaction causing the rapid collapse of an ice sheet (Hughes 1987). Such a scenario has been postulated albeit rather controversially, by Eyles & McCabe (1989) for the Irish Sea at the end of the Weichselian (Devensian) Glaciation. A similar effect was probably invoked in North America when sea water entered the Hudson Bay at *c.* 8500 BP, causing a final separation of the Keewatin and Labrador ice centres and resulting in a quick ice decay (Andrews 1987).

7.4 Ice in Siberia?

If the continental shelf was glaciated in western Siberia, then why not also in eastern Siberia? The answer is that the precipitation is much lower (Astakhov 2008). Consequently, in eastern Siberia there are no tills, no glacial landforms and no evidence of widespread glaciation. Almost no evidence, that is; Grosswald attributes the orientated permafrost lakes on the shores of the Arctic Ocean to the action of glaciers. In his opinion, the orientation of the lakes corresponds to the former ice advance direction. In recent years glacial striae were also detected in deep water on the Lomonosov Ridge, a submarine mountain range that stretches from the New Siberian Islands via the North Pole to Ellesmere Island in Arctic Canada (Polyak et al. 2001). Similar traces were found in water of depth 400–1000 m in the Chukhchi Borderland range, a system of submarine ridges north of the Bering Strait (Polyak et al. 2007). This means that during some of the Quaternary glaciations, ice of several hundred metres thick must have covered much of the Arctic Ocean (Polyak et al. 2010).

The east Siberian shelf areas are inaccessible to scientific investigation. The area is covered by thick sea ice throughout almost the whole year. In 1932 a Russian icebreaker succeeded for the first time to traverse the Northern Sea Route (Northeast Passage) without wintering. A year later, the Russians tried to repeat the same experiment with a brand-new merchant vessel. On 2 August the *Chelyuskin* began her voyage from Murmansk in an easterly direction. The

Figure 7.8 The end of the *Chelyuskin* in the Northeast Passage, 13 February 1934. Drawn by Petra Schmidt.

high expectations were not fulfilled, however; the ship became stuck in pack ice and sank on 13 February 1934 without having reached the open waters of the Bering Sea (Fig. 7.8).

The geologist and photographer Hinrich Bäsemann had the opportunity in 1999 to travel the Northeast Passage on board the Russian icebreaker *Kapitan Dranitsyn* (Fig. 7.9) in the same direction as the ill-fated *Chelyuskin*. He saw no traces of former glaciation and, when they landed on Wrangel Island, he saw no fresh glacial landforms. In contrast to the more westerly archipelago of Severnaya Zemlya, Subarctic Wrangel today is completely unglaciated.

Figure 7.9 Passing the Northeast Passage on board the *Kapitan Dranitsyn* in 1999. Photograph by Hinrich Bäsemann (www.polarfoto.de).

Figure 7.10 Musk oxen on Wrangel Island in 1999. Photograph by Hinrich Bäsemann (www.polarfoto.de).

Grosswald & Hughes (2002) had postulated a marine ice sheet in eastern Siberia, with its centres around the New Siberian Islands and Wrangel Island. However, field studies on Wrangel Island (Fig. 7.10) produced no evidence to support this hypothesis. Gualtieri et al. (2005) found that the island had been free of ice during the Weichselian.[10] Be and [26]Al dating of surface boulders and bedrock gave ages of well over 20 ka. Consequently, Wrangel Island was not glaciated (at least not during the last glaciations). They could not, however, determine whether there were any older glacial deposits present on the island.

Wrangel Island is the highest point on the eastern Siberian continental shelf. If it was unglaciated at *c.* 20 ka, this implies that the entire east Siberian shelf was not glaciated at that time. Even if the island had been covered by cold ice, which would have left no traces on the land surface, it would have been subjected to postglacial isostatic uplift and marine terraces should be present. There are in fact marine terraces on Wrangel Island, but the youngest of these are 73–64 ka in age. If there had ever been an ice sheet, it was not during the last cold stage (Box 7.2).

BOX 7.2 TERENCE HUGHES AND THE CONTINENT OF DOOM

There are scientists who advance research by studying things in meticulous detail, advancing slowly step by step, and there are others who are not interested in details but who just propose a big idea and let others do the legwork to either confirm their ingenious concept or disprove it. One of the latter is Terence H. Hughes. He says about himself:

My scientific career can be understood as postulating a number of theoretical mechanisms that might operate on scales ranging from local to global. Some ideas never took hold. Others did, after a lag of 10, 20, 30, and more years. Few if any took hold immediately. That is because most proposed interpretations were directly contrary to the prevailing wisdom.

In 1987, Boreas published 'Deluge II and the continent of doom: Rising sea level and collapsing Antarctic ice' [Hughes 1987; Figs 7.11–7.13]. Deluge I was the Genesis flood. Deluge II will be gravitational collapse of the East Antarctic Ice Sheet, since it would submerge coastal lowlands to a depth of up to 65 metres. That would be a proper deluge, especially if it was fast. My title was inspired by the movie, 'Indiana Jones and the Temple of Doom'. Arguing that the East Antarctic Ice Sheet could collapse rapidly is definitely topsy-turvy thinking …

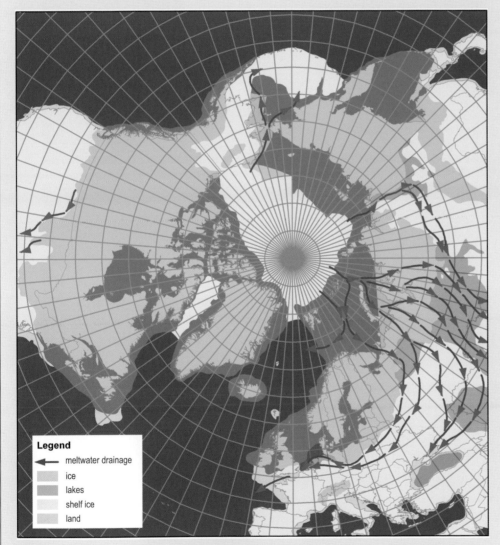

Legend

→ meltwater drainage

■ ice

■ lakes

shelf ice

■ land

Figure 7.11 Extent of glaciations in the Arctic and Subarctic as envisioned by Grosswald & Hughes (2002) and Denton & Hughes (2002). The postulated ice sheet in eastern Siberia and the Okhotsk Sea are not generally accepted.

Source: Grosswald and Hughes (2002).

(continued)

BOX 7.2 TERENCE HUGHES AND THE CONTINENT OF DOOM *(CONTINUED)*

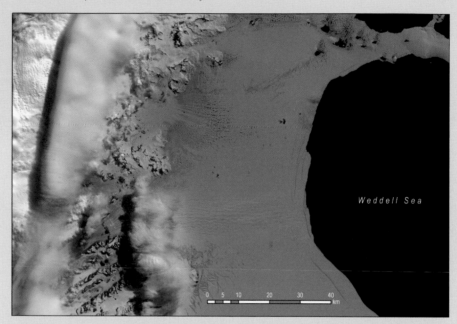

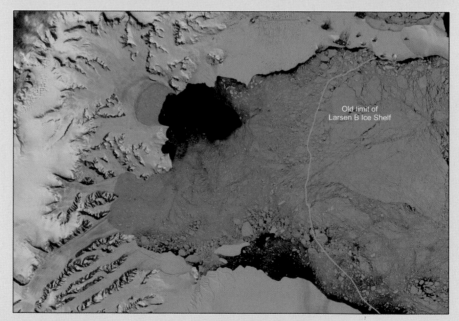

Figure 7.12 When Terence Hughes wrote his deluge article, nobody would have thought that an ice shelf could collapse rapidly. But exactly that happened later that year. In 2002 the Larsen B ice shelf on the Antarctic Peninsula disintegrated completely within months. Larsen B was a small ice shelf; the loss of large ice shelves would accelerate the deglaciation of Antarctica. Landsat 7 images of 15.12.2001 (top) and 18.12.2002 (below) document the end of Larsen B.

Source: US Geological Survey, Landsat 7 ETM, Path 217, Row 106, 13.12.2001 and 18.12.2002.

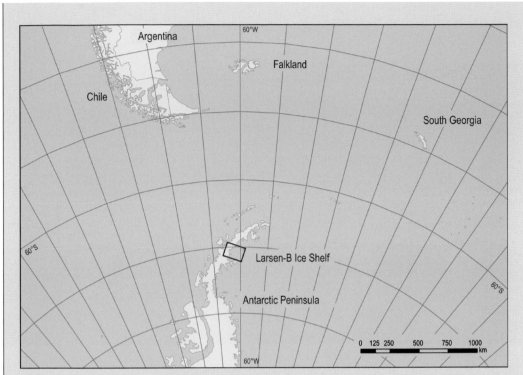

Figure 7.13 Map showing the location of the former Larsen B ice shelf on the Antarctic Peninsula.

In 2002, Quaternary Science Reviews published my paper on the dynamics of calving bays, providing a mechanism for disintegrating former ice sheets after they had been downdrawn close to sea level by surging ice streams. Calving bays migrated up ice streams into Hudson Bay and the Baltic Sea, carving out the hearts of the respective Laurentide and Scandinavian ice sheets, thereby terminating the last glaciation cycle. This 'inside-out' mechanism for deglaciation seems topsy-turvy. In the same issue, Misha Grosswald, George Denton and I improved the ice-sheet reconstructions done at the University of Maine for CLIMAP twenty years earlier. Misha and I had a field day putting 'topsy-turvy' reconstructions of ice sheets in Russia where the Establishment said there were none.

Hughes (2006)

No indications have been found of extensive ice sheets during the last glacial maximum (i.e. *c.* 20 ka) either on Wrangel Island or on mainland areas of Beringia. The vast ground ice complex in eastern Siberia, the so-called 'Yedoma complex', is not buried glacier ice but ground ice mixed with loess.

The Quaternary climatic history of NE Siberia has been investigated in a research project at El'gygytgyn Lake in Chukhotka. This lake was formed by a meteorite impact 3.6 Ma ago, and has a diameter of 12 km and a depth of 170 m. At the bottom of the lake, deposits representing the whole of the Quaternary and late Tertiary can be found. If the area had ever been glaciated, it should record the traces. However, that was not the case (Juschus et al. 2009).

The voyage through the Northeast Passage on board the *Kapitan Dranitsyn* ended with landfall at Uelen (Fig. 7.14), a small town on the northeastern tip of Chukhotka. That was

Figure 7.14 At the end of the Northern Sea Route: landing in Uelen. This is where the participants of the ill-fated *Chelyuskin* expedition also went ashore after their rescue. Photograph by Hinrich Bäsemann (www.polarfoto.de).

also the place where, after a dramatic air rescue in April 1934, the crew members of the *Chelyuskin* were returned to safety. Economically, the Northeast Passage has never attained the importance which was hoped for in the Stalinist era. The Northern Sea Route is very attractive, however; the voyage from Rotterdam to Tokyo can be reduced by some 7000 km, saving one-third of the total fuel costs. The melting Arctic sea ice makes this route economically viable. The Northeast Passage was first successfully traversed by two major carriers of the German Beluga Group from east to west in 2009. In 2013 204 ships were allowed to use the route.

7.5 Asia: The Mystery of Tibet

There are controversial ice sheets that have been proposed for various places around the world. One of the largest is located in Central Asia, the part of the Earth with the most extensive high ground, including the Himalayas (Mount Everest, 8848 m), the Trans-Himalayas (Ningshin Kangsha, 7223 m), the Hindu Kush (Tirich Mir, 7708 m), the Pamir (Kongur, 7719 m), the Karakoram (K2, 8611 m) and the Kun Lun Shan (Liushi Shan, 7167 m). Surrounded by these highest mountains on Earth is the Tibetan plateau at an average altitude of 5000 m above sea level; this plateau is 200 m higher than the summit of Mont Blanc, the highest mountain in the Alps. Mont Blanc is glaciated today and the Alps were intensely glaciated during the Pleistocene. That the highlands of Tibet were completely covered by a Pleistocene ice sheet would therefore appear to be a reasonable assumption (Fig. 7.15).

Matthias Kuhle has attempted to reconstruct the extent of the ice-age glaciers in High Asia. In some areas his calculation comes close to that of other scientists; the maps for Afghanistan show that the glacial limits drawn by Kuhle (2004) and Porter (2004) differ only slightly (Fig. 7.16). In other areas, however, the deviations are fundamental.

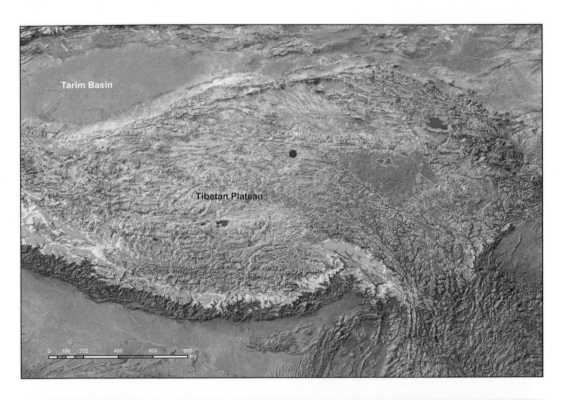

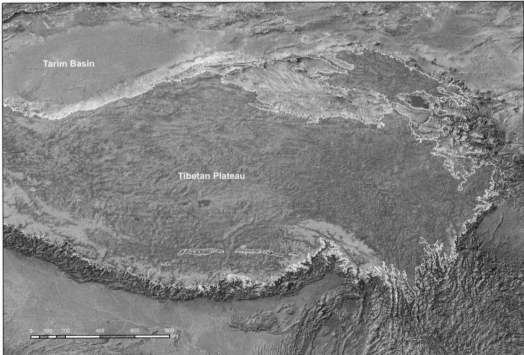

Figure 7.15 Ice in Tibet, according to Shi et al. (1991) (above) and after Kuhle (2004) (below). Red dot: location of the 'Jäkel Stone' (Fig. 7.17).

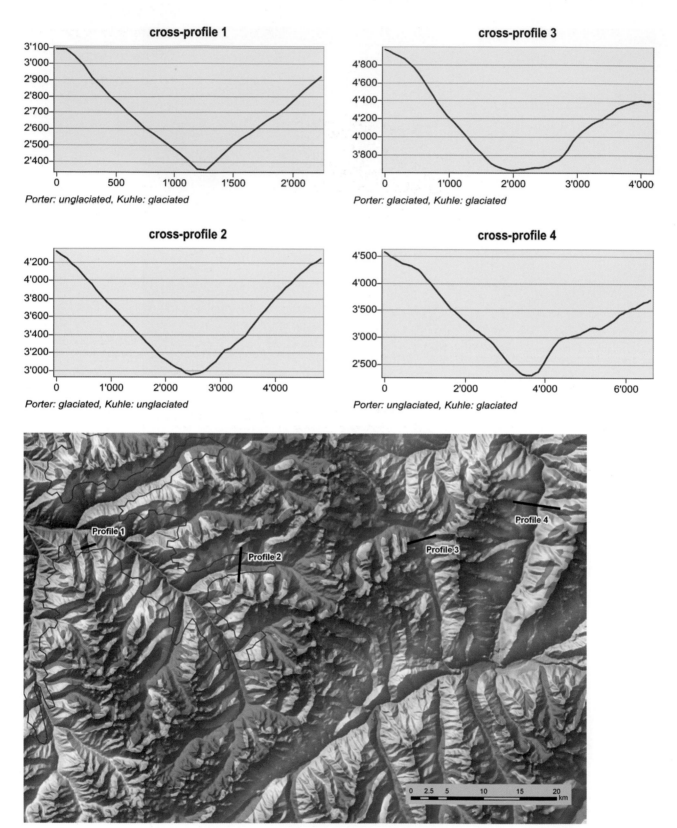

Figure 7.16 Extent of glaciation in northern Afghanistan according to Porter (2004) (blue) and Kuhle (2004) (red) against the backdrop of the SRTM elevation model. The differences are small; the GIS-generated cross-sections through valleys illustrate the different interpretations.

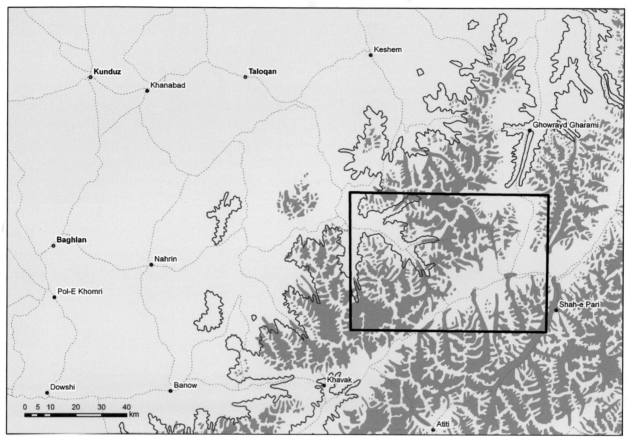

Figure 7.16 *(Continued)*

In 1976 and 1977 Kuhle visited the southern edge of the highlands of Tibet for the first time. He found that moraines and glacial striations on the south side extended right down to altitudes of 1100–1200 m, and on the north side of the Himalayan main range the glacial landforms reached down to 2580–2800 m, well below the level of the Tibetan Plateau. If those observations were correct, the ice-age glaciers must have had a much greater extent than previously thought. Investigations by Heuberger & Weingartner (1985) confirmed these results. On an expedition led by Professor Hövermann in 1981, Kuhle eventually reached the northeastern edge of the highlands of Tibet. Again, he found traces of an extensive piedmont glaciation at surprisingly low altitudes. In another expedition in 1984, Kuhle came to the conclusion that glacial ice had once covered all of Tibet (Kuhle 1986). Later, he went even further, and suggested that the uplift of Tibet and the glaciation of the plateau might have been the cause of the Quaternary Ice Age (1989).

Kuhle's ideas met instant opposition. Many workers rigorously maintain that Pleistocene glaciations over the Qinghai–Tibetan Plateau were much more restricted (including Shangzhe et al. 2011). However, the disagreement between Kuhle and others was not just concerning the extent of the largest glaciation but its timing as well. Kuhle has persistently questioned and argued against the geochronological methods used by others to date the last glaciations in Tibet. In their review, Lehmkuhl & Owen (2005) concluded that the highlands of Tibet were never completely glaciated. They regarded the map by Shi et al. (1991) as having provided the best reconstruction of glacial limits. The map shows the last glaciation as largely limited to the marginal mountain ranges. For the penultimate glaciations, a somewhat more extended glaciation is envisioned including a small (400 × 500 km) ice sheet in the northeast part of the plateau. It is unclear whether their 'penultimate' glaciation represents either the Saalian or an Early Weichselian Glaciation. Shi and his colleagues wrote in their explanatory text: "So far, no evidence was found that the plateau has born once great ice sheets, and there are no eskers, no groups of drumlins, no glacial lakes, and no large valleys, through which the ice might have flowed. The present lakes on the plateau owe their existence not to the glacial erosion, but to tectonics."

Kuhle's ideas have increasingly become a minority view, especially the concept that the Qinghai–Tibetan Plateau was almost entirely covered by an ice sheet during the Last Glacial Maximum (LGM). Many of the apparent contradictions stem from disputes regarding the age of the largest glaciation, which is now very clearly *not* the product of the LGM (e.g. Heyman et al. 2011). Nevertheless, there are some indications of a more extensive former glaciation at some point during the Pleistocene, although the extent of ice cover is debated and the timing of the maximum glaciation is poorly constrained (Heyman et al. 2011).

Figure 7.17 Striated boulder on the Tibetan plateau in a place that, according to accepted wisdom, should not have been glaciated. Photographs by Robert Hebenstreit.

Dieter Jäkel from Berlin wrote (Fig. 7.17):

On 25/05/1998 together with Robert Hebenstreit at 35° 19' 500 m North, 92° 40' 750 m East I found a 61-pound grey metamorphosed sandstone, which at one edge showed glacial striation. It lay on the southern flank of a rounded hill. According to my Thomen altimetre that was at 4700 m a.s.l., at 4660 m according to my GPS. The mound was composed of a coarse reddish sandstone. The metamorphic block differed significantly from the local sandstone bedrock and was therefore easy to find. We found five more of those rocks… I found out later that such a grey metamorphic sandstone is outcropping at two places, one of them 100 km north of the Kunlun Shan and the other in a mountain chain about 100 kilometres to the west of the site. As between the two possible source areas and the place where the erratics were found there are extensive depressions, the stones could not have got there by fluvial transport. The only possible explanation to me was that they had been transported by ice.

Zens (1998)

This plain report gives no hint of the difficulties encountered in that expedition. At one point almost everything had gone wrong. The journalist Joseph Zens who interviewed Jäkel for the *Berliner Zeitung* wrote:

After a car accident, the group had more than nine hours to struggle through inhospitable terrain until they reached the road and could stop a truck. Darkness and a snowstorm had hampered their progress, and all that at an altitude of more than 4500 metres above the sea. I was at the end of my physical strength, says Jäkel. I had given up hope to get out of this expedition alive.

Zens (1998)

But why did Jäkel find erratics on the plateau, where other researchers had looked for them in vain? There is a simple answer: most people have visited Tibet at the wrong time. In summer, when the external conditions are most favourable for such a trip, thick grass covers the ground and the glacier-transported stones are hard to find. The place where Jäkel found the erratics is clearly outside the glaciated area shown on the map of Shi et al. (1991). The same applies to several of the features interpreted by Matthias Kuhle as 'certain traces of glaciation' (Kuhle 2004). There have been far too few field investigations however, especially to the huge areas away from the main accessible highways. It is clear that much more fieldwork is required to solve the 'Tibetan Mystery'.

The extent of Pleistocene glaciation in the Tien Shan and Altai region a little further to the north is also unclear. Grosswald & Kuhle (1994) claimed that the LGM glaciers were much more extensive than today, and assumed that they extended right into Lake Baikal (Fig. 7.18). Lehmkuhl & Owen (2005) pointed out that Zech et al. (1996) and Heuberger & Sgibnev (1998) had presented clear evidence that glaciers did not reach the foothills as low as the lake, but were restricted to the higher cirques and valleys. However, in the mountains bordering the south side of Lake Baikal, there is superficial evidence from satellite imagery that ice may have extended close to the lake edge; further work is needed to resolve this issue in all valley areas.

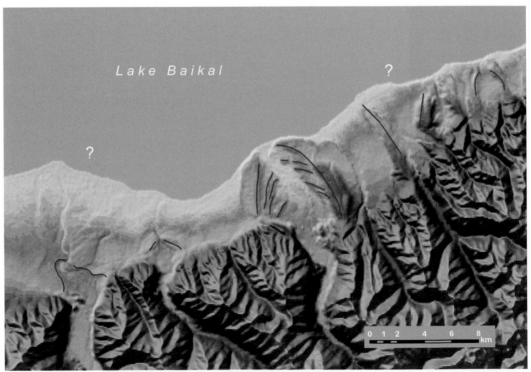

Figure 7.18 Moraines on the southern shore of Lake Baikal. Top: Landsat 7 satellite image, bottom: interpretation.

Source: US Geological Survey, Landsat 7 ETM, Path 193, Row 24, 20.1.2002.

7.6 South America: Volcanoes and Glaciers

The first scientist who grappled with the traces of the Ice Age in southern South America was Charles Darwin. He was 26 years old when he visited the coasts of Patagonia, Tierra del Fuego and the island of Chiloe aboard the HMS *Beagle* in 1834. Charles Lyell had supplied him with a copy of the first edition of his *Principles of Geology* for the journey, and asked him to pay particular attention to the occurrence of erratic blocks in those areas. On his trip along the Rio Santa Cruz, the river that drains Lake Argentino to the Atlantic Ocean, Darwin saw the expected erratic blocks. He also described outcrops for the first time, in which layers of basalt were intercalated in the sequence of glacigenic layers; these would later provide the key to the extensive Argentine Quaternary stratigraphy (Strelin & Malagnino 2009). That was much later, however; when Darwin published his geological observations in 1842, they were still in support of Lyell's drift theory. While recuperating from an illness a few decades later in 1865, Agassiz noted traces of the Ice Age in Brazil (Box 7.3).

BOX 7.3 PUSHING AGASSIZ FROM HIS PEDESTAL?

Louis Agassiz was already very famous during his lifetime. As early as in 1840, when he was only known for his research on fossil fishes, a mountain in Switzerland was named after him, the 3942 m high Agassizhorn in the Bernese Alps. The American poet Henry Wadsworth Longfellow wrote a poem to commemorate the 50th birthday of the esteemed scientist (1857). It concludes with the lines:

And Nature, the old nurse, took
The child upon her knee,
Saying: "Here is a story-book
Thy Father has written for thee."
"Come, wander with me," she said,
"Into regions yet untrod;
And read what is still unread
In the manuscripts of God."

The naturalist had become a living monument. In winter 1865 Louis Agassiz fell into poor health. A change of climate was thought useful in bringing relief, but where to? Europe was ruled out as there he would have no choice but to take part in all kinds of scientific debate. He needed rest, so chose South America. The Emperor of Brazil had shown great interest in his scientific investigations, and had not pledged his support for Agassiz' efforts to establish a major Zoological Museum in his country. Agassiz travelled with his wife, but despite all good intentions it was not a pleasure trip.

(continued)

BOX 7.3 PUSHING AGASSIZ FROM HIS PEDESTAL? (CONTINUED)

Elizabeth Cabot Agassiz wrote after a particularly stressful day:

The hospitality of our excellent hosts repaid us for all the fatigues of our journey, and our luggage being still on the road, their kindness supplied the defects of our toilet, which was in a lamentable condition after splashing through muddy water two or three feet deep. Mr. Agassiz, however, could not spare time to rest; we had followed a morainic soil for a great part of our journey, had passed many boulders on the road, and he was anxious to examine the Serra of Monguba, on the slope of which Senhor Franklin has his coffee plantation, and at the foot of which his house stands. He was, therefore, either on foot or on horseback the greater part of this day and the following one, examining the geological structure of the mountain, and satisfying himself that, here too, all the valleys have had their glaciers, and that these valleys have brought down from the hillsides into the plains boulders, pebbles, and debris of all sorts.

Agassiz & Agassiz (1868)

Yes, Louis Agassiz saw traces of the Ice Age even in Brazil. After having inspected the sediments of the Amazon lowlands, he noted:

It is my belief that all these deposits belong to the ice-period in its earlier or later phases, and to this cosmic winter, which, judging from all the phenomena connected with it, may have lasted for thousands of centuries, we must look for the key to the geological history of the Amazonian Valley. I am aware that this suggestion will appear extravagant. But is it, after all, so improbable that, when Central Europe was covered with ice thousands of feet thick; when the glaciers of Great Britain ploughed into the sea, and when those of the Swiss mountains had ten times their present altitude; when every lake in Northern Italy was filled with ice, and these frozen masses extended even into Northern Africa; when a sheet of ice, reaching nearly to the summit of Mount Washington in the White Mountains (that is, having a thickness of nearly six thousand feet), moved over the continent of North America, is it so improbable that, in this epoch of universal cold, the valley of the Amazons also had its glacier poured down into it from the accumulations of snow in the Cordilleras, and swollen laterally by the tributary glaciers descending from the table-lands of Guiana and Brazil?

Agassiz & Agassiz (1868)

In science every great discovery is matched by at least one big mistake. Errors are not sufficient to topple monuments, however; a major earthquake is required for that. At the Zoology Building of Stanford University, Louis Agassiz' statue stood in marble together with those of Alexander von Humboldt, Benjamin Franklin and Johann Gutenberg. During the earthquake of 1906, he was the only one who fell off his pedestal (Fig. 7.19). He survived the crash without major injury (only his battered nose had to be replaced).

Another quake which threatened to shatter Agassiz' reputation finally occurred in 1980, more than 100 years after his death, when Stephen Jay Gould in his book *The Panda's Thumb* (1980) casually brought to light the other major fault of Louis Agassiz: his racism. When Swiss historian Hans Fässler undertook a closer inspection in 2005 of the great naturalist's unacceptable statements, Agassiz came under fire at home.

The Swiss *Nationalrat* for the Canton of Geneva, Carlo Sommaruga (Suisse Socialist Party), took the case before the Parliament on 22.06.2007. His interpellation 07.3486 read: 'Get Louis Agassiz from his pedestal'. The Bundesrat, however, was not ready to topple Agassiz. They answered a few months later:

Figure 7.19 The toppled Louis Agassiz statue after the earthquake in San Francisco, 1906.

Source: GP 2514: Statue of Louis Agassiz, Historical Photograph Collection Stanford, Stanford University Archives.

Louis Agassiz was a great geologist and zoologist, for that he must certainly be recognized. Apart from that he held racist views that went well beyond the usual racial paradigm of his time. There is no doubt that today's Bundesrat condemns his racist thinking... [but] [t]he social values of every historical period have also to be compared its by critical confrontation with one's ancestors, but that should not result in a posthumous condemnation of a person's entire lifetime of achievements.

Source: http://www.legalanthology.ch/americanization/3-contributions/
interpellation-07-3486-von-carlo-sommaruga/

After the British, the Swedes had their moment of fame. On 19 February 1925 the Swedish newspaper *Dagens Nyheter* proudly proclaimed: 'Swedish scientists unravel the mysteries of the Ice Age.' One of them was Carl Caldenius. Caldenius was a pupil of De Geer, and it was originally planned that he should try to correlate the varved clays of Argentina with the Swedish varve chronology (Lundqvist 1991).

The varves could not be correlated; Caldenius attempted it all the same. He also mapped what he thought to be the deglaciation history of southern South America, applying the Scandinavian terminology of Daniglacial, Gotiglacial and Finiglacial to the moraines he recognized (Caldenius 1932). 'Initioglacial' was the term given to an additional glaciation in which the glaciers had reached their greatest extent, which he found at a distance of over 100 km from the foot of the Andes. The associated ice-marginal deposits east of Lago Buenos Aires consisted of, as Caldenius wrote, 'colosales morenas terminal' (huge terminal moraines; Heusser 2003). The morphological boundaries Caldenius drew were correct and are – with minor corrections – still valid today. What has changed, however, is the age of those moraines.

The terms that Caldenius used originated from his teacher De Geer and were originally related to the Scandinavian deglaciation. In the Daniglacial phase, the ice of the last glaciation melted back from its maximum extent in Denmark and northern Germany to Scania. In the Gotiglacial the ice melted back to central Sweden, until in the Finiglacial the active ice margin retreated to Jämtland. The outermost moraines which Caldenius termed Initioglacial have no equivalent in De Geer's terminology, and it is unclear whether Caldenius regarded them as an extra-wide Weichselian ice advance or the advance of an earlier glaciation.

Caldenius' mapping of the glacial limits ended at 41° 20′ S. Flint & Fidalgo (1964) extended the mapping to 39° 10′ S. They only found traces of a single glaciation, which they correlated with the Wisconsinan in North America (i.e. the Weichselian Glaciation). Denton et al. (1999) revised the glacial limits on the Chilean side of the Andes to 40° 35′ S, that is, to the north of Puerto Octay. The latest update was provided by Harrison & Glasser (2011). For the northern end of the southern Andean glaciation region, there is still no precise mapping available; the generalized map of Hollin & Schilling (1981), which in many areas is based on a mere estimate of the former snow line, is the only option.

Darwin had discovered the key to unravelling the glacial history of South America. Feruglio (1944), who mapped the geology of Patagonia and Tierra del Fuego, pointed it out once again: in various places the glacial deposits were intercalated with volcanic strata of unknown age. The latter could eventually be dated.

The INQUA meeting was held in the Southern Hemisphere in 1973 for the first time, in Christchurch, New Zealand. Here John Hainsworth Mercer presented the results of his research to a wider audience (it had previously been published in a short article in *Science*; see also Mercer et al. 1975). His contribution revolutionized knowledge of the course of the Ice Age. Together with colleagues, he had dated the Argentine basalts using the potassium–argon (K–Ar) method and had found that the oldest glaciation of Patagonia had occurred as early as 3.5 Ma. The most extensive glaciation (Caldenius' Initioglacial) was not 20 ka, but 1.2 Ma in age!

In the *Extent and Chronology* project, Coronato & Rabassa (2011) and Martínez et al. (2011) presented the best ever map of the glaciations of southern Argentina. They differentiate 13–14 individual glaciations, the oldest glacial deposits dating back to the late Miocene (7–5 Ma). The most extensive glaciation is known as the Great Patagonian Glaciation (GPG), and its age has been determined by $^{40}Ar/^{39}Ar$ and K–Ar dating to 1.2–1.0 Ma. Since then, the

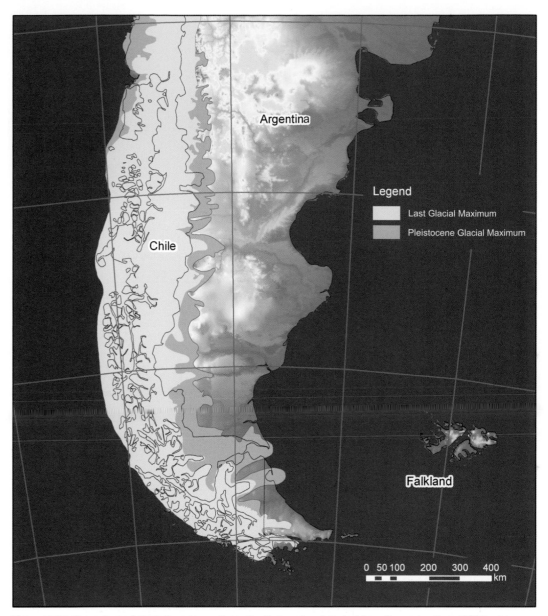

Figure 7.20 The extent of glaciations in southern South America. Adapted from Coronato et al. (2004a, b).

area has been affected by at least three more extensive glaciations (Fig. 7.20; Coronato et al. 2004b).

7.7 Mediterranean Glaciations

Many of the Mediterranean mountains supported glaciers during the Pleistocene, from Morocco in the west to the Lebanon in the east (Hughes and Woodward 2009; Hughes 2012; Fig. 7.21). In the Maritime Alps north of the Côte d'Azur, Pleistocene ice caps were contiguous with the main Alpine Ice Sheet which covered a total area of

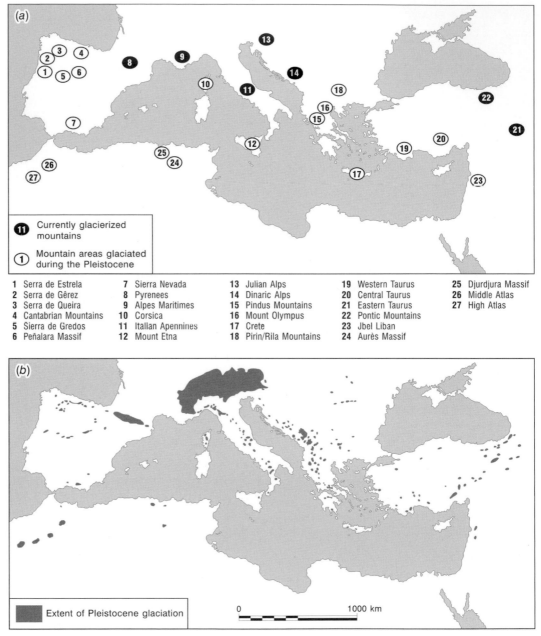

Figure 7.21 Glaciers of the Mediterranean mountains showing both (a) modern distributions and
(b) approximate distribution of Pleistocene glaciation.

Source: Hughes & Woodward (2008). Reproduced with permission of John Wiley & Sons.

126,000 km^2 during the last cold stage (Ehlers 1996). Ice caps also formed over the
mountains of NW Iberia and also over large areas of the western Balkans. Glaciers even
formed as far south as Crete during the Pleistocene. A small number of glaciers still
survive in the Mediterranean region; most of these are restricted to the highest areas,
such as the Pyrenees, the Maritime Alps and the mountains of Turkey where the highest

summits exceed 3000 m asl. However, several glaciers also exist in lower mountain areas, such as central Italy and in Montenegro and Albania (see Hughes 2012, 2014). Many of these glaciers have retreated significantly in the past 100 years from more extensive positions that were reached during the Little Ice Age. Glaciers are now restricted to the highest cirques and often survive in localities that are strongly influenced by local climate with shading and windblown and avalanching snow exerting considerable influence over glacier mass balance. For example, in the mountains of Montenegro and Albania, small glaciers are fed by snow accumulation that is several times that supplied by direct precipitation (Hughes 2008, 2009a).

There is a long history of glacial research in the Mediterranean mountains (Fig. 7.21). Early glacial research was undertaken by some of the biggest names in academic geography, including Jovan Cvijić, Albrecht Penck and Emmanuel de Martonne. Jovan Cvijić was one of the first geographers to note and map glacial landforms in Europe and worked extensively in the Balkans in the late nineteenth and early twentieth centuries (e.g. Cvijić 1917). Albrecht Penck was another early scholar who worked in the Mediterranean mountains. His research covered a range of different areas including the Pyrenees (France/Spain) and the Dinaric Alps (Balkans; Penck 1900). Emmanuel de Martonne was another widely travelled geographer, and undertook glacial research across the Mediterranean lands from Morocco to Romania (Martonne 1924).

Glacial research in the Mediterranean was particularly active in the inter-war period between 1920 and 1940 (Hughes 2012). After a break during World War II, glacial research continued apace. Bruno Messerli's (1967) classic review of glaciation illustrated the extent of glacial research in the Mediterranean region, with 364 publications cited in his article. The Messerli publication represented a watershed in Mediterranean mountain glacial geology (Hughes et al. 2006c; Hughes 2012).

This was, however, followed by an extended period of stagnation in the development of ideas with, arguably, few papers providing significant advances knowledge in the three decades which followed. One reason for this was that it was difficult to date glacial deposits and landforms and add new knowledge (many of the glaciated areas had already been identified and the extent of glaciation was relatively well known). Scholars such as Messerli focused their attentions on other subjects, and the study of Mediterranean glaciations was in the doldrums. In some areas the political situation meant that several mountain areas were out of bounds or researchers were subject to restrictions, especially in mountain areas close to national borders. This situation was exacerbated for a decade in the Balkans by the wars in the former Yugoslavia.

However, in the past decade there has been a boom in the numbers of glacial studies including: Morocco (Fink et al. 2012); Spain (Palacios et al. 2012; Serrano et al. 2013); the Pyrenees (Pallàs et al. 2010; Delmas et al. 2011); the Maritime Alps (Federici et al. 2012); Corsica (Kuhlemann et al. 2008); the Italian Apennines (Giraudi et al. 2011); Montenegro (Hughes et al. 2011); Greece (Woodward & Hughes 2011); and the easternmost Mediterranean in Turkey (Akçar et al. 2008, 2014; Sarıkaya et al. 2009; Çiner et al. 2015). These publications represent only a small selection of recent articles. The large increase in interest reflects the availability of techniques for dating glacial landforms (especially cosmogenic exposure dating) and also renewed stability in many of the countries.

7.8 Were Africa, Australia and Oceania Glaciated?

The Snows of Kilimanjaro is the title of a short story by Ernest Hemingway. On Mount Kilimanjaro there is not only snow but bare ice (Thompson et al. 2006). There were Pleistocene glaciers in Africa, but they were limited to the highest peaks. Glaciers occurred in the Atlas Mountains of Morocco and Algeria, and there were glaciers on the highest peaks of Ethiopia and on the summits of Mount Kenya, Ruwenzori, Elgon and the Aberdare. Kilimanjaro was glaciated, of course, and it still supports a small ice cap. Of all these mountains, the Ruwenzori had the largest number of glaciers during the ice ages with at least 78 former glaciers identified by Mark and Osmaston (2008).

In the south, there have been reports of small niche glaciers in the Drakensberg in Lesotho (South Africa; Mills et al. 2009). The scale of glaciation was very limited in this region and largely restricted to the Drakensberg, which reach a peak altitude of 3482 m. The glaciation of the Eastern Cape Mountains (near Port Elizabeth to the south) is rather unlikely (Hall & Meiklejohn 2011), despite reaching 3001 m at Ben Macdhui.

Some of the largest African glaciers of the ice ages formed in Ethiopia. In the Bale Mountains, which reach over 4000 m altitude, an ice cap covering *c.* 700 km^2 is thought to have covered the plateau, with individual glaciers exceeding 250 m in thickness (Mark & Osmaston 2008).

Snow patches and rock glaciers even formed in the middle of the Sahara Desert during the Ice Age. In the Hoggar (Algeria) and Tibesti (Chad) Mountains, nivation and periglacial features are present even below altitudes of 2000 m asl. Rock glaciers can be observed in the Hoggar and in the Tibesti above 2000 m, indicating very intense frost activity and possibly permafrost (Messerli & Winiger 1992) during the Pleistocene. Frost-weathered landforms occur in the Hoggar at 1100–1400 m (Rognon 1967) and are found above *c.* 1800 m in the Tibesti (Messerli 1972). At the northwestern boundary of the Sahara, the Atlas Mountains supported large glaciers during the Pleistocene. In the Toubkal Massif, containing the highest peak in North Africa, glaciers reached several kilometres in length and descended to altitudes as low as 2000 m asl. (Hughes et al. 2011). Glaciers existed throughout the Atlas although little is known about the extents of glaciers in other parts of the High and Middle Atlas. Even less is known of the glacial history of the Algerian mountains to the north. Barbier & Cailleux (1950) reported evidence of extensive glaciation in the Djurdjura Mountains. They noted that glacial deposits extend to exceptionally low altitudes for this latitude, reaching as low as 750 m asl on the northern slopes and 1270 m asl on the west. Very little (if any) glacial research has since been undertaken in this area, although glacial landforms have been noted further south in the Aurès massif of the Saharan Atlas (Ballais 1983).

The high volcanoes (>4000 and 5000 m asl) in Indonesia and Papua (New Guinea) were glaciated during the Pleistocene (Fig. 7.22). There is also evidence of glaciation on Mount Kinabalu (4101 m asl) in Sarawak, Malaysia (Prentice et al. 2011). On the Australian mainland, however, there is only one place that was glaciated: Mount Kosciuszko (2228 m asl) in the Snowy Mountains (Colhoun & Barrows 2011). In contrast, there was a Pleistocene ice sheet in Tasmania, and the mountains of the South Island of New Zealand were heavily glaciated during the Quaternary (Barrell 2011).

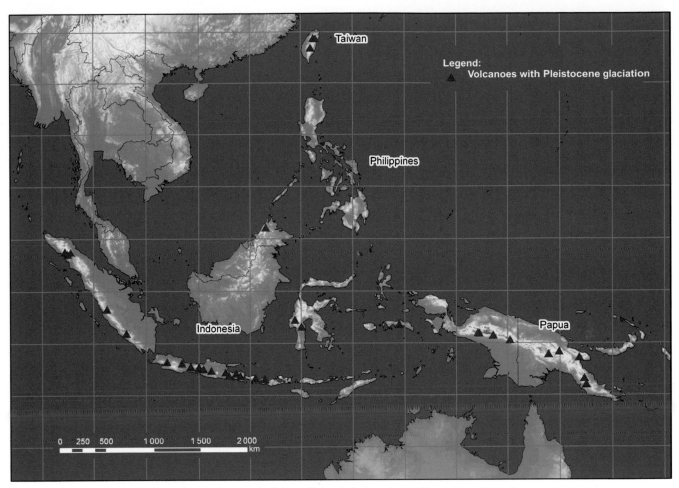

Figure 7.22 Map of the formerly glaciated volcanoes in Taiwan, Malaysia, Indonesia and Papua (New Guinea).

7.9 Antarctica: Eternal Ice?

I asked Captain Nemo if he had already discovered the Pole.

"No, sir," he replied, "we will discover it together. Where others have failed, I will not fail, although with my 'Nautilus', I have never advanced so far south. But I tell you, it will go even further."

"I believe you, Captain," I replied sarcastically. "Forward, ho! For us there's no obstacles! We shatter the ice! Or - if we do not succeed, then 'Nautilus' probably spreads her wings, and we fly over it?"

"Over it? Sir," replied Captain Nemo calmly, "we will not get over, but under it."

Verne (1870)

The Antarctic had only been discovered in 1820 after a long fruitless search for the southern continent. In 1870, when Jules Verne published his *20,000 Leagues Under the Sea*, the interior of the continent was still completely unknown. Was it even a continent? At the first

German Geographers' Congress in 1865, August Petermann had maintained that, behind a barrier of sea ice, the Arctic Ocean was ice-free right to the North Pole (Felsch 2010); why should the same not also apply to the South Pole? Verne let his imagination run wild; his heroes dived in their submarine under the ice until they came close to the South Pole, from where they covered the remaining small distance on foot.

Today we know that Antarctica is composed of two very differently structured parts. While the ice sheet in East Antarctica rests on solid land, like Greenland, West Antarctica, the area west of the Transantarctic Mountains, is covered by a marine ice sheet. Here the ice is only anchored to a group of islands, between which there are reportedly water depths of over 2500 m (Roland 2009). By passing under the Ross Ice Shelf, a foolhardy Captain Nemo might in fact (at least theoretically) reach the vicinity of the pole. (But whether it would have been possible for him to break through the thick ice cover there is another question.)

The Antarctic ice has a volume of 25–30 million km^3, but even this huge ice sheet is far from 'eternal ice'. The Antarctic Ice Sheet formed only about 40–35 Ma ago. It is believed that the ice of East Antarctica started to grow at the transition between the Eocene and Oligocene. The West Antarctic Ice Sheet may also have been present during the latest Miocene (Ingólfsson 2004). The Pleistocene glaciations, which were eventually to cover much of the Northern Hemisphere, began on the opposite side in the Antarctic.

While the South Shetland Islands and some coastal areas of Antarctica became ice free during the interglacials, the extent of the Antarctic Ice Sheet changed only slightly. In fact, based on 15 erratics on the flanks of Grove Mountain situated 800 km inland from the coast, Lilly et al. (2010) found that the thickness of the East Antarctic Ice Sheet was possibly smaller at the LGM than it is today. Similarly, Mackintosh et al. (2007) found that exposure ages from mountain tops in Mac. Robertson Land, East Antarctica, showed that there had been little change in ice sheet thickness since the last Ice Age. However, there is no guarantee that it will stay like that forever. Large quantities of icebergs are currently calving from the Antarctic ice streams into the sea. Whether the overall mass balance of Antarctica is positive or negative cannot be said with certainty. Melting of the East Antarctic Ice Sheet would result in a global sea-level rise of 59 m, flooding Washington, London and parts of Berlin and Paris.

Antarctica is still one of the least-known parts of the world (although it is not the least accessible; see Box 7.4). Julian Dowdeswell, director of the Scott Polar Research Institute in Cambridge, said in a recent interview:

> There is a lot about the ice that we don't know yet. The big drainage basins – half a million square kilometres or more – are almost completely unknown, even in terms of their basic shape. If you ask us what the shape of the rock underneath is, it's like trying to explain the shape of Britain with just two or three transects across it. You've got no idea. And, in fact, it's true to say that a lot of these basins in east Antarctica are a lot less well mapped than either the dark side of the moon or Venus. You can send up satellites and there's nothing in the way when you want to use radar to look at the shape of the moon. But you've got between 2 and 4.7 km of ice obstructing your view of what the bed of the Antarctic Ocean is like.

Jolin (2010)

BOX 7.4 DO YOU LIVE ON BOUVET ISLAND?

It is not only the Antarctic which is difficult to access. Other parts of the Earth are also covered by thick ice, for example Bouvet (Fig. 7.23). While the name Bouvet was virtually unknown to the public until recently, it is now familiar to all who have ordered something online and had to select their place of residence from a list of possibilities. One choice would be Bouvet Island (in

Figure 7.23 Bouvet Island seen from space.

Source: NASA's Earth Observatory, Astronaut photograph ISS017-E-16 161 of 13 September 2008.

theory, at least). Bouvet (in Norwegian: Bouvetøya) is considered the most inaccessible island in the world. It consists of a shield volcano, which is almost completely covered by ice. Only in some steep cliffs, such as at the high Kapp Valdivia on the north coast for example, does any dark rock surface from under the ice cover.

The island was discovered by the French captain Bouvet de Lozier in 1739, but nobody wanted it. The abandoned island in the South Atlantic was finally annexed in 1927 by a scientific expedition from Norway that visited the island. Apart from occasional visits by scientists from the Norwegian *Polarinstitutt*, Bouvet is uninhabited and virtually uninhabitable. A raft was discovered on the beach in 1964, but there was no trace of its occupants.

Solifluction over permafrost on the banks of the Aldan River, Yakutia. Photograph by Jürgen Ehlers.

Chapter 8
Ice in the Ground: The Periglacial Areas

8.1 Definition and Distribution

The term 'periglacial' refers to 'the conditions, processes and landforms associated with cold, non-glacial environments' (Ballantyne & Harris 1994, p. 3). French (2007) defined the 'periglacial' areas as those which experience a mean annual temperature of less than 3°C in areas where frost action and permafrost-related processes dominate. In a broader sense, however, it also includes any environment where cryogenic processes play an important role in the landscape and can include the effects of snow (nival processes) as well as the effects of ice and frozen ground.

One criterion for classifying periglacial areas is based on the development of permafrost. This is defined as ground in which the temperature remains below 0°C for at least two consecutive years (Ballantyne & Harris 1994). Most deposits of permafrost have an upper active layer that is 1–3 m thick. This active layer is subject to a cyclic thaw during the summer season. Permafrost can be continuous, discontinuous, sporadic or isolated. The spatial distribution of permafrost (Fig. 8.1) is influenced by altitude and latitude, with the largest continuous areas of permafrost being located in the high-latitude land areas of the Northern Hemisphere in Alaska, Canada and Russia. Mountain areas also support permafrost, often in the form of sporadic or isolated patches in areas where the topoclimate favours frozen ground, such as in shaded cirques and valleys. However, large areas of mountainous

The Ice Age, First Edition. Jürgen Ehlers, Philip D. Hughes and Philip L. Gibbard.
© 2016 John Wiley & Sons, Ltd. Published 2016 by John Wiley & Sons, Ltd.
Companion website: www.wiley.com/go/ehlers/iceage

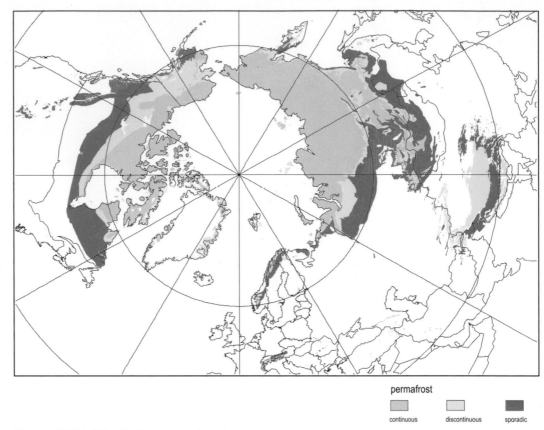

Figure 8.1 Global distribution of permafrost.

Source: Nelson et al. (2004). Reproduced with permission of John Wiley & Sons

terrain can be characterized by extensive tracts of permafrost even at low latitudes, such as in Tibet and Xinghai Provinces, China (Fig. 8.2).

During the summer the uppermost portion of the permafrost thaws. This seasonal thaw sets in when the insulating blanket of snow has melted. The layer that is affected by the seasonal freeze–thaw cycles is called the active layer. Thawing continues until the end of the summer season until the ground starts to freeze again. It is not only the thawing that starts at the surface, but also the freezing. If there is water in the soil, the water migrates to the interface between frozen and unfrozen ground where it forms ice lenses, called segregation ice. The freezing water expands and the ice lenses raise the floor. As the ice lenses are not evenly distributed, the process results in irregular uplift and subsidence areas. The freeze–thaw cycles and the associated processes of frost heave and the subsequent subsidence alter the soil structure, and can lead to a mixing or sorting of the material (cryoturbation).

Permafrost reacts to climate change. A warmer climate results in a thicker active layer above the permanently frozen ground. In NW Canada, for example, traces of an earlier permafrost table are found at a depth of *c.* 1.5 m below the current land surface, characterized by large numbers of truncated ice wedges and abundant segregation ice. This layer formed during the Holocene climate optimum some 8000 years ago, when the tree line was *c.* 100 km further north than today (Burn 1997).

Figure 8.2 The melting surface 'active' layer of permafrost close to the border between Tibet and Xinghai provinces in China. Photograph by Phil Hughes.

The largest permafrost areas are located in northern Asia. The area east of the Ob River, Russia is almost completely covered by permafrost (Kotlyakov & Khromova 2002). The frost layer in the north of Siberia is several hundred metres thick, with maximum values of *c.* 1500 m. Such strong permafrost requires millennia to develop; it also takes thousands of years to thaw in a warmer climate.

Human interference can significantly accelerate this process. Extreme caution is therefore required for structures in permafrost regions (Figs 8.3, 8.4). New buildings, such as in Yakutsk (Fig. 8.5), are set on concrete piles so that air can circulate freely between the sole of the building and the ground, preventing warming of the surface.

In North America, the continuous permafrost extends from Alaska to the southern edge of the Hudson Bay. In the east it ends on the northern tip of the Ungava Peninsula. The southern boundary of the discontinuous permafrost is at the southern edge of James Bay. In the Mackenzie Delta, permafrost thickness is c. 700 m (Burn & Kokelj 2009). On Greenland, the third major subpolar landmass of the Northern Hemisphere, continuous permafrost is restricted to the areas north of 66° N. It does not reach as far south as in America and Asia, because the ice-free parts are all coastal areas and thus moderated by oceanic climatic influence.

Due to the effects of the Gulf Stream, there is almost no permafrost in Europe despite the continent having a latitude similar to the vast frozen areas of North America. In Europe, permafrost

Figure 8.3 A four-story house in Chersky, Kolyma district partially collapsed on thawing permafrost in 2001. The lettering on the overturned boat in the foreground says 'progress'. Photograph by Vladimir E. Romanovsky.

Figure 8.4 Thermokarst sinkhole next to the building of the Geophysical Institute of the UAF in Fairbanks, Alaska. The construction of the car park led to the thawing of permafrost. Photograph by Vladimir E. Romanovsky.

is restricted to the highest mountains and occurs at a range of latitudes from southern Spain to northern Scandinavia. This means that the European mountains are at the margins of the periglacial realm, and slight increases in air temperature can have profound effects on frozen ground (Harris et al. 2003).

The present distribution of permafrost in Europe is limited to the north of Scandinavia and to small parts of the high mountain regions (Harris et al. 2009). The local distribution of recent periglacial phenomena outside the subpolar regions is controlled by altitude, lithology (rock type), vegetation cover, aspect and slope. Altitude and vegetation cover strongly influence the severity of the winter frost. While the lower limit of the recent periglacial areas in Scotland is usually at 600–800 m, it may drop to values below 400 m asl in cases where the natural vegetation cover is disturbed. Rock differences determine the type of periglacial overprint. In quartzite, microgranite and granulite block fields are formed; sandstone and granite result in a coarse, cohesionless, sandy matrix with scattered stones; and on mica schist, clay and silt as well as on lavas a fine-grained diamicton usually forms. The exposure primarily affects the extent of aeolian overprint, which is highly dependent on the prevailing wind direction. Mass movement on slopes leads to the formation of stone stripes, turf terraces and solifluction sheets (Ballantyne 1987).

Figure 8.5 Modern buildings in permafrost areas, such as in Yakutsk, are set on concrete piles so that air can circulate freely between the floor and the ground surface. Photograph by Jürgen Ehlers.

8.2 Extent of Frozen Ground during the Pleistocene

During the Quaternary cold stages, the permafrost extended much further south and (in the unglaciated areas) also increased in thickness. At that time, thanks to the lowered sea levels permafrost extended to areas that are now submerged by the sea, notably in the north of Siberia (New Siberian Islands). Central Europe was in the permafrost zone during the great Quaternary glaciations right to the southern edge of the Alps. The permafrost boundary can be traced eastwards to the northern coast of the Black Sea and in the west to the French Atlantic coast in Brittany (Fig. 8.6). On the east coast of the United States, cold-stage permafrost reached south to at least Delaware (French et al. 2009).

As soon as an area is covered by an ice sheet, the frost penetrates no further into the ground. The base temperature of a glacier is near the pressure melting point all year round. The permafrost below the insulating layer of ice is gradually degraded by the geothermal heat from below.

While the former distribution of permafrost is quite accurately known, due to the presence of ice-wedge pseudomorphs and other periglacial phenomena, its thickness can only be estimated in a few cases. It would usually have amounted to several tens of metres.

Today, permafrost is in decline globally (Fig. 8.7). In the Mackenzie Delta area, the annual mean temperature has risen by 2.5°C in the last 40 years. The thickness of the active layer has increased by 8 cm from 1975 to 2008. In Europe, mountain permafrost is rapidly disappearing with an increase in ground temperatures of 1°C occurring in Svalbard (Norway) over the period of the twentieth century (Harris et al. 2003).

Figure 8.6 Cryoturbations in Quaternary deposits at La Mine d'Or, Brittany, France. Photograph by Uta Ehlers.

Figure 8.7 Excursion on the Aldan River, Yakutia. Summer temperatures are well above 20°C. The permafrost ice can be seen in the river bank. Photograph by Jürgen Ehlers.

8.3 Frost Weathering

When water freezes, its volume increases by 9%. When the water completely fills the cavities in the rock and then freezes, it produces a pressure which is able to break rock (Fig. 8.8). This process primarily takes place in the area a few centimetres below the rock surface where the pores are filled completely or almost completely with water, and where this water freezes quickly. Mineral grains and rock fragments are broken off in this way (Box 8.1). The expansion of freezing water can also widen fractures and result in frost heave. The formation of ice lenses plays a major role in frost weathering (Walder & Hallet 1986).

Large temperature differences between day and night, for example, wear the rock itself. Rock expands when heated and contracts when cooled. As the rock's ability to expand in the horizontal direction is limited, it will eventually break. This can take several freeze–thaw cycles (fatigue), but a rapid change in temperature may also result directly in frost damage. Rapid temperature changes of more than 2°C per minute can occur if the clouds tear open during the day, and the rock is suddenly exposed to solar radiation. Chemical weathering and biological processes additionally contribute to the decomposition of the rocks in the periglacial areas.

Frost shattering is particularly prevalent at high altitudes and many mountain summits in areas once covered by ice sheets display evidence such as tors and frost-riven bedrock. However, whether such summits stood above these past ice sheets as nunataks or were simply buried beneath inactive cold-based ice is open to debate. In Scandinavia, frost-shattered

Figure 8.8 Examples of frost-shattered rocks in Iceland. Photographs by Jürgen Ehlers.

BOX 8.1 FROST ACHIEVES WHAT CHARLEMAGNE COULD NOT

When Charlemagne fought against the Saxons, it is said he once suffered a defeat. He retreated with his army to a forest south of Hamburg (Fig. 8.9). The emperor went to rest next to a large stone and, on pain of death, forbade his men to wake him. He had hardly fallen asleep when the Saxons drew near. No one dared to wake the emperor. Finally, someone threw the Emperor's favourite dog on the sleeper. He jumped up angrily, but when he saw the approaching enemy, he grabbed his sword and shouted: "So certainly as I can cleave this stone, so certainly will I defeat the Saxons!" The sword penetrated deep into the rock, after which Charles rode over the stone, the dog followed suit, and both left their traces in the rock. When the Franks saw this miracle, they took fresh courage and defeated the Saxons.

An impressive performance, but although Charles thrust a deep notch into the stone it was by no means broken apart. Admiration for this achievement pales considerably if even the dog's paws left deep marks in the rock. In the general confusion at that moment, there was of course no time to wonder about such details. The horse was obviously very confused (note the unusual position of its feet when it went across the stone).

Figure 8.9 *Karlstein* in the Rosengarten, south of Hamburg. Photograph by Jürgen Ehlers.

So much for the legend; in contrast to the emperor, frost really can split large stones (Fig. 8.10). Frost damage in the Arctic and Subarctic can literally result in the bursting of major blocks, releasing a sound like a gunshot. Whether this explosive release of tension by thermal shock is caused by the expansion of freezing water or by the formation of ice lenses is unclear (Mackay 1999). In contrast to this dramatic process, however, most weathering processes occur very slowly.

Figure 8.10 Block destroyed by frost shattering on Iceland. Photograph by Jürgen Ehlers.

landforms pre-dating the Holocene are present within the limits of the last ice sheet, and their survival suggests minimal glacial erosion (Hättestrand & Stroeven 2002).

Frost shattering often produces upstanding rocky outcrops such as tors. The origin of tors in granite has long been the subject of debate. Based on the classic tors of Dartmoor, SW England, Linton (1955) proposed a two-stage formation where differential weathering in the subsurface is followed by the stripping of weathered regolith leaving a tor outcrop. This model invokes deep weathering of the granite during tropical climates of the Tertiary. An alternative hypothesis was proposed by Palmer & Neilson (1962), who suggested that tor development was a single-phase process occurring under periglacial conditions. This was influenced by their work in the Pennines (Palmer & Radley 1961). However, in the Dartmoor area there is clear evidence of deep weathering of the granite in quarries and pits.

It is unlikely that one single model can explain the tors that are observed in different lithologies and/or different settings around the world. For example, not all granite tors are necessarily the product of frost-shattering or indeed deep tropical weathering. In coastal areas, cliff erosion can produce upstanding rock formations. On the island of Lundy in the Bristol

Figure 8.11 Tors on the west coast of Lundy, an island off the coast of SW England. This tor, part of a collection of tors known as *The Cheeses*, is typical of the coastal fringes of the island and the tors only appear on the coastal escarpment edge. Their formation is polygenetic and involves coastal erosion in producing the stacks, although frost heave is likely to promote joint widening. An ice sheet has also moulded some pre-existing tors in this area (Rolfe et al. 2012). Photograph by Phil Hughes.

Channel area of SW England, tors are widespread around the edges of the coastal escarpment. Many of the joints in these granite tors display evidence of frost heave, although the absence of tors away from the coastal escarpment edge suggests that cliff erosion is the primary mechanism for the production of these tors (Fig. 8.11). Tors are therefore a classic example whereby similar landforms can be formed by a range of different processes, a situation known as equifinality. That said, frost shattering (Fig. 8.12) does produce some fine tors and such landforms are often emblems of periglacial weathering.

8.4 Cryoplanation

Weathering and subsequent erosion processes eventually lead to levelling of the terrain. Planation under the influence of periglacial processes is called cryoplanation. Frequent freeze–thaw cycles lead to a rapid decomposition of solid rock into coarse frost rubble. In this way, block fields and block slopes are formed on level or gently sloping terrain.

Comparative studies within and outside the reach of the last ice advance in Scotland, the 'Loch Lomond Re-advance' of the Younger Dryas, have shown that the formation of frost weathering and block fields occurred predominantly during the rigid permafrost conditions of the Weichselian Pleniglacial. During the Younger Dryas, there has only been a slight additional periglacial overprint of the ground surface. Slopes and plateaux are covered with a debris mantle in the Scottish Highlands and the highest mountains of England and Wales (Fig. 8.13), the depth of which generally lies at 0.5–1 m. The absence of erratic material

Figure 8.12 Frost-shattered bedrock on the summits of the Glyder Mountains in Wales. These summits are thought to have been frost shattered during ice sheet retreat when they stood above the ice as nunataks (McCarroll & Ballantyne 2000). However, the well-developed frost shattering may have required multiple cold stages to form and frost shattering on these summits may have been preserved under cold-based ice during more than one glaciation. Photograph by Phil Hughes.

Figure 8.13 A level area of blockfield (cryoplanation surface) on the summit ridge of the Glyder Mountains in Wales. Photograph by Phil Hughes.

Figure 8.14 Granite blocks weathered by granular disintegration, Cairngorms, Scotland. Photograph by Jürgen Ehlers.

indicates that it is not a former moraine, but the result of *in situ* frost weathering. The frost heave often results in a concentration of blocks at the land surface, while the fines are enriched below (Ballantyne 1984).

Mechanical frost weathering often results in angular rock fragments. This depends on the type of parent material, however; under certain conditions even round shapes can arise, such as in the weathering of granite in the Cairngorms area in Scotland (Fig. 8.14; Ballantyne 1998).

Patches of snow lead to increased weathering which results in the sorting and removal of rock material (nivation) over the course of freeze–thaw cycles. This kind of erosion can produce slope-parallel terraces, which are referred to as cryoplanation terraces. Cryopediments are also to be found on the lower slope area.

The formation of the terraces takes place in three phases: (1) nivation results in the formation of a trough or step; (2) a terrace forms traversed by frost wedges, and enhanced solifluction and slopewash occur on the terrace step (inclined by about 7°); and (3) further retreat of the step will eventually create a hilltop planation, on which hardly any material is redeposited. Frost sorting and blow-out (deflation) may also play a role (Washburn 1979).

8.5 Rock Glaciers: Glaciers (Almost) Without Ice

Tongue-shaped accumulations can be found in the frost-debris zone of many high mountains, the surface shape of which is reminiscent of a glacier. These features are usually several hundred metres long, 100–150 m wide and 40–50 m thick, and are referred to as rock glaciers. They are not glaciers, however, but a mixture of rock debris and ice that has formed under permafrost conditions. True rock glaciers (Figs 8.15, 8.16) are not 'normal' glaciers buried under debris but have formed from scree or pre-existing moraine debris, in which ice has accumulated in the interior. They are consequently real permafrost features (Barsch 1996). Active rock glaciers move slowly down the valley, usually at a rate of a few centimetres to tens of centimetres per year. During the forward movement, the rock glacier moves over its own debris deposited at the foremost margin (Kääb 2007). Inactive rock glaciers still have a core of ice and debris but no longer move. Relict rock glaciers are ice free; however, they are witnesses of past climatic conditions. Unlike real glaciers, the outermost edge of the glacier tongue cannot recede during climatic warming so the rock glaciers remain in place. Relict rock glaciers are therefore often well preserved in the landscape and can provide useful palaeoclimatic information. This fact has been exploited in the Pindus Mountains, Greece, where the presence of relict rock glaciers has been used to demonstrate that temperatures were 8–9°C lower during the Late Pleistocene than today (Figs 8.17, 8.18; Hughes et al. 2003).

The lower distribution limit of active rock glaciers is roughly equal to the distribution of discontinuous permafrost. This relationship can be used to reconstruct the permafrost

Figure 8.15 Rock glacier at Green Lake, Idaho, overview. Photograph by Jürgen Ehlers.

Figure 8.16 Rock glacier at Green Lake, Idaho, detail. Person for scale (on the right). Photograph by Jürgen Ehlers.

Figure 8.17 The relict Tsouka Rossa rock glacier in the Tymphi Massif, Pindus Mountains, Greece. Hughes et al. (2003) argued that this feature must be Late Pleistocene in age. Photograph by Phil Hughes.

Figure 8.18 The relict Tsouka Rossa rock glacier in the Pindus Mountains, Greece, viewed from above. The arcuate transverse ridges on the lobate mass of boulder debris are characteristic of rock glaciers. Photograph by Phil Hughes.

distribution during previous glaciation phases (e.g. Patzelt 1983). The lower limit of rock glaciers is also nearly identical to the 'climatic snow line', that is, the theoretical limit on level ground above which (over the long-term average) snowfall is not completely exceeded by snow melt. With increasing continentality however, the gap between the snow line and the rock glacier distribution widens. The origin of rock glaciers is not only dependent on climate, but also on the substrate and the rate of debris supply (Humlum 1998). They are preferentially formed in rock types that are weathering to coarse blocks such as granite, basalt and sandstone (Karte 1979). Without boreholes, real rock glaciers are difficult to distinguish from normal glaciers buried under debris (debris-covered glaciers).

8.6 Involutions

Every farmer knows that stones grow in fields. Even if they are picked up and completely removed one year, new stones will have turned up the following spring. Since stones do not actually grow like mushrooms, there must be other causes of this phenomenon. The re-emergence of the stones is a result of frost heave. Upon freezing, needle-like ice crystals (pipkrake) grow perpendicular to the cooling surface, uplifting soil substrate and stones by their growth. When warmed by the sunshine, the ice below stones thaws later than in the surroundings so that fine material moves in, and the stone is uplifted relative to its environment. For this gradual redistribution of sediment, multiple freeze–thaw cycles are required but no permafrost.

Irregular folds and deformations of near-surface layers under the influence of freeze–thaw cycles are called cryoturbations after Edelman et al. (1936; Fig. 8.19). They occur mainly on

Figure 8.19

Cryoturbations in soliflucted ground at New Wulmstorf, Lower Saxony, Germany. Photograph by Jürgen Ehlers.

level or gently sloping terrain in stratified sediments of different grain size. In cases where the origin by freeze–thaw cycles has not been clearly established, the features should be referred to as involutions (Vandenberghe 2007). They include festoons and sag features. Three different interpretations have been offered for their origin (French 2007):

1 Some authors assume that the formation of involutions occurs in unfrozen soil layers between the permafrost table and the seasonal frost at the beginning of the winter, starting from the surface. Grain size differences and hence different water contents freeze at different rates, so that there are already frozen soil pockets to be found between unfrozen material. The freezing soil increases in volume, causing sediment layers to bend and deform horizontally and vertically. Although laboratory studies have shown that such processes may occur, the necessary cryostatic pressure could not be proven in field experiments (e.g. Mackay & MacKay 1976). Instead, in the unfrozen zone between the permafrost and the refreezing active layer, dehydration and overconsolidation occur due to the migration of water to the frost layers above and below (Mackay 1979, 1980). Refreezing of the active layer is therefore characterized by the absence of a viscous, semi-liquid soil mass, which could be deformed in the manner described above.
2 In many cases cryoturbations have been interpreted as the result of moisture-induced density differences between sediments of different grain size. They are therefore genetically close to load casts. Particularly favourable conditions for the formation of such features would have been present in the ice-rich environment of thawing permafrost (Eissmann 1981). Excessive water saturation of the ground in the late summers by thaw of ice-rich sediments and heavy rain may also cause this effect at the base of the active layer. This can lead to sediment liquefaction, and inversion of the density gradient can lift underlying sediments up to the surface.
3 Cryoturbations can also be caused by sediments churned by frost heave and the formation of segregation ice. This process has been confirmed by laboratory tests and field measurements (Washburn et al. 1978; Mackay 1980). The widespread occurrence of sorted polygons is attributed to this mechanism.

Caution is recommended when it comes to the palaeoclimatic interpretation of involution horizons. Through his research on Banks Island, French (1986) demonstrated that cryoturbations and load casts have a great similarity in appearance although they form under very different conditions (presence or absence of permafrost).

Most periglacial disturbances of the ground reach only a few decimetres deep. However, from the lignite mines of Central Europe much larger and strongly deformed layers are known, culminating in the emergence of whole series of diapirs (Fig. 8.20). In formerly glaciated regions such features were initially attributed to glaciotectonics. As equivalent forms also occur outside the glaciated area, it gradually emerged that they are large periglacial features (mollisol diapirs; Eissmann 1978). Under favourable conditions, several generations of such features can be distinguished. Eissmann proved that, during glacial periods of the Pleistocene, diapirs were formed in the Leipzig lowlands. Here, the pre-Elsterian features were found to be much weaker than the diapirs of the Elsterian, Saalian and Weichselian glaciations. From the depth of the buoyant forms, the minimum thickness of the former permafrost table can be estimated. During the Elsterian Glaciation

Figure 8.20 Brown coal diapirs in the opencast Schwerzau mine (Sachsen-Anhalt, Germany). Photographs by Stefan Wansa.

Figure 8.21 Drop soil in Weichselian meltwater sands at Earith, East Anglia, UK. Photograph by Jürgen Ehlers.

it was 18 m, in the Early Saalian Glaciation 30 m, in the Saalian glacial maximum 40 m and in the Weichselian Glaciation 50 m (Eissmann 1981; Fig. 8.21).

8.7 Solifluction

The slopes of the periglacial zone are areas of strong sediment movement. Apart from slopewash, solifluction (a slow, downslope mass movement; Fig. 8.22) plays a crucial role. Solifluction can occur under certain conditions (high water saturation, steep slopes, swellable clay minerals, etc.) even without a periglacial climate. However, the term solifluction is mostly used in connection with periglacial solifluction; non-periglacial slope movements are referred to as soil creep.

Periglacial solifluction is triggered by the high water content of the active layer. This leads to a reduction of the shear strength of the substrate and, after exceeding the yield point, the material moves slowly down the slope. At an angle of *c.* 2°, the active layer may begin to flow. However, the threshold value is dependent on the substrate. According to Ballantyne (1987), recent periglacial solifluction in Scotland requires a slope of 5–7.5°. Where the vegetation cover is missing or disturbed, it can result in rapid mass movement. Sorted

Figure 8.22 Solifluction in solid rock; the outcropping strata are dragged down the slope (outcrop bending). Exposure south of Ehrenbach near Idstein (on Taunus, a mountain NW of Frankfurt, Germany). Photograph by Jürgen Ehlers.

Figure 8.23 Solifluction terraces (terracettes) on Harris, Outer Hebrides, Scotland. Photograph by Jürgen Ehlers.

polygons are distorted to stone stripes by moving downslope. In extreme cases, the moisture content is so high that a highly mobile mud forms which flows very rapidly.

Depending on the vegetation cover, a distinction is made between free and bound solifluction. The latter results in a number of characteristic landforms such as terracettes (Fig. 8.23) and flow lobes. An overview of the features which occur was given by Karte (1979). During the Pleistocene cold stages there was widespread solifluction in Germany. The resulting landforms have often been altered or destroyed afterwards, but the solifluction sediments are still detectable in many places (Box 8.2).

BOX 8.2 SOLIFLUCTION IS BROUGHT TO COURT

On the mild autumn day of 20 October 1923, the young geologist Karl Gripp led an excursion to the old morainic landscape south of Hamburg. When he pointed out a series of deeply incised dry valleys near Klecken, he told his audience that these distinct landforms were caused by groundwater erosion. Present on the excursion was Professor Siegfried Passarge, who had a different view. At a discussion meeting later that month, he maintained that those valleys were not formed by water but by solifluction. Gripp adopted this interpretation later in his article about the outermost boundary of the last

(continued)

BOX 8.2 SOLIFLUCTION IS BROUGHT TO COURT (CONTINUED)

glaciation in northwestern Germany, and also used it in his application for research funds for an expedition to Spitsbergen.

Passarge was outraged at the apparent plagiarism, or was it? By then solifluction had been known of for a long time; Andersen had introduced the term in 1906. It was pointless to argue over who had used this word in a particular area first. The participants saw it differently, however, and the conflict grew quickly out of control. Colleagues and students were drawn in, the tone turned offensive and finally the argument ended in court.

On 2 June 1932 the *Landgericht* in Hamburg returned the following verdict: Professor Siegfried Passarge was prohibited under threat of a fine of unlimited size or the penalty of imprisonment for up to 6 months in the case of the infringement claim, to repeat that the plaintiff had borrowed his ideas and used them as the basis for his travel application to Spitsbergen, and that his scientific reputation was built on an unfair, unscholarly action. Passarge had to bear the costs of litigation.

Passarge went to the appeal, but the sentence was confirmed. However, the cards were reshuffled soon after. The Nazis had seized power, solifluction was no longer relevant in the dispute between the two scientists and court rulings no longer counted. In April 1933, Passarge wrote to the new Senator, "Mr. Gripp is an outspoken leftist, probably a democrat... Because of my well-known political attitude towards Jews and Marxists I was hated to the utmost by the past rulers... " In 1934, 43-year-old Gripp was forced into retirement (Krause et al. 1991).

8.8 Periglacial Soil Stripes

In flat areas of East Anglia where chalk is present just beneath the surface, often covered by only a thin layer of aeolian sand, polygons have formed which merge at a certain angle to about 7 m wide stripes. Under favourable conditions, these features can be seen on arable land from the differences in ground colour. They are also easy to detect from the differences in vegetation cover. At the INQUA meeting in Birmingham in 1977, the latter was demonstrated at Grimes Graves in East Anglia (Fig. 8.24). Heather (*Calluna vulgaris*) and gorse (*Ulex europaeus*) were growing on the sandy stripes, while the intervening chalk surfaces were covered with grass.

At that time, nobody knew how widely distributed these periglacial frost-patterned ground patterns really were. This became obvious when an analysis of aerial photographs showed that large areas of East Anglia are covered with such soil stripes (Fig. 8.25). They are only visible when moisture differences are reflected in the vegetation, reflecting the composition of the shallow subsurface. Periglacial patterned ground is not only found in England but in most areas that formerly belonged to the periglacial zone.

(a)

(b)

Figure 8.24 (a) Periglacial soil stripes at Grimes Graves in East Anglia, UK (TL 810 902). (b) Excavation of a periglacial soil stripe at Grimes Graves, East Anglia, UK. Photographs by Jürgen Ehlers.

8.9 Frost Cracks and Ice Wedges

A characteristic feature of periglacial areas is frost-patterned ground. On flat surfaces and in areas with very gentle slopes circular or polygonal patterns prevail. Large parts of northern Siberia are covered by such a honeycomb pattern. We distinguish between non-sorted and sorted polygons (Fig. 8.26). The non-sorted polygons are essentially a result of desiccation of the soil (dry cracks) and/or ice wedge formation, whereas sorted polygons are caused by frost dynamic ground movements (fine material concentrations in the centres). The sizes of the polygons vary from very small (Fig. 8.27) to over 100 m in diameter.

Wedge-shaped veins of ice (Fig. 8.28), which reach several metres depth into the ground, are found in recent periglacial areas. The formation of those ice wedges begins in the form of thin veins that grow over the years to ever greater thickness. Large ice wedges form preferentially in fine-grained or peaty substrate. In pure sand, forms of more than 0.1–1 m width are rarely seen (Eissmann 1981). Large ice wedges can reach down to depths of more than 30 m. They can expand laterally so much that the soil in between is compressed into soil columns with the ice volume significantly exceeding the volume of soil. The ice may finally form coherent ice masses of more than 80 m thickness, as for example on the New Siberian Islands; this type of ground ice is referred to as 'yedoma'.

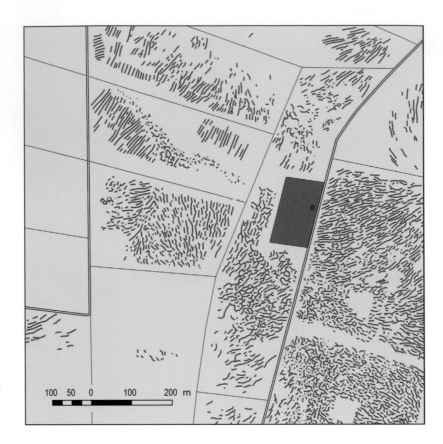

100 50 0   100   200 m

Figure 8.25 Periglacial soil stripes mapped from aerial photographs southeast of Warham, East Anglia, UK.

Figure 8.26 Sorted polygons at Juvasshytta, Norway, diameter *c.* 6 m. Photograph by Jürgen Ehlers.

Figure 8.27 Small polygons on Iceland, diameter *c*. 30 cm. Photograph by Jürgen Ehlers.

The formation of ice wedges is strongly dependent on the substrate. On the basis of investigations in recent periglacial areas of the Soviet Union, Romanovsky (1985) reported that ice wedges form at an average annual temperature of −5.5°C in sand and gravel; at −2.5°C in clay; and −2.0°C are required in peat. Consequently, ice wedges are formed in loess at higher temperatures than in sand. Such figures can only be taken as a rough guide, however. Conclusions about the climate that led to the formation of certain periglacial phenomena remain problematic. The difficulty lies in relating the former soil temperature to that of the air; there can be large differences depending on the thickness of the winter snow cover (Pissart 1987).

Low temperatures are a prerequisite for the formation of ice wedges, but they are not the only deciding factor. Especially important is a sudden decrease in temperature within the soil in the depth range of 5–10 m. This is most easily achieved if an insulating ground cover is lacking (e.g. vegetation or snow), and if the active layer is only of small thickness in the summer (Black 1976).

Ice wedges are often sediment traps filled in by slumped or windblown sand when the ice thaws. The resulting ice-wedge pseudomorphs (Fig. 8.29) are found in numerous outcrops in former periglacial areas. Fossil ice wedges are not limited to the present ground surface; occasionally several generations of ice wedges are found. Systematic investigations have been made in the Leipzig area, especially by Eissmann (1981).

Figure 8.28 Recent ice wedges in Yakutia.
Photograph by Jürgen Ehlers.

Figure 8.29 Ice-wedge pseudomorphs in Delitzsch-SW opencast mine, Germany. Main terrace gravel of the Mulde River overlying Lower Elsterian till. Photograph by Jürgen Ehlers.

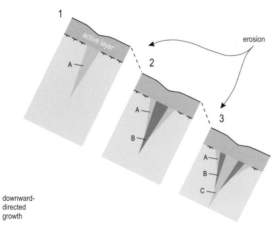

Epigenetic ice wedges

Syngenetic ice wedges

upward-directed growth

accumulation

Anti-syngenetic ice wedges

erosion

downward-directed growth

Ice wedges are normally subdivided into three groups (Mackay 1990; Fig. 8.30):

1 Epigenetic ice wedges are younger than the surrounding material. Because ice wedges usually tear in the middle and since they are V-shaped, an epigenetic ice wedge grows over the course of its development in width, but by very little in depth (Mackay 1974). Due to the increase in ice volume, the sediment layers between the ice wedges are gradually bent upwards.

2 Syngenetic ice wedges occur in permafrost areas with recent accumulation. If the sediment supply and the extent of frost-crack formation are in balance, they grow simultaneously with the raising of the ground surface. They are formed in fluvial sediments, peats and in solifluction deposits. Syngenetic ice wedges are as old as the surrounding sediment. They often have the appearance of several wedges which have been stacked within each other. The oldest ice is found in the deepest part of the ice wedges.

3 Anti-syngenetic ice wedges form when a land surface is lowered under continuing permafrost conditions. Accordingly, the ice wedges grow (in absolute terms) in depth. The oldest ice is found at the upper outer edge in an anti-syngenetic ice wedge.

Figure 8.30 Schematic representation of epigenetic, syngenetic and anti-syngenetic ice wedges. For each type, three growth stages are shown. The oldest ice is located: in the flanks of an epigenetic ice wedge; in the outer, lower part of a syngenetic ice wedge; and in the outer upper part of an anti-syngenetic ice wedge. Adapted from Mackay (1990).

In Central Europe, relict and fossil soil frost phenomena are widespread. Former ice-wedge networks in till plains can be identified under favourable conditions in aerial photographs by the light humidity differences between the infill and the surrounding material reflected in vegetation differences.

Under particularly favourable conditions, former ice-wedge networks can even be reconstructed in outwash plains. Since the humidity difference between wedge infill and surroundings is very low in this case, such networks are not visible from a single aerial survey; some small sections can be identified, however. Seven out of nine aerial photographs of the Harksheide Sandur (north of Hamburg; imagery available from the Hamburg *Landesvermessungsamt*) show part of an extensive ice-wedge network, allowing the character of the original network to be largely reconstructed (Ehlers 1990). The same shapes are found throughout any particular ice-wedge network; pentagonal and hexagonal shapes dominate, with a maximum diameter of about 20 m. Rectangular features are found sporadically.

Weichselian ice wedges were not limited to the area outside the ice boundary. Ice wedges may occur even within younger moraines (but mainly smaller forms). From detailed studies in Jutland, Svensson (1984) confirmed that large ice-wedge networks are limited to the area beyond the most recent glaciation. The rapid warming at the end of the Weichselian Glaciation prevented further ice-wedge formation; it was only during the cold reversal of the Younger Dryas that smaller frost cracks could once again form (Böse 1991).

8.10 Pingos, Palsas and other Frost Phenomena

8.10.1 PINGOS

The term 'pingo' comes from the language of the Inuit in the Mackenzie Delta, meaning 'small hill' (Washburn 1979). In the Russian language the same features are referred to as 'bulgunnyakh'. Pingos are individual conical hummocks with a round to oval shape. Their diameter ranges from a few metres up to a maximum of 1200 m and their height up to a maximum of 100 m (although their diameter is usually 20–300 m and height 5–70 m); the flanks are relatively steep (up to 35°). In addition to such eye-catching landforms, there are transitions to smaller bulges with a different outline. The core of pingos consists of frozen soil material with a very high percentage of pure ice. The overlying heterogeneous soil layer is 1–10 m thick (French 2007), and protects the core from summer thaw. Pingos are therefore perennial forms, in contrast to simple frost heave features (Karte 1979).

According to their genesis two types of pingos can be distinguished, named after their main distribution areas:

1. Mackenzie-type pingos (Fig. 8.31) form in a closed hydrologic system. They occur in shallow lakes, under which an 'island' of unfrozen soil (talik) occurs within the permafrost zone as a result of the insulating effect of the water. If the lake drains (e.g. by tapping) or falls dry by sediment infill, the insulating effect of the water surface is reduced so that the volume

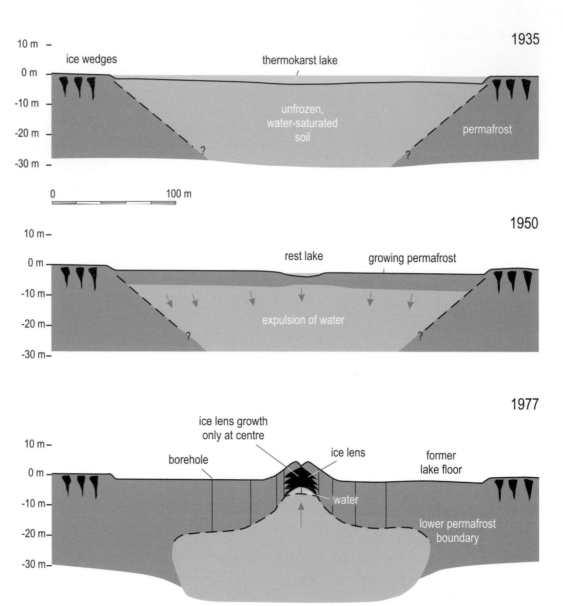

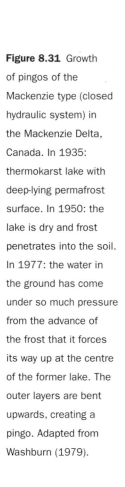

Figure 8.31 Growth of pingos of the Mackenzie type (closed hydraulic system) in the Mackenzie Delta, Canada. In 1935: thermokarst lake with deep-lying permafrost surface. In 1950: the lake is dry and frost penetrates into the soil. In 1977: the water in the ground has come under so much pressure from the advance of the frost that it forces its way up at the centre of the former lake. The outer layers are bent upwards, creating a pingo. Adapted from Washburn (1979).

of the unfrozen soil located under the lake gradually reduces from all sides. The hydrostatic pressure of the trapped water eventually rises to the extent that a breakthrough occurs, the water freezes and the overlying soil is lifted up. The rate of pingo growth in its early stages is *c.* 1.5 m a^{-1}. It later decreases continuously until the talik has been entirely replaced by permafrost.

2. East-Greenland-type pingos (Fig. 8.32) form in an open hydrologic system. They occur in areas with discontinuous permafrost, where water seeps in from the flanks. This water comes under hydrostatic and cryostatic pressure at the lower slope or the valley floor, so that an injection into the permafrost occurs or a body of segregation ice forms. Such pingos often occur in small groups. In the subpolar regions of the Arctic, they are often found on south–southeast-facing slopes (French 2007).

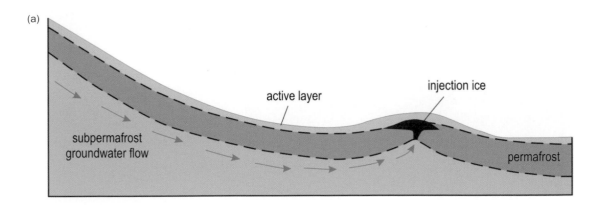

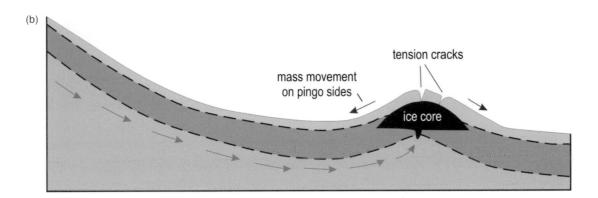

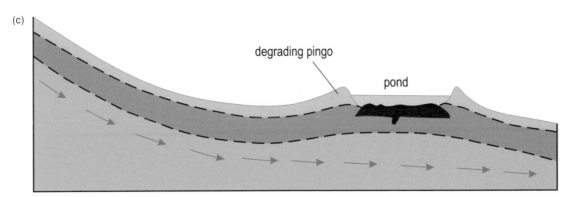

Figure 8.32 Formation and decay of pingos of the East Greenland type (open hydraulic system). (a) Below the permafrost ceiling, circulating groundwater is injected at a weak spot into the permafrost and an ice core is formed. (b) As the ice core increases in size, the overlying sediment layers are bent upwards until they crack and start sliding down the flanks. (c) A central crater is formed and the unprotected ice core starts to melt. Adapted from Harris & Ross (2007).

Figure 8.33 Pingo in Yakutia. The edge of the former thermokarst lake is visible in the background. Photograph by Jürgen Ehlers.

Some pingos are very young forms, others have been dated by radiocarbon to 4500-7000 BP. It is believed that many pingos formed during the cooling phase at the end of the Atlantic period. Both from Siberia and North America, however, cases are known where pingos have formed only during the last decades (Washburn 1979).

If excessive growth of the ice lens tears the protective sediment cover on the pingos open, the sediment starts to slide down the flanks (Fig. 8.33). This eliminates the insulation and the ice core melts, leaving behind a small lake which may be surrounded by a low annular rampart. Remnants of Pleistocene pingos have been described from many areas of central and southern Germany. A number of those landforms may, however, also have formed due to other causes.

Worldwide there are about 5000 pingos. The largest cluster of pingos is found in the Mackenzie Delta. On the Tuktoyaktuk Peninsula alone, there are over 1300 of these frost mounds. The Mackenzie Delta became free of glacial ice at *c.* 12 ka, and peat bogs have spread in the subsequent warming. It is only in recent centuries, while the climate has been cooling, that pingos have developed. Most North American forms are approximately 4000–7000 years old. Pingos can be identified in satellite imagery acquired during the winter due to their shadows; the summer image clearly shows the permafrost area penetrated by thaw lakes (Fig. 8.34). Pingos are mainly found in the Canadian Arctic, Siberia, Greenland and Spitsbergen (Fig. 8.35).

8.10.2 PALSAS

Local bulges of ice lenses also occur within the bogs of the periglacial areas. These palsas are similar to pingos, but much smaller. They usually reach a height of only about 1–10 m and are usually 10–30 m wide and 15–150 m long. In essence, they consist of permanently frozen

(a)

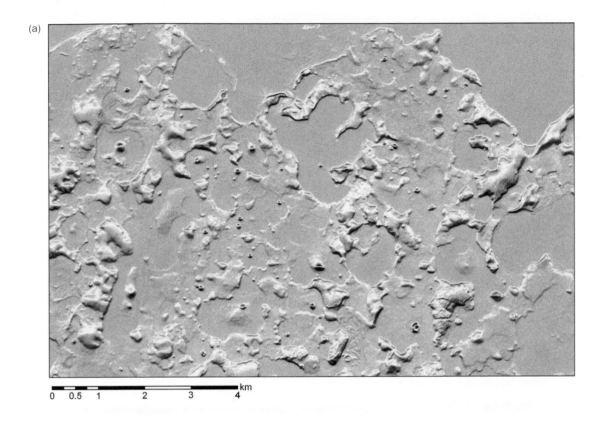

(b)

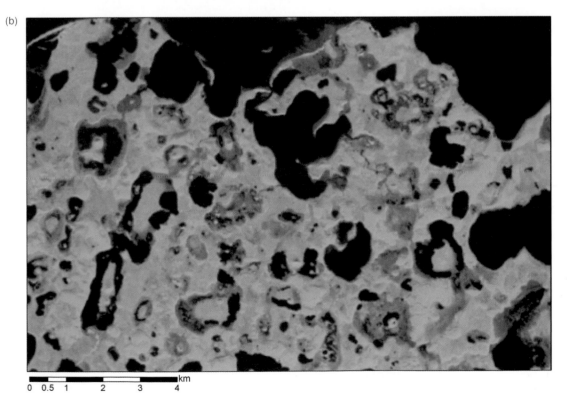

Figure 8.34 Landsat 7 satellite images of pingos in the Mackenzie Delta: (a) winter, (b) summer.

Source: US Geological Survey, Landsat 7 ETM, Path 64, Row 11, 4.2.2003, and Path 63, Row 11, 12.6.2000.

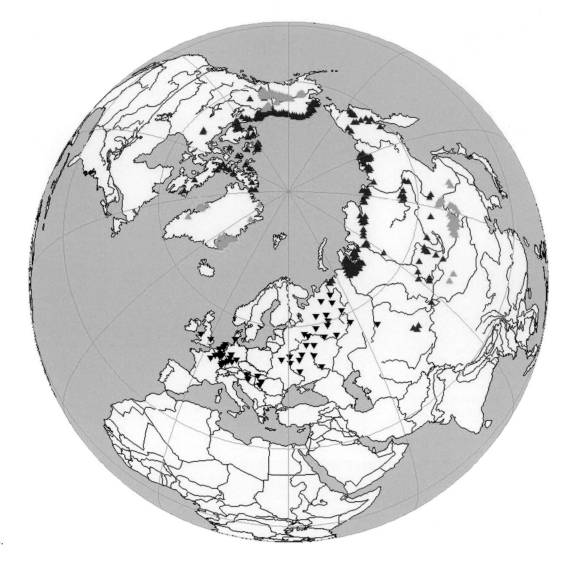

Figure 8.35 Global distribution of pingos. Red: Mackenzie type; green: East Greenland type; and black: Pleistocene pingo scars.

peat and/or silt with lenses of segregation ice and small ice crystals. Palsas owe their origin to a non-uniformly distributed winter snow cover. In places with a thin snow cover, the frost penetrates deeper into the ground. This leads to an accumulation of ground ice, including the sediment layers beneath the bog. Numerous ice lenses of a few centimetres thickness form, bulging up the peat. The bulge reduces the future snow cover, so that the process is self-reinforcing.

Palsas typically occur in areas of discontinuous permafrost. While on Iceland the southern border of palsa distribution roughly matches the 0°C annual mean temperature, in Sweden it is at about the 2–3°C annual mean temperature (Washburn 1979). The presence of palsas in Scandinavia today is largely confined to areas north of the treeline. The southernmost occurrence is found on the Dovre Fjell in Norway (Sollid & Sørbel 1998).

Palsas are relatively young landforms; their formation only began after the postglacial thermal maximum. They are only able to form if a sufficiently thick layer of peat exists, required for the insolation of the ice lens in summer (c. 0.7–0.8 m).

Figure 8.36 The southernmost palsas in Norway at Furuhaugli, showing traces of ice decay. Photograph by Jürgen Ehlers.

The age of palsas has been determined by ^{14}C dating of the overlying peat. When the frost mound begins to rise above its surroundings, the moisture conditions change and a different kind of peat is formed. While the original bog formed from *Sphagnum* (peat moss), *Carex* (sedges) and *Eriophorum* (cotton grass), on palsas *Ericales* (Heath), *Bryales* (mosses) and *Cladina* (lichen) can be found. For dating purposes, samples are taken immediately above and below the contact surface between the different types of peat. Studies in Lapland have shown that all palsas are <1000 years old. Even today, new palsas are being formed.

At present, however, the tendency to melt predominates (Seppälä 2005a, b; Fig. 8.36). The collapse of palsas begins with rupture of the protective peat blanket. The ice core starts to melt and a thermokarst lake is formed. No permanent traces remain in the soft organic matter over the long term. Consequently, glacial palsas have not yet been detected in Central Europe.

8.10.3 STRING MIRES

String mires (Aapa mires; Fig. 8.37) are special types of peat formation which occur at the edge of the northern distribution limit of the raised bogs. Raised bogs only form islands above the blanket bogs; on sloped terrain, the raised bogs create up to 2 m high and several metres wide ridges ('strings') of peat instead of islands. The distances between the ridges are usually about 10–50 m. The uneven surface of the Aapa mires is caused by differences in snow cover. Frost penetrates into the ground, preferentially in the ridges; seasonally, they contain ice lenses. By frost heave, the ridges gradually rise up out of the influence of the groundwater, attaining an extreme high moorland character in the process. The peat growth finally comes to a halt, and lichens begin to spread. The ridges are often arranged approximately parallel to the

Figure 8.37 Small string mire in Sweden. Photograph by Jürgen Ehlers.

slope contours or slightly curved. For steeper slopes, real terraced Aapa mires occur. At even greater inclinations, the ridges are arranged perpendicular to the contours and the distance between them is reduced to 3–4 m (Hallik 1975). The ridges are not stationary, but move under the influence of frost–thaw cycles. Long-term measurements have shown that they can move some 2.5–7.5 cm a^{-1} (Koutaniemi 1999).

The string mires reach significantly further south beyond the current permafrost distribution. Their origin is not bound to permafrost, but only to harsh winter frost. Radiocarbon dates of various Aapa mires have shown that the formation of the ridges began at *c.* 2–3 ka (Seppälä 2005b).

8.10.4 THUFUR

Thufur are a special type of frost-patterned ground (e.g. Fig. 8.38). These soil lumps reach a height of 0.1–0.5 m, and are formed mainly in fine-grained substrate with a minimum thickness of 0.3–0.4 m. The name comes from the Icelandic language, since thufur are most widespread in that country. Thufur are not restricted to Iceland, however, but occur in periglacial environments around the world. They only emerged after the Holocene climatic optimum.

Van Vliet-Lanoë et al. (1998) concluded that the occurrence of volcanic loess on well-drained ground, such as outwash, is of crucial importance for the origin of thufur. The internal structure of the hummocks suggests that cryoturbation plays an essential role in their formation. Thufur can be considered as a special form of sorted patterned ground. The sorting is evidently caused by a differentiated intrusion of ground frost depending on the vegetation

Figure 8.38 Thufur in Iceland. Photograph by Jürgen Ehlers.

cover. The insulating effect of the vegetation contributes to the preservation and further development of the thufur.

Similar small-scale relief is found on the *Buckelwiesen* (hump meadows) in the Alpine area. Most *Buckelwiesen* are found in the larger valleys of the Limestone Alps in Central Europe (e.g. at Mittenwald, in Bavaria, Germany), where they have developed on calcareous, glacial till and gravel deposits as well as on carbonate rocks with a thin sedimentary cover. They are also found in the high mountains of the Balkans as far south as Greece and in formerly glaciated limestone mountain areas across Europe. Their earliest possible age of formation is therefore the end of the last Ice Age, after the melting of the large glaciers. The *Buckelwiesen* are mainly a karstic phenomenon found on soluble rocks; cryoturbation has played only a minor role (Zech & Wölfel 1974).

Rancho La Brea, Los Angeles, California, USA. Photograph by Jürgen Ehlers.

Chapter 9
Hippos in the Thames: The Warm Stages

9.1 Tar Pits of Evidence

In the middle of Los Angeles, just north of Wilshire Boulevard, there is a small pond with the statue of a sinking *Mammuthus columbi* (Figs 9.1, 9.2). Only the occasional rise of gas bubbles indicates that this is not a normal pond. Tar was previously mined at this point, which is also the most important Pleistocene mammal site in North America. The natural asphalt deposits at this point, Rancho La Brea, had already been excavated for some time. It was only at the end of the nineteenth century that someone noticed that the asphalt did not contain cattle bones as previously thought, but the bones of extinct animals. Until 1912, various excavations were conducted by the University of California and later by the Museum of Los Angeles County (LACM). Excavations continue to this day. While the attention of early archaeologists centred on the bones of large mammals, today small fossils, including diatoms, are also included in the investigations.

Rancho La Brea has stimulated the imagination of researchers for a long time. Stock (1929) described the asphalt ponds as:

> ... *downright cancerous tumors in the face of nature, where the death drive with the comparative calm of life, his practical jokes ... The shouting and the agony of the wounded animals, the stench of decay and the desperate struggle of those who had not yet been swallowed by the swamp may have turned the event into a nightmarish morass of its horrors and injustices of the progress of geological time now cast a merciful veil.*

> Stock (1929)

The Ice Age, First Edition. Jürgen Ehlers, Philip D. Hughes and Philip L. Gibbard.
© 2016 John Wiley & Sons, Ltd. Published 2016 by John Wiley & Sons, Ltd.
Companion website: www.wiley.com/go/ehlers/iceage

Figure 9.1 Mammoth reconstruction. Source: Zimmermann (1885).

Figure 9.2 Mammoth skeleton in the *Naturkunde- und Mammutmuseum* in Siegsdorf, southeastern Bavaria. Photograph by Karl Stankiewicz.

Reality was not quite so dramatic, however. Although numerous animal skeletons have been recovered, it must be borne in mind that they accumulated over a period of about 25–30 ka. For the 3400 large mammals found (coyote or larger), that means that every 8 years one animal was killed in the tar swamp. Since not all the skeletons have been preserved, this is a minimum number. Assuming that the findings represent only 10% of the actual victims, the result is still an average interval of 10 months between the individual events (Marcus & Berger 1984).

It is interesting that the proportion of carnivores (meat eaters) among the large mammals recovered is unusually high. In a comparable modern environment (in Africa), the carnivores account for about 4% of mammals; in Rancho La Brea, they amount to some 85%. The animals caught in the asphalt had apparently attracted a large number of predators, which themselves in turn became caught in the deadly trap (Marcus & Berger 1984).

9.2 Development of Fauna

The large mammals of the Ice Age have long caught the imagination of people. It was known as early as in the eighteenth century that in Siberia large quantities of prehistoric mammal bones were found. What did the animals whose bones were found look like? Various mammoth reconstructions were attempted in the nineteenth century, but the question of whether the tusks were bent upwards or downwards had not yet been finally resolved.

With every change from cold to warm climate and back, the plants and wildlife were forced to react as well. Because of the short periods of the climate changes, they mostly resulted not in evolutionary adaptation processes but rather in shifts in distribution. During the early cold stages, the Arctic forms would migrate from the northeast into Central Europe, whereas the interglacial plants and animals died out. At the beginning of the next interglacial, however, the species migrated back from their refugial areas, such as the Mediterranean, to the north (avoiding the Alps).

The special situation of Central Europe, separated in the south by the Alps, meant that the differences between cold and warm flora and fauna were particularly pronounced. In addition to the differences in temperature, the plants and animals felt the alternation between the strongly continental climate of the cold stages and the more maritime climate of the warm stages. An exchange between cold continental fauna and warm maritime fauna took place repeatedly. The composition of the species that returned to their old habitat was not always the same, however. For example, during the Holocene fallow deer were originally absent from Central Europe although, during the earlier warm stages, it regularly migrated into the area (von Koenigswald 2002, 2007).

The woolly mammoth (*Mammuthus primigenius*) has become the archetypical symbol of the Ice Age. It decorates the emblem of the German Quaternary Association (DEUQUA). Individual bones or teeth are frequently found in sand pumped for beach nourishment on the North Sea coast, for example (Fig. 9.3); complete mammoth skeletons are extremely rare, however. The first mammoth skeleton found in Germany was discovered in 1903 in a clay pit south of the Cottbus–Forst railway line (Fischer 2008). One of the most complete skeletons was found in 1975 at Siegsdorf in Chiemgau (Bavaria), the age of which was 44 ka. The mammoth is now the centre of the Ice Age exhibition of the *Naturkunde- und Mammutmuseum* in Siegsdorf. This specimen had a shoulder height of 3.60 m and an estimated live weight of over 6 tonnes.

Figure 9.3 Mammoth bone found on the beach of Texel, The Netherlands. The bone is from the bottom of the North Sea and brought ashore through beach replenishment. Photograph by Jürgen Ehlers.

Apart from the mammoth, the typical cold-stage species of the so-called 'mammoth steppe' would have included the woolly rhinoceros (*Coelodonta antiquitatis*), reindeer (*Rangifer tarandus*) and, occasionally, the musk ox (*Ovibos moschatus*). Among the small mammals the lemmings (*Lemmus* and *Dicrostonyx*) are good indicators of permafrost.

The fauna of the interglacial periods includes wild boar (*Sus scrofa*) and roe deer (*Capreolus capreolus*), but also the forest elephant (*Palaeoloxodon antiquus*), rhinoceros and the forest rhinoceros (*Stephanorhinus kirchbergensis*). A special immigrant from the Mediterranean area is the hippopotamus (*Hippopotamus amphibius*). Its remains have been found in the last interglacial (Eemian or Ipswichian) deposits of the River Thames beneath Trafalgar Square, in central London (Gibbard 1985); near Cambridge, from where a complete female skeleton is known; and also in Germany (von Koenigswald & Löscher 1982; Stuart 1995) but not further east because the Atlantic impact decreased towards the continental interior. The hippopotamus spread irregularly along the rivers and coastal regions. In the Early Pleistocene, the hippos once even reached central Germany as well as northern East Anglia (Stuart 1995; Kahlke 1997–2001). Hippos did not reach northern Europe in the Holsteinian however, although that was warmer than the Eemian Interglacial during which the hippo appeared again.

Predators such as lion (*Panthera leo*) and hyaena (*Crocuta spelaea*) are less tied to particular climatic conditions, so they occur in both cold and warm periods. Their closest relatives live in Africa today. There is proof that lions and reindeer have occurred simultaneously. In an alluvial loam discovered in 1992 at Bottrop, the footprints of many reindeer are found together with those of two wild horses and a large ox (bison or aurochs), together with traces of wolf and lion. These animals crossed the site within a few hours or days of each other about 35 ka ago, before the maximum of the last glaciation (von Koenigswald & Sander 1995). The traces were preserved because they were covered with sand soon after.

For the subdivision of the Quaternary, both small mammals and large mammals are used. Small mammals such as voles have the advantage of often being preserved in large quantities, because

Figure 9.4 Bones from Neuland dredging lake in Hamburg-Harburg, Elbe glacial valley. Photograph by Jürgen Ehlers.

they were enriched by owl pellets. Statistical analyses of small mammals are therefore much easier to perform. On the other hand, the small teeth of voles are much easier overlooked than a mammoth bone. From some sites, such as the Neuland dredging near Hamburg, only large mammalian remains were recorded. Of the 457 examined bones from Neuland (Fig. 9.4), Glüsingen and neighbouring dredging lakes, 160 are steppe bison (*Bison priscus*), 137 mammoth (*Mammuthus primigenius*) and 57 of reindeer (*Rangifer tarandus*) and other large mammals (Kopp 2000).

The stratigraphical superzone, named Villafranchian (after the important Villafranchia d'Asti site in Piedmont) in the previously used large mammalian biostratigraphy, is called Villanyium in the small mammals stratigraphy and covers the period 2.65–1.8 Ma. Index fossils are deer and the 'southern mammoth' (*Mammuthus meridonalis*).

The next superzone, the Biharian, includes the rest of the Early Pleistocene and the older part of the Middle Pleistocene. In the Biharian, the steppe mammoth (*Mammuthus trogontherii*) replaces the more antique southern *Mammuthus meridonalis*. At this zone the voles *Microtus* and *Mimomys* occur together. The base of the Biharian corresponds to the first occurrence of *Microtus*, and the upper limit corresponds to the first occurrence of the great vole (*Arvicola*). Together with *Arvicola* the forest elephant (*Palaeoloxodon antiquus*) and man (*Homo heidelbergensis*) appear for the first time in the European Pleistocene.

The latest superzone known as the Thoringian ranges from the Middle Pleistocene to the Holocene. This period is characterized by the common occurrence of *Arvicola* and *Microtus*. It encompasses the three major northern glaciation cycles and intervening interglacials. The first occurrence of *Arvicola* is above the Brunhes–Matuyama magnetic reversal and provides a critical marker in the early Middle Pleistocene sequence. The cold stages are characterized by mammoth (*Mammuthus primigenius*), whereas the warm stages by forest elephant (*Palaeoloxodon antiquus*). The faunal composition provides a further subdivision. The youngest of these, the *Arvicola terrestris* fauna, spans the Eemian Interglacial, the Weichselian and the Holocene (von Koenigswald & Heinrich 2007). At the boundary between the Pleistocene and the Holocene, numerous large mammals became extinct. No mammoth steppe habitat has survived in either Asia or in North America until today. The reason for this is not yet clear; the rapid and dramatic ecological changes in the environment were surely important but human hunting may have been sufficient to 'tip the scales', causing the decline of species to critical levels.

9.2.1 THE SMALL ANIMALS: BEETLES, SNAILS AND FORAMINIFERA

The animals that are locally found in Quaternary deposits in large numbers include the remains of beetles (Coleoptera) and other insects. It can be assumed that the ecological requirements of each species experienced no significant changes during the Quaternary. Because of their mobility, beetles can respond quickly to climatic changes and even venture into areas where the soil development has yet to permit growth of any higher plants (Morgan & Morgan 1990). It is therefore unsurprising that the conclusions drawn from fossil coleopteran faunas do not always coincide with those from palynological investigations. On the basis of his studies in Britain, Coope (1977) concluded that one of the Middle Weichselian interstadials must have been much warmer than previously thought, at least during the summers. While the entire seasonal climate (especially the duration of the growing season) is of critical importance for the flora, the coleoptera faunas primarily reflect the July temperatures. Here, the significant differences in annual variation in temperature between maritime and continental areas have to be taken into account. In the area of the Pole of Cold in eastern Siberia, for example, high July temperatures regularly occur. This aspect is important when attempting to assess climatic change during the Late glacial. In the Older Dryas there might have been fairly high temperatures in July, but also very cold winters (Lemdahl 1988).

An important method for the palaeontological study of Quaternary marine deposits is foraminifera analysis. Foraminifera are unicellar organisms that live almost exclusively in ocean waters, either as plankton or as bottom dwellers. When they die, their shells or tests fall to the sea floor and are preserved in large numbers. With the help of foraminifera it has been possible, for example, to clarify the ecological changes in the transitional period from the Saalian (Wolstonian) through the Eemian (Ipswichian) to the Weichselian (Devensian) in northern Jutland. This period is completely represented there by marine sediments (Knudsen & Lykke-Andersen 1982). As with pollen analysis, the changing foraminiferal composition allows the period to be subdivided into a sequence of different biozones, allowing the correlation of even incomplete sequences. Due to the small size and large number of foraminifera, only small samples are needed so the method is ideally suited to correlate well drillings. With the help of foraminifera analyses it has been possible, for instance, to show that the transgression at the end of the Elsterian Glaciation in northern Germany and Denmark began well before the beginning of the Holsteinian Interglacial (Knudsen 1987).

In the investigation of loess sequences in Czechoslovakia, Ložek (1964) pointed out the importance of molluscs. Through malacological investigations, it is possible to reconstruct the environmental conditions during loess deposition in Central Europe. The basic type of loess molluscan associations are the *Pupilla* fauna. While the species involved today live mainly in open landscapes such as wet meadows and steppes or warm rocks, the loess sequences characterize a cold and dry steppe climate. Apart from *Pupilla muscorum* and *Pupilla loessica*, it is particularly characterized by *Vallonia tenuilabris*, a snail that today occurs in the cold regions of northern Asia. This fauna differs from that of the *Columella* fauna, which indicate a cold but humid climate, and the *Striata* fauna, which represent the transition of the arid steppe to early glacial and interglacial fauna. The investigation method using loess molluscs is summarized by Rousseau (2001). Because of the more continental conditions and the thick layers of loess, malacological investigations in Czechoslovakia and Hungary can be performed with greater success than in Germany and further west, where large parts of the loess profiles are decalcified.

As well as the loess profiles, mollusc faunas can also provide valuable information about the ecological development. The few studies published so far include, for example, works on the Eemian Interglacial sequence at Gröbern (Fuhrmann 1990) or on the travertine of

Ehringsdorf (Mania 1973, 1975) in eastern Germany. An overview of the Middle Pleistocene molluscan fauna in the east German Saale-Elbe region was provided by Mania & Mai (2001).

9.3 Development of Vegetation

The climatic changes of the Pleistocene affected different parts of the world in different ways. While the polar regions have remained largely unaffected by climate fluctuations, the mid-latitudes experienced a repeated alternation of warm and cold periods; even in the mountains of the tropics a significant change in climatic conditions occurred. Through the eustatic variations in sea level, the distribution of land and sea changed considerably. Large areas of the North Sea and the English Channel were dry during the Weichselian, the same as the northern part of the Bay of Biscay, which together with the development of sea ice in the neighbouring North Atlantic Ocean caused large parts of Europe to experience a more continental climate.

The climatic fluctuations of the past are not directly detectable, but must be reconstructed from proxy records. A particularly effective indicator is vegetation. Apart from plant macrofossils, pollen and spores preserved in large numbers allow the reconstruction of the vegetational development of the warm and cold stages. However, although pollen grains are very resistant in themselves, in most substrates they are destroyed by oxidation. Favourable preservation conditions occur mainly in organic-rich sediments including peat and muds, that is, in lake deposits and in bogs.

9.3.1 POLLEN ANALYSIS: OPPORTUNITIES AND MISTAKES

Originally, vegetation of past warm periods was investigated by analysis of plant macrofossils. Leaves, seeds and wood residue gave a limited idea of the former plant communities. A true reconstruction of the vegetation development during the warm periods was only made possible with the introduction of pollen analysis. The Swede Lennart von Post (1916) was the first to reconstruct the vegetational history from a peat bog by quantitative pollen analysis. Initially pollen analysis (palynology) was limited to the investigation of postglacial bogs. Firbas (1927) and Jessen & Milthers (1928) widened the scope by including older deposits.

Whenever possible, single samples are not investigated in pollen analysis but whole series of samples. Not only are the different plant taxa recorded (Fig. 9.5), but also their sequence of appearance and eventual disappearance. Ideally, the vegetational development of a warm stage ranges from a treeless tundra of an outgoing glaciation through the gradual reforestation to the optimal spreading of warmth-loving trees and, towards the end of the warm stage, through the development of coniferous forest back to the tundra. The details of the vegetational history, which are characterized by the prevalence of certain plant communities, are distinguished as distinct pollen (bio-)zones in the pollen diagram.

The identification of pollen grains is performed under the microscope (Fig. 9.5). While scientists originally had to compare their pollen with templates or photographs from a light

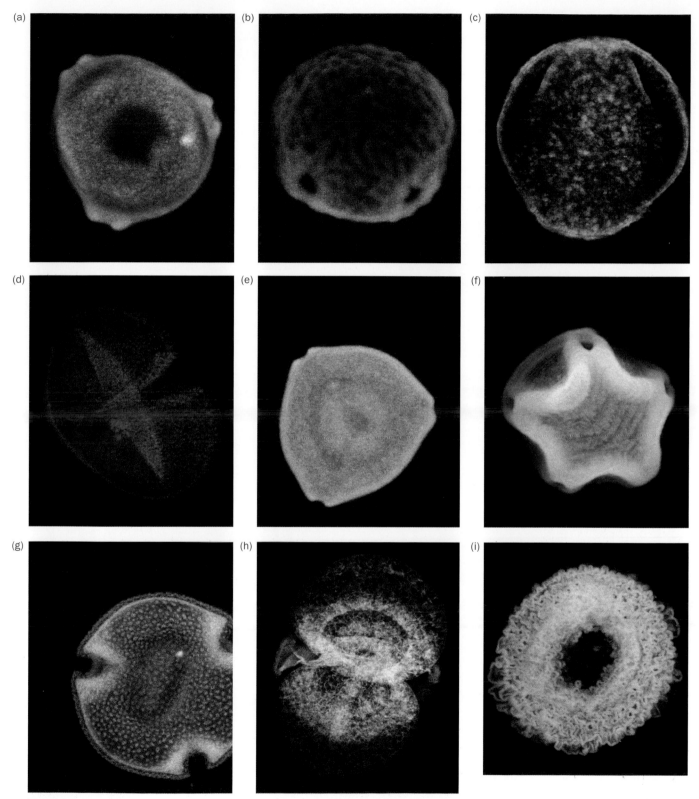

Figure 9.5 The pollen of (a) birch (*Betula*), (b) elm (*Ulmus*), (c) oak (*Quercus*), (d) spruce (*Picea*), (e) hazel (*Corylus*), (f) alder (*Alnus*), (g) lime (*Tilia*), (h) fir (*Abies*) and (i) hemlock (*Tsuga*); images by confocal laser scanning microscope. Source: Thomas Litt.

microscope, now images are also available which were made with an electron microscope, providing an almost three-dimensional picture of the different pollen grains.

The comparability of pollen profiles is complicated by the fact that the pollen flora is influenced by local factors. Deposits from a small pond, located in an oak forest, will naturally be rich in oak pollen of local origin, while a pond in an open steppe landscape will contain abundant far-transported and redeposited pollen and spores. For regional studies, deposits from large bogs and lakes are therefore preferred. The pollen grains vary in shape and size. Some (e.g. *Pinus sylvestris*) have special air sacs that improve their flight characteristics.

The vegetation is very dependent on climate as well as other environmental characteristics. However, a pollen diagram (e.g. Figs 9.6, 9.7) is not a climatic curve. Since plants are not themselves mobile, the vegetation takes significant time to respond to positive changes in climate. While a drop in temperatures at the onset of cold climate can destroy sensitive plant species over a large area in one event, the re-immigration is much slower because the seeds and fruits of many species are only transported over very short distances (e.g. pine or beech) and it takes decades for the individual plant to produce the first fruits that can be carried on again.

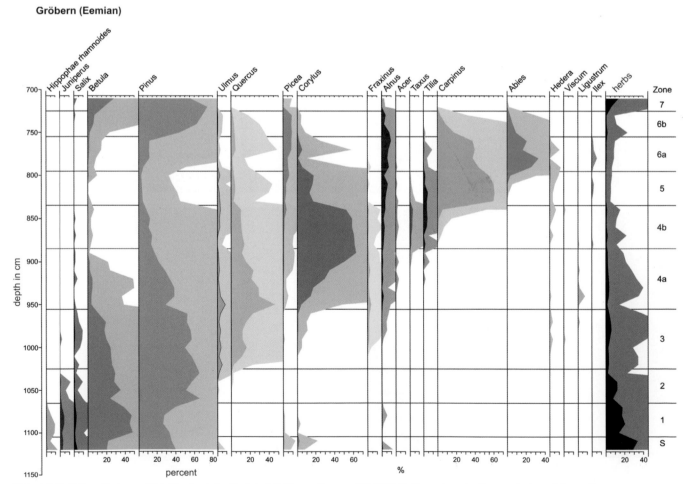

Figure 9.6 Pollen diagram of the Eemian Interglacial at Gröbern. Source: Litt (1994). Reproduced with permission of the author.

Gröbern-Schmerz (Holsteinian)

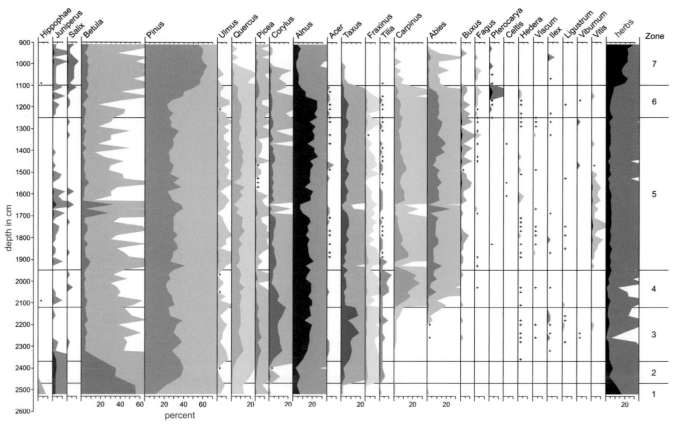

Figure 9.7 Pollen diagram of the Holsteinian Interglacial at Gröbern-Schmerz.

Source: Kühl & Litt (2007). Reproduced by permission of Elsevier.

It has been debated whether certain parts of the vast formerly glaciated areas might have remained unglaciated, and allowed at least some individual plant taxa of the plant communities to survive the cold climate. In most cases, however, no such refuge seems to have existed, and the plants have migrated gradually forwards and backwards in the course of a few millennia. Even in the case of an island such as Iceland, this was made possible by long-distance transport by driftwood or by birds.

The gradual re-immigration of plants meant that during the course of an interglacial a specific sequence of plant communities appeared. This is partially a result of the differing rate with which individual taxa migrate as well as the distance from their individual refugia. This gave rise to a differing sequence of plant arrivals during different interglacials. Moreover, not all taxa returned to the glaciated regions in every interglacial. These differences of relative timing and assemblage composition have been widely used, especially in Europe, for the stratigraphic classification of the deposits. However, many interglacial deposits are incomplete because either the lake or river channel silted up too early or peat formation started too late or terminated too soon, so that from the last part of the interglacial no pollen has been preserved. In such cases, the age determination based on palynology is rather difficult.

The typical vegetational succession of an interglacial can be illustrated by the example of the Eemian (Ipswichian) Interglacial. For this, the Eemian sequence from Gröbern is a suitable example.

- Zone I: Birch period. The first pollen zone is characterized by the dominance of *Betula*. Typical light-loving plants (heliophytes) of the late Saalian Glaciation (*Helianthenum*, *Hippophaë*) no longer occur. *Juniperus* and *Artemisia* are still present in small, steadily decreasing numbers. The proportion of *Pinus* is gradually increasing. Menke & Tynni (1984) draw the line between this zone and Zone II at the decline from the *Betula* maximum and the beginning of the *Ulmus* curve. Zone I is missing in many pollen diagrams, as the oldest Eemian deposits are often incomplete.
- Zone II: Pine–birch period. This zone is characterized by a relative maximum of *Pinus* with generally rapidly decreasing shares of *Betula*. Pollen of thermophilous trees occurs regularly, if only in small amounts at first. The start of the decline from the *Pinus* maximum marks the limit to the subsequent pollen Zone III.
- Zone III: Pine–oak forest phase. In the pollen diagram *Pinus* continues to dominate and the proportions of *Betula* decrease, while the proportion of mixed-oak forest pollen (mainly *Quercus*) is increasing. *Acer*, *Hedera* and *Viscum* are present at low levels. Around the beginning of Zone III, *Fraxinus* also appears. Menke & Tynni (1984) place the upper limit of Zone III at the spread of *Corylus* (hazel). Zone III is often poorly developed and difficult to separate from Zone II.
- Zone IVa: Mixed-oak forest-hazel phase. The zone is mainly marked by the rapid rise of *Corylus* (and *Alnus*), while the proportion of *Pinus* further decreases. The first *Ilex* pollen occurs and the pollen curves of *Taxus* (yew) and *Tilia* (lime) set in. The upper limit of this zone is located at the first significant drop from the *Corylus* maximum and the steep rise in the *Tilia* curve.
- Zone IVb: Hazel–yew–lime phase. After the *Corylus* maximum, hazel values first decrease rapidly to about half and then usually remain approximately constant. The *Taxus* and *Tilia* curves reach their maxima. The upper part of Zone IVb begins with the rise of *Carpinus* and *Picea* values (hornbeam and spruce). The border to Zone V is where the *Picea* content exceeds the percentage of *Carpinus–Tilia*.
- Zone V: Hornbeam–spruce phase. The older part of Zone V is usually characterized by a more or less strong dominance of *Carpinus*. The percentage of *Corylus* falls off sharply and *Picea* increases gradually. With the decline of *Carpinus* in the upper part of Zone V, the *Quercus* share usually takes up significantly. At the same time the younger *Pinus* rise begins. To distinguish it from Zone VI, the rapid rise of *Abies*, the upper limit of *Corylus* or the rise of *Pinus* can be used. The boundary is usually at about the middle of the younger *Quercus* maximum.
- Zone VI: Pine–spruce–fir phase. While the proportion of *Pinus* increases, the percentages of warmth-loving trees decrease. *Abies* (fir) values rise in the lower part of Zone VI (in Western Holstein to up to 10%), but then also decline. The frequency of *Picea* decreases and *Pinus* increases. As a boundary with Zone VII, Müller (1974a) has proposed the decline of pine pollen to under 1%.
- Zone VII: Pine phase. This zone is characterized by a clear dominance of *Pinus*. The *Picea* share usually ends at the turn of VI/VII, or shortly thereafter. After a pronounced maximum of *Pinus* in the higher part of the zone, the importance of *Betula* increases.

The end of the Eemian Interglacial is characterized by the retreat of closed forest and the transition to Subarctic vegetation of the Weichselian cold period. It marks the transition from an interglacial dense vegetation cover that protected the ground to the stronger redeposition of sediment in a cold climate environment.

An interglacial site with this vegetational succession could therefore be relatively easily classified as Eemian (Ipswichian); the pollen diagrams are not always so readily correlated, however. Based on palynological studies, various additional interglacials were postulated in the past but were later refuted once re-examined. The interpretation must always take into account the stratigraphical position and the palaeogeographical situation.

Comparing the vegetational development of the three last interglacials in northern Germany, it is found that the spread of individual trees varies greatly. In the Holsteinian Interglacial, fir and spruce arrive relatively early. Spruce appears much later in the Eemian and fir is of little importance. In the north German Holocene however, fir is only found in the Thuringian Forest (Overbeck 1975) and spruce is limited to the Harz Mountains and their foothills and to the Lüneburg Heath. In other areas it has only recently been introduced by the forestry industry. The European beech (*Fagus sylvatica*) only plays a minor role in the Holsteinian Interglacial in northern Germany, and it is completely absent from the Eemian. In the second half of the Holocene however, beech has spread naturally. The European beech is dominant in recent natural forest communities of Central Europe today, while oak predominates on dry and wet sites.

The warm phases of the Ice Age can be subdivided into interglacials (warm stages) and interstadials. During the warm stages, climate in central Europe was comparable to or even warmer than today's climate; during the interstadials warming was weaker. In Central Europe the interstadial warm phases were characterized by vegetation and climate conditions not reaching those of the Holocene postglacial optimum, in contrast to the interglacials (Fig. 9.8). If this definition is applied, the warmer phases of the Early Weichselian are not to be regarded as interglacials (Behre 1989).

The weakly developed interstadials (Oerel, Glinde) at the beginning of the Middle Weichselian are partly rich in Ericales (heath) that characterize acidic soils. This suggests that until that time there had not been much renewal of the soil substrate through periglacial processes. The same applies to the substrates in which the fossil soils at Keller and Schalkholz (Schleswig-Holstein) developed. Here the strong periglacial activity began only after the end of the youngest soil formation. In contrast, the organic deposits of the Dutch mid-Weichselian interstadials were apparently formed after the onset of periglacial processes. Consequently, the pollen diagrams of the Moershoofd, Hengelo and Denekamp interstadials are characterized by a lack of Ericales (Behre 1989).

Of considerable importance for climatic conditions as well as palaeogeography was the lowering of the sea level during the glacial periods. At the beginning of the Early Weichselian contact between fresh and salt water in the southern North Sea in the Brown Bank area occurred at a

Figure 9.8 During the Early Pleistocene interglacials some warmth-loving plants grew in central Europe which are no longer found there naturally today, including the wing nut (*Pterocarya*). Photograph from Schaffhausen, Switzerland. Photograph by Jürgen Ehlers.

water depth of 40 m (Zagwijn 1989). Later during the Weichselian Glaciation, the sea level never again exceeded that level. There was no mid-Weichselian high sea-level event. That meant a significant increase in continentality (colder winters, warmer summers) for today's coastal areas of northern Germany, the Netherlands and much of the British Isles. In North Denmark, on the other hand, the marine influence remained. Until shortly before the start of the Late Weichselian Glaciation, the northern part of Jutland down to about the line between Aalborg and Hanstholm was still flooded by the sea (Houmark-Nielsen 2004).

It is not always easy in long pollen profiles to distinguish interglacial or interstadial deposits from cold-stage strata. Redeposited and far-travelled windblown pollen can sometimes contaminate deposits of a treeless period with substantial frequencies of tree pollen; this is especially true for lake and river deposits. To distinguish these from true records, it is therefore particularly important to include non-tree pollen. In addition to the overall increase in non-tree pollen in relation to tree pollen, the presence of light-loving plants (heliophytes) can also give a clear indication of cold (treeless) conditions. These species include *Artemisia* (wormwood), *Plantago* (plantain) and *Armeria* (thrift). Examples of this are the pollen diagrams from Rederstall in Schleswig-Holstein (Menke & Tynni 1984), Oerel in Lower Saxony (Behre & Lade 1986) and many British lowland localities of the period (West 2011).

In general, undisturbed sequences such as peat or lake sediments are used for palynological studies to ensure a continuous sedimentation. Pollen analysis of marine sediments is also possible. The pollen diagram of the Holsteinian Interglacial from Hamburg-Dockenhuden is based on marine deposits to a large extent (Linke & Hallik 1993). Shallow marine deposits often include much stronger reworking and sorting of palynomorphs however, so that conclusions about the vegetational development are more difficult to make.

The Pleistocene warm interglacials and interstadials were not of equal length. For example, counting the varves in the diatomite from Bispingen, Müller (1974a) determined the duration of the Eemian Interglacial period (Table 9.1).

While previously pollen analysis was largely limited to the reconstruction of the vegetational history of the warm stages, efforts are also being made to obtain information about

TABLE 9.1 SEDIMENTATION RATE OF THE EEMIAN DIATOMITE AT BISPINGEN, RECONSTRUCTED FROM POLLEN ANALYSIS AND VARVE ANALYSIS.

Pollen zone after Müller (1974a, b)	Pollen zone after Menke & Tynni (1984)	Thickness (cm)	Varve years after Müller (1974a)	Sedimentation (cm/century)
IVa + b	5	240	4000 (estimate)	6.0
IIIc	4b	96	1100 (partly counted)	8.7
IIIb	4a	59	700	8.4
IIIa		36	450	8.0
IIb	3	47	450	10.4
IIa	2	21	200	10.5
I	1	7	100	7.0

Source: Kühl & Litt (2007). Reproduced by permission of Elsevier.

the climatic history. For plant growth, both temperature and precipitation are crucial. It is not the average values of temperature which are important but the seasonal cycles, which in turn depend strongly on the proximity or distance to the sea (continentality). The January and July temperatures provide information about the degree of continentality. For example, deciduous trees are now growing well in areas where the January temperature drops below −10°C if the July temperatures rise to about 18°C, and deciduous trees thrive in Britain with its mild winters (average January temperatures slightly above 0°C), even at July temperatures below 15°C. Only in the north of Britain, with July temperatures around 12°C, are the Atlantic dwarf shrubs found instead.

Present plant distribution should be compared with the current climate data. Kühl & Litt (2007) have done this for a grid of 0.5 × 0.5°. As an example, the values for *Ilex* (holly) and *Carpinus* (hornbeam) are plotted against temperature in Figure 9.9. The green points indicate the temperature conditions under which *Carpinus* and *Ilex* grow today.

Comparing the present values with the corresponding plant distributions from pollen diagrams, information about the temperatures of past warm climate events can be deduced. For example, investigations have shown that during both the Eemian and the Holsteinian temperatures were similar to today. In the Eemian Stage the temperatures increased rapidly at the beginning of the interglacial and then remained high throughout the warm stage. This result is consistent with

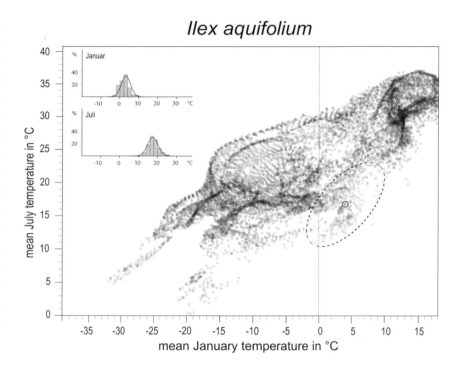

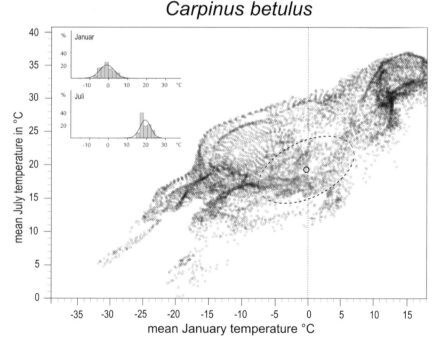

Figure 9.9 Presence (green) and absence (red) of *Ilex aquifolium* and *Carpinus betulus* at present-day mean January and July temperatures in North Germany. Source: Kühl & Litt (2007). Reproduced by permission of Elsevier.

temperatures reconstructed from oxygen isotopes in Gröbern (Böttger et al. 2000). Conse-quently, the vegetational development in the Eemian was not a result of temperature fluctu-ations, but rather controlled by rainfall and soil development.

In the Holsteinian Interglacial there was also a rapid temperature rise at the beginning of the interglacial, and (as in the Eemian) a rapid drop in temperature at the end of the warm period. The temperatures in the Holsteinian were higher than in the Eemian, at least in the second half of the interglacial. They generally increased throughout the Holsteinian, while throughout the Eemian they gradually decreased. In some pollen diagrams of the Holstein-ian, three dramatic setbacks have been recognized (in zones VIII, XI and XIIb/c, after Müller 1974b), but they were probably of short duration meaning that the vegetation could recover rapidly. In the pollen diagram of Gröbern-Schmerz, the second and third of those setbacks are seen in the sharp decline of the *Carpinus* curve. The resolution of the temperature diagram (Fig. 9.10) is not sufficient to cover these short-term events.

A similar event, termed the high non-arboreal pollen (NAP) phase, is found in British Hoxnian pollen sequences, such as at Marks Tey in Essex (Turner 1970). Changes in precipi-tation and therefore moisture availability may also have occurred through the warm periods. For example, incomplete sequences in shallow lake or pond basin infills in lowland Britain indicate changes in the regional water table during this period. This suggests that precipi-tation varied through the interval with initial relatively higher precipitation in the early part of the interglacial. Later in the zones Ho I, II there were relatively low water levels, which were again replaced by higher water levels and therefore moisture availability in Ho III, IV (Gibbard et al. 1977).

9.4 Weathering and Soil Formation

Climate fluctuations not only cause changes in the vegetation and animal kingdoms, but also have an effect on the weathering of rocks and on soil formation. Temperature and precipita-tion play a central role in soil acidification, for instance. The development of the soils in the course of the warm stages is an irreversible process that leads to a progressive depletion of the substrate. In the following warm stage, the relict soils are overprinted or further developed. Pedogenesis can only restart in the next warm stage if the old soil was removed, and fresh substrate has been brought to the surface.

During the cold stages, soil development in northern Germany under permafrost con-ditions did not go beyond the raw soils of Arctic periglacial patterned ground (cryosols). It was only with the onset of warming that differentiation into soil horizons began. In a typical till, with a carbonate and clay content of about 20% as is found widely distrib-uted in the formerly glaciated areas, soils developed as follows (according to Scheffer & Schachtschabel 2010):

1 In the early phase of the interglacial, the unfolding vegetation yielded litter and scattered plant remains. As a consequence of the intense activity of soil organisms, organic matter was mixed with the substrate and a relatively thick humic topsoil above the unweathered, calcareous parent material of an Eutric Regosol with a loose crumb structure developed.

Gröbern (Eemian)

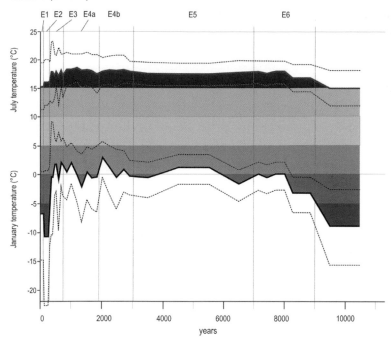

Gröbern-Schmerz (Holsteinian)

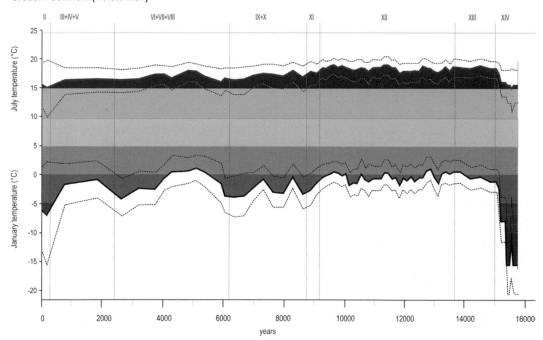

Figure 9.10 Reconstruction of the July and January temperatures for the Eemian and Holsteinian interglacials after Kühl & Litt (2007). Reproduced by permission of Elsevier. Pollen zones for the Eemian after Menke & Tynni (1984) and for the Holsteinian after Müller (1974b).

2 The leaching of the carbonates resulted in a continuous deepening of the soil profile by acidification and weathering of silicates. The rate of leaching decreased with increasing depth, because the topsoil was more exposed to seepage water than the underground. Decalcification was followed by formation of a Bw horizon. Mainly through the weathering of mica and feldspars, clay minerals were formed. In this way, a Cambisol developed.

3 In the later stage of the Cambisol, neo-formed clay was translocated from the topsoil to the subsoil. A particularly intense clay migration took place, especially during the warm and humid Atlantic period (Holocene), which apart from the clay also included the fine silt fraction. This led to the formation of the argillic Bt horizon, which characterizes a Luvisol. Holocene Luvisols developed from tills of the last glaciation are today often decalcified to a depth of 1–2 m. In tills of older glaciations, decalcification reaches down to depths of >4 m. These depths depend very much on the originally lower carbonate content of the substrate, however.

3 The previous increase in porosity of a Cambisol as a result of decalcification was greatly reduced again by compaction and clay accumulation in the Luvisol. Under the influence of decreasing temperature and increasing precipitation during the Subatlantic period, soil moisture increased due to impeded drainage. Tn this period the Haplic Luvisols developed to stagnic Luvisols and Stagnosols with typical mottling of the deeper stagnic horizon and formation of concretions of Fe and Mn oxides in the upper subsoil.

4 The leaching of bases and nutrients led to strong acidification of the topsoil and therefore to a decrease in biological activity. Due to the lack of bioturbation and inhibited decomposition of organic matter, a humus layer formed on top of the mineral soil which was the first stage of the formation of a Podzol. Characteristics of weak podzolisation are now known to be widespread in soils from glacial till under forest.

This sequence of stages of soil development has been repeated on till in a similar way throughout all Quaternary interglacial periods in a temperate, oceanic climate. The end stage of a Podzol was not always reached in interstadial and interglacial periods, because of the relatively short time of climatic warming between the cold stages. Further, pedogenesis did not always proceed undisturbed. Climatic setbacks (e.g. at the end of the Late Weichselian Bølling and Allerød periods) led to a disruption of soil formation and to the renewed periglacial mixing of the substrate.

A humic topsoil could form more rapidly on meltwater sands and gravel because of the greater permeability of the substrate. Since the carbonate content was usually lower than in the tills and more quickly dissolved and leached, the soils left the stage of an Eutric Regosol relatively fast and transformed to a Cambisol. Within the cambic horizon at pH values of 5–7, the newly formed clay minerals from the weathering of mica were partially leached and precipitated as dark brown bands of clay enrichment in the C-horizon. Later, under the cool, moist climate of the Subatlantic period, soil acidification and nutrient depletion increased. This resulted in a strong podzolisation, in which iron and aluminium oxides were dissolved and leached as metal-organic complexes from the topsoil into the subsoil where they reprecipitated in the spodic horizon (which partly forms a solid hard pan).

On loess, which was deposited far from the inland ice in periglacial areas, the first stage of a Regosol developed into a Chernozem during the early Holocene (Preboreal and Boreal) under a forest steppe. The steppe soil formation ended with the spread of a closed forest cover at the start of the Atlantic period however, when the more oceanic climate favoured the spread of forest. The Chernozems in Germany are therefore regarded as a relict of the early Holocene soil formation. Typical interglacial loess soils in Central Europe are Haplic and stagnic Luvisols. In east and southeast Europe, however, the formation of Chernozems lasted longer as a result of the more pronounced continental climate; sometimes it was only interrupted when the steppes were cultivated. In western Europe, which had been under forest and the influence of a oceanic climate since the Atlantic period, an increasing degradation of the Chernozems with a transformation into Luvisols occurred. This development was delayed in areas where humans have farmed since the Neolithic (from *c.* 4 ka ago). In addition, Chernozem relicts are preserved where, after the onset of a humid climate, waterlogging and a high water table resulted in the formation of a stagnic or gleyic Chernozem. The waterlogging resulted from poorly water-permeable layers in the deeper underground (e.g. tills or Mesozoic clays). In the Hildesheimer Börde area of Germany, this resulted in very large areas of relict Chernozems lacking advanced decalcification and weathering of silicates.

The development of many bogs began in the Late glacial and Preboreal with the formation of subhydric soils. The gyttjas in kettle holes and meltwater channels can reach thicknesses of up to 20 m in extreme cases. Lakes silted up under the warmer climate of the Boreal period, and bogs were formed from fens. More bogs developed in the humid climate of the Atlantic. Bog formation was favoured in northern Germany when the postglacial sea level caused a significant rise in groundwater levels.

A special case of interglacial soils are the Histosols, which consist of peat with more than 30 wt% organic matter. Peat is formed in ponds when a lack of oxygen prevents the organic litter from decomposing. Subhydrically formed fens (topogenic Histosols) are distinguished from raised bogs that formed independently of the groundwater table (ombrogenic Histosols) and transitional bogs. In topogenic Histosols, organic matter from mosses is accumulated on top of an acid Podzol or a former topogenic Histosol which is poor in nutrients. This process is favoured by a cool climate and nutrient-poor water, when the activities of humus-decaying soil organisms are reduced.

Histosols of fens form in stagnant water. The starting material is primarily common reed (*Phragmites*), cattails (*Typha*) and/or sedges (*Carex*). When the peat growth has reached the mean water table, the marsh plants are complemented by the appearance of alder (*Alnus*) and willow (*Salix*). The alder swamp forest is the final member of the fen development (Overbeck 1975).

If the fen grows out of the water table, typically birch and pine spread. This type of bog in transition from a fen to a raised bog is called a transitional bog. On further peat growth, a different vegetation appears whose diet is adapted to the nutrient-poor rainwater. The alder forest disappears and pod grass (*Scheuchzeria*), cotton grass (*Eriophorum*) and peat moss (*Sphagnum*) spread. The residues of *Sphagnum* in particular now form the raised bog. A bog is often arched upwards like a watchglass. Where the rainwater cannot drain from the central parts of the bog, marsh ponds and lakes are formed (so-called 'bog eyes').

9.4.1 SOILS AND PALAEOSOLS

The degree of soil development is widely used to estimate the relative ages of glacial sequences. It can be integrated within morphostratigraphy as illustrated in Figure 9.11. In Greece and Montenegro, Hughes et al. (2006b, 2011) quantified soil development on moraine sequences using the Harden Profile Development Index. This was important because it allowed moraines that were sometimes close in proximity to be grouped together in age or separated. For example, differences in soil development enabled recessional moraines that are close in age to be differentiated from moraine successions that were widely different in ages (Fig. 9.11). Examples of soils formed on Greek and Montenegrin moraines are shown in Figures 9.12 and 9.13. Often

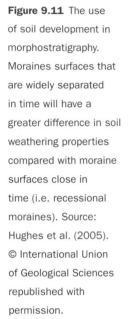

Figure 9.11 The use of soil development in morphostratigraphy. Moraines surfaces that are widely separated in time will have a greater difference in soil weathering properties compared with moraine surfaces close in time (i.e. recessional moraines). Source: Hughes et al. (2005). © International Union of Geological Sciences republished with permission.

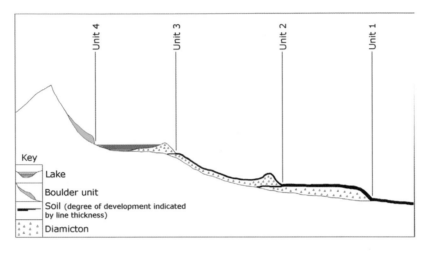

Figure 9.12 Well-developed soil on Middle Pleistocene moraines in the Orjen Massif, Montenegro. This soil has developed over multiple glacial cycles. The thickness of the soil is promoted not only by its age but also the abundant silty-clay composition of the moraine matrix. This results in soils that are much thicker on moraines compared with those on coarse fluvial gravel terraces of similar age in the same area (Adamson et al. 2014). Photograph by Phil Hughes.

these soils are still forming today and represent polycyclic soils that have formed over multiple glacial–interglacial cycles. Some are relict however, such as the buried soil in Greece in Figure 9.13. In these cases the soils are called palaeosols.

Relict and fossil soils (Box 9.1) from pre-Holocene or even pre-Pleistocene geological periods can be found on previous land surfaces. Under the hot-humid climatic conditions of the Tertiary, a weathering mantle (regolith) formed that can be divided into two distinct genetic units, on the Rhenish Massif.

The solum has developed near the land surface by soil-forming processes. It has soil horizons and a characteristic soil structure. Below this solum, the saprolite was formed by profound weathering and neo-formation of silicates and leaching, but still shows the undisturbed rock structure. Solum and saprolite combined can still be up to 150 m thick (Felix-Henningsen 1990).

The lower Tertiary soil cover was probably already eroded and largely removed throughout the middle–late Tertiary, meaning that at the beginning of the Quaternary the saprolite was found near the surface over wide areas. During the cold periods of the Quaternary, the Rhenish Massif was a periglacial area. As a result of tectonic uplift, increasing fluvial erosion in combination with solifluction on the valley slopes resulted in a progressive removal of the weathering mantle and regressive dissection of the peneplain, starting from the existing main rivers. By cryoturbation and frost shattering, parts of the kaolinitic saprolite were converted into a structureless, loamy substrate which was subjected to slopewash and solifluction. Redistributed remnants were left behind as 'Graulehm' on top of the

Figure 9.13 A buried soil formed on a moraine surface in the Pindus Mountains, Greece. The soil is buried by colluvium and is no longer developing (now a palaeosol in this locality). The presence of this soil on the underlying moraine is a distinctive marker for the oldest Middle Pleistocene moraine surfaces in the Pindus Mountains. Photograph by Phil Hughes.

BOX 9.1 A FOSSIL SOIL

Students found humic layers in a sand pit near New Wulmstorf (Lower Saxony) in 1972. These soils did not represent an interglacial however, but just redeposited lignite. Hamburg geologist Friedrich Grube, to whom they had demonstrated their discovery, found something much more fascinating in a different exposure of the pit: within a Weichselian, predominantly sandy valley fill, there was a fossil soil. What looked at first like a buried Podzol turned out to be a far more complex palaeosol on closer inspection. At Grube's suggestion the students mapped the profile (Fig. 9.14; Grube 1981):

- 0.0–0.5 m: modern lamellic Luvisol–Podzol with solid iron enrichment horizon (topsoil mostly removed) and numerous root channels and clay enrichment bands in the underlying sand;
- 2.5–3.0 m: Weichselian niveofluvial sand with thin clay and gravel layers; and
- 0.5–1.0 m: fossil soil on Pleistocene sand, with underlying Weichselian sandy slopewash.

(continued)

BOX 9.1 A FOSSIL SOIL *(CONTINUED)*

Figure 9.14 Investigation of the palaeosol at New Wulmstorf, Lower Saxony: (a) overview and (b, c) exposing the profile and sampling. Photographs by Jürgen Ehlers.

According to borehole information, the present groundwater level is at 12 m below the ground surface.

The ^{14}C Laboratory of the *Niedersächsisches Landesamt für Bodenforschung* in Hannover radiocarbon-dated organic matter from the charcoal-bearing top layer of the fossil soil. The ^{14}C model age of the extracted humic acids was $36,300 \pm 2200$ ^{14}C years before 1950, with the charcoal yielding $22,200 \pm 750$ ^{14}C years. Contamination of the fossil humic acid fraction by inwashed younger humic acids from the present surface soil could not be excluded, but the data show that the soil had been formed prior to the advance of the Weichselian ice into northern Germany.

The fossil soil had apparently been formed in several stages:

1 Formation of a flat surface with relatively uniform, medium to fine sand in a periglacially infilled depression.
2 In a warmer climate (probably under conifer vegetation; tree root tubes) there was a relatively intense humus accumulation, acidification and podzolisation of the former topsoil (Ah-E-Bs-profile development) and (in the deepest part of the depression) gleyification (Bg-horizon formation) with iron and manganese accumulations in the subsoil, possibly over a former, impermeable permafrost table, in today's layered sand of the C-horizon (soil type: gleyic Podzol).
3 During the cold phases of the Weichselian (perhaps temporarily under tundra vegetation) the gleyic Podzol was then affected by cryoturbation and, at the margins, overprinted by solifluction. The outermost Weichselian ice margin was only some 30–40 km northeast of this location.
4 Just before the start of the coldest part of the Weichselian Glaciation *c.* 22,200 ^{14}C years ago, fire affected the contemporaneous vegetation. Since no remains of trunks or branches were found, it was perhaps a shrub vegetation. The charred remains of vegetation had been relocated or washed in (or blown in?). The orange-brown colour in the upper part of the podzolic B horizon originated from progressive microbial decomposition of humic matter in the Holocene.
5 During the Weichselian maximum (and also in the Late glacial?) the palaeosol was overlain by niveofluvial sands. During cold periods, syngenetic ice wedges formed in the frozen sands; in some cases their yellow-brown sand fillings reached down to the lower part of the fossil podzolic B horizon.
6 A lamellic Luvisol was formed in the postglacial (Holocene), initially under deciduous or mixed forest. The brown clay enrichment bands pervade both these sands as well as all the horizons of the fossil gleyic Podzol, including the tree root tubes and syngenetic ice-wedge fillings.
7 The final pedogenesis was a recent podzolization of the lamellic Luvisol after the plant cover changed to conifer and/or heath vegetation.

saprolite, buried in the aftermath by a loess–loam-rich solifluction sheet. It is overlain by an aeolian sediment cover of tephra-bearing loam, which was deposited during the Younger Dryas Stadial.

Figure 9.15 shows an exposure in the saprolite at the Oedingen kaolin pit (Oberwinter, south of Bonn). This completely soft, white kaolinitic saprolite (with distinct primary structures of the parent rock still visible) has been mined. The kaolinite-rich fine substance was extracted at the Oberwinter plant and concentrated for the ceramic industry. The colour zoning shows the transition between the white oxidation zone (where the primary organic substance of the slate has been degraded by oxidation) and the black zones characterized by prevailing reducing conditions in the lower parts of the saprolite (where

Figure 9.15 Saprolite in the Oedingen kaolin pit, Oberwinter (south of Bonn). Photograph by Peter Felix-Henningsen.

the primary organic substance was maintained by air exclusion). The zones are interlinked through a zone of transition of several metres thickness, where the black zones in the saprolite originating from clay slate due to the poorer air circulation and reducing conditions were better preserved. A better air flow and oxidative conditions existed in the saprolite from silt slate and sandstone, due to the wider pores. The fact that the black, kaolinitic saprolite turns white in the oxidation process of the organic matter (and in burning) is evidence that silicate weathering, desilication, leaching of bases, neo-formation of kaolinite and the removal of the silicate-bound iron occurred prior to oxidation in an acidic, warm migrating groundwater. Had the iron remained, a brown saprolite would have formed, which is widespread at higher elevations in the Rhenish Massif. Presumably, the groundwater level was lowered at the onset of the tectonic uplift but also during phases of arid climate in the late Miocene.

In the cooler climates of the Ice Age, soil formation could not reach the intensity of Tertiary pedogenesis. During the Early–Middle Pleistocene warm stages, strong soil formation still occurred however. The latter includes the widespread Rubified Valley Farm Palaeosol in East Anglia, for example (Fig. 9.16). This palaeosol is penetrated by periglacial features which might have originated in an early phase of the Elsterian (Anglian). The Valley Farm Palaeosol marks the transition from the Early to the Middle Pleistocene. The palaeosol at Stebbing (Fig. 9.17) also consists of a combination of Valley Farm Palaeosol with a younger periglacial overprint (Whiteman & Kemp 1990).

Palaeosols of older warm stages are not often exposed. In the ancient morainic landscape, few remains of Eemian-age interglacial soils can be expected. Within the young morainic

Figure 9.16 Rubified Valley Farm Luvisol at Great Blakenham, Suffolk, England. Photograph by Jürgen Ehlers.

Figure 9.17 Valley Farm Soil at Stebbing, Essex, United Kingdom. Photograph by Jürgen Ehlers.

landscape, however, soils of the last interglacial and Weichselian interstadials are found in many places under younger Weichselian glacial deposits. Stephan (1981) lists 28 sites of Eemian-age soils in Schleswig-Holstein. The type of soil formation during the Eemian Interglacial is largely equivalent to the modern soil formation. The Eemian fossil soil at Schalkholz developed in younger Saalian till (Warthe; Menke 1992). It is overlain by sandy strata that have been partly deposited by wind and partly by snow meltwater at the beginning of the Weichselian Glaciation. Three more podzol-like soils have developed within the sands above the Eemian palaeosol, each of them slightly weaker than its predecessor. These soils represent the Brørup, the Odderade and the Keller interstadials.

The type of soil formation is not only dependent on climatic conditions but is also strongly influenced by the original substratum. It so happens that considerably thicker soils have formed in the more sandy and poor-in-clay tills of the Older Saalian Glaciation than in the clayey, chalk-rich tills of the Middle Saalian Glaciation. While weathering only reached a depth of 1.5–2.8 m in the Warthe Till in Schleswig-Holstein, the decalcification, clay illuviation and brunification in the Older Saalian Drenthe Till can exceed a depth of 10 m.

Another important factor is age. The soils in the old and young morainic areas differ in thickness and intensity of soil development. While in the young morainic areas of northern and southern Germany stagnic Luvisols with more or less strong waterlogging are common, on the more deeply weathered, old morainic landscape podzolised Stagnosols predominate. The difference arises from the fact that soil development on the young moraines was limited to the Holocene, while in the old morainic landscape pedogenesis of the last interglacial, the Weichselian interstadials and the Holocene all contributed to the soil formation. The soils beyond the last glacial maximum limit often display a number of properties that are inherited from earlier influences. Remains of ancient interglacial clay-enrichment horizons whose clay cutans were fragmented by the action of Weichselian permafrost are frequently found. Periglacial aeolian processes have resulted in an admixture of silt (loess) and sand to the soil profile. The material may have been brought in by periglacial slope movements or mixing *in situ* by cryoturbation, so that in the topsoil many new substrate layers are often found over indigenous soil relicts.

One of the most enigmatic soils in northern Germany is the 'bleached loam' (*Bleichlehm*) of Sylt (Fig. 9.18). Till of the Older Saalian Glaciation is exposed over a length of 4.5 km in the Rotes Kliff section of the Isle of Sylt, overlying the remains of some Elsterian till in places. This till has been decalcified, weathered and bleached, probably as early as during the Holsteinian Interglacial. At the foot of the cliff, Pliocene kaolin sands are exposed. At Wenningstedt in the upper part of the Older Saalian till, a 2–4-m-thick bleached horizon is present. A strong, reddish-brown-coloured iron accumulation horizon is exposed below that, from which the name 'Rotes Kliff' (red cliff) is derived.

The bleached loam might of course be the palaeosol of the Eemian Interglacial; many scientists believe that this is so. However, there are some features that do not fit. First, at the beach stairway the bleached loam of Wenningstedt is superimposed by a 5 m thick

(a)

(b)

Figure 9.18 The 'bleached loam' of the island of Sylt. Photographs by Jürgen Ehlers.

younger diamicton. Felix-Henningsen (1979, 1983) interprets this as the till of a younger (Warthe) ice advance. A strong soil has formed in this till which pre-dates the Holocene, since it has been strongly modified by cryoturbation and ice-wedge casts. The involutions (cryoturbations) that are clearly visible in the exposure have affected a relict Bt horizon in that till, which consequently pre-dates the Weichselian and is assumed to be of Eemian age.

If the soil that developed in the upper diamicton of Wenningstedt is originally from the Eemian Interglacial, the bleached loam must have been formed within the Saalian cold Stage. But under what conditions? Stremme (1979) argued that it represented a new warm stage, the so-called 'Treene Interglacial'. However, evidence of an intra-Saalian interglacial has not been found elsewhere. According to Felix-Henningsen, soil formation may have occurred under a cool, moist climate. The impressive soil of Wenningstedt probably only represents an interstadial within the Saalian sequence.

In North America, the warm stages are not defined by palynological evidence but primarily by fossil soils. The Eemian Interglacial period is represented by the Sangamon(-ian) interglacial soil, which can be traced from the Great Lakes southwards into Western Texas. The soil type varies, depending on the substrate and the local climatic conditions. The Sangamon Soil is formed both in basal till and in loess. It is a strong interglacial soil, and is often used as a marker horizon. In the west, where no deposits of the Saalian Glaciation (the Illinoian) are found beyond the Weichselian glacial limit, the Sangamon soil is often found to be directly overlying the soil of the previous interglacial period (Flint 1971). However, the Sangamonian is not equivalent to the Eemian Stage. The former covers all of MIS 5, while the Eemian is only equivalent to MIS 5e. The Sangamonian soil is a product of pedogenesis in the Eemian, Brørup and Odderade as well as the intervening cold-climate oscillations.

9.5 Water in the Desert: The Shifting of Climate Zones

Since the Weichselian glacial maximum, the climate and ecosystems of the Earth have changed drastically. Ezcurra (2006) has illustrated those changes in his *Global Deserts Outlook*, prepared for the United Nations (Fig. 9.19). During the LGM, the belt of tropical vegetation narrowed and the deserts expanded equator-wards, while in the middle latitudes grassland, bush and semi-arid steppe or open woodland expanded into what had previously been desert.

In short, in the higher latitudes the desert margins were colder and wetter then today, while towards the equator the margins were dryer. During the LGM dunes developed in the southern transition zone of the Sahara to the Sahel, while the Mediterranean coastal areas of Algeria and Morocco would have been wetter and colder. Similarly, the Chihuahua Desert in North America was colonized by pine forest with juniper and oak, while central Mexico became drier and the forests there were replaced by cactus. The subsequent warming in the early–middle Holocene (8000–5000 years ago saw an intensification of the subtropical

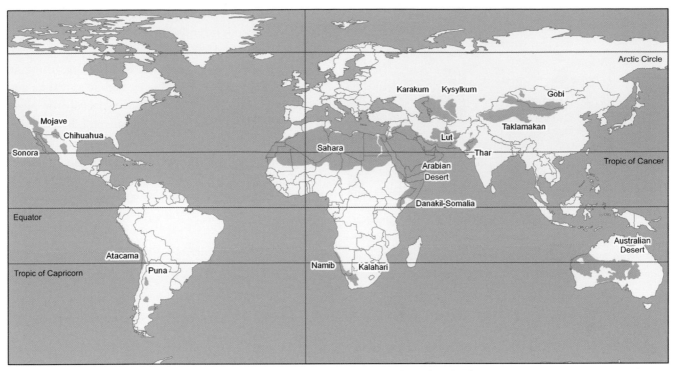

Figure 9.19 Today's deserts on Earth. Source: San Diego Natural History Museum. Reproduced with permission.

monsoon. The tropical rainforests and forests around the equator expanded to the north and south, while the deserts shifted towards the middle latitudes. During this time the southern Sahara and the Sahel were much wetter than today. There was an extensive vegetation cover, a rich animal life and many human settlements. This phase of the 'green Sahara' ended during the middle Holocene when a sudden transition to 'Sahara desert' took place, the state which we see at present.

In the Sahara a large number of endorheic basins exist which, under a more humid climate, were occupied by lakes. The latter include the pluvial lakes in the area of the Schotts of Tunisia and Algeria as well as in large parts of Mali and Chad. Many closed depressions or playas exist even in the Western Desert of Egypt and Sudan, from where river and lake deposits and multiple traces of prehistoric human activity have been found (Hoelzmann et al. 2001). The extent of the playa sediments suggests that considerable areas of water once existed, which attracted the Neolithic settlers. Many of these sediments have been radiocarbon dated and testify to the ubiquity of an early–middle Holocene wet phase, which is often referred to as the 'Neolithic pluvial'. The large Westnubian palaeolake existed in the extreme northwest of present-day Sudan. It reached its largest extent 9000–4000 years ago, and covered an area of *c.* 7000 km². Tremendous amounts of rainfall were required to sustain a lake of that size; today's annual rainfall in that area is below 15 mm.

Even more extensive was the 20,000 km² 'Lake Ptolemy' which Pachur (1997) found in northern Sudan. It was surrounded by vast swamp areas and might have been connected to a

Figure 9.20 Skeletal remains of a fish (presumably Nile pike) found in diatomite in Dogonboulo Fachi, Niger. Photograph by Detlef Busche.

palaeolake Chad. The amphibious landscape which can be derived from it extended over a distance of more than 2000 km (Busche 1998).

Four former lake phases have been identified in the Sahara. The oldest of these existed at *c.* 120 ka and the younger at 33, 11 and 7 ka, especially in Niger (Fig. 9.20), Chad (Box 9.2) and southern Libya (Busche 1998). Lake phases are also known from the Western Desert of Egypt, from northern Sudan (e.g. Pachur 1997) and from the Western Sahara (Petit-Maire 1991; Busche 1998). An even younger lake period has affected the Lake Chad basin (*c.* 5–4 ka).

BOX 9.2 SHORELINES IN THE SAND: THE HISTORY OF LAKE CHAD

With the help of remote sensing, it is now possible to reconstruct the location and extent of many former lakes; an excellent example is Lake Chad (Fig. 9.21). During the Holocene, a huge Lake Megachad (Fig. 9.22) occupied the Chad Basin. This lake covered an area of >350,000 km², which corresponds to the size of the Federal Republic of Germany. By comparison, the largest lake on Earth today, the Caspian Sea, has an extent of 386,000 km². The shorelines of Lake Megachad were preserved in the dry climate and are still visible. The lake was only discovered by evaluation of the SRTM30 digital terrain model. It revealed that a system of old shorelines was present around present-day Lake Chad. These features have been mapped (Schuster et al. 2005) with the additional use of Landsat TM satellite images (resolution 28.5 m). A number of excellently preserved landforms that formed on the banks of Lake Megachad have been identified.

Figure 9.21 Traces of a palaeolake in the Largeau region (northern Chad); recent dunes overlying diatomite. Photograph by Detlef Busche.

1 *The Angamma Delta.* A well-developed former delta at the foot of the Tibesti Mountains on the extreme northern edge of the former lake was observed in aerial photographs (Ergenzinger 1978). A closer look revealed that this ancient delta had been dissected by erosion. The fragmentation phase was followed by re-activation of the delta with simultaneous formation of beach ridges. The beach ridges on the surface of the delta were, in turn, crossed by a system of braided and meandering channels. The former underwater slope, the so-called

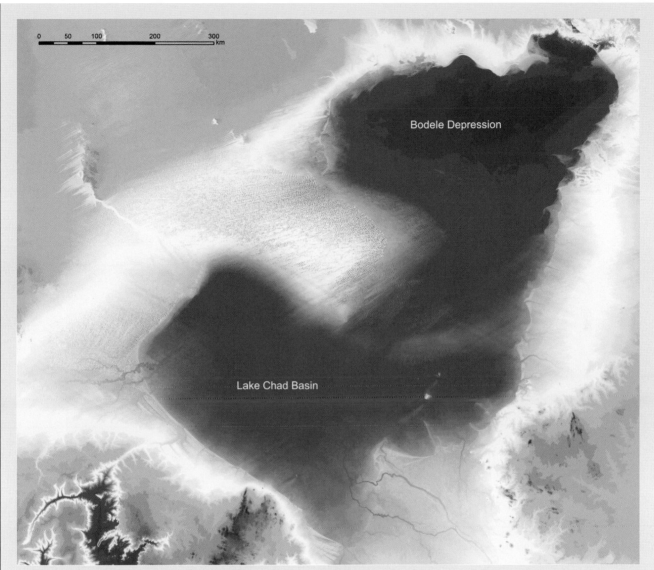

Figure 9.22 Lake Megachad some 4000 years ago. Courtesy NASA/JPL-Caltech.

Angamma Cliff, continues to the west in a large beach spit. Weakly developed beach ridges at two lower levels represent later lake phases with lower water levels (Figs 9.23, 2.24).

2 *The Goz-Kerki system of spits.* The east coast of Lake Megachad resembled today's north coast of the Azov Sea. Seven large spits are identified on a 200-km-long section of the east coast, belonging to two different lake levels. At least two phases of beach ridge formation can be distinguished: (1) small, clearly developed spits at about 325–330 m; well-developed beach ridges at this level form an almost continuous shoreline of sand ridges along the former lakeshore and the mouths of small tributaries are highly visible; and (2) another series of large spits slightly deeper (c. 310 m), which appear less distinct (Fig. 9.25).

(continued)

BOX 9.2 SHORELINES IN THE SAND: THE HISTORY OF LAKE CHAD (CONTINUED)

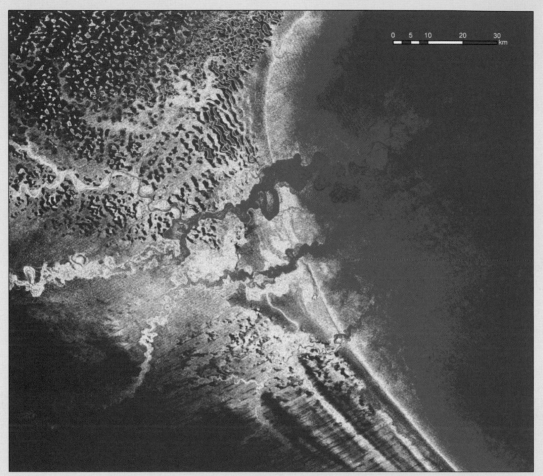

Figure 9.23 Dunes and beach ridges on the western shore of Lake Megachad. Courtesy NASA/JPL-Caltech.

3 *The Chari-Palaeodelta.* The Chari River flowed into the palaeolake from the south. A huge delta (Fig. 9.26) formed on the southeast corner of the basin where it entered the lake, occupying a surface area of *c.* 50,000 km². The delta is surrounded by younger beach ridge systems that reflect different stages of reactivation, associated with a lake level of 315–320 m. As the water level of Lake Megachad fell, the delta was abandoned. Today, the Chari River runs further to the west and is currently accumulating a new delta in today's smaller Lake Chad. The age of Lake Megachad has been determined by radiocarbon dating. The beach ridges of Goz-Kerki were overlain by a shell-rich sediment layer which yielded a radiocarbon age of 5–4 ¹⁴C ka BP.

How can the former extent of Lake Megachad be mapped? Remote sensing provides no pictures, but data that must be transformed into digital images. The SRTM terrain model consists of a multitude of altitude data, which are nothing more than a number of points with some information. In order to turn these points into an image, a programme such as ArcGIS must be used. But there are limits to image processing,

Figure 9.24 Dunes and beach ridges on the western shore of Lake Megachad. Courtesy NASA/ IPI -Caltech.

and to depict minor altitudinal differences in detail, the available colour scale is not sufficiently detailed, In ArcGIS only 32 colours can be distinguished. If they are spread evenly over the range of elevations around Lake Chad, each shade represents over 30 m of altitude, and little can be seen. Certain key altitudes must be examined. The most conspicious coastal landforms occur between 310 and 330 m.

The image of the northwest shore of Lake Megachad demonstrates that the dunes in the area of Lake Chad have not moved since the palaeolake dried up. They have been frozen in their steps for over 4,000 years. The area evaluated here covers about 1,000 × 1,000 km, i.e. one million km². It extends from the southern edge of the Sahara right to the middle of the desert. If in the southern half of the Sahara after

(continued)

BOX 9.2 SHORELINES IN THE SAND: THE HISTORY OF LAKE CHAD (CONTINUED)

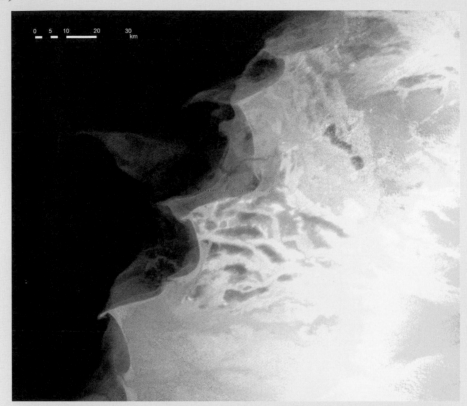

Figure 9.25 Details of the deltas and spits on the east bank of Lake Megachad. Courtesy NASA/JPL-Caltech.

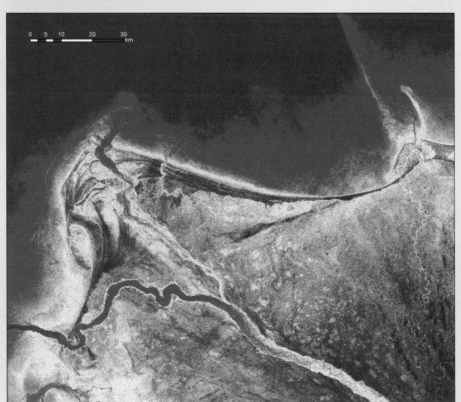

Figure 9.26 The former Chari Delta. Courtesy NASA/JPL-Caltech.

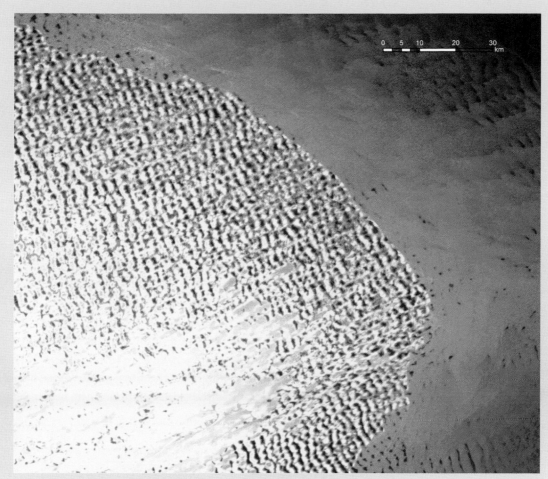

Figure 9.27 The dunes are clearly older than Lake Megachad. Courtesy NASA/JPL-Caltech.

4,000 years ago all dunes ceased migrating – was that also true of the northern half? And does it also apply to the west and east?

Satellite pictures of Mauritania show that the north-south trending dune ridges, the youngest dunes in Mauritania, can be traced south into Senegal, where they are fixed and covered with vegetation. These dunes have also been stationary for several thousand years (cf. Chapter 11).

According to their deposits, the older large lakes of the Sahara must have been even more impressive than Lake Megachad, however, their traces have been largely destroyed by the wind, their shorelines and lake sediments dissected into yardangs, visible only locally in the field or on aerial photographs. The dissection dates from a time of much stronger winds occurred, when the dunes have moved, that is why those dunes are found to cover the floors of the old lake basins. The dunes in the Chad Basin were established between the phase of the giant lakes and the formation of Lake Megachad (Fig. 9.27).

Busche (1998) describes a huge marsh and lake basin from the Ténéré. Because of their bright colour, the shorelines with beach ridges and mollusc shells can be tracked for many kilometres in aerial photographs or satellite images around the Dirkou airfield in northern Mali. These sediments are strongly affected by wind erosion, some bizarre pillars (yardangs) remain (Fig. 9.28). Most of the former lake sediments fell victim to deflation processes after drying up (Bristow et al. 2009).

It must be assumed that, before the known Late Pleistocene and Holocene lake phases, numerous older lakes existed throughout the course of the Quaternary. The diatomite sediment formed in these lakes is very susceptible to erosion, however. What sediment survived erosion by the wind would certainly be soaked and destroyed in the following lake phase (Busche et al. 2005).

Huge palaeolakes were not limited to Africa. Another lake that had grown to enormous proportions in the wet periods of the Quaternary is Lake Eyre in Australia. The lake reached its greatest extent by 125, 80, 65 and 40 ka, in times of stronger monsoon activity. The higher lake levels resulted in the formation of distinct shorelines, some of which have been known of for a long time. During the most extensive lake phase (at *c.* 125 ka, the last interglacial period), the lake level stood at about +10 m. Lake Eyre at that time covered an area of almost 35,000 km², which is more than three times the size of today's playa. Together with the Frome-Gregory lake system adjoining in the southeast, the palaeolake comprised 430 km³ of water. The largest known Lake Eyre from historic times, in contrast, only had a capacity of 30 km³. It is obvious that the former lake must have formed in a much wetter climate. Because of the incomplete preservation of the beach deposits and the

Figure 9.28 Yardang formed in diatomite lake sediment in the Kafra Depression in NE Niger, Africa. Photograph by Detlef Busche.

Figure 9.29 Sinkhole in limestone karst 30 km WSW of Doha in Qatar. Photograph by Holger Wolmeyer.

huge lake system, it was only possible to determine the former extent of Lake Eyre and the Frome-Gregory system when a digital terrain model of the area recently became available (DeVogel et al. 2004).

Traces of former aquatic systems are not only found at the land surface but also underground. Palaeokarst features point to earlier, more humid conditions, in areas which are now desert. These include the numerous sinkholes and caves in the Eocene rocks of Qatar (Fig. 9.29). According to present knowledge, the now-dry karstic landforms formed in a wet phase during the Middle Pleistocene (*c.* 560–326 ka; Sadiq & Nasir 2002).

9.6 Changes in the Rainforest

There are tropical rainforests in South and Central America, Africa, Southeast Asia and New Guinea. On a world map showing the distribution of the rainforest, Australia is mostly omitted; the rainforest areas of that continent today are rather small. They include the Daintree National Park (Fig. 9.30) about 100 km northwest of Cairns, the vegetation of which is equivalent to other tropical rainforests.

If tropical rainforest is mentioned today, it is usually in the context of radical deforestation and replacement of natural vegetation by farm land. Palynological studies have shown that the present South American rainforest is a relatively young phenomenon. In SW Amazonia (Laguna Bella Vista and Chaplin, Bolivia), there are indications that the rainforest only reached those areas *c.* 3000 years ago. While the extent of the Brazilian rainforest has

Figure 9.30 Tropical forest in the Daintree National Park, Australia. Photograph by Eva-Maria Ludwig.

increased in a southwest direction, as a result of postglacial sea-level rise, mangrove has expanded at the expense of the rainforest on the coast (Behling 2007).

It was first suggested that the area of the South American rainforest had been significantly reduced in the last glacial period when sediment cores from the deep sea, off the coast of Guyana, were analysed. The proportion of unweathered feldspars in the sediments from the last Ice Age was 25–60%, while in today's deposits it is about 17–20%. This large difference can only be explained if erosion had increased under arid conditions, during which the rainforest was much reduced in size (Damuth & Fairbridge 1970). The map of Kadomura (1995; Fig. 9.31) shows the extent of the change.

Or does it? Other scientists have disagreed, arguing that the results from the deep sea were not evidence. After all, the Ice Age sea level was lower by 120 m, and the mineral entry could have been a result of the exposed shelf areas. After studying lake deposits from the Amazonian lowlands, Colinvaux et al. (1996) countered: 'the pollen profile indicates that the Amazonian lowlands was continuously covered by tropical rainforest for the last 40,000 years. The rainforest had not been fragmented into isolated refuges, and the Amazonian lowlands had neither been the source area for dust deflation. Only the species composition differed slightly from today, reflecting perhaps a cooling of 5–6°C.'

The palynological studies of van der Hammen (1974) and van der Hammen & Hooghiemstra (2000) suggest that dry vegetation spread during the Pleniglacial. However, the localities they investigated were not in the rainforest but in the surrounding savannah areas. Extrapolation to the inner Amazonian lowlands might not have been justified.

However, there are other indications that the climate was drier than today: large areas of northern Brazil are covered with now-inactive sand dunes. The dunes along the Rio Negro can be dated to 32–8 ka, testifying to a period of enhanced aeolian activity (Carneiro-Filho et al. 2002). The dunes on the Rio Branco began migrating c. 17 ka ago (Teeuw & Rhodes 2004). However, Colinvaux et al. (2000) do not think much of dunes as climate indicators. The dunes had been active periodically since the Pliocene–Pleistocene boundary, and enhanced activity during the glacial maxima could not be proved unequivocally.

The only pollen profile from the Amazon lowlands which extends as far back in time as the last glaciation is the La Plata profile studied by Colinvaux et al.; it showed nothing but rainforest. However, However, d'Apolito et al. (2013) were able to demonstrate that the presence of a dry period at the glacial maximum is masked by drought-induced hiatuses in the studied profile.

The findings from the areas surrounding the Amazonian lowlands all seem to suggest that the rainforest was pushed back considerably during the Pleniglacial (Marchant et al.

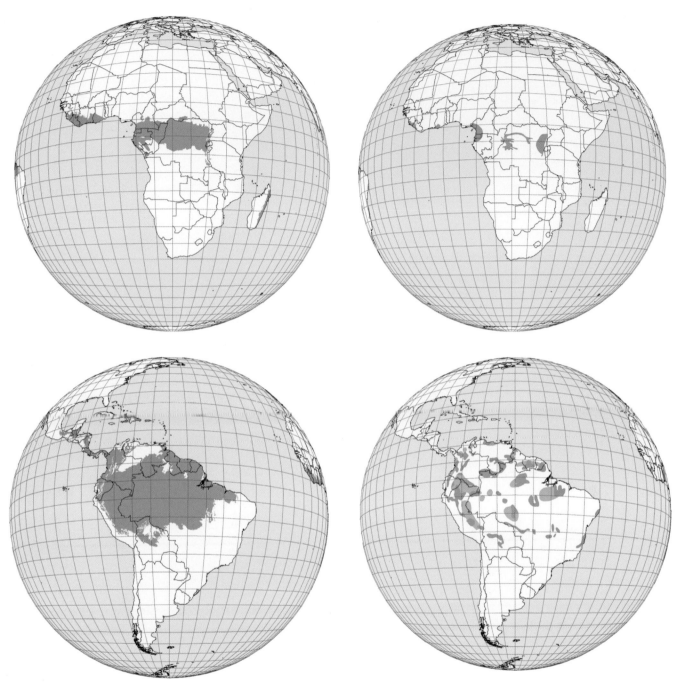

Figure 9.31 Tropical forests today (left) compared with their distribution during the ice ages (right). Adapted from Kadomura (1995).

2009). Africa also saw a strong suppression of the tropical rainforest at the Last Glacial Maximum, in which the Dahomey gap between the West African rainforest area and the Congo Basin increased significantly in size. However, as in South America, the number of sufficiently long pollen profiles is still far too low to allow for a final evaluation (Meadows & Chase 2007).

Decay of dead ice in front of the Skeiðará glacier and on the edge of the Brúarjökull, Iceland. Photographs by Jürgen Ehlers (above) and Emmy Mercedes Todtmann (below).

Chapter 10
The Course of Deglaciation

10.1 Contribution to Landforms

Ice decay and retreat produces a range of distinctive landforms. An excess of melting (ablation) over snow accumulation will eventually destroy glaciers and produce sediments and landforms associated with melting. The speed and magnitude of climate change resulting in deglaciation affects the nature of these landforms. If glaciers retreat gradually they can leave a series of recessional moraines. If climate change is rapid with sharp rises in temperature, then glaciers can stagnate *in situ*, producing a complex assemblage of waterlain sediments. Whatever the mode of the glacier decay, it almost always involves production of large amounts of meltwater (the rare exception being cold-based glaciers which lose mass by sublimation, i.e. direct evaporation of ice to gas). For most glaciers, the transition from ice to water during deglaciation results in a distinct change from glacial to fluvial and lacustrine sedimentation. This is often characterized by the presence of sand and gravel spreads and laminated lake sediments as meltwater pulses from glaciers into forelands downvalley or ponds as ice stagnates, forming kettle holes.

Deglaciation also reveals a landscape completely vulnerable to new forces. Rock walls once propped up by glaciers become open to the forces of gravity and the removal of huge pressures of ice results in rock fractures. The new dominance of gravity as the main force causes slopes to undergo considerable re-adjustment following deglaciation. This paraglacial response (Ballantyne 2002) can be just as important as glaciers themselves in removing large masses of rock from the Earth's surface. In fact, in some settings there is evidence that glaciers simply clean up rock slope failures that occur in intervening deglaciated periods. Deglaciation is therefore one of the most important mechanisms in landscape evolution.

The Ice Age, First Edition. Jürgen Ehlers, Philip D. Hughes and Philip L. Gibbard.
© 2016 John Wiley & Sons, Ltd. Published 2016 by John Wiley & Sons, Ltd.
Companion website: www.wiley.com/go/ehlers/iceage

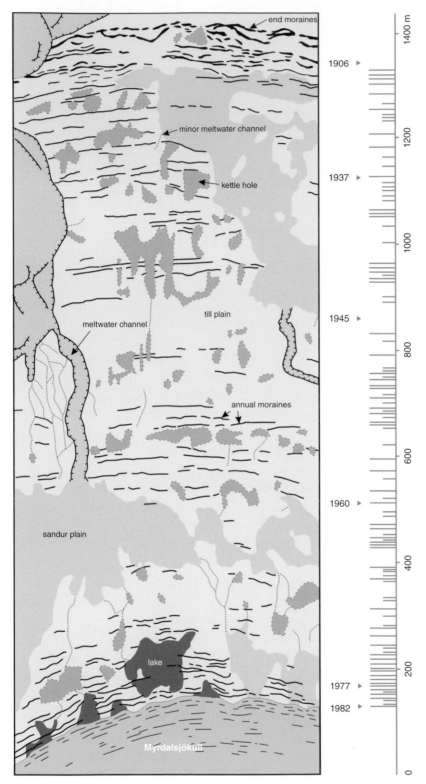

10.2 Ice Decay

The decay of ice sheets produces a distinct landscape that is often characterized by numerous lakes and morainic mounds (Fig. 10.1). This landscape can even be seen on continental-scale maps, with the widespread scattering of lowland lakes in Finland and Canada hinting at the former presence of large ice sheets over these areas. These lakes formed as ice retreated and come in a wide range of forms; one which is diagnostic of ice decay *in situ* is the kettle-hole lake.

Ice decay produces different landforms to those formed when ice is advancing or in equilibrium. The classic landscape of ice decay is 'kettle-and-kame topography' which is characterized by gently undulating terrain with gravel hillocks separating small lake basins, many of which may now be infilled. Price (1969) described such features in front of Breiðamerkurjökull in Iceland, and noted that these had formed in front of a glacier which had undergone very rapid wastage. However, Evans & Twigg (2002) noted that these features were characterized by flat-topped outwash surfaces which separated pits (kettles). They suggest that this type of terrain is therefore better termed 'pitted outwash', and that 'kame-and-kettle topography' should be reserved

Figure 10.1 Map of the proglacial area between Mýrdalsjökull (Iceland) and the terminal moraine of 1890. All clearly visible (long dash) and weakly visible (short dash) annual moraines that could be measured have been recorded. Such a landform assemblage is characteristic of an active retreat of the glacier front. Adapted from Krüger (1987). Reproduced with permission of Johannes Krüger.

for features produced by the reworking of debris by supraglacial and englacial drainage systems (e.g. Price 1973; Benn & Evans 2010).

The landforms of northern Germany often show no clear assignment to one of the landforms found commonly at the ice margin. The young morainic landscape of the Schwansen peninsula (Schleswig-Holstein) may serve as an example. The area is bordered in the north by the Schlei and in the south by the Eckernförde Bay. The internal composition of the landforms is exposed in the steep cliffs of Waabs and Bookniseck, which had been mapped by Prange (1979). The area is characterized by a large number (about 1000) of mainly circular to elliptical landforms that usually rise c. 5–10 m above their surroundings. The diameter of most of the forms is 100–150 m and only about 20% have a distinct long axis. The strike direction of those axes varies greatly, and no connection with a possible ice advance is recognizable. Geological mapping has shown that the area is largely till underlain by ice-marginal sand and gravel. Meltwater sands occur in the southwest part of the Sieseby map sheet along the valley of the Kolholmer Au, but elsewhere are limited to small areas and individual hills (i.e. southwest of Thumby). In the cliff outcrops at Klein Waabs and Bookniseck, till forms the uppermost layer. Those undisturbed basal tills are underlain by partly glaciotectonically disturbed older till pervaded by sand lenses (Prange 1979).

The undulating relief and the glaciotectonics exposed in the cliffs may have contributed to the interpretation of the Schwansen landscape as terminal moraine originally shaped by active ice. Only Prange (1979) pointed out that the thrust features seen in the underground had been formed before deposition of the upper glacial till. The present-day morphology does not allow the reconstruction of older, overridden thrust moraines.

In Figure 10.2 two of the end moraines mapped by Gripp (1954) are shown as dark lines. To map the end moraines, Gripp coloured the contours of 1:25,000 scale maps. While the areas of highest ground are still relatively easily connected to arcs of end moraines on the topographic map of scale 1:25,000, this is difficult on the larger-scale map of 1:5000 used here; the latter displays a landform assemblage that is characteristic of an ice-decay landscape: it consists of irregularly distributed hills and kettles; the hilly

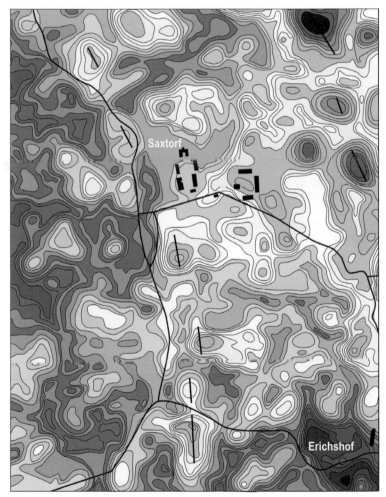

Figure 10.2 Ice-decay landscape at Saxtorf, Schwansen (Schleswig-Holstein, Germany). Cartographic basis: Topographic Map 1:5000, sheet Saxtorf.

Figure 10.3 Crevasse fillings in the forefield of Brúarjökull, Iceland. Photograph by Emmy Mercedes Todtmann.

region is covered in dead-ice landforms over a large area, and does not consist of elongated end-moraine ridges; some of the hills with a recognizable long axis strike NW–SE and others NE–SW, suggesting the influence of crevasses (Fig. 10.3).

Comparable landform assemblages were described by Gravenor & Kupsch (1959) and Colgan (2007) from North America and interpreted as ice-decay landforms (Fig. 10.4). In both cases features occur which are superimposed on older landscape elements (such as moraines or drumlins). Some of the mapped 'terminal moraines' of various authors within the dead-ice landscapes of northern Germany are likely to be explained this way. It should be noted however, that squeezing-up of underlying material can take place when an ice sheet decays (Hoppe 1952) and that the glaciotectonic features observed in the cliffs could in part be a consequence of such events.

During ice decay the crevasses of the stagnant ice are party filled by sediment washed in from above, partly by squeeze-up of underlying sediment (Colgan et al. 2003). Gripp (1929) had already described the formation of hummucky moraine through the decay of 10–15 m high 'mud walls' that had originated from former crevasses filled with till from the ice margin of the Nathorst Glacier on Svalbard. After studying glaciers on Iceland, Woldstedt (1939) concluded that the hummocky moraine of North

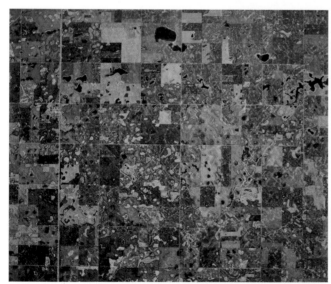

Figure 10.4 Ice-decay landscape in Bottineau, North Dakota. The higher ground of the Turtle Mountains can be seen in the northeast. Both levels were formed under the influence of thawing dead ice.

Source: US Geological Survey, CIR, NHAP image 5MGY01032_205, November 1981.

Germany owed its existence primarily to the thawing of buried dead ice. Gripp (1974) calls this form assemblage a 'meltdown landscape'. The development of such dead-ice landscapes has been studied in detail on Iceland (Kjær & Krüger 2001; Figs 10.5, 10.6).

The north German glaciated area was especially susceptible to the detachment of larger dead-ice fields because of its location on the southern edge of the Baltic Sea Basin. However,

Figure 10.5 Dead-ice hollows in front of Brúarjökull, Iceland. Photograph by Emmy Mercedes Todtmann.

Figure 10.6 Dead-ice hollow in front of Skeiðarájökull, Iceland. The ice and the meltwater sands are from the 1996 jökulhlaup. Photograph by Jürgen Ehlers.

the presence of dead-ice landforms in the Alpine Glaciation area should not be underestimated. The phenomena of ice decay are however limited to smaller areas in many cases, such as the Bachhauser Filz and Buchsee areas east of Lake Starnberg (Schumacher 1981; Jerz 1987) or the Oster Lakes south of Lake Starnberg (Meyer & Schmidt-Kaler 2002). The landform assemblages are similar to those found in northern Germany. It is not always easy to distinguish features created by subsidence over thawing dead ice from ice-pushed deposits. Extensive ice-decay landscapes are found in front of the outermost Würmian moraine of the northern Rhine glacier (i.e. Rohrsee at Bad Wurzach, northeast of Ravensburg; Schreiner 1997).

10.3 The Origin of Kettle Holes

Kettles (kettle holes) are widely regarded as a characteristic of the morainic landscape. In the British Isles, kettle holes are found in the lowlands within the limits of the last ice sheet. These features are widespread in the Cheshire and Shropshire basins for example, where radiocarbon ages from the base of lake cores show that these basins formed at the end of the Pleistocene. Kettle holes are also found in older glacial terrains, such as in East Anglia within Middle Pleistocene glacial limits. However, many of these older kettle holes have become infilled and contain important evidence of Middle Pleistocene interglacial deposits resting on top of Anglian-age till (e.g. Boreham & Gibbard 1995). It is believed that the irregular topography left behind after the thawing of the ice sheet is preserved until, under increased erosion and redeposition during the following cold stage, the endorheic depressions are partly filled and partly connected to form continuous valleys (Marcinek et al. 1970).

Old morainic landscapes have been separated from young morainic landscapes according to this concept, based largely on the presence or absence of a kettle holes. Garleff (1968) gave an overview of their distribution in Lower Saxony. Since the occurrence of kettles of several metres depth (e.g. at Geestenseth; Fig. 10.7) are especially difficult to reconcile with extensive periglacial mass movements during the Weichselian, it was assumed that some of the hollows in the old morainic landscape only emerged at the end of the Weichselian Glaciation. The emergence of such depressions can have different causes:

1 Most kettles are best explained by sagging over thawing stagnant ice. Salisbury (1892) first described sandur deposits with numerous kettle holes (pitted outwash plains) from North America. This interpretation was later extended to the majority of the endorheic depressions, including those which were found in till areas. There are different hypotheses to explain how the ice got into the ground; either ice within or covered by the till are envisioned, or sand-covered blocks of dead ice (Woldstedt 1961). The sand-covered ice need not necessarily be former glacial ice. Even meltwater ponds frozen in winter (Galon 1965) or icings (Kozarski 1975) can be covered by sediment and eventually result in the formation of kettle holes. Buried dead ice can survive in the ground over a long time. It is known, for instance, that some Late Weichselian–early Holocene moraine ridges in the Arctic still have an ice core. On the basis of structural analyses, French & Harry (1988) have shown that the ice inside the Sandhills Moraine on Banks Island in northern Canada is Weichselian glacial ice. From the boulder content of the ice, it was determined that the initial melting of a

(a)

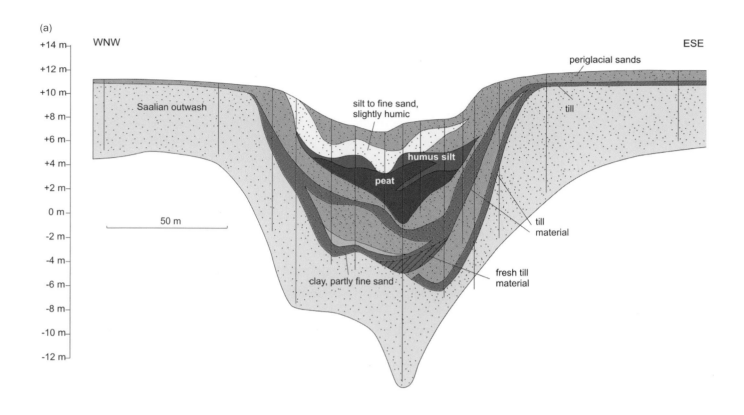

(b)

Figure 10.7 (a) Cross-section through a deep kettle hole at Geestenseth, Lower Saxony. Adapted from Lade (1980). Reproduced with permission of the author. (b) Drilling in the Geestenseth kettle hole. Photograph by Jürgen Ehlers.

c. 10-m-thick layer of ice was sufficient to produce the 2–3-m-thick blanket of ablation till, which has protected the ice from further melting. Buried glacier ice from the last glaciation has been found in various places in northern Canada and NW Russia (Harris & Murton 2005). Sugden et al. (1995) have found buried glacial ice in a dry valley in Antarctica which is 8.1 Ma in age.

2 Using the example of some hollows in the area of Neuenwalde Geest, Meyer (1973) discussed the possible formation of a kettle hole by erosion through a former glacier moulin. That process is known to form deep potholes even in bedrock, as can be seen in Norway. However, such forms in unconsolidated material seem to be relatively rare.

3 Pingos such as those found in recent periglacial zone around the Arctic can also lead to the formation of kettle holes. With progressive growth of those ice lenses, the outer sediment layers of the hill tear open above the ice core and slide down the flanks. The ice core is no longer protected from the Sun, the ice melts and a lake forms, surrounded by a small annular rampart. Maarleveld & van den Toorn (1955) have reported such forms from the Netherlands and Sparks et al. (1972) from East Anglia. Well-studied remnants of former pingos include the features on the Hohes Venn, Belgium (Pissart 2003). These cases differ from most ponds in northern Germany in the fact that no ramparts have been preserved. Most of the kettles that do not have a rampart probably have other origins. Even the remains of a rampart such as at Lake Wollingst (Lade 1980) are not certain evidence of pingo genesis.

4 Thermokarst processes can also result in the development of kettle holes (Garleff 1968), some of them larger than 1000 m in diameter. Ballantyne & Harris (1994) describe such landforms from the Fenlands (England) and pollen and macrofossil stratigraphy from sequences within these depressions shows that many of these formed during the last cold stage (Sparks et al. 1972). An overview of the processes occurring during thermokarst is given by Burn (2007).

5 Aeolian processes can also lead to the formation of kettle holes. In combination with dunes, frequently shallow blowout depressions are found. However, they can usually be easily distinguished from kettles of other origins.

Unusual events have also been invoked to explain the formation of kettle holes. In his novel *Nooit meer Slapen*, the Dutch geologist and novelist Willem Frederik Hermans (1966) lets the protagonist check the academic idea that some circular hollows in northern Norway were caused by meteorite impacts; this is not found to be case either in the book or in reality. For the majority of geologists, the so-called 'Tüttensee impact crater' in Chiemgau (Bavaria) is nothing but a harmless dead-ice hollow (Doppler & Geiss 2005).

Since a number of kettles in the old morainic landscape contain deposits of Eemian Interglacial peat, it has been demonstrated that at least these forms have survived the Weichselian glacial period. The kettles are preserved almost exclusively on flat terrain with a slope of less than 2°, where solifluction was very limited. In his study of 12 endorheic depressions in the Bremervörde-Wesermünde Geest, Lade (1980) demonstrated that in five cases formation by sagging over dead ice was probable. In four cases, formation by Weichselian ground ice seemed probable; the remaining three cases could not be clarified. Lade points out that kettle holes are often found as an extension of small dry valleys. The thawing of dead-ice blocks may in some cases have controlled the valley formation. The time at which the ice eventually thawed depended largely on the thickness of the sedimentary cover. In some cases, Weichselian stagnant ice is known to have survived into the Allerød Interstadial and longer.

The previously mentioned kettle at Geestenseth is 3 m deep and has a diameter of about 50 m. Drillings have shown that today's depression is only a small residual form of the original kettle. The true base of this dead-ice hollow is at a depth of over 16 m. The largest part of the original hollow form had been infilled by solifluction and slopewash. The formation of peat in the Eemian Interglacial and in the subsequent Brørup Interstadial has contributed to the infilling of the depression. However, material transport from the surrounding high ground was not sufficient to fill the Saalian kettle hole.

10.4 Pressure Release

As glaciers retreat, the stresses associated with the immense weight of ice become replaced by gravity. This causes rocks to release the strain resulting from the previous overburden, and glaciers often excavate rock producing steep slopes. When ice is present these steep slopes are supported by the ice itself; when the ice retreats these oversteepened slopes are prone to collapse under gravity. This process is promoted by the pressure release and rock fracturing (Figs 10.8, 10.9). The result is a massive adjustment of landscape to glacier unloading. This adaptation of a formerly glaciated area to ice-free conditions is referred to as paraglacial processes (Ballantyne 2002). The volume of rock mass remobilized by paraglacial processes can be immense. Later glaciations then return and remove this material, starting the cycle of erosion once again. In this respect, glaciers not only erode landscapes but also transport large amounts of rock that are released in the intervening non-glacial periods.

The compensatory movements of bedrock begin as soon as an area becomes ice free, and they last up to the present day. In Scotland, Ballantyne & Stone (2013) found that while

Figure 10.8 After becoming free of ice, the rock slope above Nigardsbreen Glacier in Norway becomes fractured. Photograph by Jürgen Ehlers.

Figure 10.9 Fractured rock at Nigardsbreen, Norway. Photograph by Jürgen Ehlers.

there was enhanced rock slope failure frequency during or immediately after deglaciation, this has been followed by approximately constant periodicity. While glaciation provides the pre-conditioning for rock slope failures, and these failures have greatest frequency soon after ice retreat, the timing of rock slope failures in the intervening interglacial can occur at any time.

In the Alps, Lebrouc et al. (2013) provided dating evidence that large landslides did not immediately follow deglaciation, but occurred several thousand years after ice down-wastage in the valleys. Lebrouc et al. conclude that that debuttressing is not the immediate cause of landslide initiation. They suggest that climatic mechansims may play a role, with landslides coinciding with warmer and wetter conditions in the Alps. However, they also note that seismic effects cannot be ruled out. Delayed paraglacial response can be particularly common in tectonically active high mountain areas. In the High Atlas, Morocco, a large rock slope failure situated in a glaciated valley dates from the mid-Holocene. Ice last retreated from this valley in the Late Pleistocenc, yet Hughes et al. (2014) suggest that glacial oversteepening combined with seismic activity near an active fault could have contributed to rock slope failure.

Large rock slope failures are the most dramatic landscape response to glacier retreat (Figs 10.10, 10.11). However, most movements are more subtle and slower. For example, occasional release of rock from a cliff face downslope produces talus (scree) while whole rock masses can creep downslope along local fault planes (Ampferer 1939).

Figure 10.10 A rockfall destroyed a petrol station in Dale, Norway. Note the rock on the roof (April 1976). Photograph by Jürgen Ehlers.

Figure 10.11 The interior of the ruined petrol station; note the indented rear wall of the building. Photograph by Jürgen Ehlers.

10.5 A Sudden Transition?

The end of the last cold stage was characterized by the occurrence of several interstadials. These are clearly defined in both ice-core records, such as in Greenland (Björk et al. 1998), and in vegetation records around the world.

The strongest of the Late-glacial climatic warmings is the Allerød Interstadial (equivalent to Greenland Interstadial 1 in the ice-core records). During the Allerød, a birch–pine forest grew in northern Germany and adjacent areas. Considering today's distribution of such forests, there might have been relatively low average summer temperatures of about 12°C. The palynological investigations demonstrated the presence of certain aquatic plants however, that show that the summer may have been considerably warmer (*c.* 15–16°C). The discrepancy is explained by the fact that the water plants can spread much faster than trees so the absence of species such as oak, elm and elderberry, which could grow well under those thermic conditions, should not be misinterpreted. The Allerød sediments are easily identifiable in large parts of northern continental Europe because of an ash layer (tephra) from the eruption of Laacher See volcano (Eifel region, Germany).

Many so-far unanswered questions of bio- and chronostratigraphy of the Weichselian Late-glacial have been clarified by sediment studies of the Eifel maars. In the sediments of the Holzmaar and Meerfelder Maar, which have an annual layering, the Meiendorf, Bølling and Allerød interstadials can be clearly demonstrated.

The Meiendorf/Bølling/Allerød interval defined in Germany is often simply defined as the Bølling/Allerød elsewhere in Europe. There has been a move towards defining the Late-glacial interval using high-resolution ice-core records to provide a global

TABLE 10.1 SUBDIVISION OF THE WEICHSELIAN LATE-GLACIAL IN TERRESTRIAL AND ICE-CORE RECORDS. THE TERRESTRIAL AGES ARE BASED ON EVIDENCE FROM THE EIFEL MAARS (LITT ET AL. 2003). THE GREENLAND ICE CORE AGES ARE FROM RASMUSSEN ET AL. (2006).

Eifel Maar Stratigraphy	Age from Eifel Maars	Age from Greenland ice core	Ice-core stratigraphy
Younger Dryas	12,680–11,590	12,896–11,703	Greenland Stadial 1
Allerød Interstadial	13,350–12,680	14,075–12,896	Greenland Interstadial 1abcd
Older Dryas	13,540–13,350		
Bølling Interstadial	13,670–13,540	14,692–14,075	Greenland Interstadial 1e
Oldest Dryas	13,800–13,670		
Meiendorf Interstadial	14,450–13,800		
Pleniglacial (Oldest Dryas)			Greenland Stadial 2a

stratigraphical scheme for correlation. The Meiendorf/Bølling/Allerød interval is correlated with Greenland Interstadial 1 (van Raden et al. 2013). The Weichselian Late-glacial stratigraphy is shown in Table 10.1 and the terrestrial chronology from Eifel Maar in Germany is compared with the high-resolution ice-core chronology from Greenland. Confusingly, the Oldest Dryas occurs within the Meiendorf/Bølling/Allerød interval in Germany, whereas elsewhere in the world the Oldest Dryas precedes this interval and is correlated with the Pleniglacial in the German sequence (Fig. 10.12).

The Older Dryas, between the Allerød Interstadial and the Bølling Interstadial, is only a comparatively minor deterioration. The birch forest of the Bølling was only replaced by a more open birch forest. Following the Allerød Interstadial, the Younger Dryas interval led to significant re-advances of the last northern ice cap, forming major end moraines in Norway (Ra-moraine), Sweden (Central Swedish End Moraine) and Finland (Salpausselkä; Mangerud 1987). In Scotland this ice advance is represented by the terminal moraines of the Loch Lomond Re-advance (Fig. 10.12). Further south in the British Isles, small cirque glaciers formed in the uplands of England and Wales (Fig. 10.13).

The Younger Dryas is an extensively studied interval and is especially pronounced in the lands bordering the North Atlantic. Various theories have been put forward to explain this 1200-year-long cold phase, the last of the Pleistocene. One suggestion is that it relates to the diversion and discharge of meltwater from former proglacial Lake Agassiz in North America from a southerly route into the Gulf of Mexico to an easterly

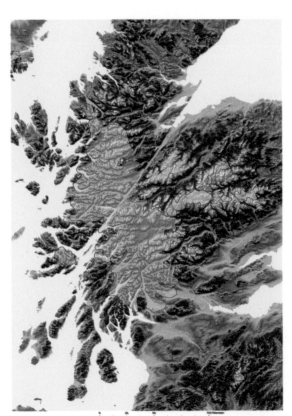

Figure 10.12 Extent of the Younger Dryas glaciers in Scotland (Loch Lomond Re-advance).

Figure 10.13 Cirque moraines bounding the lake of Llyn Cwm Llwch in the Brecon Beacons, South Wales. Small glaciers occupied many similar sites across Wales during the Younger Dryas Stadial (Hughes 2009b). Photograph by Phil Hughes.

route into the North Atlantic via the St Laurence Seaway (Broecker et al. 1989). The ensuing freshwater pulse into the North Atlantic caused the shutdown of the North Atlantic conveyor ocean current and dramatically reduced air temperatures in the region, especially over NW Europe (Isarin et al. 1998). The highly controversial view that the Younger Dryas cold reversal might have been caused by an asteroid impact (Firestone et al. 2007) has been refuted (Meltzer et al. 2014).

Further controversy has arisen concerning the extent to which the Younger Dryas is represented in records around the world. In New Zealand, Denton & Hendy (1994) and Ivy-Ochs et al. (1999) used radiocarbon and cosmogenic exposure-age dating, respectively, to argue that glaciers in New Zealand advanced during the Younger Dryas interval. However, a more recent study has shown that while glaciers were present they were in fact retreating and not advancing during the Younger Dryas Stadial (Kaplan et al. 2010). The hemispheric synchroneity of a climatic event correlative with the Younger Dryas is in doubt, and key evidence of a Younger Dryas cold reversal in the Southern Hemisphere has been dismissed (Green et al. 2013). It would therefore appear that the Younger Dryas Stadial is very much a Northern Hemisphere event, specific to the North Atlantic region.

It had long been assumed that the transition from the last glacial period to the current warm stage occurred in a globally similar way; today we know that this was not the case, however. Although climatic variations during the transition phase have also been detected on the southern continents, the comparison between the NGRIP ice core from Greenland (Rasmussen et al. 2006; Lowe et al. 2008) and the EPICA Dome C ice core from Antarctica (Jouzel et al. 2007) shows marked differences. The fluctuations in the Northern Hemisphere were much stronger than in the Southern Hemisphere, and the climatic changes in both hemispheres were not simultaneous (Hoek 2008).

While the Holocene is often thought of as a stable period of climate following the last ice age, there have been significant climatic events and more gradual fluctuations in climate. The large continental ice sheets took time to melt and the Laurentide Ice Sheet over North America did not retreat to the current ice extents in the Canadian Arctic until *c.* 6000 years ago; this means we are more than 5000 years into the current interglacial. One legacy of the last ice age is the 8.2 ka event when, as with the Younger Dryas, a pulse of meltwater associated with the decaying Laurentide Ice Sheet released into the Atlantic Ocean caused the shutdown of the North Atlantic conveyor. This plunged the North Atlantic region into several hundred years of cooling (Alley et al. 1997).

The subdivision of the Holocene commonly used in Europe today is based on studies conducted by Axel Blytt and Rutger Sernander at the end of the nineteenth century in Norway and Sweden. Norwegian Blytt introduced the concept of Boreal, Atlantic, Subboreal and Subatlantic periods, originally used to characterize the current distribution of the flora in Norway (Blytt 1876). After a number of other works by Blytt and Swede Sernander, these terms were included in Sernander's dissertation as names for the postglacial vegetation development phases (Sernander 1894).

Sernander was well aware of the problems of his division, and wrote: 'It is obvious that a real coincidence in time between an aspen-birch-horizon in Scania and a similar zone in Jämtland must be regarded as a priori very doubtful' (Sernander 1894, cited in Mangerud 1982). The immigration of plants over distances of thousands of miles must have taken a long time. Sernander therefore proposed that the age determination be based on indicators such as climate variability or changes in sea level. Since these were initially un-dateable however, vegetation history was used as a classification scheme for the lack of a better option.

After the International Geological Congress in Stockholm in 1910, the terminology of Blytt–Sernander was adopted all over Northern Europe. The system was originally based on the analysis of macrofossils found in peat. As in the 1930s, the attention of Quaternary scientists shifted towards lake deposits and, as macro analysis was succeeded by pollen analysis, the meaning of the original terms changed and they were used to distinguish the newly formed pollen zones.

With the advent of radiocarbon (^{14}C) analysis since the 1960s, it became possible to date organic deposits directly. The contrast between the biostratigraphic and chronostratigraphic interpretation of the terms that Sernander had pointed out was now obvious. Mangerud et al. (1974) proposed a redefinition of the established chronological terms which closely followed the

TABLE 10.2 GEOBOTANICAL SUBDIVISION OF THE HOLOCENE.

Period	Time span cal. years
Subatlantic	2,400–present day
Subboreal	5,660–2,400
Atlantic period	9,220–5,660
Boreal	10,640–9,220
Preboreal	11,590–10,640

Source: Litt et al. (2001).

existing practice on the one hand, but refrained from any geological definition on the other. Unfortunately, this proposal came at an early phase of ^{14}C dating. The method has been significantly improved through the advent of accelerator mass spectrometry (AMS) dating and calibration (comparison with tree rings), and the timescale of Mangerud et al. (1974) is no longer valid. The current figures (Table 10.2) were provided by Litt et al. (2001).

A formal threefold subdivision of the Holocene has recently been proposed (Walker et al. 2012). These authors advocate the 8.2 ka cooling event as the primary guide for the Early–Middle Holocene boundary, with a suggested GSSP in the Greenland NGRIP1 ice core. The event is expressed in several Greenland ice cores by a pronounced, brief (~150 year) interval of heavier δ^{18}O values, together with a decline in ice-core annual layer thickness and deuterium excess, a minimum in atmospheric methane and subsequent CO_2 increase. A distinct acidity peak, probably representing an eruption from an Icelandic volcano, was proposed as the marker horizon and gives an age of 8,236 years b2k. The 8.2 ka event is most conspicuous in mid- to high latitudes, but is a near-global phenomenon. It is recognized in marine (benthic and planktonic) as well as lacustrine, speleothem and ice-core records. In the South Atlantic and Southern Ocean, it may be expressed by a brief warming phase (Walker et al. 2012).

Walker et al. (2012) also proposed the 4.2 ka aridification event as the primary guide for the Middle–Late Holocene boundary. This event has been linked to cultural disturbance and collapse in North Africa, the Middle East and southern Asia. It is represented by a speleothem record, with a five-year sample resolution, from Mawmluh Cave in Cherrapunji, Meghalaya, north-east India. This record is based on δ^{18}O, the variability of which is used as a proxy for monsoon strength. It is constrained by U/Th dating, and spans 3.6 to 12.5 ka ago. The heaviest δ^{18}O values are between 4071±18 and 3888±22 years B.P. (midpoint 3.98 ka ago), representing an interval of 183 years, and signal a brief shift in Indian monsoon dynamics. This pronounced excursion is almost synchronous with the widely recognized low-latitude aridity event.

10.6 The Little Ice Age

Glaciers started to melt worldwide at the beginning of the Holocene as a result of general warming. It was originally assumed that the present-day glaciers are small remnants of the great ice sheets of the last ice age; this is not the case, however. Today, we know that a large number of mountain glaciers had completely disappeared by the middle Holocene. The mountains of Norway were ice free for most of the current warm stage, and the glaciers had mostly melted in the Alps; it was not until later cooler and/or humid climate phases that new glaciers formed. This process is commonly referred to as 'neoglaciation'. There are different views regarding the number, extent and age of the Holocene ice advances. These can be dated using radiometric dating techniques (see Chapter 2), but because of their young age biological proxies such as dendrochronology and lichenometry can also be used (Boxes 10.1 and 10.2).

BOX 10.1 DENDROCHRONOLOGY: WHAT TREE RINGS REVEAL

Dendrochronology is concerned with the dating of tree rings. Trees grow in thickness in the cambium, the growth layer between wood and bark. Cell division in the cambium is suspended during the winter and spring and sets in with increasing spring temperatures. Under favourable growth conditions, thin-walled cells are formed next and appear lighter in a cross-section through a tree. Heat and drought lead to a decline in cell division, and there are smaller cells with thicker walls. Even Leonardo da Vinci was aware that tree rings represent annual layers.

Without human interference, most trees would grow to a few hundred years old. The natural maximum age of the hazel (*Corylus*) is 80 years, of maple (*Acer*) 150 years, of white fir (*Abies alba*) 600 years and oak (*Quercus robur*) c. 800 years. Individuals may be significantly older (about 1800 years). The natural maximum age of the Chilean Araucaria (*Araucaria araucana*) is 1300 years and the California redwoods (*Sequoiadendron giganteum*; Fig. 10.14) live 2000 years. Where wood is grown commercially, the most advantageous age (the so-called rotation) is applied; this is 90–130 years for the white fir and 180–300 years for oak.

Figure 10.14 *Sequoia* wheel in the Hamburg Groß-Flottbek botanical garden; the tree was over 1000 years old. Photograph by Jürgen Ehlers.

In order to obtain reproducible results from dendrochronological analyses, it is advisable to examine several samples per trunk and several trunks from each locality. In this way, non-specific deviations are most likely eliminated (Smith & Lewis 2007).

Scientists at Hohenheim University have succeeded in obtaining a seamless oak chronology from trunks found in the alluvial deposits of the rivers Main, Rhine and Danube, which dates back to the end of the Boreal period when oak first re-immigrated to Central Europe. The oldest oak in the Hohenheim collection dates from the year 10,429 before present (Kromer 2009). The approximate age of 'floating' chronologies can be calculated by using the ^{14}C method to get a rough age prior to matching them with the Hohenheim curve. Dendrochronology has the advantage that a large number of wood samples, often whole tree trunks, can be processed in a relatively short time. They also result in a better age determination of river terraces than dating individual samples, where there is always the danger of sampling redeposited older material.

The growth of annual tree rings depends not only on climate. Trees that grow at the edge of their natural range have stronger fluctuations in growth than others which have grown under more ideal ecological conditions. Local peculiarities such as pest infestation, so-called 'cockchafer years', can stop the growth and make cross-regional correlation difficult.

On a field trip, Jürgen Ehlers once came across such a disaster:

In southern Sweden, we had lunch in a light beech wood. Someone said, 'Oh, look, here's a caterpillar!' A small, yellowish creature with thin black stripes and a red 'spike' at the back end. 'Cute!' 'Here's another!' We went a few steps further, and came very soon to a part of the wood that was completely devoid of summer foliage. The reason was not hard to see: on each of the trunks thousands of small yellow caterpillars were busy trying to crawl up in search of food. Once in the crown, they would find that all the leaves had already been eaten by their predecessors. A very light breeze was blowing, and this meant that a gentle rain of caterpillars trickled down on us. These caterpillars set out immediately for the nearest tree trunk and crept up again.

The affected part of the forest was not very big (a few thousand square metres), but in that area in that year the growth rings would be very different from the rest of southern Sweden (Figs 10.15–10.17).

Figure 10.15 Pale Tussock (*Calliteara pudibunda*) caterpillar on beech leaf. Photograph by Jürgen Ehlers.

(continued)

BOX 10.1 DENDROCHRONOLOGY: WHAT TREE RINGS REVEAL *(CONTINUED)*

Figure 10.16 Caterpillar invasion in a beech forest in southern Sweden. On the trunk section in the photograph, about 1250 caterpillars can be seen. On the back of the tree and in the crown there would have been at least an equal number, so that in this beech tree there were about 4000–5000 caterpillars. Photograph by Jürgen Ehlers.

Figure 10.17 Several Pale Tussock caterpillars. Photograph by Jürgen Ehlers.

BOX 10.2 LICHENOMETRY: AGE DETERMINATION USING LICHEN GROWTH

In newly deglaciated areas, the first plants to spread after about 15 years are the lichens. When looking at the lichens of moraines of different age at the front of melting glaciers in Scandinavia or the Alps, it may be expected that lichens found on the most recent end moraines are the smallest and on the oldest moraines the largest. This fact can, for example, be used to estimate the ages of moraines or landslides. This method is called lichenometry.

More accurate results are obtained when the study focuses on a particular species of lichen. Map lichen (*Rhizocarpon geographicum*; Fig. 10.18) is ideally suited for this purpose. The diameter of its circular or elliptical thalli in the first decades of its growth increases rapidly. This is followed by a phase of steady growth, which can last several

Figure 10.18 Map lichen (*Rhizocarpon geographicum*) on a boulder in Sweden. Photograph by Jürgen Ehlers.

centuries. However, the growth of lichens will eventually come to a halt. With increasing age of the thalli, they often lose their middle parts. New thalli settle there and blend with the old lichen body, so that they are no longer distinguishable. Dating by means of lichens is therefore generally limited to a period of *c.* 300 years (McCarthy 2007).

The ice-free period varied in length. The largest ice mass in Scandinavia, the Jostedalsbreen, had a relatively short ice-free period (*c.* 7–6 ka ago), while small glaciers in Jotunheimen or the Nordre Folgefonna were melted completely over much longer periods of time. The strong warming in the Atlantic period meant that the elevations of vegetation shifted upwards significantly. During the thermal optimum of the Holocene, *Ulmus* (elm) grew in southern Norway up to a height of 700 m asl (which is now the height limit of birch; Nesje et al. 1991).

It is now generally assumed that large-scale neoglaciation in Scandinavia began *c.* 6–5.3 ka before present, at the transition from the Atlantic to the Subboreal period. The number and extent of the glacial advances differ from glacier to glacier. A total of 17 phases of glacier advances can be distinguished in the Holocene in Europe (Mathewes 2007). Glaciers were widespread even in the mountains surrounding the Mediterranean Sea, as far south as southern Spain and Morocco (Hughes 2014).

Many glaciers were melting during the early Middle Ages. Around 1030–1080 AD, however, they began to advance again. From historical sources we know that during 1650–1680 a further significant deterioration of climate occurred that eventually culminated in the maximum extension of glaciers around the middle of the eighteenth century (Nesje et al. 1991). This most recent glacial advance, the largest of the Holocene, is known as the 'Little Ice Age' (LIA).

Today, most glaciers on Earth are melting; it is possible that they will disappear altogether from mountains such as the Alps and Pyrenees within a few hundred years.

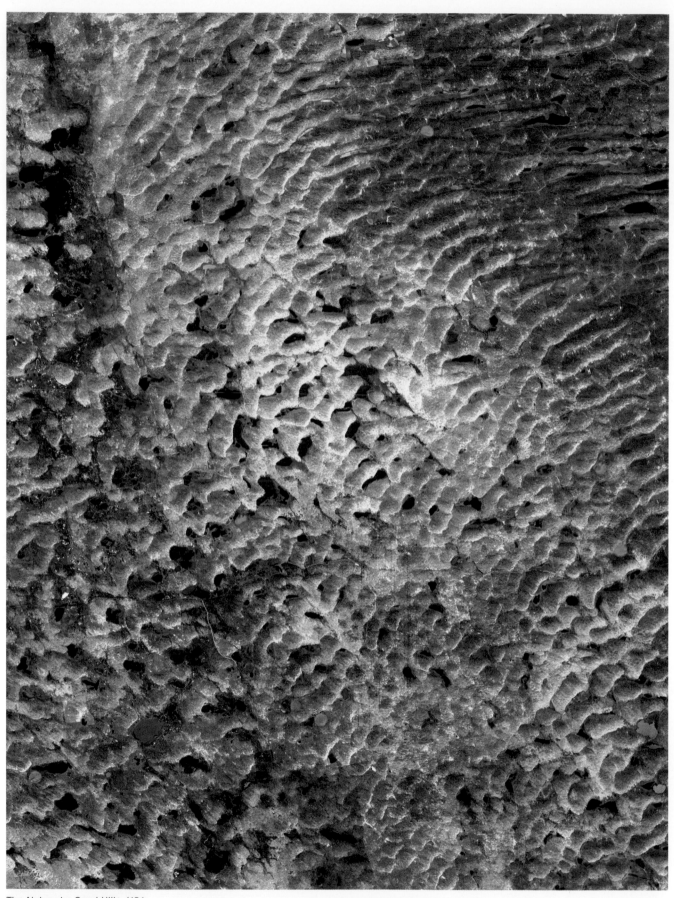

The Nebraska Sand Hills, USA.

Source: NASA ASTER satellite image of 10 September 2001.

Chapter 11

Wind, Sand and Stones: Aeolian Processes

11.1 Dunes

When dunes are mentioned, an image of either the coast or the large sand seas of the Sahara, the Namib or the Atacama spring to mind. The wind can only move large quantities of sediment if it can attack the substrate directly, and that is the case for vegetation-free zones (i.e. on a beach or in a desert). The deserts include not only the dry belt of the subtropics, but also the cold subpolar deserts. The latter extended to the middle latitudes during the Pleistocene cold periods, replacing the temperate forests of the interglacials. Today the cold subpolar deserts are confined to the extreme northern edge of the northern continents, and only in a few of those areas is there enough dry sand available for recent dune formation.

The dune fields of the subpolar region usually reach a size of no more than a hundred square kilometres. The sediment supply is limited, as at present in many areas the land surface consists either of the bare rock or the ground is covered by swamps and bogs. The exceptions in North America include the Athabasca sand dunes of northern Saskatchewan (1900 km^2).

In the middle latitudes there are relict dune areas of a much larger extent. In the Quaternary cold stages, large tracts of land were devoid of vegetation. Sand could be freely blown away from the wide sandur plains and river valleys, and even outside of those bare patches the ground was usually covered with no more than a steppe vegetation which did not constitute a complete protection against sand drift. Most of the resulting dune areas are no longer active. In England, this includes the relict dunes of the Breckland in East Anglia. Such dune areas are much more extensive

The Ice Age, First Edition. Jürgen Ehlers, Philip D. Hughes and Philip L. Gibbard.
© 2016 John Wiley & Sons, Ltd. Published 2016 by John Wiley & Sons, Ltd.
Companion website: www.wiley.com/go/ehlers/iceage

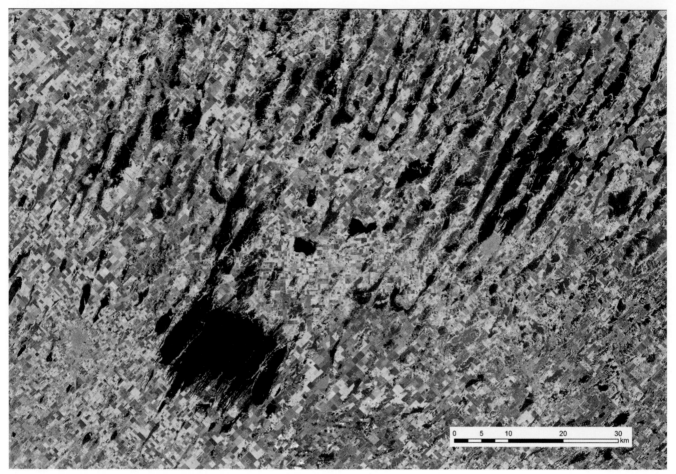

Figure 11.1 Dunes in the Argentine pampas.

Source: US Geological Survey, Landsat 7 ETM, Row 96, 10.2.1999.

in Germany where dunes are found on the edge of the ancient river valleys, such as Boizenburg (Elbe), and in the Upper Rhine Valley a belt of relict inland dunes stretches from Rastatt as far as Mainz. Even larger dune fields are found in the Americas, such as the western Nebraska Sandhills (covering an area of over 50,000 km²) and the vast dune areas of the Argentine pampas.

In the Pampas, stabilized longitudinal dunes of over 100 km in length and 2–5 km in width are found, arranged in a NE–SW-orientated slight curve (Fig. 11.1). They are now covered with vegetation and used for agriculture. Because of the dune relief, drainage is difficult. Some of the dune valleys are covered by lakes all year round. The majority of the sand dunes were apparently formed in the last (Weichselian) cold stage, during the cold and dry periods coinciding with MIS 4 and 2. Some of the dunes, especially the parabolic dunes further to the southeast, may still have been active in dry periods of the early Holocene.

Even in the great deserts of the Earth, the active phase of dune formation was in the distant past. With the exception of the aeolian landforms, the large-scale relief of the Sahara dates back to the Tertiary. The largest aeolian-shaped surfaces are covered with wind ripples (Fig. 11.2). Ripples are not only found on dunes, but also in the sand layer of a few centimetres thickness that covers different types of old relief in whole or in part. While the ripples

Figure 11.2 Wind ripples covering dunes, Tassili, Algeria. Photograph by Sigrid Wegner.

move, the dunes are almost stationary. Dunes of different shape and size may occur in patches at obstructions hindering the wind flow. The bulk of the Saharan sand, however, is concentrated in large basins. These are the sand seas, the ergs (Fig. 11.3), which are usually built of up to over 100 m high dunes (*draas*) which do not move, but are merely crossed by low dunes on the flanks. How the great sand seas were formed exactly is unknown. It is assumed that the Murzuq Basin was cut off from the fluvial drainage network as early as in the late Tertiary by the formation of the surrounding escarpments. The alluvial sands must have been transformed

Figure 11.3 Large sand dunes in an erg, Tassili, Algeria. Photograph by Sigrid Wegner.

Figure 11.4 Sandstone moulded by the wind into a bizarre sculpture, Tassili, Algeria. Photograph by Sigrid Wegner.

into ergs in a later phase of strong aeolian activity. The present-day aeolian overprint is comparatively minor (Busche 1998).

The reorganization of the relief by wind also includes the older hard rock. Wind erosion has transformed loose pebbles to ventifacts, solidified lake sediments were cut into yardangs and hard rocks carved into bizarre sculptures (Fig. 11.4).

The most active dune elements are the barchans (Fig. 11.5). These are crescent- to sickle-shaped dunes which migrate towards their concave side. Their height ranges from a few centimetres to tens of metres. A smaller amount of sand is transported in the marginal parts of barchans than in the centre, resulting in the lateral arms of the landforms travelling faster than the cores, accentuating the crescent shape. The migration rate of barchans depends on their size. A comparison of aerial photographs shows that the barchans south of Doha (Qatar), for instance, move at a rate of 5–10 m a^{-1}. Busche (1998) has reported barchans from southern Egypt which migrate at a rate of 20 m a^{-1}. In combination with water, aeolian sand can be transformed into quicksand. However, it is not quite as dangerous as some movies suggest (Box 11.1).

Figure 11.5 Barchans south of Doha, Qatar. Photographs by Holger Wolmeyer.

BOX 11.1 THE SANDS OF DEATH

Who has not heard of the dreaded quicksand? Anyone who has watched the film *Lawrence of Arabia* will certainly remember the scene in which Daud (one of Lawrence's servants) sinks in the quicksand in sight of his master while on his way from Aqaba through the Sinai Desert. The film is based on fact. Farraj and Daud, the two servants, are real characters that can be found in T.E. Lawrence's autobiographical account *The Seven Pillars of Wisdom*. However, there is no reference to Daud's death in the quicksand. No-one in the book perishes in the sand, and the term quicksand does not occur anywhere in the whole volume.

Is it all fiction? Not quite: the German geologist Berendt described his experience with 'Triebsand' on the Kurische Nehrung (Curonian Spit), close to the former village of Carwaiten (Karvaiciai, Lithuania, about 2 km north of Palanga; Fig. 11.6):

Figure 11.6 Horse caught in the quicksands of the Curonian Spit.

Source: Das Buch für Alle. Illustrierte Familien-Zeitung. Chronik der Gegenwart. 1889, issue 13.

But as soon as we had gone a few steps on the flat, dry ground, the horses began to break in. The whip whizzed and - in the next minute, the dangerous place had been passed. I bent over the edge of the coach at that moment and saw what I never thought possible, that the ground, without bursting, bent upwards 12–14 inches high between and in front of the wheels, so that with our rapid driving (...) the ground was moving in waves, more than a foot high.

But so easy we were not to get away. Again the horses broke in, again the whip whizzed and did its best, while already wet sand was splashing about, and in an instant the horses had sunk chest-deep into the sand.

Fortunately, the sand surface carried the burden of a person very well, so we could move around safely and could quickly pull back the coach, only the front wheels of which had sunk 6 inches deep into the ground without breaking through. (...) Our combined efforts succeeded, at least to get the lead horse out of the sand by pulling it at head and tail that it got its feet free, so that it could be dragged away a few steps, where we might hope that it would not break in again. With the other two animals, however, the sand had settled so firmly around their bodies that they were immured, they had immediately broken in right to the chest at the first moment, and now they still sank slowly, but noticeably, and if no help was available, they would both be lost.

Quoted in Solger et al. (1910)

The horses were finally rescued. Wet quicksand, as described by Berendt (1879), is often found in nature on tidal flats, for example. It is not possible to sink completely into it; the main danger is drowning in the rising tide. Dry quicksand on the other hand – the Sinai example – has been little studied although it is possible to produce it under laboratory conditions. It is thought to be extremely rare however, so the risk for travellers in desert areas should be very low.

TABLE 11.1 DUNE GENERATIONS IN MAURITANIA.

Width (m)	Trend	Age (ka)
5000	NE–SW	15–25
500	NNE–SSW	10–13
50	N–S	<5

In the past the dunes of the deserts were more active. The great climatic changes since the end of the last cold stage have also resulted in a shift in the wind belts of the Earth. Several generations of dunes can therefore be distinguished in many places. Lancaster et al. (2002) identified three different dune systems in Mauritania (Table 11.1, Fig. 11.7). The oldest system is composed of large linear dune ridges which are aligned NE–SW. The oldest system is overlain by smaller linear dunes, which run NNE–SSW. They are overlain in turn by even smaller N–S-trending linear dunes, which are still active. The three directions represent repeated changes of the wind regime. The wind has not turned slowly, but there were several phases of dune formation separated by periods without significant sand movement. Using optically stimulated luminescence (OSL) dating, it could be demonstrated that the oldest

(a)

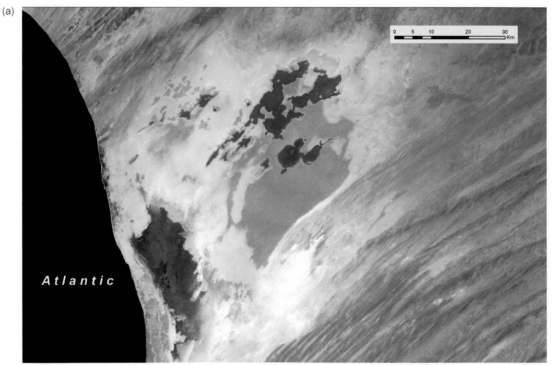

Figure 11.7 Generations of dunes in Mauritania. (a) Large NE–SW-aligned dunes dominate the picture, overlain by smaller NNE–SSW-oriented dunes.

(b)

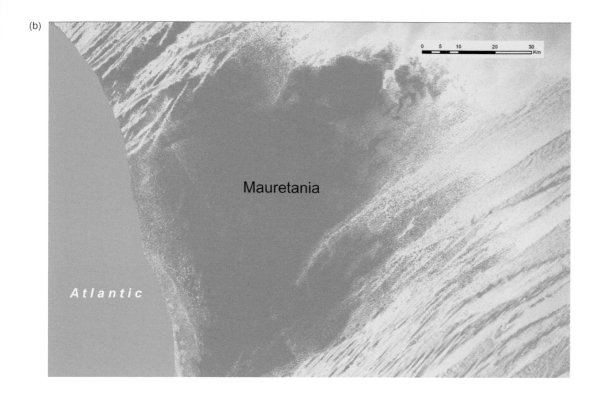

(c)

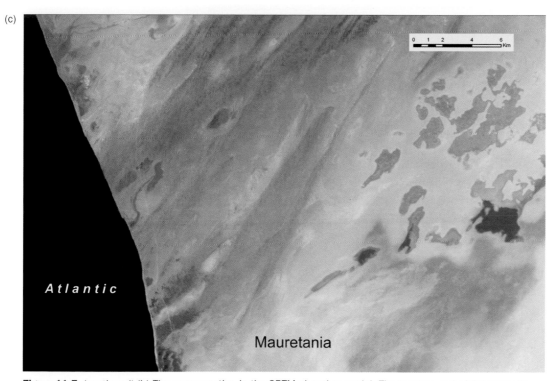

Figure 11.7 *(continued)* (b) The same section in the SRTM elevation model. The western end of the dunes is cut off by the Atlantic Ocean. (c) In detail, we see that the two older generations of dunes are overlain by small dunes aligned north–south.

(d)

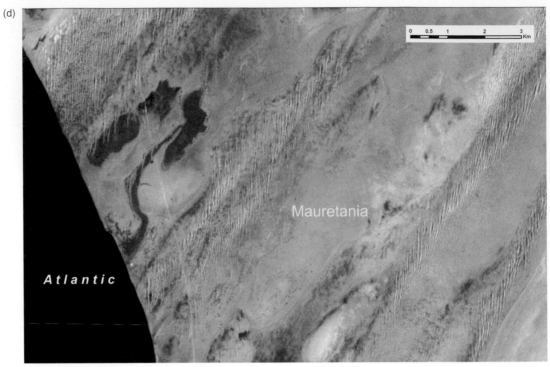

Figure 11.7 *(continued)* (d) Detail from (c).

Source: US Geological Survey, Landsat 7 ETM, Row 205, 04.05.2003.

generation of dunes were formed 25–15 ka ago; the middle generation formed 13–10 ka ago and the youngest 5 ka ago.

If we trace these dune systems to the south, we find that the middle generation extends far beyond the borders of the present-day Sahara southwards into the Sahel. Here the dunes are stabilized by vegetation. The age of these dunes in NE Nigeria is *c.* 18–13 ka and in Mali about 12.7 ka (Lancaster 2007). The southern border of today's active dunes in the Sahara is around 14° N, which corresponds to the 250 mm rainfall contour. The passive dunes extend into an area that now receives nearly 1000 mm of precipitation per year. These vegetated dunes today form a west–east-trending belt *c.* 400 km wide. It is therefore clear that the Sahara once extended much further south than today.

The external appearance of dunes is decisively influenced by the source material. While in the Sahara red to yellow colours dominate, the White Sands dune field in southern New Mexico (>700 km^2) is almost white. Its very bright appearance is caused by the source material: gypsum (Fig. 11.8). White Sands is the largest gypsum dune field in the world. OSL dating of the base of the dunes has yielded an age of 7 ka. The dune field is located in the Tularosa Basin, and the sediments from which the dunes have formed are the deposits of former Lake Otero. The lake dried up and salt flats remained to be deflated by the wind. The dune field is still active (Kocureka et al. 2006).

The sand which leads to the formation of dunes can be transported over long distances, even across wide stretches of water. The dunes on Lanzarote and Gran Canaria (Canary Islands; Fig. 11.9) do not owe their existence to deflation from the relatively narrow island beaches.

Figure 11.8 Gypsum dunes at White Sands, New Mexico, USA. Photographs by Klaus Stribrny.

Figure 11.9 Dunes on Gran Canaria; the sand for the dunes of the Canary Islands is transported from the Sahara. Photograph by Sigrid Wegner.

Figure 11.10 Sandstorm over the Western Sahara and the Canary Islands (in the upper part of the picture). The Cape Verde Islands can be seen in the lower left part of the picture.

Source: NASA ASTER satellite image of 4 March 2004.

Instead, large dust storms blow sand from the Sahara across the sea over a distance of more than 200 km. The transfer of Saharan dust offshore to the Canary Islands is promoted by the prevailing trade winds (Fig. 11.10).

11.2 Aeolian Sand

Weichselian–early Holocene aeolian sands are widespread in the lowlands of western Europe. Mostly they are found in the form of blanket-like deposits, often referred to as cover sands (Koster 1982). While cross-bedding is most frequent within the dunes, horizontally layered sediment prevails in the sand blankets (Schwan 1988).

A regional differentiation occurs within the Weichselian sand belt. The cover sands of the west are replaced eastwards by dunes that were formed simultaneously. The cause of the different kind of aeolian deposition was probably the aridity increasing towards the east (Böse 1991).

In some cases it is difficult to determine whether sandy deposits that do not occur in the form of dunes are fluvial or aeolian in origin. However, aeolian and fluvial sediments can be distinguished on the basis of the grain surfaces. Cailleux (1942) has shown that aeolian-transported glacial sand grains are generally characterized by a matte finish. It should be noted that chemical processes can also lead to matting, however, so that each case has to be carefully examined.

The sands in NW Europe have been remobilized since the beginning of the Neolithic. Apart from historical sources, these processes can also be dated by pollen analysis and by radiocarbon dating of sand-covered peat layers. A first major phase of sand drift started when the digging of sod was introduced (about 750–1200 AD). The resulting 'young dunes' are distinguished from the older dunes of the Weichselian Glaciation and the early Holocene.

11.3 Loess

The term 'loess' originated in the Upper Rhine Graben in Germany where it was used as a term for the widespread fine soil. Lyell (1834) introduced the term into the Anglo-American literature. The origin of the loess has long been controversial; studies by von Richthofen (1877) in China have shown that the loess was an aeolian sediment and that it was deposited in a steppe environment. The loess is very well sorted with a pronounced grain-size maximum in the coarse silt fraction. Loess consists mainly of quartz dust and limestone, with a minor proportion of feldspars, clay minerals and mica. The heavy-mineral composition may allow the assignment to specific areas of origin.

Loess deposits are found widespread in Central Asia, the northwest and central United States, central Europe and Argentina. Most loess is of Pleistocene age and has been deposited under arid to semi-arid conditions. The raw material comes from deserts, from periglacial deposits and from the deposits of the large meltwater streams. Loess usually lies like a blanket over the landscape. Because it consists of fine-grained, slightly weathered minerals, it provides the source material for very fertile soils in humid to semi-humid areas (black earth).

There are thick loess sequences in Tajikistan and Uzbekistan, but the thickest loess deposits are found in central China. There, the loess covers a wide dissected plateau (the Loess Plateau), the largest part of which is located in the northwards-pointing arc of the Yellow River (Huang He). This area is dissected by deep valleys in the steep walls of which the alternating layers of yellow loess deposits and reddish brown palaeosols can be seen (Fig. 11.11). In most loess areas the palaeosols represent periods in which the loess deposition was interrupted, so that the soil could develop undisturbed. In China it is different; a continuous loess sedimentation took place there, which was slowed down during the soil formation phases but not interrupted.

For the subdivision of the sequence of loess and palaeosols in the Loess Plateau and for the correlation of different profiles, lithological differences have been used (e.g. colour, relative thickness, grain size, mineralogy) as well as pedological and magnetic properties. Magnetic susceptibility has been found to be particularly useful. The intensity of the magnetic signal is due to the activity of soil bacteria that produced magnetic particles (maghemite) in the fine-grained material. The magnetic susceptibility reaches the highest values in the interglacial deposits since the activity of bacteria was highest under the influence of a warm interglacial climate.

Figure 11.11 Loess profile with palaeosols in the Bad Soden brickyard pit, Taunus, Germany. Photograph by Jürgen Ehlers.

Figure 11.12 Loess profile at Longhua, China, with numerous palaeosols. Photograph by Frank Lehmkuhl.

In the Chinese loess stratigraphy, loess layers are designated with the letter L and the palaeosols with the letter S. The youngest loess (L1), the loess of the Weichselian Stage, is known as Malan loess. The underlying soil of the Eemian Interglacial (S1) is a palaeosol complex which consists of three palaeosols separated from each other by thin layers of loess.

A certain problem is the dating of the strata. Radiocarbon dating can only be used down to the upper Malan loess (i.e. for the last 50 ka). Thermoluminescence (TL) dating is applied for layers up to about 200 ka in age. Layers which are any older must be dated by OSL, cosmogenic isotopes ([10]Be) and palaeomagnetic properties.

Study of the Chinese loess profiles (Fig. 11.12) has shown that the sequence can be correlated without major problems with the marine oxygen-isotope stages of the deep-sea stratigraphy. In the loess profiles as well as in the deep-sea stratigraphy, the earlier 40 ka cycle of cold and warm stages is replaced by a 100 ka cycle at *c.* 800 ka.

In the western and central part of the Loess Plateau, the loess reaches a thickness of 300–400 m. It not only covers the whole of the Quaternary climatic cycles, but the layer sequence extends into the Tertiary. The loess is underlain by the so-called 'Red Clay', which contains numerous mammal bones. More detailed studies have shown that the Red Clay is a sequence of reddish aeolian silt and calcareous dark red palaeosols, similar to the overlying sequence of loess/palaeosols. The actual loess deposits are nearly 300 m thick in the central Loess Plateau, while the red clay reaches a thickness of only *c.* 125 m. The rate of deposition of the red clay was about half that of the loess layers, and its formation began at *c.* 7.2–6.5 Ma (late Miocene). The onset of loess deposition therefore extends far back into the Tertiary (Porter 2007).

Since the loess spans a relatively long period of time with a rather uniform sedimentation, the resolution of pedostratigraphy is very high. This also applies to the central European loess deposits. In a well-developed Weichselian loess sequence, 5–10 palaeosols can be found. These soils are distinct from the pristine loess that was little influenced by soil formation by their colour and texture. The colours are usually given according to the Munsell colour scale.

Whilst the palaeosols within the loess sequences indeed vary in intensity from place to place, their stratigraphic sequence remains essentially the same. The variations have several causes. Individual soils (especially in the upper slope area) may therefore be thinned by solifluction or completely eroded. Other soils (especially in a lower slope position) can be represented two or three times by slopewash. Local small-scale factors have resulted in a greatly variable appearance, especially in the more differentiated soil formations. In addition, deep soil formation by clay migration (especially in the interglacial Luvisols) can alter underlying older soils beyond recognition (Ricken 1983).

The Rhine Falls near Schaffhausen, a result of glacial relocation of the Rhine. The metal sign marks the spot up to which the river was frozen in the winters of 1880 and 1963. The distance to the nearest recent glaciers is c. 100 km. Photograph by Jürgen Ehlers.

Chapter 12

What Happened to the Rivers?

12.1 River Processes and Landforms

A river may either cut into the subsoil or accumulate sediment. In any given segment of its course, usually one of the two processes outweighs the other over the long term. However, throughout its history the river may experience repeated changes in which process is dominant. The question of whether a river switches from erosion to accumulation depends largely on two factors: changes in discharge and the supply of sediment (climate); and changes in slope ratio (tectonics). Whether a change in sea level may also have a significant impact depends on the initial relief. If the level falls some 100 m, in the case of the Thames for instance, only the lower reaches of *c.* 200 km are affected. Where repeated changes of erosion and accumulation occur, during the accumulation phase the valley is filling in to a certain level and during the subsequent erosion phase these deposits are partially removed again, resulting in the formation of terraces.

The term 'terrace' is often used for both land surfaces as well as for deposits. In order to avoid confusion, the Commission of Terraces and Erosion Surfaces of the International Geographical Union decided that the term 'terrace' should only be used for the landform and not for the sediment body (Howard et al. 1968). The use of the same term 'terrace' for both the landform and the sediment body is based on the idea that both represent a single accumulation cycle. This is frequently not the case, however; the body of the terrace may have a very complex structure. More important than the terrace surface, which has been modified by subsequent erosion, is the base of the sedimentary body.

The Ice Age, First Edition. Jürgen Ehlers, Philip D. Hughes and Philip L. Gibbard.
© 2016 John Wiley & Sons, Ltd. Published 2016 by John Wiley & Sons, Ltd.
Companion website: www.wiley.com/go/ehlers/iceage

Traditionally, terrace surfaces were often correlated with each other according to their height. This approach is problematic, since the surfaces represent breaks in the sedimentary sequence which may have lasted much longer than the accumulation of the sediment body. No information on the climatic conditions of that period is available from terrace heights; in contrast, the preserved sediments provide clear indications of the sedimentary conditions at the time of their deposition. The core of any terrace investigation should therefore consist of an analysis of the sediments themselves. To distinguish between the different sedimentary bodies, petrographical–sedimentological parameters can be used such as grain size, texture, lithological composition and the degree of weathering of the gravel or sand fraction. Changes in the catchment area are reflected in the composition of the clasts and in the heavy mineral spectra. The spatial relationship of the individual sediment bodies to one another can be used to confirm lithostratigraphic matches and to support correlations.

Important evidence comes from the internal stratification and the position of the terrace base. The age determination of the layers can be facilitated through intercalated organic beds, which can be bio- or chronostratigraphically correlated with other layered sequences. In addition, molluscs or mammal remains provide age information. Archaeological finds help to date the sediments, especially in the younger terraces. In this way, a terrace stratigraphy can be established even where the sequence of terraces is masked by a thick cover of younger sediment (Gibbard 1985). Today, it is possible to determine with luminescence dating the accumulation age of terrace bodies: OSL dating of quartz and feldspar can be used on materials of age up to 100 ka; and infrared radioluminescence (IR RL)/infrared radio fluorescence (IR RF) can determine ages of up to several 100 ka.

Where the lack of exposure prevents the sediment body from being sampled, the gradient of the terrace body or of its surface can (with care) be utilized for their correlation, if it has not been modified by tectonics. This is possible in some mountain areas such as the foothills of the European Alps, especially with the proximal parts of the accumulations, because the largest differences in height between the terraces of different age are found in the vicinity of the ice margin. For correlations over long distances it is important to remember that the gradient of the terrace body may differ from the gradient of the modern valley floor, and that tilting of the accumulation levels may have occurred due to differences in uplift or subsidence. For this reason, it is also important that the height of the terraces is given in relation to sea level, not their relative height above the recent valley floor.

Unfortunately no uniform system for terrace terminology was agreed upon in the past, so almost every author uses their own nomenclature and clarity suffers. Originally, river terraces were commonly subdivided on the basis of their altitude above the river, for example 110 foot, 50 foot, 25 foot, etc. (in England) or sometimes simply as High Terrace, Middle Terrace and Low Terrace (in Germany). The most prominent members of a terrace sequence were referred to as 'Main Terrace'. This basic type of subdivision is present in many rivers, but the meaning of the terms varies from case to case. The '100-foot Terrace' of the River Thames in England was thought to have been deposited during the Hoxnian Interglacial, while the '25-foot Terrace' of the Thames was formed in the Ipswichian (Last Interglacial). The number of terraces recognized soon failed to fit this simple scheme. This finally led to terrace names being supplemented or replaced by numbers. However, this process is not very appropriate to take into account complex stratigraphic conditions. Difficulties occur whenever a new terrace has to be added or erroneously distinguished terraces have to be combined. It is therefore recommended, as elsewhere in the stratigraphy, to work with local names and type localities. Terrace terminology today has been brought into line with other stratigraphical schemes by assigning the sediments to lithostratigraphical units and the surfaces to morpho- or allostratigraphical terms.

Where rivers cut laterally into solid rock, this leads to the formation of rock (strath) terraces. Rock terraces are also formed under the influence of periglacial climate on the coast and on lakeshores. Dawson et al. (1987) have evaluated this process using the example of a former ice-dammed lake in southern Norway. Terrace forming acts primarily through frost shattering at the lake level. Winter lake ice and drift-ice transport play an important role in removing the detritus (Dionne 1981). Dawson et al. (1987) found an average horizontal growth of the terrace of 3–4 cm a^{-1} (maximum 7.07 cm a^{-1}) in the case of the Norwegian lake.

In Glen Roy in Lochaber (Scotland), three rock terraces are found on the valley flanks. According to ancient legend, the giant Fingal used those 'roads' for hunting. Darwin (1839) was the first to map the terraces in Glen Roy and the neighbouring valleys. Based on his experience in South America, he believed that they were marine terraces. The so-called Parallel Roads (Fig. 12.1) at 260, 325 and 350 m altitude represent three former lake levels of an ice-dammed lake, however. The terraces are up to 12 m wide and their landward rocky cliff is up to 5 m high (Sissons 1978).

During the Loch Lomond (Younger Dryas) Stadial, glaciers re-advanced into Lochaber. The drainage through Glen Spean to the west was blocked by the ice. An ice lake was created which finally drained over a pass at 260 m to the east. As the ice advanced further, the connection between Glen Roy and Glen Spean was also cut off and the lake level in Glen Roy rose to 325 m. At this altitude, overflow was possible through another pass into Glen Spean. Eventually, the ice advanced further and also blocked this discharge route so that the lake level rose to 350 m. The drainage then went from the north end of the valley eastwards into Glen Spey. During the melting of the ice, the drainage routes opened again in reverse order (Gordon 1993).

Figure 12.1 The 'Parallel Roads of Glen Roy', Scotland: a series of rock terraces formed by ice-dammed lakes. Photograph by Jürgen Ehlers.

12.2 Dry Valleys

The ancient glacial and periglacial landscapes of southern Britain, northern France and northern Germany are dissected by numerous valleys. Many of them contain no flowing water today. These dry valleys originally formed under the influence of permafrost; the water would only have drained on the surface (following the existing gradient) and not seeped into the permeable ground. Under a periglacial climate, solifluction occurs on slopes steeper than 2° (see Section 8.7; Fig. 12.2). The water-soaked active layer flows downhill on the permafrost. In the transition zone to the undisturbed layers of the substrate, there is a downslope drag of the material. Most sediment is transported off along the valley; coarse gravel and stones remain as a lag.

In the periglacial landscape, erosional processes have occurred repeatedly during several cold stages. On the present slopes, however, only traces of the last (Devensian/Weichselian) redeposition are found. The gravelly diamicton that flowed down the slopes of the northern European dry valleys is often covered by fine-grained aeolian sediments (sandy loess). In fact, in countries such as Belgium this forms the dominant soil substrate. The presence of loess in the dry valleys of northern Europe (e.g. Fig. 12.3) suggests that cold arid conditions prevailed towards the end of the glaciations. At the valley floors, however, multilayered valley fill sequences with gravel beds testify that repeated periglacial overprint took place.

In the uplands, beyond the glacial limits, the Quaternary periglacial climate also led to significant material reworking. The periglacial sediments form a significant element of the landscape in most upland areas today, where they may range up to several metres in thickness at the foot of slopes (Ballantyne & Harris 1994).

The following sections examine the Ice Age history of three of Europe's most iconic rivers: the Rhine and Elbe in Germany and the Thames in England.

Figure 12.2 Solifluction layer on the flank of a dry valley in Appelbüttel near Hamburg. The uppermost decimetres of the cross-bedded meltwater sands were captured by the movement and bent downhill. Photograph by Jürgen Ehlers.

Figure 12.3 Floor of the dry valley in Appelbüttel near Hamburg. Gravel lag overlain by slopewashed sandy loess. Photograph by Jürgen Ehlers.

12.3 The Rhine: Influences of Alpine and Nordic Ice

The Rhine can be divided into several sections (Fig. 12.4), the geological evolutions of which are rather different. The headwaters of the Rhine (the Vorderrhein and Hinterrhein) unite in the Swiss Alps at the village of Reichenau on the Rhine. The Rhine then flows north and empties at Fußach into Lake Constance. The following section from Stein am Rhein to Basel is referred to as the Hochrhein (High Rhine). Further downstream, the 'Upper Rhine' covers the Upper Rhine Graben and the Mainz Basin. The Middle Rhine is the section from Bingen to Bonn, where the river crosses the Rhenish Slate Mountains. The Lower Rhine (Niederrhein) lies further north, followed at the German–Dutch border by the Rhine delta. Here, the Rhine splits into the rivers Waal, Niederrhein-Lek and IJssel (Westerhoff 2009). During Pleistocene lowstands, the Rhine Delta continued far into the North Sea.

The Rhine has taken meltwater from both the Alpine glaciations and the glaciations of northern Europe. Both glaciation areas have forced the Rhine to change its course. In recent geological history, the Rhine has considerably expanded its catchment area. Until the late Pliocene, the Aare flowed into the Danube. It was then initially deflected towards Saône/ Rhône, just like the Reuss and Limmat. It eventually became diverted to the west during the Donau – Günz glacial periods to the present Rhine, and has drained ever since through the Rhine valley into the North Sea (Villinger 2003). The basin of Lake Constance was excavated through the course of the Alpine glaciations, which meant that fluvial Rhine gravel from the high Alpine area could not under any conditions be transported beyond this sediment trap. The glaciation of the Alps and the Black Forest forced further changes to the drainage system.

The Upper Rhine Graben began forming at c. 35 Ma during the late Eocene, creating the conditions for the development of today's Rhine. First the sea entered the newly formed depression from the south, however. In the north, large parts of the present-day mainland were flooded. Until the middle Miocene the sea had retreated to a line joining Venlo, Geldern and Wesel. It was during this period that a river existed for the first time that might be described as the forerunner to today's Rhine. It crossed the Lower Rhine Basin from south to north. The sea continued to retreat slowly, and the about 135 m thick Lower Rhine brown coal was deposited. The river followed the tectonically determined Rhine axis. Further south, the corresponding sediments are missing. The heavy mineral composition of the sands suggests, however, that the catchment area extended far to the south. The headwaters of the Rhine at that time would have been located in the area of the Kaiserstuhl volcano, which formed at c. 19–15 Ma (Wimmenauer 2003). The more southern regions at that time still drained via the Burgundy Gate into the catchment area of the Rhône.

The Vallendar layers, of which only remnants are preserved, were deposited during the late Eocene–Oligocene (Schnütgen 2003). Clearly identifiable Rhine sediments were also formed along the Middle Rhine during the late Miocene–Pliocene. Due to their content of fossiliferous and petrographically characteristic 'Kieselooliths' of the Muschelkalk and Jura of Lorraine, it is assumed that the course of the ancient Rhine which had accumulated this 'Kieseloolithschotter' largely followed the valley of the Mosel. In the Rhine Valley at that time a tributary was found (Boenigk 1981). The Rhenish Slate Mountains were uplifted at that time, giving rise to the formation of a first terrace staircase of three different levels. In the

Figure 12.4 Map of the Rhine.

Lower Rhine Basin, however, subsidence continued so that a staircase of terraces was built, including two or three separate Kieseloolith accumulations.

In the latest Pliocene, the Swiss Jura and the Molasse Basin were uplifted. The drainage was then directed to the north, and the Rhine catchment area widened into the foothills of the Alps. This change can be detected in the clast and heavy-mineral composition of the Upper Rhine sediments (Hagedorn & Boenigk 2008). While the older deposits were characterized by stable minerals, the proportion of unstable minerals (sphene, epidote, garnet, green hornblende) increased significantly (Kemna 2005). During this time, the current Rhine became the main river and the Mosel its tributary.

While between the Middle Rhine (Rhenish Slate Mountains, Bingen–Bonn) and the Lower Rhine (Bonn–mouth) a rough correspondence of the stratigraphic sequence is found, the correlation with the Upper Rhine poses considerable difficulties. The river section between Basel and Bingen is characterized by tectonic activities in the area of the Upper Rhine Graben and the Mainz Basin. The alternating deposition of coarse gravel and fine-grained sediments in this section of the river was at least partly due to tectonic movements. Several gravel bodies can be distinguished from the results of boreholes. The Upper Rhine Graben is divided into a southern and northern section, separated from each other by the Karlsruhe threshold. The southern part is characterized by rather coarse clastic sediments, whereas the northern Upper Rhine Graben contains more fine-grained sediments. The research boreholes at Heidelberg, Ludwigshafen and Viernheim were drilled in the centre of subsidence in the Heidelberg Basin (Hoselmann 2008; Weidenfeller & Knipping 2008). The thickness of Pleistocene deposits in the Heidelberg basin is 225–300 m, possibly (after the results of pollen studies) as much as 500 m.

Until recently it had been assumed that the uppermost gravel body was accumulated during the last glaciation, and that it was underlain by layers of the Eemian Interglacial. Palynological studies of the Mannheim borehole have however shown that the interglacial deposits of the Ludwigshafen Formation (formerly referred to as the 'upper intermediate horizon' or OZH), which were encountered at less than 40 m depth in the Viernheim research drilling, contained pollen of *Fagus* (beech), *Celtis* (hackberry) and *Azolla* (a fern), which did not occur in the Eemian Stage. Malacological investigations have confirmed that these strata must have a much greater age. They are now correlated with the Cromerian Complex Stage (Knipping 2008). Below the over-100-m-thick gravel bodies in the eastern part of the graben that are primarily interpreted as cold-stage deposits, and therefore as an expression of Alpine glaciation, sandy-silty sediments of the Lower Pleistocene have been preserved. A more detailed age determination has not yet been possible.

A key role in the Rhine stratigraphy is played by the 'Mosbach Sands', which yield a rich fossil record. At Wiesbaden-Biebrich the fluvial sediments of the Rhine and Main are exposed, and the Rhine sediments of the Grey Mosbach (e.g. Hauptmosbach Subformation in LithoLex; Figs 12.5–12.7) probably correspond to the deeper sands of the northern Upper Rhine Graben ('Rhenish facies'). The Middle Rhine equivalent of the Mosbach Sands are the 'Hönningen Sands' that are found in the older Main Terrace (Fig. 12.8).

By the uplift of the Rhenish Slate Mountains, the incision of the Rhine caused the formation of a terrace staircase (Fig. 12.9). In the subsiding Lower Rhine area, on the other hand, stacked 'terrace deposits' were formed, (Fig. 12.10). Up to 14 different terraces are

Figure 12.5 Sands of the Hauptmosbach Subformation (Grey Mosbach) in the Dyckerhoff quarry in Wiesbaden-Biebrich. Photograph by Jürgen Ehlers.

Figure 12.6 *Hydrobia* layer from the carbonaceous Tertiary deposits from the Mainz Basin, underlying the Hauptmosbach Subformation. Dyckerhoff quarry in Wiesbaden-Biebrich. Photograph by Jürgen Ehlers.

Figure 12.7 Sands of the Hauptmosbach Subformation in the Dyckerhoff quarry in Wiesbaden-Biebrich. Photograph by Jürgen Ehlers.

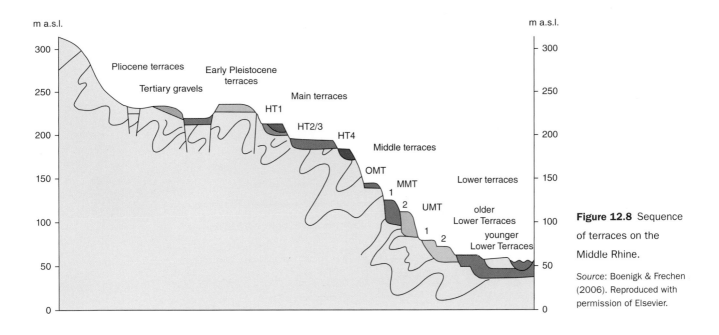

Figure 12.8 Sequence of terraces on the Middle Rhine.

Source: Boenigk & Frechen (2006). Reproduced with permission of Elsevier.

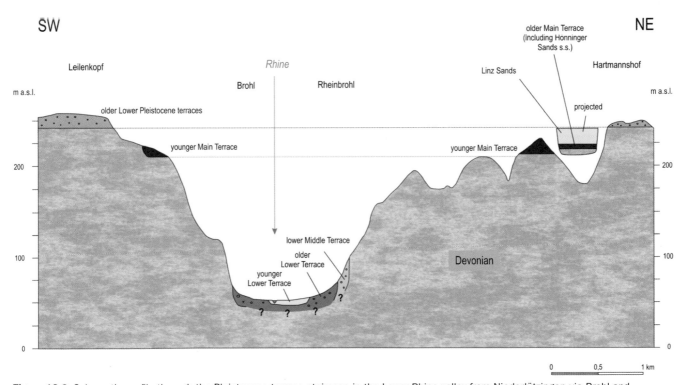

Figure 12.9 Schematic profile through the Pleistocene terrace staircase in the Lower Rhine valley from Niederlützingen via Brohl and Rheinbrohl to Hartmannshof (10× height exaggeration).

Source: Hoselmann (1996). Reproduced with permission of Zeitschrift der Deutschen Geologischen Gesellschaft.

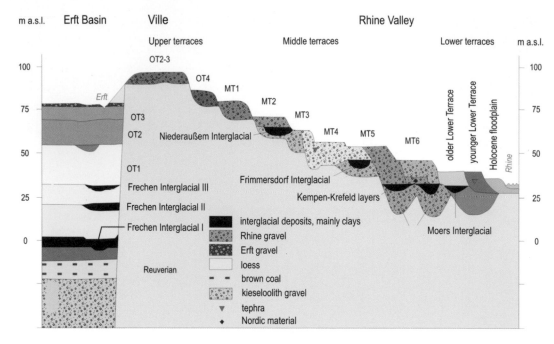

Figure 12.10 Sequence of terraces on the Lower Rhine.

Source: Boenigk & Frechen (2006). Reproduced with permission of Elsevier.

distinguished at the Middle Rhine, two of which are Pliocene and twelve of which are placed in the Pleistocene. The Lower Rhine has at least eleven different terraces. There are very few datable chronological marker horizons, including:

- the Brunhes–Matuyama boundary that lies between Upper Terraces 1 and 2/3 (the 'Main Terrace' of the older literature);
- the first appearance of volcanic minerals from the East Eifel volcanic field (Box 12.1) in the lower level of the younger Main Terrace and then amplified in the Upper Middle Terrace (UMT or MT1) prior to *c.* 600 ka; or
- the first appearance of Nordic pebbles in the Lower Rhine (after the Holsteinian Interglacial, MT6 or LMT2; Boenigk & Frechen 2006).

To characterize the altitude of the terraces, only the lower limit is of importance. The differences between the terrace levels are, however, extremely small. It is possible that the base of an older terrace gravel is found at the same level as a later terrace. This is partly because the Rhenish Slate Mountains have risen by *c.* 200 m since the deposition of the Lower Pleistocene terrace sequence. Not all parts have been equally uplifted; the cross-section (Fig. 12.9) shows an example of differences in height (see base of Younger Main Terrace) and similarities (base of the Lower Pleistocene Older Main Terrace). Correlation is hampered by the fact that only a few layers are present in each profile of the model terrace staircase for the Middle Rhine (Hoselmann 1996).

The Scandinavian Ice Sheet also advanced into the Lower Rhine area. Unfortunately, the link between the glacial deposits and the terraces of the Rhine is so far not clearly understood (Schirmer 1990), and the stratigraphic position of the layers is partly disputed. There seems to be no evidence of Elsterian-age ice. Saalian till was found interlocked with deposits of the

BOX 12.1 THE YOUNGEST VOLCANOES IN GERMANY

There are a number of young volcanic areas in Germany. Most of the volcanoes in the Rhön, Westerwald, Vogelsberg and Hegau were active during the Miocene. The volcanoes of the High Eifel (e.g. Hohe Acht and Arensberg) also belong to this group. In addition, the Eifel region also contains the only Quaternary volcanic area of Germany. It actually consists of two volcanic fields: West Eifel and East Eifel.

The West Eifel volcanic field consists of about 240 individual volcanoes, some of which are still visible in the terrain as ash or cinder cones or maar lakes (Fig. 12.11). According to studies from the 1990s, it is likely that all the West Eifel volcanoes were created in the Brunhes Chron, that is, they are younger than 780 ka. A first peak of volcanic activity seems to have occurred at *c.* 550–450 ka. A second active phase, in which the maars of the West Eifel were formed, is estimated to be younger than 50 ka. The Meerfelder Maar is 45 ka in age and the Ulmener Maar (the youngest volcano in Germany) possibly less than 10 ka (Litt et al. 2008).

Figure 12.11 The Weinfelder Maar in the Eifel region, Germany. The maar has a diameter of 500 m and is 51 m deep. Photograph by Jürgen Ehlers.

The smaller East Eifel volcanic field includes about 100 volcanoes. Here, three well-dated main phases of volcanism are differentiated. The oldest phase started at *c.* 460 ka BP. The culmination of that phase saw the formation of the Rieden volcano complex, dated to *c.* 430–360 ka BP. After a period of rest, another series of volcanic eruptions

(continued)

BOX 12.1 THE YOUNGEST VOLCANOES IN GERMANY (CONTINUED)

followed in which the Hüttenberg Tephra or Wehr Tephra originated. The third phase involved the formation of the Laacher See system. According to recent dating, the age of the Laacher See eruption is estimated at about 13 ka BP. The ash from the Laacher See eruption can be found hundreds of kilometres to the south (northern Italy) and north (Sweden). It is an important marker horizon which, due to its chemical composition, can be easily distinguished from the Late Glacial Icelandic tephra which is also present in the bogs of continental Europe (van den Bogaard & Schmincke 2002).

The eruption of the Laacher See volcano released enormous amounts of tephra which temporarily blocked the drainage of the Rhine. A natural dam of pumice and ash formed at the mouth of the Brohlbach creek into the Rhine, behind which a lake of at least 18 m depth was dammed. The lake was covered with a blanket of drifting pumice, which largely stayed behind after the eventual catastrophic drainage of the lake. The traces of the flood released when the dam was breached can be traced down to Bonn (Park & Schmincke 2009).

It is often assumed that the volcanoes of the Eifel were extinct. This assessment, however, is not based on valid arguments. The Eifel volcanic fields rest today; they have rested for tens of thousands of years before and then suddenly erupted again. Future eruptions in both volcanic fields of the Eifel are to remain possible. If this happens, new crater lakes and cinder cones could be created (Schmincke 2010).

Younger Middle Terrace 2 at Hamminkeln, now below the Younger Middle Terrace 3 (Jansen 2004). It seems that three Saalian ice advances have reached the Lower Rhine, and that all of them were part of the Older Saalian Glaciation (Skupin et al. 1993).

The Lower Rhine was forced by the Early Saalian Glaciation to change course (Fig. 12.12). When the glaciers were advancing up to Düsseldorf, the Rhine flowed south of the Reichswald through the Niers valley. After the melting of the ice sheet, the Rhine moved north into the IJssel valley in the Netherlands, and turned only at Zwolle towards the northwest. In the tongue basin of the overdeepened IJssel valley it initially formed a 50 km long lake. This lake was already completely filled with sediments during the Saalian Glaciation, but the Rhine retained its course during the Eemian Interglacial and into the Early Weichselian. It was not until the formation of a glacio-isostatic forebulge, probably of 5–10 m altitude during the Weichselian Glaciation maximum, that the Rhine was forced to a more southerly and westerly course (Busschers et al. 2007). In the Netherlands this was also a phase of strong incision. Initially the Rhine flowed north of Montferland through the Liemers towards the west. The present-day river path through the Gelderse Poort formed a little later. This last change only occurred after the Brørup, as demonstrated by peat in the Gelderse Poort. During heavy floods in the Late Weichselian both the Niers and especially the IJssel valley continued to take a portion of the Rhine drainage. Another recent development is that today's upper IJssel became part of the Rhine river estuary. The section between Deventer and Westervoort was only formed at about 600 AD (Makaske et al. 2008).

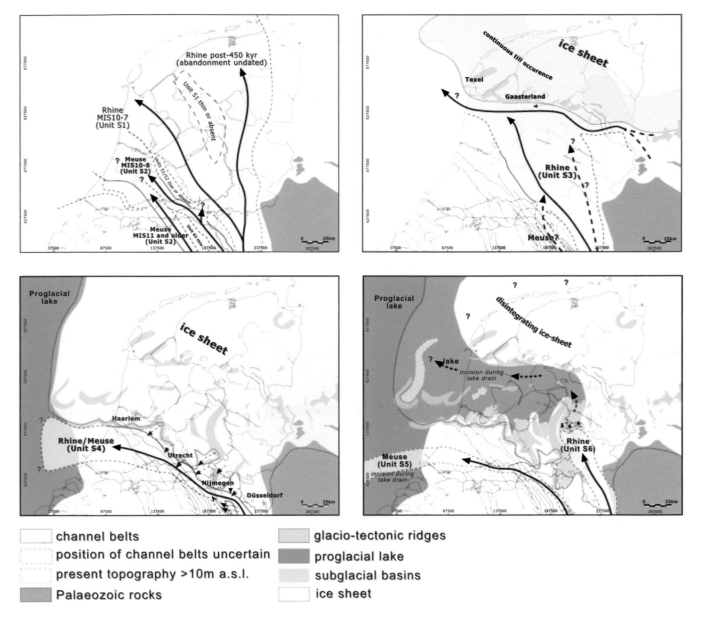

Figure 12.12 Development of the Rhine estuary. Adapted from Busschers et al. (2008). Reproduced with permission of John Wiley & Sons.

The most spectacular example of glacial erosion in the area of the Alpine Rhine Glacier is Lake Constance (Bodensee). Its rock floor is lower than the North Sea. In the melt phase of the last glaciation, the lake was still much larger. It extended over 70 km further upstream beyond Chur. This extension consisted of several sub-basins which were backfilled with sediment at the end of the last glaciation. The former lake was also deeper than that today. At Koblach the base of the Quaternary is found at over 200 m below the present sea level (Keller & Krayss 1993).

The 25 m high Rhine Falls near Schaffhausen are not the highest but the most water-rich waterfall in Europe. They are geologically very young. Until the Rissian, the Rhine flowed north through the Klettgau. Only then did the advance of Alpine ice push the river to the south; the new channel went through Neuhausen. During the Weichselian Glaciation, the

Figure 12.13 The Rhine Falls near Schaffhausen developed during the Weichselian ice age. Photograph by Jürgen Ehlers.

Rhine was again forced to move to another more southerly course. Since then, it flows across the Malm escarpment of today's Rhine Falls (Hofmann 1987; Fig. 12.13).

New terraces were formed in the Middle Rhine and Lower Rhine during the Weichselian glacial period. We can distinguish two Lower Terraces: the Laacher See tephra is first found in the alluvial loam on the older Lower Terrace and then redistributed in the younger Lower Terrace. Since the eruption of the Laacher See volcano could be dated to the Allerød Interstadial, it is clear that the younger Lower Terrace must have been deposited after the Younger Dryas Stadial. A threefold subdivision of the lower terrace has even been found at Düsseldorf and at one point in the Neuwied Basin. The additional ('Middle') Lower Terrace was deposited before the Bølling period (Schirmer 1990).

12.4 The Elbe: Once Flowed to the Baltic Sea

The sea also retreated from Saxony and Thuringia towards the end of the Tertiary. Fluvial-lacustrine sediments of the Miocene and Pliocene were deposited in shallow trough valleys. In the transition area to the north German lowland, the valleys terminated in broad alluvial fans. The first indications of a cool climate are found in the Rauno Formation which was deposited at the transition between the Miocene and Pliocene. Cryoturbations, drift blocks and small frost cracks indicate the influence of winter frost. Central Bohemian stone blocks, which are found in these sediments north of Dresden (Senftenberg Elbe) and in Lower Lusatia, were likely transported by drift ice (Eissmann 1997).

The history of the Elbe (Fig. 12.14) begins with the river occupying this Senftenberg Elbe course during the late Miocene–Pliocene. Early in the Quaternary the Elbe flowed from Pirna via Bautzen to the northeast (Bautzen Elbe). Its course shifted several times even in this Early Pleistocene period, and eventually flowed to the west and northwest (Schildau Elbe, Schmiedeberg Elbe, Early Elsterian Streumen Elbe; Eissmann 1975; Wolf 1980). How these courses continued to the west is unclear. The Elbe drained temporarily through the Netherlands, with the Saale, Mulde and Weser forming its tributaries. Between Dresden and Riesa, the Elbe used the same valley as that today for the first time during the Elsterian Stage. By the end of this cold stage, it ran via Jüterbog and Berlin to the contemporaneous Holsteinian-age Baltic Sea (Berlin Elbe). The advance of the Saalian ice sheet forced the river to change its course again. The development during the Drenthe Advance is not known. During the

(a)

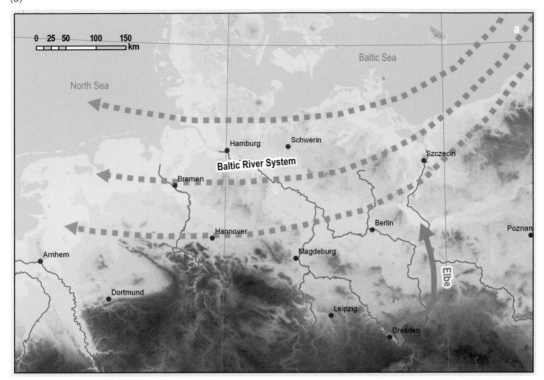

(b)

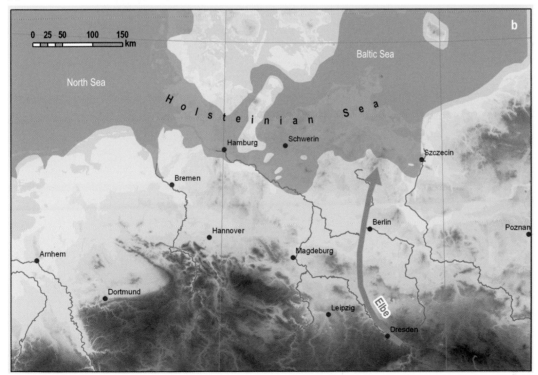

Figure 12.14 Development of the Elbe river. (a) In the Early Pleistocene the Upper Elbe was a tributary of the Baltic River System; (b) in the Holsteinian Interglacial it flowed via Berlin to the Baltic Sea (Berlin Elbe)

(c)

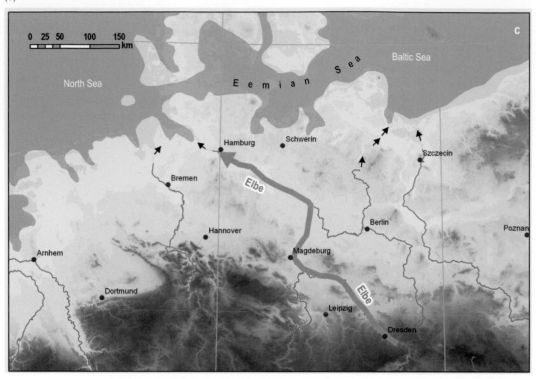

(d)

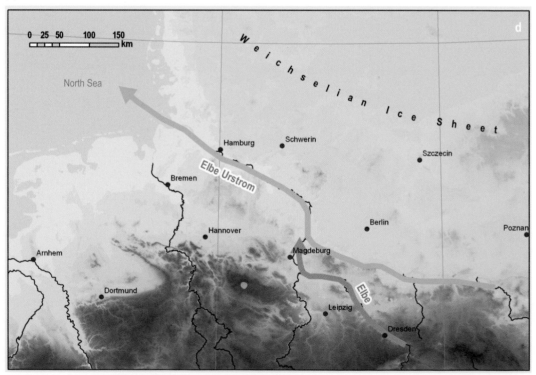

Figure 12.14 (*Continued*) (c) during the Eemian Interglacial it adopted its present course; and (d) during the Weichselian glaciation the Upper Elbe was a tributary of the Elbe Urstrom.

Warthe Advance the Elbe first drained through the Breslau–Magdeburg–Bremen glacial valley (urstromtal) via the present-day Aller and Weser rivers. It was only at the end of the Saalian Glaciation that it occupied its present course (Litt & Wansa 2008).

The history of the middle and lower reaches of the Elbe was heavily influenced by the Pleistocene glaciations. The history of this part of the river during the Early Pleistocene and its relocations during the Elsterian Glaciation are still poorly known. First references to the development of the Lower Elbe are only found in the Saalian deposits. A river with a southerly source area was present between Gorleben and Lauenburg after the Holsteinian Interglacial period and before the Younger Saalian ice advance. It deposited milky quartz-bearing gravelly sand. Deposits of this system, however, are not tied to today's Elbe Valley (Lübbow, Woltersdorf). The gravel spectrum of those sands lacks the high frequencies of brown sandstones and quartzites and the Thuringian Forest porphyries, characteristic of the Weichselian sediments of the Elbe glacial valley in this section of the river (Schröder 1988).

The morphological development of the Elbe through the sequence of the various Weichselian glacial valleys was described in Chapter 5. In this section, we discuss some geological and sedimentological findings. The basis of the Weichselian 'Lower Terrace' in the Lower Elbe between Hamburg and Gorleben is deeply incised in various channels into the substratum. Schröder (1988) demonstrated through his study of the gravel fractions that the Weichselian sands and gravels of the Lower Elbe differ from older Quaternary sediments by a blend of southern components derived from what is now the Elbe River Basin. They include the Thuringian Forest porphyry and brown sandstones and quartzites. The body of the Weichselian accumulation was deposited in two parts. At the transition to the younger gravel layer, the proportion of southern components

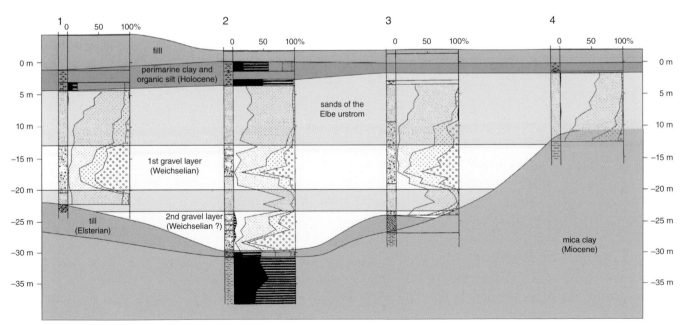

Figure 12.15 Particle size distribution of four boreholes through the Elbe glacial valley sediments in Hamburg-Georgswerder. The distribution of the gravel layers is tied to former channels. Adapted from Ehlers (1990).

suddenly increases. While the southern components in the Gorleben area are clearly detectable, their share is significantly lower in the west. Ehlers (1990) could not prove their presence in the 3.15–5 mm fraction in the Hamburg area; however, Meyer's investigation of 2–6 cm Weichselian Elbe gravels west of Hamburg still yielded southern material (personal communication, 1990). In the Gorleben area both gravel bodies overlie an occurrence of the Eemian Interglacial deposits, so their Weichselian age can be regarded as proven (Schröder 1988). The low-lying Eemian deposits in the tributary valleys of the Elbe in Hamburg (Seeve, Alster) show that the Elbe was already in its present-day position during the Eemian Interglacial.

The Weichselian erosion in the Lower Elbe valley was subdued. Essentially, an infilling took place of Late Saalian river channels; the gravel layers are restricted to the respective channels (Fig. 12.15). The high altitude of the Eemian peats on the banks of the Elbe in Schulau and Lauenburg, however, demonstrate that the valley was widened during the Weichselian Glaciation. Weichselian sands and gravels of the Elbe glacial valley are completely buried by younger (Holocene) sediments in the Hamburg area.

12.5 The Thames: Influence of British Ice

The Thames system is the largest drainage basin in Britain. For convenience, it can be divided into three regions that reflect bedrock and river form (Fig. 12.16). The Middle and Lower Thames occupy the London Basin, a syncline of Cretaceous chalk and overlying Tertiary sands and clays. Here the Thames is a broadly west–east-aligned stream axial to the basin, with tributaries entering from both the northern and southern margins. Upstream of Reading however, the river enters the London Basin from the Upper Thames catchment. In the latter region it traverses a series of gently dipping Jurassic limestone and clay strata. North of Oxford, three main streams converge to form the trunk River Thames. The catchment here includes the Cotswold Hills and the south English Midlands. In the extreme east a large estuary occurs where the river and its drowned tributary valleys enter the North Sea.

The deposits of the Thames and its tributaries occur from the tops of the highest hills on the basin margin to below sea level in the Thames Estuary. The earliest Thames deposits, the so-called Pebble Gravel Formation, represent a fragmentary series of gravels composed predominantly of local materials, particularly flint. They postdate marine sands (Red Crag = ?late Pliocene) that also occur up to 180 m asl on the margins of the London Basin, the occurrence of which indicate relative uplift of the western end of the basin during the Pleistocene (Gibbard 1988; Mathers & Zalasiewicz 1988). The Pebble Gravels therefore represent the Thames and tributaries established immediately following regression of the sea in the Early Pleistocene (Wooldridge & Linton 1955).

A profound change in gravel lithology is observed in the next-youngest units, which form a series of terrace remnants that are characterized by their content of rocks which are exotic to the present Thames catchment. These Kesgrave Formation units can be traced from the Upper Thames, where they are aligned parallel to the river Evenlode, the modern

Figure 12.16 The lower reaches of the River Thames in London, England. Top: the Thames in central London. Buckingham Palace and its grounds are visible on the left of the image. Bottom: moving eastwards downstream (at same scale) near the O2 Arena and London City Airport, with the Thames Barrage visible in the middle of the image.

Source: GoogleEarth.

Thames tributary downstream through the Middle Thames Valley. Here they diverge from the modern course and pass through Hertfordshire, parallel the shallow Vale of St Albans and enter East Anglia, where they form a terrace-like system mostly buried beneath tills of the Anglian Glaciation (Hey 1976, 1980; Rose et al. 1976; Rose & Allen 1977; Whiteman 1992; Whiteman & Rose 1992).

Glaciation in the Anglian overrode the drainage system in east and central England, damming the Thames and its southbank tributaries north of London. This resulted in the river adopting a new course through London (Gibbard 1977, 1979, 1985, 1994; Bridgland 1983, 1988). Subsequent evolution of the Thames system has been marked by the cyclic development of a sequence of gravel and sand aggradations under periglacial climates during the Middle and Late Pleistocene. These aggradational members

(Maidenhead Formation) show a marked reduction in exotic material, accompanied by an equivalent increase in local lithologies. This change marks a severing of the headwater catchment in the West Midlands.

The Thames Valley also includes many important interglacial fossiliferous sequences that provide both stratigraphical control and palaeoenvironmental evidence. East of London, the valley was repeatedly invaded by the sea so that, as today, a substantial estuary developed during periods of high eustatic sea level (interglacials). Submergence of the valley system has meant that offshore of SE Essex, a drowned course of the Thames and its tributaries occurs aligned towards the east and southeast (Bridgland 1988; Bridgland & D'Olier 1995). Upstream in the estuary, a thick wedge of Holocene marsh and mud sediments has accumulated (Devoy 1979). Both in and upstream of the estuary, floodplain alluvium overlies Late Devensian gravels.

In the last three decades of the 20th century the Thames sequence underwent a reappraisal from its pre-existing subdivision largely based on morphological evidence to one based substantially on geological sequences and their three-dimensional relationships. This change reflects the explosion of developments in Quaternary science in general. Today the sequence is in a phase of consolidation and although substantial changes are still possible, they are less likely than some 20 years ago. Much remains to be determined however, particularly in terms of dating and finer stratigraphical resolution of the unfossiliferous parts of the sequence.

Of the problems remaining, the most obvious is the disagreement over the stratigraphical status of individual terrace aggradations. The lithostratigraphical approach of Gibbard (1985, 1988, 1989, 1994) has been adopted in the following. In this system individual sediment bodies are assigned member status as this is thought to be the most appropriate hierarchical level; it is also compatible with neighbouring areas and with other unit classifications. The term 'formation' has been used to refer collectively to members with broadly unified lithological characteristics. This scheme has since been adopted by the British Geological Survey for mapping fluvial sequences in the Quaternary throughout Britain (Fig. 12.17). This is contrary to Bridgland (1988, 1994) who considers that terrace aggradations should be assigned formation status. Likewise, Bridgland favours group status for some of the formations defined below, considered necessary because the complexity of the Pleistocene sequence requires the use of all available hierarchical levels.

More problematic is the means by which chronostratigraphical correlation of individual temperate character deposits is achieved both between sites and between the conventional terrestrial and marine isotope stages. Conflicting results have arisen from the application of conventional biostratigraphical techniques, particularly palynology, and geochronology, particularly amino-acid racemization. This has led to the Thames' sequence being subdivided chronostratigraphically using systems that stress different elements of the sequence (Gibbard 1977, 1985, 1994; Bridgland 1994). The scheme adopted here is an attempt to integrate both approaches as far as possible.

General publications on the Thames system stratigraphy and evolution include Gibbard (1977, 1985, 1994), Gibbard & Allen (1994), Bridgland (1988, 1994), Whiteman & Rose (1992) and Rose (1995).

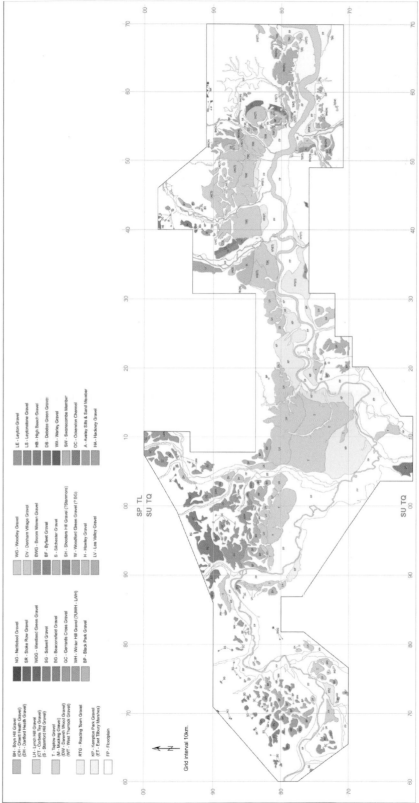

Figure 12.17 Map showing the Quaternary geology of the Middle and Lower River Thames, including the London area. Map provided by Phil Gibbard.

Raukar: individual rock pillars on the beach of Langhammars, Fårö, Sweden. Photograph by Jürgen Ehlers.

Chapter 13
North and Baltic Seas during the Ice Age

C arl Linnaeus described the limestone pillars at the beach on the island of Fårö near Gotland in 1741 with the words:

Between Strandsriddaregården and the lime kilns of Kyllei at Hafsviken there was a gently sloping hill, on which many quite large and massive limestones, 4–6 fathoms in height, were arranged in rows, looking like ruins of churches and castles, those at the foot of the slope being higher than those above, so that their heads had the same height. From a distance, they looked like statues, busts, horses or whatever kind of ghosts.

von Linné (1745)

Linnaeus had been commissioned by the Swedish Riksdag to explore the islands of Öland and Gotland. Linnaeus described everything that seemed remarkable: not only the plants and geological features, but also the customs of the inhabitants. He caused some distrust by these activities, as Sweden and Russia headed directly towards another military conflict (after the Great Northern War of 1700–1721). Was this strange scholar in fact a Russian spy? Linnaeus was lucky; despite suspicions he escaped unscathed.

The Raukar are rocks over 10 m in height which had survived erosion in the surf better than the surrounding limestone, and can be found in Gotland, Fårö, Lilla Karlsö and Öland (at Byrum). Most Raukar are found in the beach area, although some are still in the water.

The Ice Age, First Edition. Jürgen Ehlers, Philip D. Hughes and Philip L. Gibbard.
© 2016 John Wiley & Sons, Ltd. Published 2016 by John Wiley & Sons, Ltd.
Companion website: www.wiley.com/go/ehlers/iceage

As a consequence of the ongoing land uplift, however, some (such as the 12-m-high 'Maiden' in Lickershamn) are already beyond the influence of the sea and the surf.

The North Sea and the Baltic Sea are marginal seas of the Atlantic Ocean, both having achieved their present form in the course of the Ice Age. Their shape is still changing today.

13.1 Development of the North Sea

The warm stages of the Ice Age are not periods of geomorphic inactivity. On the shores of the Baltic and North seas today, major changes are underway. The North Sea, as well as large parts of Britain, the Netherlands, northern Germany, Denmark and Poland, has been part of an area of subsidence since the Late Palaeozoic. The North Sea is, in fact, older than the Atlantic, the opening of which (as a result of the drifting apart of the North American and the Eurasian plates) had not begun before the Jurassic. The general subsidence was repeatedly interrupted by phases of uplift. Although most of the recent history of the North Sea is represented by marine deposits, there are significant breaks. The red sandstone cliffs of Helgoland, for example, owe their existence to terrestrial sedimentation. In the long run, however, subsidence clearly prevailed. The thickness of sedimentary rocks in East Frisia is *c.* 4500 m and in Dithmarschen *c.* 6000 m. It is only since the Miocene that subsidence was concentrated on what is now the North Sea Basin. The thickness of Tertiary deposits in the central part of this basin is over 3500 m (Petroleum Exploration Society of Great Britain 2007).

The subsidence in the North Sea Basin continued into the Quaternary. While the thickness of Quaternary sediments on the Isle of Sylt is close to zero (a result of the high position of the Pliocene kaolin sands), whereas in the 'duck bill' region, in the middle of the North Sea, it is about 830 m (Brückner-Röhling et al. 2005).

The floor of the North Sea is extremely smooth. Very few landforms provide information about the basin's Quaternary development. For the reconstruction of the shaping of the submarine landscape, the results of geological and geophysical investigations must be relied on almost exclusively. This has both advantages and disadvantages. While the construction of the underground stratigraphy on land relies on point information (boreholes and exposures), in the marine area seismic techniques offer not only the ability to study continuously recorded profile sections but with the help of 3D seismic measurements the true spatial structure of the strata can also be determined. The resulting seismostratigraphic units, however, cannot be directly correlated with the continental litho- and chronostratigraphy. This is only possible via the evaluation of specifically planned boreholes, which can be used as guide profiles within the seismostratigraphic framework.

The surface of the North Sea floor has a small number of morphological features, including a few incompletely filled remnants of Weichselian-age subglacial channels (e.g. the Outer Silver Pit) and the large glacial thrust-moraine zone of the Dogger Bank. Some gravel-rich areas, such as the Borkum Riffgrund, are also interpreted as terminal moraines. Seismic studies have also shown strong glaciotectonic deformation in some places. Borth-Hoffmann (1980) described such structures north of Helgoland, and Andersen (2004) discovered a thrust sequence west of the Danish island of Mandø. In both cases, the glacier had exerted pressure from an E–ENE direction. The thrust slices in the Danish Wadden Sea were found to dip at an angle of 10–40 degrees and extend down to a depth of 200–360 m. The thrust features can be traced in a north–south direction for over 5 km (Andersen 2004). However, most areas of the North Sea floor display undisturbed stratification.

The Quaternary deposits underlying the North Sea provide further evidence of the basin's evolution. The Early and Middle Pleistocene deposits of the central and northern North Sea consist almost exclusively of marine sediments. The most striking feature in the seismic profiles is a major unconformity, which was probably formed during the Elsterian (Anglian) glacial period. In the southern North Sea, this unconformity is underlain by Early and Middle Pleistocene sediments of up to 550 m thickness. They comprise delta deposits of the large rivers, notably the Rhine and the Meuse, but also the substantial Baltic River System (also called the Eridanos River). During the cold phases of the Early Pleistocene, when sea levels were low, the deltas reached far into the North Sea. They form the seaward end of the glacial river terraces. Above the unconformity, predominantly glaciomarine deposits are found which, in contrast to the subhorizontal stratification of the older glacial sediments, are in many places dissected by deep channels, eroded by meltwater under the ice.

Investigation of 3D seismic data from the central North Sea has demonstrated that much of the early Quaternary sedimentary sequence contains buried iceberg plough marks. Their source was an early Quaternary Scandinavian Ice Sheet (Dowdeswell & Ottesen 2013). The oldest till unit of the North Sea area dates from c. 1.1 Ma. This so-called 'Fedje Till' was found at the base of the Norwegian channel immediately overlying Oligocene deposits (Sejrup et al. 2000).

There are indications that a major glaciation affected the North Sea before the Elsterian Glaciation. In the evaluation of 3D seismic data from the Witch Ground in the central North Sea, Graham (2007) found glacial striae produced by icebergs within the Aberdeen Ground Formation at a depth of 130–170 m. Palaeomagnetic investigations of BGS borehole 77/02 suggest that the features might be of Cromerian age. The system of striae is traversed by subglacial troughs, which are at least of Elsterian age.

In the southern North Sea, thrust features in Middle Pleistocene sediments in the Brown Bank area in the Dutch sector suggest a first (probably Elsterian) ice advance to at least 52° 20′ N. In East Anglia the Anglian (=Elsterian) tills and small buried channels reach south as far as the area of Ipswich (52° N). It is assumed that the British and Scandinavian ice sheets were confluent during the Elsterian (Long et al. 1988). The distribution of tills at the bottom of the North Sea is extremely discontinuous. Where till is found, an exact age determination is often difficult. Consequently, Beets et al. (2005) just refer to two 'Middle Pleistocene' till strata drilled in the Dutch sector at 54° N and 5° E.

The most ancient channels of the southern North Sea are substantially filled with subglacial and glaciolacustrine deposits of the Elsterian Glaciation. In the upper part of the channel fill, Holsteinian (=Hoxnian) marine deposits are also found. In the central North Sea, the situation is different. Here the Elsterian channel fill is largely absent and thin Holsteinian sediments near the channel base are overlain by Saalian glaciomarine strata. In this part of the North Sea, a large number of channels were not infilled during the Weichselian. They include the Devil's Hole, a group of up to 150 m deep channels about 200 km east of Dundee. The channels are 1–2 km wide and 20–30 km long (Long & Stoker 1986).

With the help of 3D seismics the filled channels on the floor of the North Sea can be reconstructed quite accurately. The number of channel generations exceeds that of the number of known glaciations. Kristensen et al. (2007) distinguish seven major phases of channel formation, which they attribute to ice-marginal oscillations of multiple glaciations.

In the Netherlands, the ice of the Saalian Glaciation advanced to the south of Amsterdam. It had previously been assumed that the Scandinavian and British ice met in the North Sea (Rappol et al. 1989). In the Dutch sector of the North Sea, the corresponding till can be traced from the coast to only *c.* 40 km offshore (Joon et al. 1990). Saalian till is absent from most of the central and northern North Sea. However, the lack of till does not necessarily mean that no contact existed between British and Scandinavian ice in the southern North Sea. The Saalian ice movement directions in the Netherlands and northern Germany (Ehlers 1990) and the distribution of subglacial channels are suggestive of an extended Saalian Glaciation of the North Sea Basin (Graham et al. 2007). Equivalent glaciation in the British Isles has been identified in the Fenland region of East Anglia and areas immediately to the north under the North Sea (Gibbard et al. 2009).

Throughout the North Sea region the sedimentary reconstructions indicate that the ice dammed the river systems of the Rhine, Meuse, Thames and Scheldt to form a massive glacial lake that persisted throughout much, if not all, of the Anglian/Elsterian Glaciation (Gibbard 2007). The rivers and meltwater streams (Gibbard et al. 1996, 2008; Leszczynska 2009; Gibbard & van der Vegt 2012) formed deltaic accumulations (including the Cromer Ridge in Norfolk), built out into the lake through the glacial episode. Overspill from this lake gave rise to the initial breach of the anticlinal Weald-Artois bedrock ridge to initiate the Dover Strait (Gibbard 1988, 1995; Cohen et al. 2005).

The pre-existing glacial landscape largely controlled the extent of the interglacial transgressions. The Holsteinian Sea invaded large fjords, which followed the course of the Elsterian tunnel valleys far into the interior of Denmark, northern Germany, the Netherlands and Belgium. In East Anglia marine deposits are found in the Nar Valley and near Peterborough, about 20 km from the present coast. The transgression was however accompanied by a strong reworking of sediment that led to the removal of high-altitude till plains and a filling of the deep channels.

The expansion of the last precursor of today's North Sea in the Eemian (=Ipswichian) Interglacial is relatively well known. The transgression of the Eemian Sea did not usually reach the maximum extent of the Holocene North Sea. The major Holocene embayments, however, are each characterized by Eemian precursors (Zuider Zee, Lauwerszee, Dollard, Jade Bay). In Schleswig-Holstein a temporary connection between the North Sea and Baltic Sea existed (approximately along the present-day Kiel Canal; Streif 2004).

However, coastlines are dynamic and this general information tells us nothing about the actual alignment of the coastline at any given time. Few details are known about the development of the Eemian coast. In some cases sea caves and beach deposits indicate the elevation of former sea levels (Fig. 13.1). Unlike the far-reaching inland fjords of the Holsteinian Sea, the bays of the Eemian Interglacial sea were less extensive; they essentially followed the former

Figure 13.1 Former highstands (or lowstands) of the sea can be deduced from the corresponding landforms. They include sea caves (left: Elgol, Isle of Skye) or marine terraces (right: cemented gravels of the so-called 'Patella beach', Gower Peninsula, Wales). Photographs by Jürgen Ehlers.

river valleys. For example, in Britain a substantial estuary was established in the Thames Valley, comparable to that today, as far upstream as central London. As the duration of the Eemian Interglacial was similar to the Holocene, it must be expected that a barrier similar to the current coastline formed in the present-day Wadden Sea, even if the sea level was lower than today. The existence of a coastal barrier is supported by the faunal composition of Danish North Sea coastal deposits (Konradi et al. 2005). Evidence for the barrier itself has not yet been detected. It was probably located further seaward and fell victim to Holocene marine erosion.

While Eemian deposits are found at a much higher level on the English eastern and southern coasts (in a tectonically 'stable' zone, sea level reached a maximum of about +7.5 m), the surfaces of the Holsteinian and Eemian marine sediments in northern Germany are much lower than the present sea level. This has been attributed to a general subsidence trend of the German North Sea coast.

During the Weichselian Glaciation, when sea level was lowered by *c.* 110–130 m (Shackleton 1987), the ice-free areas on the North Sea and English Channel floors were subject to periglacial transformation, just like today's landmass. Traces of ice wedges have been found, at the bottom of the southern North Sea, together with aeolian sand.

Periglacial landforms have also been recorded on the Channel floor. The asymmetric filling of a number of channels may be due to periglacial solifluction. A continuous till sheet (Bolders Bank Formation) can be traced from the East Anglian coast *c.* 100 km to the northeast. Glaciolacustrine clay and diamicton have also been found south and east of the Dogger Bank. Farther north, however, the occurrence of Weichselian tills is usually limited to a strip less than 100 km wide east of the British coast. Recent discoveries have shown that British and Scandinavian ice met during the Early Weichselian glacial period (MIS 4), and also during the Late Weichselian glacial maximum (Graham et al. 2009). Older hypotheses that parts of Scotland (Caithness) remained free of ice during the Weichselian have now been refuted (Ballantyne & Hall 2008; Ballantyne 2010). The English Channel was occupied by a substantial Channel River, and was joined by modern streams that extended their courses on to the continental shelf. The drainage of the resulting ice-dammed lake towards the south, that also entered the Channel River, is clearly visible in the sediments at the foot of the continental slope in the Bay of Biscay (Toucanne et al. 2010).

The initial relief of the North Sea, prior to the Holocene transgression, was primarily shaped by the Pleistocene glaciations. In principle, it can be assumed that the Pleistocene landform assemblages, which are found on the mainland today, also extended into the realm of the North Sea Basin. In addition, the Elbe river valley can be traced on the floor of the North Sea NNW to the White Bank area (Figge 1980).

At the end of the Weichselian, sea level rose relatively rapidly. Dated peats show that the Dogger Bank was still dry land in the Boreal. The transgression reached the −45 m contour 9000 years ago (Vink et al. 2007). The North Sea coast was struck by a major tsunami *c.* 8150 years ago (Box 13.1). Before the sea reached the present-day coastline in the Atlantic period, the changed drainage conditions led to increased waterlogging of the coastal area and extensive peat growth occurred. This resulted in formation of the so-called *Basaltorf* (the lowest member of the Holocene sequence), which is regularly found in boreholes throughout most of the German coastal area.

Under the influence of the mild climate in the Atlantic period, the remains of the circumpolar ice caps melted. By 5000 years ago the last remnants of the Laurentide Ice Sheet in North America were gone. This is also when the postglacial sea-level rise slowed down considerably. The opening of the English Channel resulted in a change in the tidal conditions, thereby influencing sedimentation along the coasts of the southern North Sea (Streif 1986). Peat bogs developed later in the coastal area as a result of increasing freshwater influence. This phase of regression came to an end along the Dutch coast at *c.* 1700 BC. At the German and Danish coasts, the transgression resumed somewhat later. Since then, apart from an interruption during the 'Little Ice Age', it has continued to this day.

During the first centuries of the Holocene sea-level rise was very rapid. This was reflected not only in the vertical but also in the horizontal transgression. A recorded increase of *c.* 3.3 cm a^{-1} at *c.* 8500 years ago may not seem very much, but with the low gradient of the North Sea floor it resulted in significant changes in the configuration of the coastline. Bäsemann (1979) calculated that the sea then advanced landwards at a rate of 267 m a^{-1}. Under those conditions, there was not enough time to build up a coastal barrier system such as the present Wadden Sea.

BOX 13.1 NORTH SEA TSUNAMI

The continental shelf falls sharply towards the Atlantic Ocean. During the glaciations, a thick sediment body formed on the continental slope off the Norwegian coast which was not very stable. At 8150 years ago, 100 km seaward of Ålesund, 2400 km³ of sediment slipped down the continental slope in a massive landslide (Storegga Slide). Model calculations show that it travelled at a speed of 25–30 m s⁻¹ and resulted in a tsunami wave of 20 m height. The sediments that were swept away by this wave were deposited as a sand and gravel layer in the bog and marsh areas of the adjacent coasts. In areas of progressive peat growth, it has subsequently been covered by peat again. The traces of the Storegga tsunami have been detected in Norway, Scotland and on the Shetland Islands (Fig. 13.2; Bondevik et al. 2003, 2005).

Figure 13.2 Tsunami deposits (bright layer) in peat at Sullom, Shetland Islands. A large landslide on the continental slope off-Norway more than 8000 years ago triggered a tsunami wave that affected the coasts of the northern North Sea. Photograph by Jürgen Ehlers.

The situation changed at *c.* 8000 years ago, when the horizontal transgression almost came to a halt. This was the birth of the modern coastal barrier. A beach barrier system formed along the Dutch coast, allowing lagoonal deposits and mud sediments to develop in great thicknesses under its protection. Dunes soon formed on the beach ridges. Those so-called 'old dunes' can still be identified from south of The Hague to north of Alkmaar. In this first phase of barrier formation, sea level was still rising rapidly (at a rate of 0.1–0.06 cm a⁻¹) and sediment supply was not sufficient to fill up the resulting tidal flats. At *c.* 5000 years, sediment supply increased to such an extent that the beach ridges were extended to form a continuous coastal barrier in whose protection lakes, freshwater marshes and swamps could proliferate (Beets et al. 2003).

Sea-level changes in Britain were strongly influenced by isostatic land uplift, which reached up to 1.6 mm a^{-1} in central and western Scotland, while in southwest England the land subsided at a rate of up to 1.2 mm a^{-1}. Sediment consolidation in areas of thick Holocene deposits added an extra *c.* 0.2 mm a^{-1} to the subsidence rate. Changes in tidal range have to be taken into consideration when trying to recalculate the coastal dynamics (Shennan & Horton 2002).

The development of the North Sea coast was not always synchronous. To some extent, large-scale trends were obscured by local palaeogeographic features (Streif 1990). A correlation with the well-known Dutch subdivision of postglacial sea-level rise in a Calais and Dunkirk Transgression (each with 4–5 individual transgression phases) to other coastal areas is therefore seldom possible.

The formation of the modern coastal dunes (the 'young dunes') began on a large scale in the twelfth century and intensified later, especially in the fifteenth and sixteenth centuries, when strong coastal erosion provided large amounts of sand for aeolian transport. The shape of the coastline changed greatly. Parts of the old dunes and beach ridge systems were eroded. Due to large-scale siltation tendencies, others were relocated far inland (Netherlands, Dithmarschen, Eiderstedt). From historical sources it can be shown that the main phase of the young dune formation on the Dutch mainland coast was completed by the seventeenth century (Zagwijn 1984). The same applies in principle to the dunes on the barrier islands.

Today's barrier islands of the Wadden Sea are very young features. The existence of Langeoog island can be traced back to no more than 3000 years (Barckhausen 1969), and the oldest saltmarsh sediments (Groden layers) on Juist have yielded a ^{14}C age of 1965±130 years. The development of the dune islands was a result of changes in tidal processes and increased sand supply.

The barrier islands of the North Sea (like other barrier islands such as on the east coast of the United States) adapt to changes in sea level. A rising sea level results in a landward shift of the barrier. Under natural conditions, this process takes place under erosion of the sea-facing dunes and aeolian sand transport to the interior of the island, with the dune ridges being breached and 'washover processes' transporting sand into the lagoon. Fixing the dunes and implementing a rigid defence of the current coastline runs counter to this natural adaptation, and requires increasing efforts to keep the shoreline at its current position (Ehlers 2009).

Juist is one of the few East Frisian Islands which was favoured by nature in the last hundred years. As on the neighbouring islands, in 1913 a sea wall and groynes were built to protect the settlement. This turned out to be an unnecessary measure, however. The groynes and sea wall disappeared soon after in freshly deposited immense masses of sand, and have not since reappeared (Fig. 13.3). The positive sand balance at the central beach is deceptive, however; in recent years the western part of the island has experienced a significant loss of dunes. Over 20 m were lost during the winter of 2006–2007, and breaching of the dune ridge could only be prevented by beach nourishment.

Figure 13.3 A wide beach, as here on Juist, is an indication of a positive sand balance. Photograph by Jürgen Ehlers.

Coastal erosion along almost the entire North Sea coast (e.g. Fig. 13.4) is largely a result of sea-level rise. Worldwide sea-level rise is currently assumed to be *c.* 20 cm per century (IPCC 2007). Apart from short-term regional differences, this value can be significantly lower or higher. Extrapolations from short time series are always problematic.

Figure 13.4 Eroding cliff coasts. Erosion destroyed a road at Holmpton, Holderness Peninsula, Yorkshire, England (left) and destroyed beach stairs at Ristinge Klint, Langeland, Denmark (right). Photographs by Jürgen Ehlers.

The current rise in sea level can no longer be explained with the further melting of glaciers; it must therefore be assumed that the thermal expansion of sea water also contributes significantly to the current transgression. Sea-level rise will consequently continue and the exposed parts of the coasts of North Sea and Baltic Sea will continue to be eroded.

On the British North Sea coast, the cliffs around Happisburgh (Norfolk) are being strongly eroded. Detailed measurements have shown retreat rates of almost 10 m a^{-1}. Erosion increased considerably after the wooden coastal defences were abandoned. However, sea defences do not solve the problems. Locally they may slow down erosion, but they result in undernourishment of the downdrift coastal areas (Ohl et al. 2003; Poulton et al. 2006).

13.2 Development of the Baltic Sea

The early history of the Baltic Sea (prior to the Weichselian Glaciation) is largely unknown. At the time of the Baltic River System (from the Tertiary to the Waalian), river sediments of Scandinavian origin were poured into northern Germany and the Netherlands, including the Loosen Gravels in Mecklenburg and the Kaolin Sands of Sylt. During this period the Baltic Sea Basin had yet to come into existence; instead it was a substantial river valley. The further development from the Menapian to the Elsterian is unknown. Whether early glaciations had produced a proto-Baltic Sea, and to what extent, cannot be clarified. There is, however, no doubt that glacial scouring played a major role in sculpting the Baltic depression and that the Baltic Sea Basin increased considerably in depth throughout the Quaternary. The low proportion of flint and chalk in the Elsterian tills in the north of Germany suggests that at the time, the Tertiary sediment cover over the chalk areas of the western Baltic Sea was still largely intact (Meyer 1991).

First clear evidence for the existence of a Baltic-precursor is known from the Holsteinian interglacial period. Marine Holsteinian deposits have been described from the Kaliningrad district and from Latvian sites (Marks & Pavlovskaya 2003; Zelčs et al. 2011). The Holsteinian sites in Estonia, however, are terrestrial (Kalm et al. 2011). No evidence has yet been found for the sea reaching further north than the Riga region of Latvia during this time. In Poland, Holsteinian marine deposits are only found in the extreme northeast on the border with the Russian exclave of Kaliningrad. In Germany, the Holsteinian transgression (from the North Sea) extended into eastern Mecklenburg-Vorpommern and, in a brackish facies, southwards to the northwest of Brandenburg (Müller 2004a). Several sites with marine sediments in western Mecklenburg and NE Lower Saxony suggest a connection at the time between the North and Baltic Sea via the Lower Elbe embayment.

That the glaciers of the Older Saalian crossed the Baltic Sea basin without being significantly deflected (Meyer 1991) does not necessarily imply that a Holsteinian precursor of the Baltic did not previously exist. The same phenomenon can be observed in the Weichselian, at a time when the Baltic Sea Basin was definitely in existence. Ice-movement directions seem to have been controlled by glacier dynamics rather than topography (Ehlers 1990).

The Eemian Sea filled much of the Baltic Sea Basin (Fig. 13.5). Marine Eemian deposits have been identified in Denmark and northern Germany (Schleswig-Holstein and

Figure 13.5 Maximum extent of the Baltic Sea during the Eemian Interglacial. A connection existed between the North Sea and Baltic Sea in Schleswig-Holstein. Large parts of Finland and NW Russia were inundated. Compiled by Jürgen Ehlers.

Mecklenburg-Vorpommern). Marine Eemian deposits found in boreholes suggest that a narrow connection existed along the route now followed by the Kiel Canal between the Eemian North and Baltic seas (Kosack & Lange 1985). In Mecklenburg-Vorpommern, the Eemian Baltic Sea extended tens of kilometres inland in several bays. The largest of those bays extended from Rostock Laage about as far south as Bützow (Müller 2004b). Two transgressions of the Eemian Sea can be distinguished in Poland. In the lower Vistula Valley a wide bay reached from Gdansk far inland (Makowska 1979; Knudsen et al. 2012). In Sweden, brackish-marine sediments of probable Eemian age have been found at Bollnäs, north Sannäs, Skulla and Nyköping (Robertsson 2000). The Eemian transgression not only reached the Finnish coast, but also affected parts of northern Russia (as indicated by marine Eemian sites of Mga and

Figure 13.6 Mollusc-bearing marine clay from the Eemian Interglacial, Ristinge Klint, Langeland, Denmark. Photograph by Jürgen Ehlers.

Petrozavodsk). The far eastward-reaching marine influence is explained by the greater extension of the Saalian Ice Sheet and the resulting stronger isostatic depression (Forsström et al. 1988). The model of Lambeck et al. (2006) shows the ice distribution at the end of the Saalian Glaciation and the subsequent encroachment of the sea (see Chapter 14). The Eemian marine transgression (Fig. 13.6) was the last marine foray into the Baltic Sea prior to the Holocene. Many new insights into the extent and history of the Eemian transgression in the Baltic Sea have been obtained from the BALTEEM project (Knudsen & Gibbard 2006; Kristensen & Knudsen 2006).

It has been known from the early eighteenth century that the Baltic Sea has undergone significant changes in sea level. De Geer (1888–90) constructed the first map of the highest sea levels detected in Sweden. A few years later, De Geer (1896) was able to explain the basic principles of postglacial development of the Baltic Sea and also convincingly demonstrate the effects of isostasy for Scandinavia. The influence of eustasy was still unknown at the time; another quarter-century passed before Nansen (1922) and Ramsay (1924) demonstrated its effect.

The development of the Baltic Sea since the end of the last glaciation can be divided into four main phases:

1 Baltic Ice Lake: to 10,200 years BP;
2 Yoldia Sea: 10,200–9300 BP;
3 Ancylus Lake: 9300–8000 BP; and
4 Litorina Sea: 8000 until today.

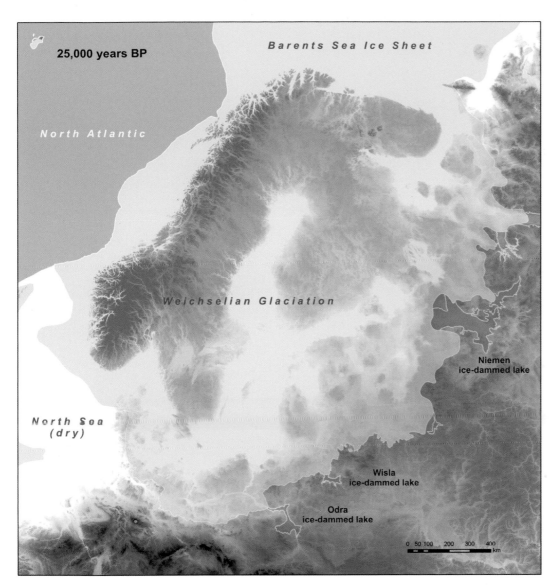

25,000 years BP

Barents Sea Ice Sheet

North Atlantic

Weichselian Glaciation

Niemen ice-dammed lake

North Sea (dry)

Wisla ice-dammed lake

Odra ice-dammed lake

0 50 100 200 300 400 km

Figure 13.7 During the Weichselian glacial maximum the Baltic Basin was completely filled with ice. Large ice-dammed lakes formed in the river valleys at its margins. Compiled by Jürgen Ehlers.

Of the four main stages of the Baltic Sea development, the Baltic Ice Lake was the last to be discovered (Figs 13.7, 13.8). At the end of the nineteenth/beginning of the twentieth century it was still generally assumed that the Baltic Sea Basin was filled with ice until the Yoldia Sea broke finally through. Munthe (1902) was the first who adopted an ice lake stage prior to the formation of the Yoldia Sea in his description of the geological map sheet Kalmar. Today we know that the Baltic Ice Lake gradually evolved from a large number of local ice-dammed lakes, which coalesced until the ice-free Baltic Sea Basin could combine to create a single lake basin (Kvasov 1978). However, the Baltic Ice Lake did not spread into NE Finland as Sauramo (1958) had assumed. Moreover, the idea that the Baltic Ice Lake was preceded by a Late-glacial marine incursion ('late-glacial Yoldia Sea'; Sauramo 1958) has not been confirmed (Hyvärinen & Eronen 1979).

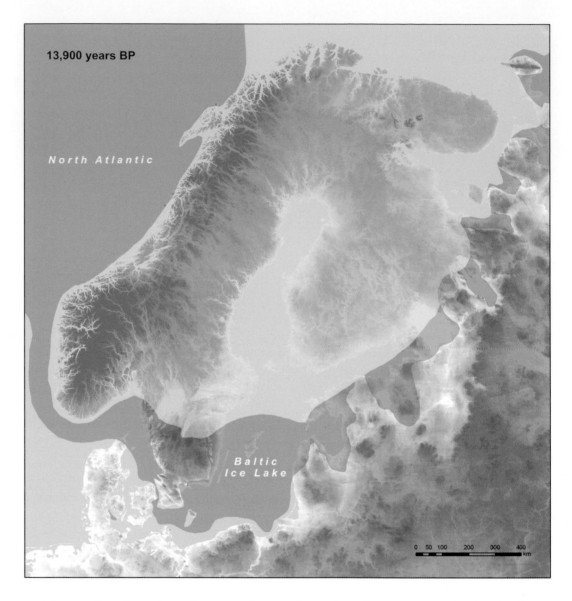

Figure 13.8 At
c. 13,900 years BP
the ice had retreated
from the southern
part of the Baltic Sea
Basin. The German and
Polish Baltic coast was
in isostatic uplift and
therefore not inundated
by the Baltic Sea at
that time. Compiled by
Jürgen Ehlers.

The overflow from the Baltic Ice Lake was directed to the west, since the lake level was above that of the ocean. The western edge of the lake in the Baltic Sea Basin was located at the Darss Sill. The shape and size of the lake changed rapidly, controlled by the further melting of the ice and the continued uplift. Björck & Digerfeldt (1984, 1989) showed that the final drainage of the Baltic Ice Lake took place over two phases. An interruption of several hundred years occurred when during the Younger Dryas Stadial the ice sheet pushed vigorously forwards again. The final drainage of the Baltic Ice Lake occurred when the ice margin in Sweden had retreated beyond Mount Billingen. This event is coeval with the transition from the Younger Dryas to the Preboreal, that is, with the Pleistocene–Holocene boundary. The zero-year of the Finnish varve chronology refers to the date on which the level of the Baltic Ice Lake dropped to the global sea level (i.e. the Yoldia Sea; Fredén 1979).

Curiously, the traces of this event around Mount Billingen in Skövde, between Lakes Vättern and Vänern (Fig. 13.9), are somewhat ambiguous. Högbom (1912) had discovered a

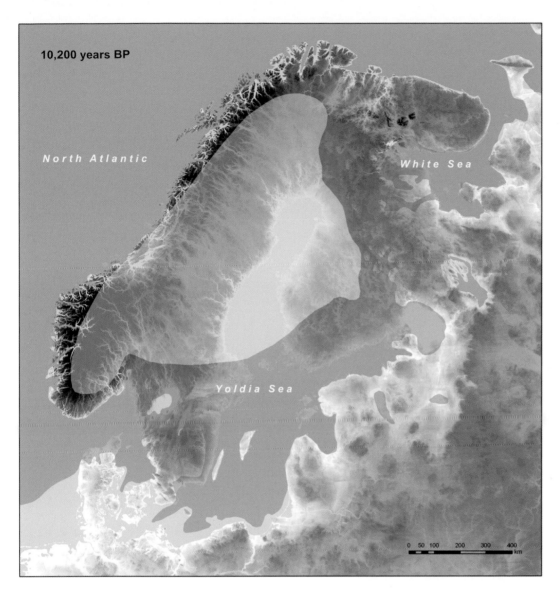

Figure 13.9 At 10,200 BP the Närkesund (north of Lake Vättern) connected the Baltic Sea to the Atlantic Ocean. The Yoldia Sea covered parts of southern Sweden and extended over large parts of Finland. Compiled by Jürgen Ehlers.

gully cut in Cambrian sandstone near St Stolan (15 km north of Skövde) and interpreted it as an outlet of the Baltic Ice Lake. The feature is relatively small, however, compared to the mass of water that must have been released by the emptying of the giant Baltic Ice Lake. The height difference between the levels of the ice-dammed lake and the ocean at Billingen was still 26 m (Fredén 1979). Offshore submarine evidence of the discharge event has yet to be identified. Drainage probably followed the Norwegian Channel.

The High Arctic saltwater mollusc *Portlandia arctica* (formerly *Yoldia arctica*) characterizes the next stage of development of the Baltic Sea, the Yoldia Sea. The prevailing view is that the salt water penetrated eastwards as soon as the ice had retreated from Billingen and the level of the Baltic Sea had adapted to the global sea level. The saltwater influence, however, may have been much lower than originally anticipated. De Geer (1940) had already suggested that the living conditions in the Stockholm area were favourable for *Portlandia arctica* only for a short time (barely 100 years). How far saline water actually penetrated to the east is controversial. For example, Florin (1977) found no saltwater diatoms in Yoldia phase deposits south of the Mälar Valley (Fredén 1979). On the other hand, the Yoldia Sea has been demonstrated in Finland on the basis of diatom floral assemblages. Eronen (1974) has, however, shown that the faunas found do not represent reliable indicators for a saline environment; some are reworked forms from Eemian Interglacial deposits. Abelmann (1985) only found brackish diatoms of the Yoldia phase in drill cores from the Karlsö basin (at Gotland). In the Gotland Basin, Bornholm Basin and the Gulf of Gdansk areas, only freshwater diatoms have survived from the 'Yoldia phase'.

As a consequence of the ongoing land uplift, the shallow Närkesund gradually closed until saline water could no longer penetrate into the Baltic Sea Basin. That was the end of the Yoldia Sea period. The name Ancylus Lake (Fig. 13.10) for the next freshwater phase of the Baltic Sea was introduced by de Geer (1888–90), named after the freshwater snail *Ancylus fluviatilis*. The beginning of the Ancylus stage in Sweden is a couple of hundred years before

Figure 13.10 At c. 9300 BP the land was uplifted to such an extent that the connection between the Baltic Sea and the Atlantic Ocean was cut off again; this period is referred to as the Ancylus Lake phase. Compiled by Jürgen Ehlers.

the first appearance of alder (*Alnus*), that is, *c.* 9200–9300 BP. The overflow of such a large lake basin in the direction of the sea should hypothetically have led to the incision of a deep gully. Munthe (1927) and von Post (1928) thought they had finally found that drainage channel at Degerfors in Sweden. The traces are not convincing, however. A re-examination of the area by Fredén revealed no clear evidence of a major drainage event. The water levels in the Baltic Sea and in the ocean were probably almost the same at that time, and a saltwater intrusion was only prevented by the threshold between the two seas (Fredén 1979).

As a result of the very different isostatic uplift rates, the Ancylus Lake reached its greatest extent in the south later than the north. At the peak of the transgression, the overflow of the Baltic Sea Basin moved back to the area of the Darss Sill where a 10–20-m-deep channel was cut. Catastrophic drainage through that channel led to a fall in lake level by *c.* 20 m (Kolp 1986).

Figure 13.11 Before *c.* 8000 BP, the connection to the ocean was restored. The Baltic Sea level was still significantly higher than today. Compiled by Jürgen Ehlers.

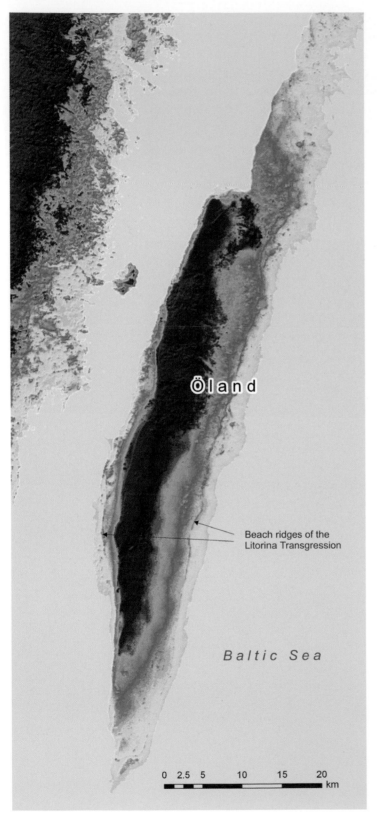

Öland

Beach ridges of the
Litorina Transgression

Baltic Sea

0   2.5   5   10   15   20
                                    km

The expanding Ancylus Lake in the west eventually reconnected with the ocean (Fig. 13.11).

The Litorina period is the final phase of the Baltic Sea development (Fig. 13.12). The current connection to the ocean, via the Sound, the Great and the Little Belt, was established during this phase. The stage is named after the snail *Littorina littorea*. The change from the Ancylus Lake to the Litorina Sea was not an abrupt one, but a gradual transition. The saltwater diatom *Mastogloia smithii* has been found in Blekinge since the late Boreal (from *c.* 8500 years BP). In contrast, in Ångermanland the transition occurred *c.* 1000 years later (Björck 1995). The Litorina phase can be subdivided into a number of subphases: Mastogloia Sea (low salinity); Litorina Sea *sensu stricto* (high salinity); Limnea Sea (brackish influence since *c.* 4000 BP); and Mya Sea (today's Baltic Sea, characterized by the sixteenth–seventeenth century immigration of the bivalve *Mya arenaria*).

The coast of Mecklenburg-Vorpommern was reached at *c.* 7800 BP by the Litorina transgression (Kliewe & Janke 1982). The beach ridges and conspicuous spits on Rügen and Usedom were formed at *c.* 6000 BP (Figs 13.13, 13.14; Hoffmann et al. 2005). In Schleswig-Holstein and Mecklenburg, there is evidence of a regression at *c.* 1000 BP. Due to the different isostatic uplift, its impact was limited to the western Baltic Sea. The change between transgressive and regressive sections decisively influenced the development of graded shorelines and the coastal beach barrier and spit systems of the Baltic Sea coast (Kolp 1982; Fig. 13.15).

Figure 13.12 The transgression of the Litorina Sea has left clear traces on the shores of the Baltic. Its beach ridges are clearly visible in the SRTM terrain model. Courtesy NASA/JPL-Caltech.

Figure 13.13 After the ice had melted and the associated sea-level rise was completed, the land uplift in the formerly glaciated areas is still continuing. On the Baltic coast here at Hudiksvall, Sweden, traces of the uplift are evident in flights of beach ridges. Photograph by Jürgen Ehlers.

Figure 13.14 Raised beach barrier on the northwest coast of Öland, Neptuni Akrar nature reserve, 3 km NNE of Byxelkrok. The individual ridges are marked by the blue-flowered Viper's Bugloss (*Echium vulgare*). The cobbly beach ridges lie directly on outcropping limestone. Photograph by Jürgen Ehlers.

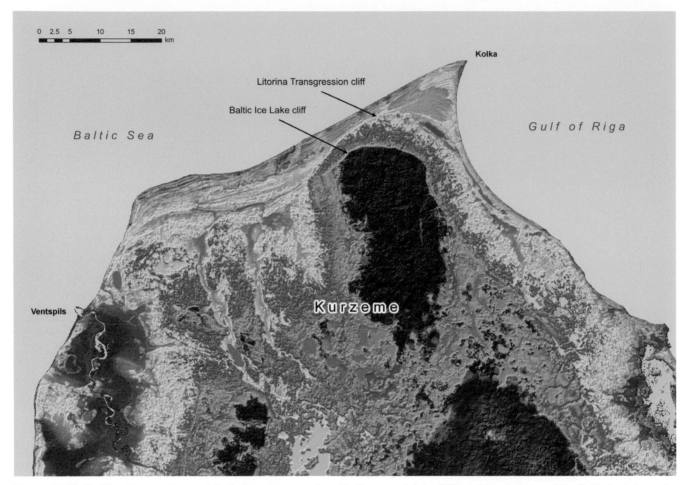

Figure 13.15 Significant traces of several Baltic Sea stages are found in Kurzeme, Latvia. Cliffs and beach ridges of the Baltic Ice Lake and the Litorina Sea are present. Compiled by Jürgen Ehlers.

Reconstructing the glaciations and their climatic effects is like trying to put together a puzzle of which many parts are missing. Photograph by Jürgen Ehlers.

Chapter 14

Climate Models and Reconstructions

14.1 Ice Cores

Ice, just like water, consists of oxygen and hydrogen. Oxygen comprises two isotopes: ^{16}O and ^{18}O. During the cold periods, more of the lighter ^{16}O was bound in the ice sheets and therefore removed from the water cycle. These changes in the isotope ratio are reflected in the composition of the glaciers and ice sheets, making it possible to reconstruct the climate history of the Ice Age from ice cores.

The first deep hole in the ice was drilled in 1966 at Camp Century, an American military base on Greenland. Its results were limited by the low resolution, the uncertainties of dating and a lack of knowledge about the ice flow behaviour at the drilling site. However, the hole reached a depth sufficient to provide a first important result. In the deepest core that was retrieved from the hole, there was ice as old as the last interglacial. This proved that the Greenland Ice Sheet did not melt completely during the Eemian Stage. The drilling had highlighted the potential of ice cores for the study of past climates in the Ice Age.

The first deep hole in the ice of Antarctica was drilled at the US Byrd Station in 1968. When the results were compared with those from the Camp Century sequence, there was a strong match between the cores from the two profiles. In 1981 the Dye 3 hole in southern Greenland was drilled. This intersected the ice from the last 90 ka and had a better resolution than any previous drilling.

The Ice Age, First Edition. Jürgen Ehlers, Philip D. Hughes and Philip L. Gibbard.
© 2016 John Wiley & Sons, Ltd. Published 2016 by John Wiley & Sons, Ltd.
Companion website: www.wiley.com/go/ehlers/iceage

New knowledge was provided by the Greenland Ice Core Project (GRIP) and Greenland Ice Sheet Project (GISP2) boreholes that were drilled in 1992 and 1993 with improved drilling techniques in the centre of the Greenland Ice Sheet with improved drilling techniques. The ice cores revealed that there were a number of significantly warmer periods during the Weichselian. Those short-term climatic fluctuations, which are called Dansgaard–Oeschger cycles after their discoverers, lasted a few thousand years and appear to reflect temperature changes of 10–12°C in less than a century. A total of 23 of those cycles have now been identified, and it is known that they occurred at intervals of 1500 years or a multiple of that time (Rahmstorf 2003).

The two deep boreholes cut through the layers of Weichselian ice and, just above the solid substrate, encountered ice from the Eemian Interglacial. Here the dramatic climate changes of the Dansgaard–Oeschger cycles appeared to continue. Or was that an artefact created by interference of the moving ice? Comparison with detailed pollen profiles from Europe, for example Gröbern in northern Germany, showed no evidence of any possible dramatic climatic deterioration within the Eemian Interglacial time.

Clarity was sought from the North GRIP borehole, which was completed in 2003. Unlike at GRIP and GISP2, clearly stratified ice was encountered to the base. The Weichselian part of the profile coincided completely with the results of GRIP and GISP2, but the lower part of the profile showed clear deviations. In the North GRIP core no indications were found of any strong climatic fluctuations during the last interglacial. What the three boreholes had in common, however, was that they all encountered Eemian-age ice. Earlier model calculations, according to which large parts of Greenland might have been ice free during the Eemian, had not been confirmed. Only the southern part of the inland ice sheet seems to have been slightly smaller than today (Johnsen & Vinther 2007).

Is it possible to find even older ice in Greenland? Geophysical surveys suggest that the best chances are in the northwest. The North Greenland Eemian Ice Drilling Project (NEEM) reached the base of the Greenland Ice Sheet at 2537 m depth in 2010. The cores obtained present a complete sequence of ice layers from the last interglacial period to the present.

While the Greenland ice cores date back no further than the Eemian Interglacial, at the other end of the Earth much older ice is found. The Russian Vostok drilling in East Antarctica was the first to penetrate strata several hundred thousand years old (Fig. 14.1). The European borehole on Ice Dome C recovered 740 ka old ice in 2005 (Taylor 2007).

The drilling in Dronning Maud Land in the framework of the European Project for Ice Coring in Antarctica (EPICA) encountered ice 800 ka old. Investigations have shown that in all the warm stages since Marine Isotope Stage 11, high levels of CO_2 and NH_4 comparable

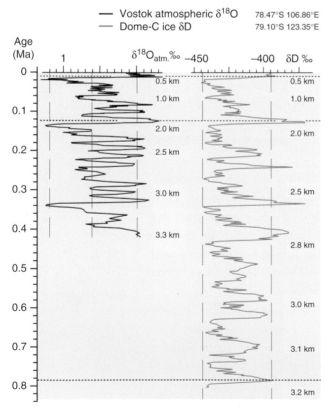

Figure 14.1 Oxygen-isotope curves from the Antarctic ice cores Vostok and Dome-C. The shape of the curves corresponds to the oxygen-isotope curves from deep-sea cores.

Source: Cohen & Gibbard (2011), http://quaternary.stratigraphy.org/charts/.

to those today were encountered. Only once, during MIS 11, was a higher concentration of N_2O recorded before the pre-industrial Holocene maximum. This value is now significantly exceeded, however (Schilt et al. 2010).

14.2 The Marine Circulation

Like the air masses in the atmosphere, sea water is also subject to a global circulation system. Its movement is driven primarily by the heating of the water surface, salinity differences and the wind, but ice cover and freshwater inflows are also significant. This circulation system is therefore called the thermohaline system.

The oceanic circulation can be likened to a giant conveyor belt that is driven by the polar Arctic water masses. The conveyor belt transports salt- and oxygen-rich water from the depths of the ocean, until it finally reaches the surface in the northern parts of the Indian Ocean and Pacific Ocean (Fig. 14.2). As the North Pacific is effectively sealed off from the cold waters of the Arctic Ocean by the shallow Bering Strait, the Arctic polar water is supplied mainly from the northern North Atlantic Ocean. The formation of North Atlantic deep water is therefore decisive for the entire oceanic circulation. Cold deep water is also produced around the Antarctic, but the existence of a circumpolar ocean in the Southern Hemisphere prevents exchanges between the Antarctic deep water and the water of lower latitudes.

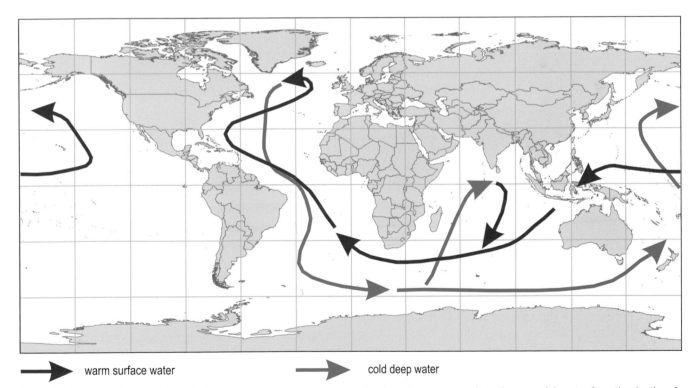

→ warm surface water   → cold deep water

Figure 14.2 The marine circulation. Arctic cold water sinks in the North Atlantic and transports salt- and oxygen-rich water from the depths of the oceans to the upwelling areas of the northern Indian Ocean and the North Pacific. Adapted from Broecker & Denton (1989). Reproduced with permission of Elsevier.

The oceanic circulation exerts a strong influence on climate. Ocean currents, such as the Gulf Stream system, transport warm water from the equatorial low latitudes to high polar latitudes. This heat is released when the water cools, until it finally sinks as new North Atlantic deep water. In winter, the amount of heat released by the deep-water formation in the North Atlantic is of the same order of magnitude as the amount of heat generated by sunlight.

The oceanic circulation depends not only on temperature but also on the salt content. If larger amounts of water are removed by evaporation from the oceans, such as was the case during the extensive glaciations, then the density of the sea water increases. A strong transportation of freshwater into the polar regions, however, reduces the salt content and therefore the formation of deep water. Broecker & Denton (1989) assume that a great flood of fresh water flowed into the North Atlantic at about 12 ka as the ice sheets melted. The inflow of the meltwater prevented the formation of North Atlantic deep water and interrupted the oceanic conveyor belt system, which triggered the cold phase of the Younger Dryas Stadial. The cooling was felt most strongly around the North Atlantic. This has been a popular idea explaining the last cold snap of the Ice Age, although it must be noted that several other hypotheses have been proposed (including asteroid impacts, volcanic eruptions as well as various different routes of catastrophic meltwater flooding).

Ocean currents are driven mainly by horizontal temperature gradients. During the cold phases the global temperature gradient between the ice-covered polar ocean (0°C or less) and the tropical waters (about 25°C) was compressed into a much narrower belt than it is today. The temperature gradient was therefore steeper, leading to stronger winds and stronger ocean currents. The changes in ocean circulation are reflected in the sedimentation. Consequently, the palaeo-oceanographic conditions can be derived from the sedimentary sequence. For example, the oxygen-isotopic composition in the calcite shells of planktonic foraminifera can be used to reconstruct the former sea-surface temperature and salinity.

The study of climate variations that occurred over a few millennia and their global correlation is complicated by the fact that a sufficiently precise chronology is not usually available. For ages >15 ka, uncertainties in the ^{14}C dating result in uncertainties of the order of 100 years. The calibration of the data is complicated by the fact that no linear relationship between ^{14}C age and the calendar age exists. Marine carbon is not exchanged with the atmosphere fast enough to ensure that the surface water in each case reflects the current state of the atmosphere. The dating of marine sediments therefore gives ages that are too old. The difference between this apparent age and the actual age of a sample is referred to as reservoir age, which varies between regions from <100 to >1000 years. As the exchange in the past has not been constant, but was subject to potential changes in ocean currents causing local fluctuations, the respective values can only be determined with the help of a 3D model (Franke et al. 2008). The ice cores from Greenland are very similar to each other, and show the same characteristic sequence of Dansgaard–Oeschger cycles. Each Dansgaard–Oeschger cycle represents the connection of the temperature fluctuations in Greenland with corresponding changes in atmospheric circulation in the vicinity of the northern ice sheets. The sawtooth-shaped variations that are found in ice cores from Greenland are also observed in the sediment samples from the Atlantic Ocean, obtained from about 45° N. This seems to indicate that the Dansgaard–Oeschger cycles and Heinrich events are coupled with latitudinal variations of the polar front and the associated wind-driven circulation.

Cores from the North Atlantic show that a large part of the Greenland Sea and Norwegian Sea were free of ice during the warm interstadials of the last cold stage (Weichselian Glaciation), at least during the summers. On the other hand, the sea ice probably spread much further than today during stadials.

Further south in the North Atlantic between 55° and 40° N, the composition of deep-sea sediments is dominated by drift-ice events (Heinrich events). This zone, called the Ruddiman belt, is the area where the icebergs are melted on their eastwards drift along the polar front. The broad sediment bands of the Heinrich events consist of debris transported by icebergs (ice-rafted detritus or IRD; see Section 2.3 and Box 2.3). The Canadian Shield could be determined as the main area of origin of the IRD, which was deposited in the Heinrich events. The material is of age 3000–4000 Ma and has a significant proportion of metamorphic carbonates and iron minerals.

The grain size of the deposits of individual Heinrich events varies from clay particles to grains of sand and gravel. Some of the Heinrich events, such as Heinrich Event 6 and Heinrich Event 3, have a different composition. They contain more clastic material from the Arctic and western Europe due to the fact that in those periods the ice shelf there extended beyond the continental margin.

In the North Atlantic, each Heinrich event was connected to a distinct cooling of the surface waters. Evidence of these cold fluctuations is also observed in the European vegetation records in pollen found in long lake sequences (Fletcher et al. 2010). Heinrich events are also recorded in the isotope ratios of stalagmites from loess and caves, even as far away as SE Asia (Porter & Zhisheng 1995; Wang et al. 2001). Stalactites have the advantage that they can be dated by the uranium–thorium method with an error of only about 500 years up to an age of 50 ka.

Temperature fluctuations of some 1000 years duration have been found in sediment cores from the southern oceans between latitudes 40° and 55° S. The fluctuations appear to be caused by an increase in IRD from the edge of the Antarctic Ice Sheet. The currently available dates suggest that these IRD events in the southern oceans did not occur simultaneously with the Heinrich events of the North Atlantic. However, it seems that the Dansgaard–Oeschger cycles and Heinrich events have some southern counterparts. For each Heinrich event, at least one corresponding A-event can be detected in Antarctica and surrounding areas. In contrast to the Dansgaard–Oeschger cycles, those A-events are characterized by a gradual relative heating (about 10°C over several millennia) of the Antarctic, followed by a slightly longer cooling of similar magnitude.

14.3 Modelling the Last Ice Sheets

There is little accurate information about the course of the glaciations. The best-known histories of deglaciation since the last glacial maximum are from North America and NW Europe (Boulton et al. 2001; Dyke 2004). The findings of the earlier stages of development are extremely sketchy because their traces are in most cases overprinted, or have been completely removed by younger glaciations. Where deposits are still present, even the age determination is complicated by the fact that it lies beyond the limits of radiocarbon dating. The thickness of the former ice sheets is largely unknown. While morphological indicators (limit of glacial polish) allow the reconstruction of the maximum ice thickness in the case of mountain glaciations, they are

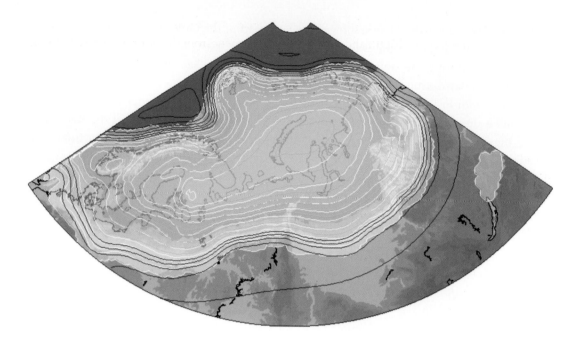

Figure 14.3

Palaeogeographical reconstruction of the Late Saalian glaciation maximum at 140 ka.

Source: Lambeck et al. (2006). Reproduced with permission of John Wiley & Sons.

missing in the area of extensive lowland glaciations such as the Laurentide Ice Sheet in North America or the glaciation of Russia, the Barents Sea and Kara Sea. With regard to isostatic movements, only the postglacial rise of Scandinavia and parts of North America are roughly known, however, such information is unavailable for older periods and also for areas outside the actual ice sheet (forebulge).

For such model calculations, only some boundary conditions are known. The maximum extent of the ice sheets during the Saalian Glaciation (Fig. 14.3) of northern Europe is relatively well mapped (Ehlers & Gibbard 2004a; Kjær et al. 2006a, b). However, reliable data on past sea levels are much harder to obtain. Due to the viscosity of the Earth, the sea level at any point is not only a function of the global ice volume but is also determined by the current and previous ice load (isostasy). The former ice load has an effect because the compensatory movements are much slower than changes in sea level. Subsequent isostatic rebound determines whether deposits are found above or below the present sea level. In sea-level modelling, the glacial conditions that precede the interval under consideration must therefore be taken into account (Dawson & Smith 1983).

To model the evolution of the northwest European glaciated area, the fluctuations in the more distant large ice sheets in North America, Greenland and the Antarctic cannot be neglected. Former shorelines that can be incorporated into the model refer primarily to high sea levels. The development of the Eemian coastlines, the timing and degree of connection between the Baltic and White Sea and Early–Middle Weichselian shorelines in the areas of the North Sea and the Taimyr Peninsula can only be calculated.

There are a number of models in which the glaciation of northern Europe during the last cold stage has been reconstructed. They include works by Boulton et al. (2001) and Arnold et al. (2002). Most extensively, however, Kurt Lambeck and his colleagues have dealt with the reconstruction of the northern European ice sheets. After first having established a model of the best-documented period from the glacial maximum up to the present (Lambeck et al. 1998), they also modelled the period from the Late Saalian glaciation to the onset of the Weichselian Glaciation (Lambeck et al. 2006), and finally the development from the Early Weichselian to the glacial maximum (Lambeck et al. 2010). The results are summarized in the following sections.

14.3.1 SAALIAN GLACIATION–EARLY WEICHSELIAN

At the time of greatest ice extent (*c.* 140 ka), the ice of the Late Saalian glaciation reached its greatest thickness. The model shows that there were *c.* 4500 m of ice over the Kara Sea and *c.* 4000 m above the Gulf of Bothnia. The maximum isostatic depression of the crust in the Late Saalian glaciation peaked at *c.* 1100 m in the Kara Sea and 1000 m in Finland, so that the highest part of the ice sheet was *c.* 3500 m above the current (or former) sea level. The ice thickness and the resulting isostatic depression are larger than the eustatic sea-level changes by an order of magnitude.

Much of the subglacial topography from the North Sea to the Taimyr Peninsula was below sea level at the end of the Saalian Glaciation (Fig. 14.4). As a consequence, the melting of the ice resulted in a rapid transgression. Also, large areas of water formed on the southern edge of the European part of the ice sheet. These lakes were initially separated from the North Sea and the Atlantic Ocean (at *c.* 136 ka). A connection to the sea was later established, first in Denmark and, upon further retreat of the ice margin, via southern Sweden. In western Siberia, the outflow of rivers flowing north was blocked by the ice. An ice-dammed lake formed between the Urals and the Putorana Massif, into which the rivers Ob and Yenisei drained. A potential overflow existed via the rivers Irtysh-Toboj and Turgay into Kazakhstan and the Aral Sea. Once the ice had melted far enough, drainage shifted to the west, using the lower passes through the northern Urals such as the Sob Pass. This was aided by the still significant depression of the crust after the retreat of Saalian ice. By 135 ka the Turgay Pass was probably *c.* 15 m higher than the Sob Pass in the northern Urals.

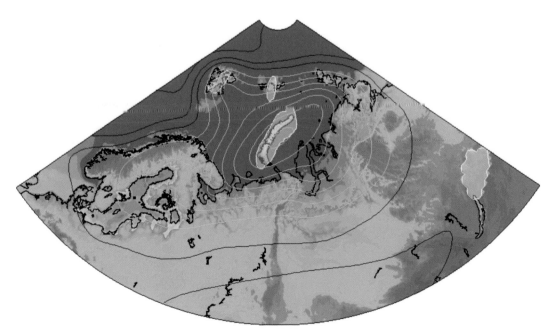

Figure 14.4

Palaeogeographical reconstruction for the period of the Late Saalian transgression at c. 129 ka.

Source: Lambeck et al. (2006). Reproduced with permission of John Wiley & Sons.

The modelling of the Early Eemian Interglacial period can be based only on a few concrete observations. Eemian lakes and shorelines are largely unknown in Russia. This is partly because many of the older records were destroyed by the ice of the Weichselian Glaciation and by periglacial overprint. Mangerud et al. (2001) however found beach sands and gravels at the Upper Pechora that were deposited prior to the Eemian Interglacial; they were overlain by Weichselian lake sediments and yielded an OSL age of 141±15 ka. Those beach pebbles are now at an altitude of *c.* 72 m asl and correspond to the calculated model sea level for 136–134 ka.

On further ice retreat, the entire northern Siberian lowlands were affected shortly after 135 ka by a transgression that reached up into the Khatanga Valley. This extensive flooding of the Arctic lowlands persisted over the entire warm stage. The lowlands of the Ob and Yenisei were still flooded at the end of the Eemian as the isostatic land uplift was still incomplete. At the onset of Saalian ice sheet melting, the western part of the former ice sheet was affected by a transgression from the North Sea and western Baltic Sea and much of low-lying northern Denmark was under water. There was a connection from the Atlantic to North Jutland and the Baltic Sea through the Danish Belts. According to the model calculation, there was initially no connection between the German Bight via Schleswig-Holstein to the Kiel Bay. It must however be borne in mind that the present barrier is just 20 m high and that possible post-Eemian changes in topography by subsequent Weichselian ice advances could not be considered in the calculations.

The marine connection through southern Sweden was significant and lasted until after 134 ka. The sea then entered from the Atlantic to northern Europe once the ice had disappeared from those areas, while further north in the Barents Sea and Kara Sea there were still significant masses of ice.

Until 135 ka the Scandinavian ice withdrew from northern Finland, Sweden and Norway, but the resulting uplift was not sufficient to prevent large-scale flooding of the lowlands. The transgressions into the Baltic Sea occurred from both the Atlantic and the Arctic Ocean, with the transgression from the north being suppressed as long as an ice tongue extended over the Murmansk region and the Kola Peninsula. This ice barrier had no effect on the extent of the flooding, but it directed the drainage of the Kara Ice Sheet into the Baltic Sea. The deposition of sediments with a cold freshwater fauna, found at some sites in Ostrobothnia underlying boreal Eemian deposits, occurred around the same time as when marine or brackish-water deposits were formed further to the west. This suggests that during the earliest part of the interglacial the White Sea had no connection to the Arctic Ocean.

After 134 ka the remaining ice was limited to the present Kara Sea and the Arctic islands, and there was a connection from the Atlantic Ocean via the Baltic Sea to the Barents Sea and via the northern Taimyr Peninsula to the Laptev Sea. However, the Strait of Karelia was narrowing quickly and, by 132 ka, was already strongly limited by the isostatic land uplift. The formation of the Karelian watershed occurred soon after 129 ka. At that time the shape of the Baltic Sea already resembled its present appearance, but parts of the lowlands around the Gulf of Finland and on the southern edge of the Baltic Sea were still flooding. The most comprehensive reconstruction of the Eemian-age connection from the Atlantic through the Baltic to the White Sea was undertaken by Funder et al. (2002). Lambeck's model, which predicts a narrow marine connection between the Baltic Sea and White Sea for the time of pollen zone 2b–4a, is consistent with their their conclusions.

The model predicts: (1) a relatively long time had passed between the ice retreat and the beginning of the Eemian Interglacial; and (2) the lowlands of Scandinavia and Russia during the earliest phase of the Eemian Interglacial were subjected to a significantly larger marine transgression than during the climatic optimum. Consequently, considerably higher sea levels must be adopted for this region than for the subsequent warm stage. It must be assumed that the short-term and narrow link between the Baltic Sea and the White Sea in the Early Eemian was preceded by a much longer and extensive marine transgression of cold water, which began when the ice retreated from the Karelian watershed. There is little field evidence for this transgression, however. The sea-level data for this early part of the Eemian Interglacial in Europe are limited to a few early interglacial marine terraces on Spitsbergen. Furthermore, we only know that the warm phase of the Eemian Interglacial in Ostrobothnia (Finland) was preceded by a cold-water phase and that during the Late Saalian Glaciation brackish cold-water conditions prevailed in the southernmost parts of the Baltic. Lambeck et al. (2006) suggest that there was insufficient time for a clearly identifiable early interglacial fauna and flora to develop.

The pre-Eemian connection between the Atlantic and the Arctic Ocean via the Baltic extended initially to the Urals and then continued to the Taimyr Peninsula. The connection lasted for some 1000 years. The penetration of relatively warm Atlantic water in those northern areas in the early warm stage facilitated the rapid spread of boreal vegetation and interglacial marine fauna in the region. A temperature rise was recorded everywhere in northern Europe at that time. In contrast to the Eemian, at the end of the last glaciation there was no connection between the Baltic and the Arctic Ocean and there was no widespread transgression in northern Russia. During the Eemian Interglacial forest grew in northern Russia in areas that are now occupied by tundra. The Nordic Eemian Sea is referred to in the Russian literature as the 'boreal transgression'.

If the sea level during the Eemian Interglacial was c. 5 m higher than today, then correspondingly more water must have been available. This additional water could only have been derived from a stronger ablation of the remaining large ice sheets. The East Antarctic Ice Sheet is regarded as relatively stable, but the West Antarctic Ice Sheet may have undergone change. Changes in the Greenland Ice Sheet may have been even larger; while most areas in Greenland did not become free of ice during the Eemian, the ice thickness may have declined sharply. Tarasov & Peltier (2003) modelled the behaviour of the Greenland Ice Sheet and concluded that the melting of the Greenland ice probably contributed to about half of the sea-level rise at that time (c. 2.7–4.5 m).

14.3.2 EARLY–MIDDLE WEICHSELIAN

The model suggests that the Barents–Kara Sea Ice Sheet reached the northern Urals during MIS 5d and that there was a connection to the ice sheet on the Putorana Massif so that, by 113 ka, a large west Siberian Ice Lake developed between these two glaciation centres. The lake consisted of a northern and a southern basin, separated by an east–west-trending topographic high area only dissected by the Ob and the Taz. The northern basin is similar to that which had already existed during the Late Saalian. On the other hand, the second lake extended much further south into the region of the present-day cities of Omsk and Tomsk. It had an overflow through the Turgay Depression into the basin of the Aral Sea, from which the water flowed into the Caspian Sea and through the Manych Depression into the Black Sea (Fig. 14.5).

The difference between the two glaciations probably resulted from the fact that the ice was much thicker at that time, and that during the Late Saalian the northern Urals had been isostatically lowered by a greater amount than during MIS 5d. The watershed between Sob

Figure 14.5 Discharge of the west Siberian ice-dammed lake through the Turgay Depression towards the Aral, Caspian and Black seas.

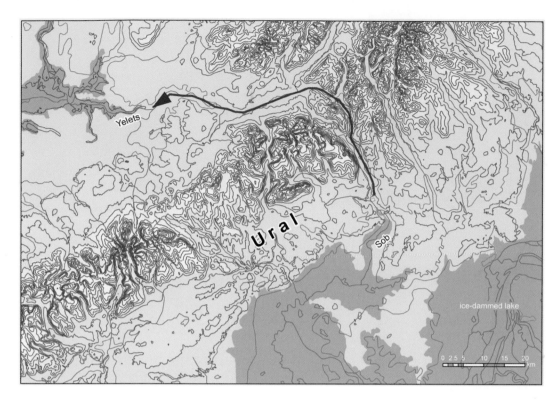

Figure 14.6 Discharge of the west Siberian ice-dammed lake over the northern Urals towards the west.

and Yelets rivers in the northern Urals is currently at +154 m, located west of the Urals on a moraine ridge which was accumulated by a local glacier (Fig. 14.6). At MIS 5d the sea level was *c.* 31 m below the present level; in order to drain westwards at that time, the water at the Sob pass would have had to stand at *c.* 185 m above sea level. The level of the Turgay Depression was a few metres lower. Consequently, the natural drainage went through the Turgay Depression. Only the current ground level was used in the model calculation and not the height of the rock threshold which, in the case of the Turgay Depression, is *c.* 55 m asl (*c.* 500 km north of the present watershed; Mangerud et al. 2004). If the height of the solid rock controlled the lake level, then it definitely had to drain through the Turgay Valley.

This pattern was repeated in the cold phase of the MIS 5b Substage (Fig. 14.7). At *c.* 113 and 95 ka the west Siberian lowland was filled by two vast ice-dammed lakes which drained (at least temporarily) to the south towards the Aral, Caspian and Black seas. This means that during MIS 5d and 5b the catchment area of the Black Sea and therefore also of the Mediterranean Sea had expanded to the east into the Lake Baikal region (Fig. 14.8).

The ice in the Kara Sea, which blocked drainage to the north, was quite similar during both substages, and the residual isostatic influence of the Saalian Glaciation was probably already very low at that time. It is therefore likely that the ice-dammed lakes in both glaciations had a similar extent. This implies that the older river terraces were overprinted by the younger ice lake so there are no OSL data for MIS 5d, while data for MIS 5b and MIS 4 are commonly found. Mangerud et al. (2004) estimate that during MIS 5b the lake level in the Upper Ob valley would have been *c.* 60 m depth. That would have been *c.* 122 m above sea level, meaning that a runoff was easily possible through the Turgay Depression over the bedrock threshold.

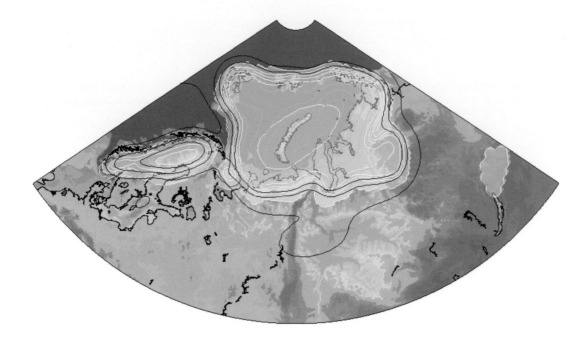

Figure 14.7

Palaeogeographical reconstruction for c. 113 ka (MIS 5b).

Source: Lambeck et al. (2006). Reproduced with permission of John Wiley & Sons.

During interstadials MIS 5c and MIS 5a the ice retreated largely to the area of the Kara Sea. The topography of west Siberia looked much like today, with the exception that the northern parts of Yamal and Gydan were still below sea level. During MIS 4, the southern margin of the Middle Weichselian Glaciation stayed in the Kara Sea north of the Urals and the Putorana Uplands, and the extent of the Siberian lakes was controlled by the location of the

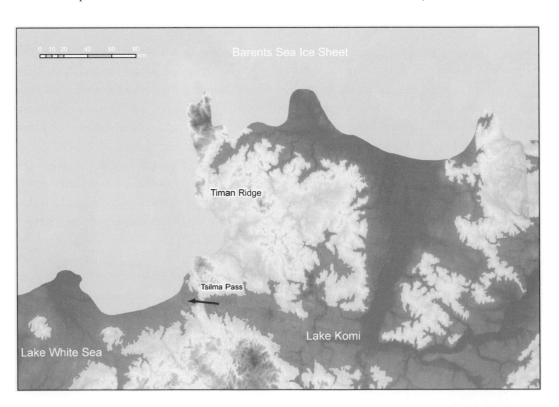

Figure 14.8 Drainage of Lake Komi to the west.

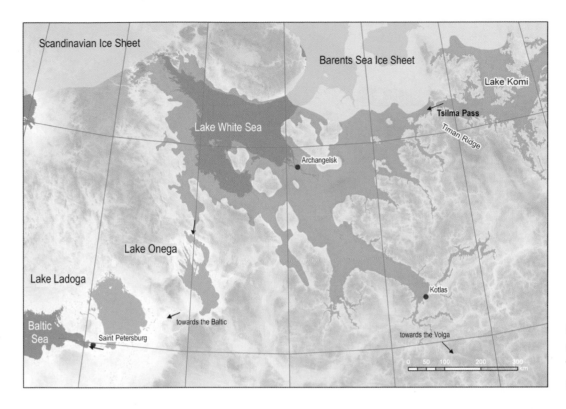

Figure 14.9 Drainage of Lake Komi across the Timan Ridge (via Tsilma Pass) to the west.

ice margin on the Taimyr Peninsula and in the Pechora Lowland. For this period, the model suggests extensive inundation of the Ob, Yenisey and Taz lowlands; the maximum lake level was not reached during the earlier stages however, because deeper ice-free overflow passes were available at the northern edge of the ice sheet, allowing drainage to the west.

West of the Urals, the formation of ice-dammed lakes depended mainly on the position of the ice margin in the area of the Pechora and Mezen-Arkhangelsk basins, which is not known. The model therefore cannot provide a definitive answer to the question of the former drainage between the Timan Ridge and the White Sea. For a detailed analysis, large-scale terrain data would be required. The predictions of the model provide similar results for the two glaciation phases of MIS 5. An ice lake between the Urals and the Timan Ridge is predicted in both cases, the level of which was controlled by the Tsilma Pass (the lowest pass in the Timan Ridge; Fig. 14.9). The resultant lakes are equivalent to 'Lake Komi' of Mangerud et al. (2001). Another ice-dammed lake is predicted for both periods for the White Sea and Archangelsk area, dammed by an ice barrier which extended from the Pechora Sea to the Kola peninsula (MIS 5b) or by a thick ice mass which extended to Scandinavia (MIS 5d). This lake corresponds to the 'White Sea Lake' of Mangerud et al. (2001). The outflow of this lake was either via Karelia or south through the valley of the Dvina and via the Keltma Pass towards the Volga. In the case of smaller ice sheets on the Kola Peninsula, drainage via the Kola River would have been possible.

The presence of these ice-dammed lakes depends heavily on whether the drainage routes were open. Larsen et al. (2006) are of the opinion that there was an ice-free corridor between the ice of the Barents Sea and the Scandinavian Ice Sheet. If this was the case, a southern drainage of the lakes was impossible. The question of whether a White Sea Lake existed or not cannot be answered by the model.

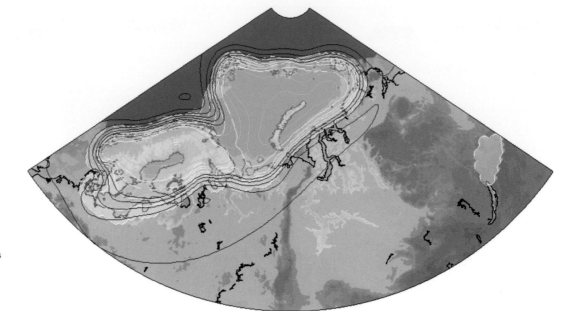

Figure 14.10

Palaeogeographical reconstruction for *c.* 64 ka (MIS 4).

Source: http://people.rses .anu.edu.au/lambeck_k/ index.php?p=research. Reproduced with permission.

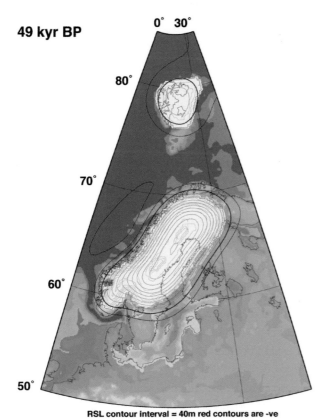

49 kyr BP

RSL contour interval = 40m red contours are -ve

Figure 14.11 Palaeogeographical reconstruction for *c.* 49 ka (MIS 3).

Source: http://people.rses.anu.edu.au/lambeck_k/index .php?p=research. Reproduced with permission.

The isostatic response to a glaciation depends on the rheology of the Earth and the geometry of the ice load. The position of the ice margins is essential to calculate the ice load, at least for some periods of time. These may be adapted from Svendsen et al. (2004), and individual contributions from Ehlers & Gibbard (2011a). The changes in sea level are included in the calculations. It has to be taken into account that the sea level after the Weichselian maximum still reflected inherited effects of the glaciations prior to the Weichselian maximum. The glacial history of the other large ice sheets (especially in North America) is certainly part of the equation for sea level.

The calculations of Lambeck et al. (2010) showed that the maximum ice thickness in the Gulf of Bothnia area was significantly lower than that assumed by Denton & Hughes (1981). Consequently, the ice centre was not above the Baltic Sea but over Sweden, and further south than previously thought at Ångermanland (rather than Norrbotten).

Very few data are available for the extent of the glaciers during the Early Weichselian. By 64 ka the ice in the eastern part of the glaciation area had almost reached its maximum extent, which led to the formation of large ice-dammed lakes (Fig. 14.10). In the southwest the ice filled the entire Baltic Sea basin. In Denmark, the Ristinge Klint till was deposited.

By 49 ka, the ice on Svalbard and Scandinavia was still limited (Fig. 14.11). For this period the model shows an expansion of the precursor of the Baltic Sea, which extended

41 kyr BP

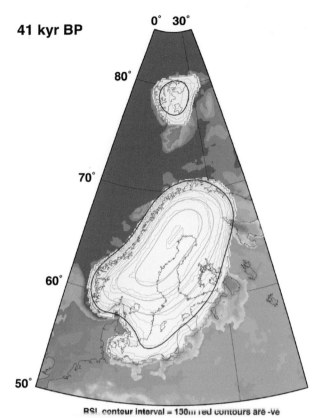

RSL contour interval = 150m red contours are -ve

Figure 14.12 Palaeogeographical situation for *c.* 41 ka (Jæren–Skoghelleren Stadial).

Source: http://people.rses.anu.edu.au/lambeck_k/index .php?p=research. Reproduced with permission.

35 kyr BP

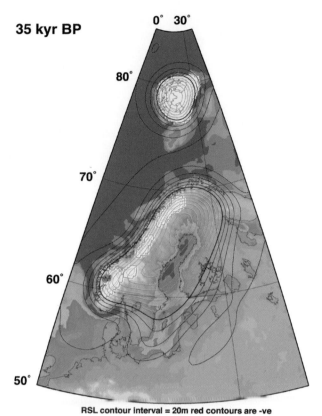

RSL contour interval = 20m red contours are -ve

Figure 14.13 Palaeogeographical reconstruction for *c.* 35 ka (Ålesund Interstadial).

Source: http://people.rses.anu.edu.au/lambeck_k/index .php?p=research. Reproduced with permission.

southwards across the area of today's Baltic Sea. The North Sea was almost completely dry. The outflow of the Baltic Sea occurred via an ice-dammed lake in central Sweden.

The first major ice advance occurred by 41 ka (Fig. 14.12). This advance during the old Baltic Jæren–Skoghelleren Stadial followed the Baltic depression and reached the area of the Danish islands. A large number of extended ice-dammed lakes formed on the southern edge of the ice sheet, giving the impression of a Baltic Sea outside the modern Baltic Sea Basin.

In the following Ålesund Interstadial (by 35 ka), the glaciers were confined to the mountainous regions of Scandinavia (Fig. 14.13). The new glacier advance began at *c.* 30 ka. At 28.5 ka it resulted in an ice advance towards the west across the North Sea, where Scandinavian and British ice met.

The model shows that by 28 ka a large ice lake formed in the southern Baltic, which probably drained via Schleswig-Holstein into the North Sea (Fig. 14.14). This was followed by the decay of the westwards ice advance at *c.* 27 ka and another great ice advance across the Baltic Sea into northern Germany and Poland at *c.* 25 ka (Fig. 14.15). With further growth of the ice sheet, the ice-dammed lake moved to the southeast of the present-day Baltic Sea. The drainage was now directed via the Elbe Urstrom, which first originated in large lakes in its upper reaches.

By 23 ka there was an ice retreat. The area of the Danish islands became icefree, and glaciolacustrine silts were deposited in the ice-dammed lake. Such deposits underlie the Langeland Till of

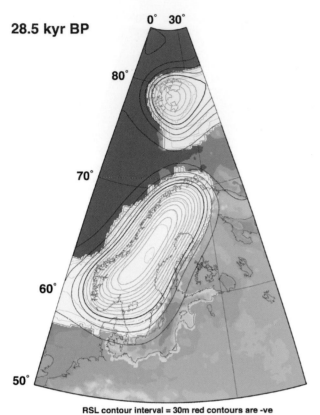

28.5 kyr BP

RSL contour interval = 30m red contours are -ve

Figure 14.14 Palaeogeographical reconstruction for *c.* 28.5 ka.

Source: http://people.rses.anu.edu.au/lambeck_k/index
.php?p=research. Reproduced with permission.

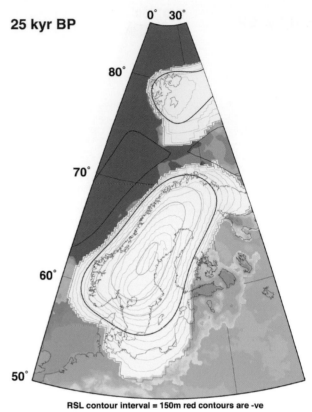

25 kyr BP

RSL contour interval = 150m red contours are -ve

Figure 14.15 Palaeogeographical reconstruction for *c.* 25 ka
(beginning of the LGM).

Source: http://people.rses.anu.edu.au/lambeck_k/index
.php?p=research. Reproduced with permission.

the last ice advance, observed in the cliff sections on Langeland. With the renewed ice advance at *c.* 21 ka, the drainage was directed through the glacial valley (*urstromtal*). The ice advance culminated by *c.* 20 ka during the Weichselian Glaciation maximum (Fig. 14.16).

14.4 Modelling Glaciers and Climate

Glaciers have a close relationship with climate, and geomorphological and geological evidence provides the basis for palaeoclimatic reconstructions. Large ice sheets such as those described in the previous section are not best for reconstructing climate because their sheer size means that these ice masses can be out of phase with prevailing climates by thousands of years. Smaller valley and cirque glaciers respond more rapidly to climate changes, and these are best suited to climate reconstructions. At the other extreme, tiny niche glaciers are unsuitable because these can be decoupled from the regional climate and are often influenced by strong local topoclimatic controls such as shading, avalanching and windblown snow.

The limits of former valley glaciers are often well constrained by geomorphological evidence; such evidence can be used to build the former ice surfaces by superimposing ice-surface contours on the former glaciers. The shape of ice contours should follow that observed on modern glaciers, where ice-surface contours are concave towards the top of the glacier and convex

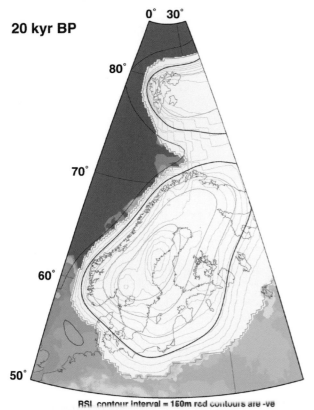

Figure 14.16 Palaeogeographical reconstruction for *c.* 20 ka.

Source: http://people.rses.anu.edu.au/lambeck_k/index.
php?p=research. Reproduced with permission.

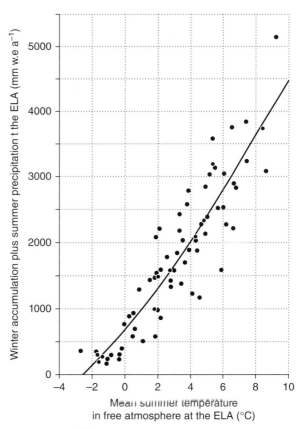

Figure 14.17 Relationship between winter balance + summer precipitation and summer temperature (June–August) observed on 70 glaciers around the world.

Source: Hughes (2011). Reproduced with permission of Springer Science and Business Media.

towards the snout. Once the ice surface is reconstructed, the equilibrium line altitude (ELA) can be reconstructed. The ELA of a glacier is the altitude at which accumulation and ablation are equal, and knowledge of its location is crucial for palaeoclimate reconstruction. It is also a critical concept in the understanding of glacier dynamics, and there is a very close relationship between the ELA and local climate (Ohmura et al. 1992; Braithwaite 2008). Several methods have been applied to calculate the ELA of former glaciers and the choice of method depends on the former glacier morphometry, former climate regime and the nature of the geomorphological evidence (Benn & Lehmkuhl 2000; Pellitero et al. 2015). Once the ELA is known, various modelling approaches can be applied to reconstruct climate at this ELA. Hughes (2011) summarized three different types of approach, briefly outlined in the following section.

14.4.1 EMPIRICAL, CURVED OR LINEAR RELATIONSHIPS

The relationship between glaciers and climate is fundamental in understanding glacier behaviour. There are well-established relationships between annual precipitation (or accumulation) and summer mean temperature at the equilibrium line altitude (ELA) on glaciers around the world (Ohmura et al. 1992; Fig. 14.17). Various forms of this relationship have been applied to reconstructions of Pleistocene glaciers, and the relationship provides one of the most useful

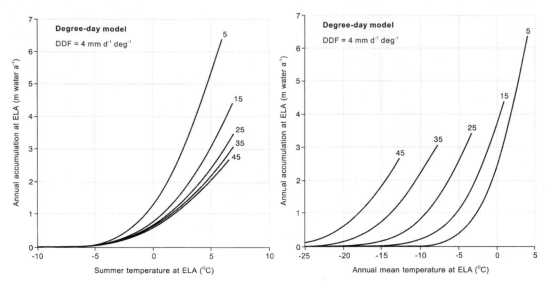

Figure 14.18 Annual accumulation at the equilibrium line altitude (ELA) as a function of summer mean temperature (June–August) and mean annual temperature (January–December). A degree-day model was used to calculate the annual accumulation that is required to balance annual melt under annual temperature ranges of 5, 15, 25, 35 and 45°C.

Source: Hughes & Braithwaite (2008). Reproduced with permission of Elsevier

approaches for palaeoclimate reconstruction using glacial evidence. However, the utility of such a relationship is constrained by the need to isolate one of these variables in order to derive the other. For example, summer temperatures may be derived from palaeoecological evidence from lake sediments near a glacier site. Chironimids (non-biting midges) are especially useful in this regard (e.g. Heiri et al. 2014).

Empirical relationships such as that of Ohmura et al. (1992) are widely used for palaeoglacier–climate reconstruction, but they have their limitations. Braithwaite (2008) used a degree-day model (described in the following section) to demonstrate that the data underlying the single empirical curve of Ohmura et al. (1992), relating summer temperature and winter balance + summer precipitation, is in fact better described by multiple curves depending on the region. These different regional relationships are effectively determined by different annual temperature ranges (Fig. 14.18).

14.4.2 DEGREE-DAY MODELLING

A degree-day model can be used to calculate the amount of accumulation required to sustain the glaciers. The degree-day model is based upon the notion that glacier melting occurs when air temperatures 1–2 m above the glacier surface are above the melting point (0°C). The total melt over a period at some point is therefore proportional to the sum of positive temperatures at the same point, that is, the positive degree-day total. The annual accumulation required at the equilibrium line altitude to balance melting equals the sum of daily snow melt, using a degree-day factor. Braithwaite (2008) found that degree-day factors for snow on 66 glaciers worldwide had averages of 3.5 ± 1.4 and 4.6 ± 1.4 mm day^{-1} K^{-1} in low- and high-accumulation conditions, respectively, with an overall mean of 4.1 ± 1.5 mm day^{-1} K^{-1}.

The degree-day model is widely used for modelling the mass balance of the Greenland Ice Sheet, but has also been applied to mountain glaciers and ice caps (Braithwaite et al. 2008). Degree-day modelling has also been applied to reconstruct past climates associated with former glaciers in the USA (Brugger 2006), Greece (Hughes & Braithwaite 2008) and Montenegro (Hughes et al. 2010).

14.4.3 ENERGY-BALANCE MODELLING

This approach tends to be more complex than the previous two approaches (Sections 14.4.1 and 14.4.2) and involves computation of different energy fluxes at the reconstructed glacier surface. Plummer & Phillips (2003) used a simple energy balance model to calculate the effects of topography on shortwave radiation, the largest component of the surface energy balance. Their model calculates the distribution of snow accumulation using a surface mass and energy balance approach and calculates the resultant glacier shapes with a 2D flow model. Energy balance modelling has been used effectively to reconstruct former glaciers and climates in New Zealand by Rowan et al. (2014). However, this approach is less widely used than simpler relationships between air temperature and mass balance, such as the Ohmura et al. (1992) regression. This is probably because of problems of complexity and uncertainties involved when reconstructing the former energy balance on palaeoglaciers.

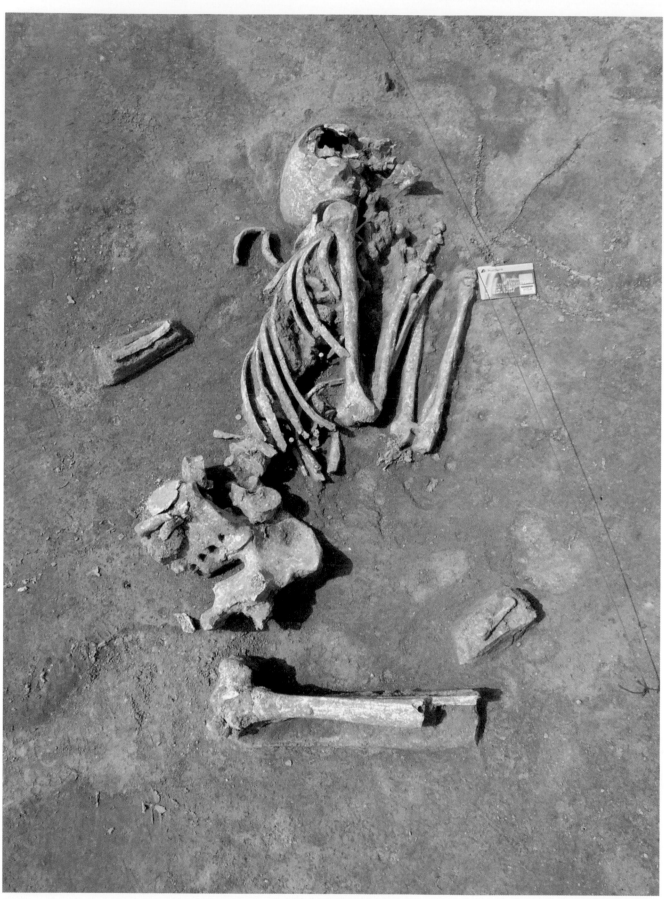

Skeleton found in an archaeological excavation at Karsdorfer Flur, north of the Unstrut River, Sachsen-Anhalt. Photograph by Jürgen Ehlers.

Chapter 15
Human Interference

Today humans are in a position to change not just their environment but also the development of humankind, for the better or worse. When asked by a reporter from the *New York Times* if it was possible to reconstruct the Neanderthal man after the latest successes of genetic research, Dr George Church said yes, that was possible.

He said he would start with the human genome, which is highly similar to that of Neanderthals, and change the few DNA units required to convert it into the Neanderthal version.

This could be done, he said, by splitting the human genome into 30,000 chunks about 100,000 DNA units in length. Each chunk would be inserted into bacteria and converted to the Neanderthal equivalent by changing the few DNA units in which the two species differ. The changed lengths of DNA would then be reassembled into a full Neanderthal genome. To avoid ethical problems, this genome would be inserted not into a human cell but into a chimpanzee cell.

The chimp cell would be reprogrammed to embryonic state and used to generate, in a chimpanzee's womb, a mutant chimp embryo that was a Neanderthal in many or most of its features.

Dr. Church acknowledged that ethical views on such an experiment would vary widely. But bringing a Neanderthal to birth, he said, would satisfy the human desire to communicate with other intelligences.

Dr. Church said he had no plans for such an experiment, but if someone were eager to supply the financing, 'We might go along with it.'

Wade (2009)

The Ice Age, First Edition. Jürgen Ehlers, Philip D. Hughes and Philip L. Gibbard.
© 2016 John Wiley & Sons, Ltd. Published 2016 by John Wiley & Sons, Ltd.
Companion website: www.wiley.com/go/ehlers/iceage

For many years we have been just a small step away from the brave new world of genetic interventions in human development. It is not so much moral concerns which have halted progress, but rather the lack of direct economic benefit.

15.1 Out of Africa: Humans Spread Out

The first humans originated in Africa. New research shows that *Homo erectus*, the first hominid who used fire and was able to hunt and run like a modern human, spread in a first wave of emigration from Africa across the world *c.* 1.8 million years ago (Ma). These early humans were only replaced by the more advanced *Homo sapiens* during the last ice age. Until recently it was unclear whether this change occurred simultaneously in different parts of the world, or whether a new wave spread from Africa to replace the old population. It is now widely believed that the latter was the case.

Until a few decades ago, the only way to find out more about the evolution of humans was the study of archaeological finds. Because radiometric dating methods often fail in the period in question, apart from biological evolution (change in bone structure, especially the skull), the only other option is analysis of artefacts. Today, research into the genetic basis of human development is key to the reconstruction of early human development.

Africa is a relatively closed continent. Only a few routes to other land masses were available for the spread of *Homo sapiens* (Fig. 15.1): (1) the Strait of Gibraltar; (2) Sinai; or (3) the Strait at the Horn of Africa (Bab el Mandeb). Two of these possible exits were actually used. By 120 ka ago a first group of people advanced northwards through Egypt and Palestine. However, adverse climatic conditions prevented any further advance. This branch of human propagation died out by 90 ka. On the basis of archaeological findings, it is now clear that the first successful migrants from Africa used the southeast exit route. As the strait never fell dry, even in periods of low sea level, boats must have been available as a means of transport from a very early date.

By 85 ka, humans began a second wave of emigration via the southern Arabian peninsula and through Persia into India (Fig. 15.1). This group formed the nucleus from which all non-African modern humans developed. First, they spread to the east. China and Southeast Asia were reached as early as 75 ka. The road to the southeast to Australia and New Guinea (which at that time formed a unit) was taken by 65 ka.

Humans achieved something that is still denied to many other species: crossing the natural border between Asia and Australia. In his studies in the Malay Archipelago, the British naturalist A. R. Wallace (1823–1913) noted the clear dividing line between the fauna of Asia and Australia. This imaginary line, which is named after Wallace, runs between Bali and Lombok and between Borneo and Sulawesi. The reason for this becomes clear when considering the depth of the sea floor: Sumatra, Java and Bali are located on the Malay shelf, a shallow sea area which fell dry during cold periods. The same goes for Australia, New Guinea and Melanesia, which also share a common continental shelf which in the cold stages became part of the mainland. These two continents remained separated along the Wallace Line by a 50-km-wide strait. It is unclear whether *Homo sapiens* used the southern route via Bali–Sumba–Flores–Timor–Celebes or the northern route, Maluku–Ceram–New Guinea. Australia was reached no later than 50 ka before present.

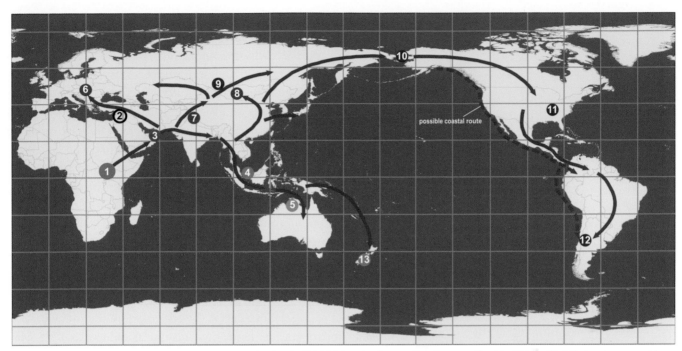

Figure 15.1 The spread of modern humans across the Earth. (1) 150 ka: the first modern humans lived in Africa. (2) A first group moved to the north before 120 ka. They crossed Egypt and Israel, but died out before 90 ka. (3) 85 ka: another group of people migrated along the southern edge of the Arabian Peninsula to India. All non-African modern humans are descended from that group. (4) 75 ka: modern humans spread into SE Asia and China. (5) 65 ka: they reached Australia via Timor. (6) Prior to 50–46 ka, *Homo sapiens* came to Europe. (7) 40 ka: another group of people migrated from Pakistan via the Indus valley into Central Asia. (8) At the same time, people migrated from East Asia along the Silk Road to the west. (9) 20–30 ka: people from Central Asia went westwards into Europe and eastwards to Beringia. (10) 25–22 ka: they eventually crossed the Bering Strait. (11) 19–15 ka: they reached Pennsylvania. (12) 12.5 ka: they migrated either by land or by sea along the American west coast to Monte Verde in Chile. (13) It was not until 800–1300 AD that Maori reached New Zealand. Adapted from Oppenheimer (2009). Reproduced with permission of Elsevier.

It is uncertain whether the continent was originally inhabited by *Homo erectus* (Fig. 15.2). Although the Australian Aborigines exhibit certain physical characteristics that seem to suggest that there might have been mixing with *Homo erectus* (pronounced eyebrow, robust figure), this is not the case. Studies have shown that the genetic material of the Australians and Papuans fully complies with *Homo sapiens*. The strong genetic differences to most other nations within the Indian Ocean may have evolved due to the isolation of the continent after the disappearance of the land bridge at *c.* 8 ka (Hudjashov et al. 2007).

Homo sapiens reached Europe at *c.* 50–46 ka. Almost all of today's Europeans can be genetically traced to this origin, and their mitochondrial DNA originated 50–13 ka ago. Humans spread into Central Asia at *c.* 40 ka. From there, they advanced east into Beringia at *c.* 30 ka. During the low sea-level stand at *c.* 25–22 ka, people crossed the Bering Strait which was then dry and, via this land bridge, came to Alaska. Here they encountered a problem: their further advance to the south was blocked by the Wisconsinan Ice Sheet. South of the ice sheet area, human settlement in America has only been proven after *c.* 15 ka ago. New Zealand was the last major land mass to be inhabited by modern humans. The Polynesian Maori occupied the

Figure 15.2 Rock paintings at Nourlangie, Arnhem Land, Australia. Photograph by Eva-Maria Ludwig.

Figure 15.3 The lower jaw of *Homo erectus heidelbergensis* is about 600–500 ka in age. Photograph by Jürgen Ehlers.

islands *c.* 800–1300 AD. Separated from Australia by nearly 2000 km of water, these islands were uninhabited until then.

The date of the earliest colonization of Europe by humans has yet to be determined. The southern Caucasus was occupied *c.* 1.8 Ma, where five *Homo erectus* skulls were found in the Dmanisi cave (Lordkipanidze et al. 2013). Further to the west, the skull fragments from Ceprano, Italy (*c.* 800 ka) and human remains from the karstic caves of Atapuerca, Spain (older than 780 ka) show that early humans had dispersed to the Mediterranean hinterland before the Brunhes–Matuyama magnetic polarity reversal (Parfitt et al. 2005).

Recent discoveries from Pakefield and Happisburgh (Britain) have provided clear evidence for an unexpectedly early human occupation of NW Europe. The sites were found in the deposits of interglacial rivers and estuaries, located on the ancient North Sea coast. The artefacts span the older and younger parts of the 'Cromerian Complex' Stage. The older of these sites pre-date *c.* 0.5 Ma based on the presence of a *Mimomys* micromammal fauna, and may be as old as 1.0–0.78 Ma. On the European continent, stone artefacts of such an age are only known from the Mediterranean region (Parfitt et al. 2010; Cohen et al. 2012).

Finds of early man (*Homo erectus*) elsewhere in Europe are known from only seven sites: Petralona in Greece (dating controversial); Arago in France (*c.* 70 remains); Ranuccio, Pofi and Castel di Guido in Italy; Vértesszöllös in Hungary; Mauer near Heidelberg; and Bilzingsleben in Thuringia. The oldest human remain found in Germany is the lower jaw which was discovered in 1907 in the Grafenrain sand pit at Mauer (Fig. 15.3). The exact age of the layer in which this *Homo erectus heidelbergensis* was found is not clear, it can only be roughly estimated as 600–500 ka (Wagner et al. 2007). A tibia from a *Homo erectus heidelbergensis* was found together with numerous artefacts at the Palaeolithic site of Boxgrove in southern England (Pitts & Roberts 1998; Roberts & Parfitt 1999).

The subdivision of the archaeological discoveries is largely based on tools. The entire Quaternary Period until the end of the Weichselian Glaciation belongs to the Palaeolithic era, the 'Old Stone Age'. The division into Early, Middle and Late Palaeolithic is based on changes in lithic technology. The ages of the different industries is not clear, for the pre-Weichselian deposits in particular, while an assignment to a certain type of industry is always possible.

The most important traces of Palaeolithic occurrence in northern Germany were found during excavations in the Schöningen opencast brown coal mine near Helmstedt. In addition to other pieces of processed wood seven pine spears were found, the largest being 2.5 m long. They postdate the Reinsdorf Interglacial and are approximately 270 ka old (Urban 2007b). Although earlier occasional finds of worked wood of probably Early Palaeolithic age had been reported, the only remains still existing are one lance or spear tip from England (Clacton-on-Sea). Apart from the wooden tools from at the excavations in Schöningen, more than 20,000 mammal bones were recovered; 90% of these are wild horse, so it can be assumed that people were regularly hunting at that time (Thieme 1999).

Middle Palaeolithic hunting places have been found more frequently. These include the Lehringen site near Bremen, where an Eemian deposit containing a forest elephant was discovered in 1948, complete with a 2.5-m-long yew lance (Adam 1951; Thieme & Veil 1985). Nearly 40 years later in 1985, another Middle Palaeolithic hunting ground was found in Neumark-Nord in the Geisel valley (Mania 1990). Two years later in 1987, a slaughtering place of Eemian forest elephants was detected in Gröbern, Kreis Gräfenhainichen (Litt 1990). In contrast to the situation on the continent, there are no traces of humans in Britain that have been dated to the Eemian. The reason for this is at that time the channel was open, separating Britain from the rest of Europe.

Palaeolithic man was not yet in a position to alter the natural landscape. He maintained himself by hunting and gathering fruit; the population density was correspondingly low.

The Neanderthals appeared in Germany around the time of the Upper Palaeolithic. The heyday of the *Homo neanderthalensis* came sometime during the Weichselian Glaciation. The naive representation by Flammarion's nineteenth century book (1886–87; Fig. 15.4) gives the impression that early humans were helpless and at the mercy of wild animals, especially since they could not escape into the trees as the monkeys could; the reality was different, however. The cave bear (Fig. 15.5) became extinct about 30 ka ago and, even if the extinction of most large mammals was not a direct result of excessive hunting (the 'overkill hypothesis'), the end of *Ursus spelaeus* was perhaps accelerated by human action. The Neanderthals, however, not only lived in caves; in Germany at that time they occupied freeland settlements. For example, a settlement from the Late Saalian Glaciation has been found at Mönchengladbach Rheindalen in the open countryside. The settlements were probably moved seasonally to provide better opportunities for hunting and food gathering. The Neanderthals were also, as far as we know, the first people to bury their dead (Bosinski 1985).

Figure 15.4 Antediluvian man (Flammarion, late nineteenth century).

Figure 15.5 Cave bear skeleton from the Geological Museum, Hamburg. Photograph by Jürgen Ehlers.

15.2 Neanderthals and *Homo sapiens*

In his novel *The Inheritors* William Golding (1955) describes a clash between a group of Neanderthals and modern humans. The naive Neanderthals have no chance. The new branch, which suddenly appears next to Lok's head, is really an arrow fired at him by the 'new companions'. The new arrivals have weapons and alcohol. They also fear the primitive Neanderthals. Before poor Lok understands what is happening around him, his family is exterminated. Did modern humans really kill off the Neanderthals? Nobody knows. The Neanderthals appeared before *c.* 400 ka in Europe. They populated Europe and advanced eastwards into Siberia and southwards to the Middle East, but *c.* 30 ka they were gone. Since this is about the time when modern humans (*Homo sapiens*) entered Europe, it has long been assumed that the Neanderthals mixed with the successful newcomers. For example, excavations at Mugharet Es-Skhul and Jebel Qafzeh in Israel have shown clear evidence of a millennia-long coexistence between these types (Bosinski 1985).

But was this 'coexistence' also a 'mixing'? Study of the mitochondrial DNA of Neanderthals initially seemed to indicate that there was no mixing. However, recent studies on DNA in cell nuclei of modern humans in Eurasia have provided evidence that an amount of genetic material (1–4%) was taken over from the Neanderthals; a slight mixing between Neanderthals and modern humans must therefore have occurred (Green et al. 2010). The population density was extremely low. It is thought that only about 10,000 Neanderthals lived throughout Europe.

Did the Neanderthals speak? Perhaps. An essential distinction between man and ape is the human language. A certain gene is held responsible for the ability to develop a language (the FOXP2 gene that is found also in primates) but, as compared to the chimpanzee, the human gene has undergone two genetic changes. The Neanderthal genes had already undergone those crucial changes; they therefore possessed the prerequisite for the development of language.

The famous cave paintings in France (e.g. Lascaux) and Spain (e.g. Altamira) were produced during the Upper Palaeolithic. The paintings of the Chauvet Cave in the Ardèche, first discovered by a French cave explorer on 18 December 1994, are older and therefore close to the beginning of the Upper Palaeolithic. Unfortunately, because of the sensitivity of the paintings and bad experiences from other caves, the Chauvet Cave is closed to the public.

15.3 The Middle Stone Age

The Middle Stone Age, also known as the Mesolithic period, began when the landscape became reforested at *c.* 11.6 ka. The woods, however, were increasingly unfavourable for human settlement. Mixed oak forest with high lime shares began to spread. There was little undergrowth in the shade of the big trees and the stocks of hazelnuts, which were previously used for food, fell sharply in the course of the Atlantic. The people of the Middle Stone Age therefore preferably settled along rivers and lakes where, apart from hunting and gathering, fishing was also possible. The distribution of Middle Palaeolithic sites reflects these natural conditions.

While only a few settlements have been found in the old morainic landscapes of Lower Saxony and Schleswig-Holstein, the lake-rich young morainic landscapes of Ostholstein, Mecklenburg and Brandenburg were densely populated (Behre 2008).

15.4 The Neolithic Period: The Beginning of Agriculture

Agriculture began only in the early Atlantic period (*c.* 11 ka) at the beginning of the Neolithic in the Middle East. This culture spread into central Europe to the area between Hungary and the Rhine at *c.* 7500 year ago. The new techniques eventually led to extensive deforestation and transformed the natural landscape into a cultural landscape. Cultural remains from the Neolithic period include, among other things, the great stone circles of Stonehenge, Avebury and Stenness (Orkney), which date from *c.* 5 ka BC.

Among the Neolithic relics on Orkney is the 'Watchstone', a >5.6 m high, 1.5 m wide and 40 cm thick slab standing next to the road between the Stones of Stenness and the Ring of Broadgar (Fig. 15.6). It is not known how deep it penetrates in the ground. The remains of a second stone of this kind were found (and removed) in 1930 during road construction (http://www.orkneyjar.com/history/monoliths/watchst.htm).

In the North German lowlands, the transition to agriculture and livestock farming took place at *c.* 6.1–6 ka. This change is clearly visible in the pollen diagrams; in the forest vegetation a rather sharp decline in the proportion of elm is noticeable. This decline in elm was caused by elm disease, and occurred virtually simultaneous throughout northern Germany and beyond. It is easy to detect in pollen diagrams and marks the transition from the Subboreal to the Atlantic period (Behre 2008).

Figure 15.6 A Neolithic 'Watchstone' on Orkney. The stone is well fixed so passing cars are not endangered. Photograph by Jürgen Ehlers.

After the 'elm decline', pollen of cereals and weeds are found on a regular basis in the north German pollen diagrams, reflecting the changing economy. Deforestation initially played no significant role and cattle were fed with leaves. This leaf-feeding business was eventually replaced by another type of economy that, after its Danish term, is referred to as the 'Landnam phase'. From then on, cattle were sent to forage in the forest all year long. Consequently, the houses of the 'Landnam phase' had no stables. Since cattle could only eat the young growth, and did this very effectively, this economy resulted in increasing deforestation after a few decades, represented in the pollen diagrams by a sharp decline of lime and oak and a proliferation of grasses and herbs. Finally, heath spread (Behre 2008).

15.5 Bronze and Iron

Metallic tools arrived in Britain at *c.* 4500 years and in North Germany at *c.* 4000 years from the south. Bronze, an easily workable alloy of copper and tin, enabled the manufacture of jewellery and tools. The end of the Stone Age had come, and agricultural techniques improved. Instead of the simple wooden plough hooks of the Neolithic which man had to drag through the soil himself, ploughs of *c.* 3 m width (so-called 'Ard') were introduced which, as we know from petroglyphs, were drawn by two oxen. As in the Neolithic period, Emmer wheat (*Triticum dicoccum*), barley (*Hordeum vulgare*) and naked barley (*Hordeum vulgare nudum*) remained the major cereals consumed (Behre 2008).

Among the most impressive works of art from the Bronze Age are the rock carvings in the vicinity of Tanum (southern Sweden) showing, among other things, hunting scenes and boats that could apparently carry at least 11 people (Fig. 15.7). To date, over 3000 such

Figure 15.7 Rock carved drawings (petroglyphs) of the Bronze Age in Fossum, southern Sweden. Photograph by Jürgen Ehlers.

petroglyphs have been found. The largest of the rock panels, Vitlyckehäll, was only discovered in 1972.

The introduction of metal was not limited to the Old World. In fact, the indigenous people of North America used copper as early as in 7000 years ago, much earlier than the Europeans, but they processed it by cold hammering and not by melting the metal (Levine 2007). Copper was used in the Andean regions of South America from *c.* 3400 years, mainly to create objects representing the high status of the owner (Bruhns 1994). Bronze was not used in pre-Columbian America.

The use of iron was not established in northern Germany until *c.* 300 BC. There were tools and weapons made of iron available, but these were imported from the south. Behre (2008) writes of the strong technological backwardness of the north as compared to the southern areas, which was only overcome during the Roman Empire.

In the pre-Roman Iron Age, agriculture changed fundamentally. Vast chambered arable fields were built in the dry old morainic areas of northern Germany which, after their British counterparts, were erroneously referred to as 'Celtic fields'. Their distribution extends as far as Sweden and the Baltic States. The beds are 10–50 m long, 8–16 m wide and rise by up to 80 cm above the surrounding ground. Their features are partly preserved to this day in forest areas, and they may often be seen on aerial photographs of arable land as a result of differences in soil colour. Agriculture at that time was highlighted by the strengthening of the still-common Ard by iron, and use of a new plough with a moldboard which could turn the clods.

15.6 The Romans

At around the time of Christ's birth in the Roman Empire, Romans advanced into southern Germany and to the Lower Rhine (Figs 15.8, 15.9). Although Romans were not present in most of northern Germany, Roman coins, pottery, bronze vessels, jewellery and military equipment have been found. Such artefacts have long been thought to reflect the intensive trade between Germany and the Roman Empire. A problem, however, was that no one could think what goods Germania might have had to trade with the Romans. It is assumed today that the majority of Roman objects found in the north are either from some tribute or military pay, or the booty of occasional raids. Conspicuously, hardly any Roman artefacts have been found from peaceful times (e.g. from the third century; Behre 2008).

The Romans considered the Germans as barbarians because they had little culture and no written language. The relapse of Germania into barbarism at the end of the Roman rule is nicely illustrated in the exposition of the museum in Xanten.

When the Roman Empire fell into decline, the Migration Period started. Its onset was around 375 AD when the Huns advanced into Europe. The end is recognized by 568 AD when the Lombards moved from Lower Austria to Italy. The extent of the migrations of the various tribes is difficult to assess. One of the causes is likely to be the attractiveness of the richer and more climatically favoured areas in the south, to which access was now freely available. The migration resulted in a significant decrease of population in Germania; however, a residual population stayed behind in most areas (Behre 2008).

Figure 15.8 Pompeii columns in front of the 'Building of Eumachia' in the forum. Eumachia was an entrepreneur and Venus priestess, and the building may have been a wool market. Photograph by Jürgen Ehlers.

Figure 15.9 Traces of the Romans on the Lower Rhine: an exhibition of archaeological finds in the *LVR RömerMuseum* in Xanten. The 'Lüttinger Boy' was discovered in 1858 by a salmon fisherman on the banks of the Rhine. Photograph by Jürgen Ehlers.

15.7 Middle Ages

A re-colonization gradually began during the early Middle Ages. When Charlemagne took the field against the Saxons, northern Germany was already densely repopulated. The population grew rapidly; forests were cleared and new villages founded. Dyking began on the North Sea coast and the marshes were protected by a continuous line of dykes as early as in the thirteenth century. Agriculture technology also changed; while the three-field system in southern Germany can be demonstrated from the eighth century, in northern Germany north of the loess limit a single-span system with the so-called Esch prevailed. The Esch was farmed land near the village, fertilized with sods (Behre 2008, 2014).

By the middle of the fourteenth century there was a serious agricultural crisis in Germany. A sharp increase in population led to a famine as early as in 1309–1318, in which many people died of 'hunger-typhus'. The plague reached Germany in 1349–50 and killed a large proportion of the population. Three more waves of the plague followed during the fourteenth century. The population decreased by one-third all over Europe; Hamburg is said to have lost half of its inhabitants in the first two years of the plague and Bremen three-quarters. The demand for cereals, the price of grain and harvests fell; farming was no longer profitable in many areas. The Medieval Warm Period lasted from c. 800 to 1350. It was followed by the 'Little Ice Age', during which many villages were abandoned. In central Germany, where the crisis was felt most strongly, half the villages were deserted in some areas. Historical place names indicate the locations of abandoned villages, and 'Celtic Fields' are frequently found under forest around those sites (Behre 2008).

The Villenhusen village to which the Wölbäcker in the aerial photograph belonged (Figs 15.10, 15.11) was located at a little creek, the Sebberbeck. The village was first documented in 1196 when it belonged to the Lords of Heimbruch. The seven farmsteads of the village were then given to the Altkloster Bendictine monastery. In the year 1263 the tithe of 'Vilhusen' went to the 'old monastery'. The village had shrunk considerably by around 1310–12, with only one farm left. The cause of the reduction in this particular case is unknown. The village was deserted from the first half of the fourteenth century.

Many of the fields of deserted villages were later put back into use when demand grew. Today, more and more areas of the world are taken into human use; it is only recently that some have begun to worry about the rate of development.

15.8 Recent Land Grab

A much discussed example of rapid overdevelopment is the destruction of tropical rainforest, widespread mainly in the Amazonian lowlands. This deforestation does not take place secretly. Not only are the latest figures regularly published, but each of us can check the progressive destruction for ourself. The satellite images are freely available and no special software is required for a visual comparison; the changes are so drastic that direct comparison is sufficient in most cases (Fig. 15.12). Brazil and China have put their own satellites into orbit (CBERS-1 and 2) to monitor the rainforest. LandSat images are also freely available

(a)

(b)

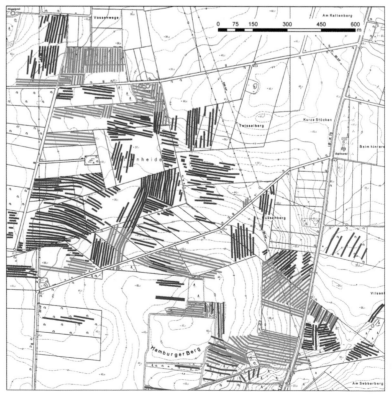

Figure 15.10 High medieval 'Wölbäcker' ('Celtic Fields') of the deserted Villenhusen village south of Buxtehude (a) in an aerial photograph and (b) mapped (aerial photo from flight 717, dated 01/05/1971, LVA Hannover). Mapping by Habermann (Buxtehude, green), Brandt (Harburg, blue) and Ehlers (red).

Figure 15.11 'Wölbäcker' beds have survived under forest, but because of the undergrowth they may be difficult to detect. The picture shows 'Wölbäcker' of the deserted Villenhusen village. Photograph by Jürgen Ehlers.

through the website of the Brazilian Space Agency *Instituto Nacional de Pesquisas Espaciais* (INPE). The South American satellite data can be found and downloaded from http://www .inpe.br/ingles/index.php.

15.9 Drying Lakes, Melting Glaciers and other Problems

The major changes in our environment are no longer hidden from view, but visible to everyone. Whoever wants to see what is left of the Aral Sea and Lake Chad can do so at any time (http://rapidfire.sci.gsfc.nasa.gov/subsets/). Whoever wants to learn about the current glacier fluctuations in Norway is no longer dependent on the literature (e.g. Winkler 2009a, b), but can find the relevant information on the Internet. Anyone who wishes to know about the current state of the ozone hole (Fig. 15.13) simply needs to visit to the appropriate website (http://ozonewatch.gsfc.nasa.gov/).

Finally, anyone interested in the latest data on sea-level rise will find the information online (http://ibis.grdl.noaa.gov/SAT/SeaLevelRise/LSA_SLR_timeseries_global.php). Sea-level rise data are traditionally determined by comparing gauge records. Errors result from the fact

Figure 15.12

Decreasing rainforest in Brazil demonstrated by satellite imagery from 07.09.1984 (below) and from 28.4.2010 (above).

Source: US Geological Survey, Landsat 5 TM, Path 225, Row 69.

that not all gauges were properly fixed vertically; some were built on unstable ground (e.g. peat, humic clay) and have moved downwards. Tectonic movements may additionally mask actual changes in sea level. Today it is possible to measure the rise or fall in sea level directly from satellites, and not just along the coasts but also in the open sea. Figure 15.14 shows the current development for the entire world ocean and for the North Sea. The sea level rises by 2–3 mm every year.

Why are there seasonal fluctuations visible in the graphs? Would we not expect any hemispherical differences be balanced by water flowing freely back and forth? The fluctuations are a consequence of the different land–sea distribution; 57% of the world's oceans are in the Southern Hemisphere. The reaction of the near-surface layers of the sea water to heating (the so-called 'steric effect') is therefore stronger in the Southern than in the Northern Hemisphere.

The different land–sea distribution makes itself felt in other ways. The Northern Hemisphere has twice as much land as the Southern Hemisphere, and most of it lies so far

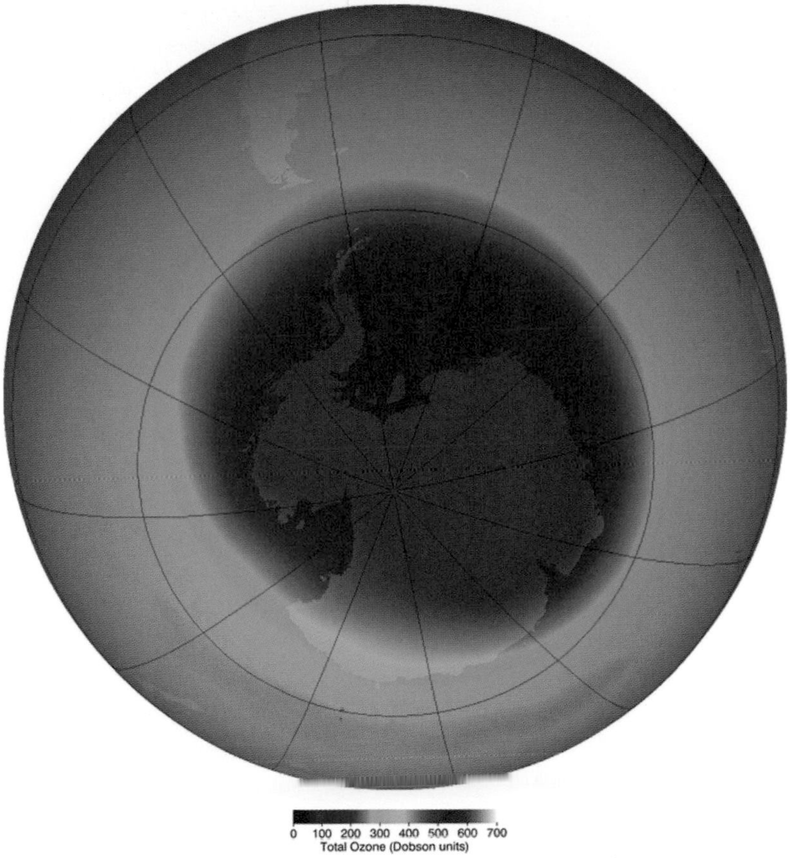

Figure 15.13 Largest expansion of ozone hole in the Southern Hemisphere in September 2014.

Source: NASA.

north that in the winter a greater proportion of rainwater is bound in the form of snow and ice. In this way, more water is removed from the northern seas in the winter than in the southern winter.

The two effects are opposite, but they do not cancel each other out completely. The differential expansion of water when heated produces a difference of 4 mm between the Northern and Southern hemispheres, while the difference in retention of the continents affects the sea level by 8 mm. A difference of 4 mm therefore remains, and the world sea level is lowest in early April and highest in late September.

The melting of glaciers can be tracked easily by comparison of satellite images. The measurement of the decay rate of ground ice is, however, much more difficult. The permafrost regions of the Earth become warmer, resulting in widespread thaw (Fig. 15.15). This leads to a change in the hydrological regime; in areas of permafrost and low gradient, the thawing of ground ice results in increased waterlogging (wet thermokarst). This process causes boreal forests to turn into swamps. On slopes and in high altitudes, however, thawing results in an improvement of the drainage (dry thermokarst). The water table falls, and the boreal forest develops into a steppe. Thawing in areas of wet thermokarst releases large amounts of CO_2 and methane; Vladimir Romanovsky has put together some figures on the warming of permafrost (Table 15.1).

(a)

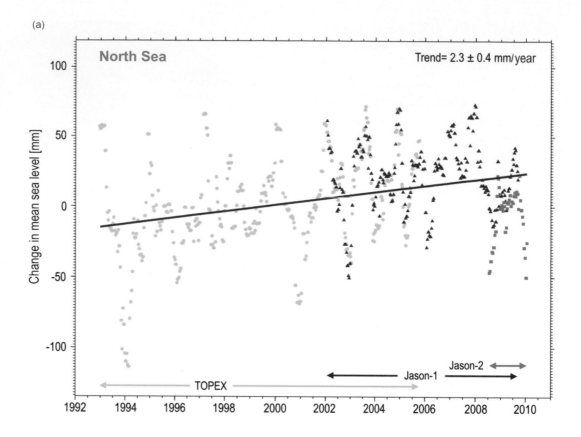

(b)

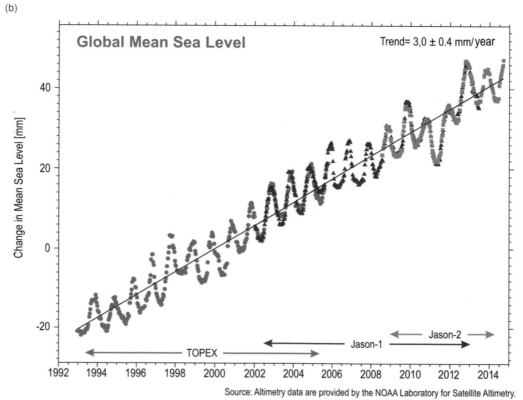

Figure 15.14 Current changes in sea level (a) in the ocean and (b) in the North Sea. The zero value in the graphs is set arbitrarily to match the sea level of 1 January 2000.

Source: NOAA Laboratory for Satellite Altimetry.

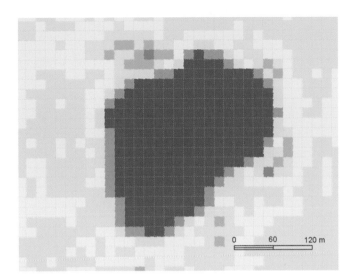

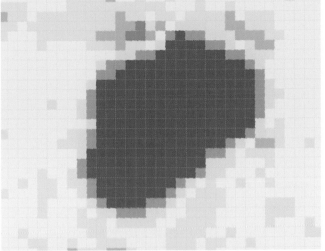

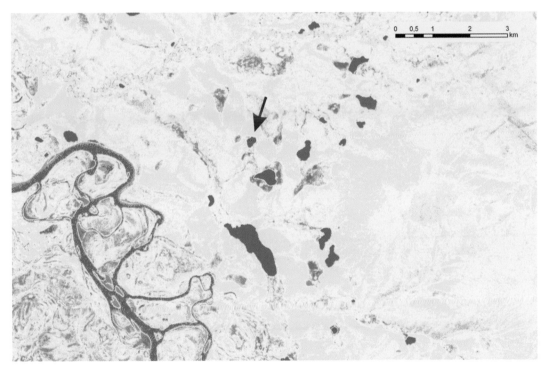

Figure 15.15 Changes in the size of permafrost lakes in the Kolyma region between 2000 and 2010, evaluation of Landsat7 ETM satellite images (Band 8). Apparent size of the lake in 2000: 236 pixels; in the year 2010: 225 pixels. The red arrow shows the location of the lake.

Source: USGS Landsat7 ETM satellite images 104/12 dated 14.7.2000 and 24.6.2010.

But to what extent is the permafrost actually thawing? When looking at a satellite image from northern Siberia or Alaska, the Emmental cheese-like landscape seems to indicate rapid ice decay. Walter et al. (2006) reported that lakes in the Kolyma region in eastern Siberia have risen sharply both in number and size. The lake areas have apparently increased by 14.5% from 1974 to 2000 (9.6–11% of the total land surface).

TABLE 15.1 TEMPERATURE CHANGES IN PERMAFROST AREAS.

Location	Depth (m)	Changes in permafrost temperature ($°C\ a^{-1}$)	Period
US Trans-Alaska Pipeline	20	+0.6 to +1.5	1983–2003
Barrow Permafrost Observatory	15	+1	1950–2003
Russia Siberia	1.6–3.2	+0.03	1960–1992
NW Siberia	10	+0.3 to +0.7	1980–1990
European North of Russia, related permafrost	6	+1.6 to +2.8	1973–1992
European North of Russia, patchy permafrost	6	up to +1.2	1970–1995
Alberta, Canada	15	+0.15	1995–2000
Northern Mackenzie Basin	28	+0.1	1990–2000
Middle Mackenzie Basin	15	+0.03	1985–2000
Northern Quebec	10	−0.1	late 1980s – middle 1990s
Juvvasshøe, southern Norway		+0.5 to +1.0	
Jansson Haugen, Svalbard		+1 to +2	
China, Qinghai–Tibet Plateau		+0.1 to +0.3	1970s–1990s
Kazakhstan, northern Tien Shan		+0.2 to +0.6	1973–2003
Mongolia and Khangai Mountains, Khentei, Lake Hovsgol	up to 50	+0.3 to +0.6	1973–2003

Source: Romanovsky (http://www.arctic.noaa.gov/essay_romanovsky.html).

Such a change should be large enough to be detected by the newer high-resolution satellite images. However, when comparing images from 2000 and 2010, instead they seem to show a slight decrease in the lake areas. The dramatic increase reported by Walter et al. (2006) is probably an artefact caused by misinterpretation of the low-resolution satellite images of 1974 (pixel size 60 × 60 m). Alleged newly formed lakes result from the fact that those small water bodies were not recognized on the old satellite images because of the low resolution. The new images have a resolution 16 times higher.

The disappearance of permafrost does not necessarily lead to an increase in lakes. On the contrary, Smith et al. (2005) found a shrinkage of the water areas in western Siberia. There, the size of the lakes in the discontinuous permafrost region has decreased by 11% from 1973 to 2000. The thawing of permafrost only results in a short-time extension of the lake areas. If the ice in the ground disappears, most of the lakes are drained. Lakes will only remain over an impermeable surface (e.g. peat).

Since 1958, the Mauna Loa Observatory in Hawaii measures the levels of CO_2 in the atmosphere. It was found that the average CO_2 concentration has risen steadily (Fig. 15.16), apart from seasonal fluctuations. It initially seemed obvious that this was an effect of the release of greenhouse gases by human activities. However, studies of ice cores from Antarctica show that a significant increase in CO_2 was also recorded in the atmosphere during the Eemian Interglacial; this cannot be attributed to human interference. Then, as in our present interglacial, the increase in CO_2 occurred at the end of the interglacial period. It therefore looked as if the CO_2 was not the cause but, conversely, a result of global warming.

This apparent conundrum has still not been resolved, however. Meanwhile, boreholes have penetrated into deeper layers of the Antarctic ice and the positions of several previous interglacial periods could be examined with regard to their gas content. It was found that the CO_2 content has never reached the level of the current rise. The highest value ever reached in the last 450 ka was about 310 ppm; in December 2014, the measured values exceeded 397 ppm. This increase was calculated from an average of between 0.86 ppm a^{-1} (1959–69) and *c.* 2 ppm a^{-1} (data from Mauna Loa Observatory).

The changes that we observe today may seem small to the observer. Whether the sea level rises this year by 2 mm or whether the CO_2 content in the atmosphere increases by 2 ppm seems insignificant, as if we have a pot of milk on the stove and the temperature rises from 49 to 51°C. With the milk we know at what point a sudden unwelcome development can be prevented by rapidly turning the switch. With regard to sea level, the CO_2 in the atmosphere and other environmental changes, that threshold is not so clear. There is no such 'off' switch with which to halt unpleasant developments however, and the effects may be catastrophic.

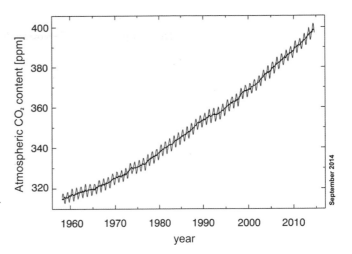

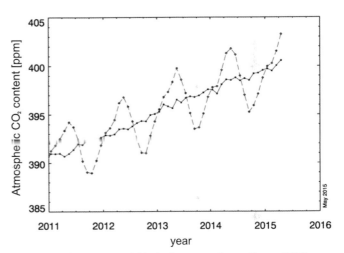

Figure 15.16 Increase of CO_2 in the atmosphere. Above: 1959–2014; below: 2006–2015.

Source: NOAA Earth System Research Laboratory.

15.10 The Anthropocene: Defining the Human Age?

In recent years the term 'Anthropocene' has become increasingly used. It was first proposed by Paul Crutzen and Eugene Stoermer (2000) to denote the present time interval in which many geologically significant processes and conditions have been, and continue to be, profoundly altered by human activities. These processes include changes in erosion and sediment transport resulting from agriculture and urbanization and changes in the chemical composition of the atmosphere, oceans and soils, with increasing variation in the cycling of elements such as carbon, nitrogen, phosphorus and various metals. Changes in environmental conditions generated by these processes include: global warming; atmospheric instability including extreme weather events; ocean acidification and spreading oceanic 'dead zones'; habitat loss, predation and species invasions; decreasing biodiversity; and a range of globally significant physical and chemical changes.

The rise to dominance of humans during the present (Holocene Series) interglacial has been marked by a progressive parallel increase in their effect on the natural environment and its processes, the impact of which has long been recognized in palaeoenvironmental reconstructions. These anthropogenic effects, which reached significant levels during the last few centuries and particularly since the Industrial Revolution in northern Europe, have resulted in marked changes to the Earth's surface; indeed, almost every aspect of the contemporary global environment is now modified, or at least influenced, by human activity (Crossland 2005; Zalasiewicz et al. 2010; Steffen et al. 2011). Within the geological community there has been a growing opinion that this accelerated human impact may be reflected in the recent geological and therefore stratigraphic record, and that the anthropogenic effect can be distinguished from natural 'background' conditions (Zalasiewicz et al. 2008). Although the term 'Anthropocene' was initially proposed informally to describe this period of expanding human influence on the global environment (Crutzen 2002), and was apparently suggested by chance (Steffen et al. 2004), it is increasingly being applied to the recent geological record. Some have suggested that the term should be used as a division of geological time of equivalent status to the subdivisions of the Cenozoic Erathem ('recent life'), that is, Pleistocene and Holocene.

An important consequence of the recognition of increasing intensity of human impact on the natural environment, and its possible representation in the geological record, has been the initiation of discussions to determine whether the stratigraphic signature of the Anthropocene is now sufficiently clearly defined to warrant its formal definition as a new period of geological time. This is currently being considered by a Working Group of the Subcommission on Quaternary Stratigraphy (SQS).

While there is no doubt that human activity is influencing the majority (if not all) of the natural environments on the Earth's surface today, the point at which this influence can be said to have surpassed what might be termed 'normal background conditions' has been increasingly debated. The original suggestion that this point was passed with the start of the Industrial Revolution in northern Europe has been met by others who point to events such as the marked expansion of industrial activity across the world in the post-Second World War years. Others highlight much earlier human impacts however, such as widespread alluviation in river systems resulting from forest clearance for agriculture, the arrival of domesticated animals or the activity of humans in hunting.

Some authors focus on the changing composition of the atmosphere as a result of human activities as a marker for the Anthropocene. The most obvious example of the latter is the increased carbon dioxide concentrations since the Industrial Revolution which some argue has caused irreversible climate change (Solomon et al. 2009). However, another hypothesis invokes the onset of early human agriculture for significant rises in atmospheric methane concentrations, another more potent greenhouse gas, since the early Holocene (Ruddiman 2003). Defining a marker for the onset of the Anthropocene based on human activities and their impacts on the natural environment can therefore be difficult. One way of defining the base of the Anthropocene has been to suggest the distinct radiogenic fallout signature from nuclear bomb testing as a marker (Zalasiewicz et al. 2015). However, some may consider this radiogenic 'spike' too arbitrary for defining the onset of the Anthropocene, given the wide-ranging timespan of human activities that impacted on the environment as described above. Whatever is decided about the formality and definition of the Anthropocene, one thing is certain: human endeavours and ideas have made huge impacts on the way we understand and affect our environment.

References

Aario, R. (1977) Classification and terminology of morainic landforms in Finland. *Boreas* **6**, 87–100.

Abelmann, A. (1985) *Palökologische und ökostratigraphische Untersuchungen von Diatomeenassoziationen an holozänen Sedimenten der zentralen Ostsee.* Berichte – Reports, Geologisch-Paläontologisches Institut der Universität Kiel 9.

Aber, J.S. (1991) The glaciation of northeastern Kansas. *Boreas* **20**, 297–314.

Adam, K. D. (1951) Der Waldelefant von Lehringen - eine Jagdbeute des diluvialen Menschen. *Quartär* **5**, 79–92.

Adamson, K. R., Woodward, J. C. & Hughes, P. D. (2014) Pleistocene glaciofluvial sediment flux and catchment decoupling in a Mediterranean mountain karst. *Quaternary Science Reviews* **94**, 28–43.

Agassiz, L. (1840) *Études sur les Glaciers.* Neuchâtel: Jent & Gaßmann.

Agassiz, L. (1841) *Untersuchungen über die Gletscher.* Solothurn: Jent & Gaßmann.

Agassiz, L. & Agassiz, E. C. (1868) *A Journey in Brazil.* Boston: Ticknor & Fields.

Aguirre, E. & Pasini, G. (1985) The Pliocene-Pleistocene Boundary. *Episodes* **8**, 116–120.

Ahlmann, H. W. (1935) Contribution to the physics of glaciers. *Geographical Journal* **86**, 97–113.

Ahrens, H. & Lotsch, W. (1976) Zum Problem des Pliozäns in Brandenburg. *Jahrbuch für Geologie* **7/8**, 277–323.

Akçar, N., Yavuz, V., Ivy-Ochs, S., Kubik, P. W., Vardar, M. & Schlüchter, C. (2008) A case for a down wasting mountain glacier during Termination I, Verçenik Valley, NE Turkey. *Journal of Quaternary Science* **23**, 273–285.

Akçar, N., Yavuz, V., Ivy-Ochs, S., Reber, R., Kubik, P.W., Zahno, C. & Schlüchter, C. (2014). Glacier response to the change in atmospheric circulation in the eastern Mediterranean during the Last Glacial Maximum. *Quaternary Geochronology*, **19**, 27–41.

Aksu, A. (1985) Climatic and oceanographic changes of the past 400,000 years: evidence from deep-sea cores on Baffin Bay and Davis Strait. In: Andrews, J.T. (ed.) *Quaternary Environments: Eastern Canadian Arctic, Baffin Bay, and West Greenland.* Springer, pp. 181–209.

Albertz, J. (2009) *Einführung in die Fernerkundung*, 4. Auflage. Darmstadt: WBG.

Alley, R. B., Blankenship, D. D., Bentley, C. R. & Rooney, S. T. (1986) Deformation of till beneath ice stream B, West Antarctica. *Nature* **322**, 57–59.

Alley, R. B., Blankenship, D. D., Rooney, S. T. & Bentley, C. R. (1987) Continuous till deformation beneath ice sheets. In: Waddington, E. D. & Walder, J. S. (eds) *The Physical Basis of Ice Sheet Modelling.* Proceedings of the Vancouver Symposium, August (1987). IAHS Publication **170**, 81–91.

Alley, R. B., Mayewski, P.A., Sowers, T., Stuiver, M., Taylor, K.C. & Clark, P.U. (1997) Holocene climatic instability: A prominent, widespread event 8200 years ago. *Geology* **25**(6), 483–486.

Alloway, B. V., Larsen, G., Lowe, D. J., Shane, P. A. R. & Westgate, J. A. (2007) Tephrochronology. In: Elias, S. A. (ed.) *Encyclopaedia of Quaternary Science.* Amsterdam: Elsevier, pp. 2869–2898.

Ampferer, O. (1939) *Über einige Formen der Bergzerreißung.* Sitzungsberichte der Akademie der Wissenschaften, Wien, Mathematisch-Naturwissenschaftliche Klasse, Abteilung I, **148**, 1–14.

Andersen, B. G. & Borns, H. W. Jr (1994) *The Ice Age World.* Oslo: Scandinavian University Press.

Andersen, K. K., Svensson, A., Johnsen, S.J., Rasmussen, S.O., Bigler, M., Röthlisberger, R., Ruth, U., Siggard-Andersen, M-L., Steffensen, J.P., Dahl-Jensen, D., Vinther, B.M. & Clausen, H.B. (2006) The Greenland Ice Core Chronology 2005, 15-42 ka. Part 1: constructing the time scale. *Quaternary Science Reviews* **25**, 3246–3257.

Andersen, L. T. (2004) *The Fanø Bugt Glaciotectonic Thrust Fault Complex, Southeastern Danish North Sea. A study of large-scale glaciotectonics using high-resolution seismic data and numerical modelling.* Ph.D. thesis. Danmarks og Grønlands Geologiske Undersøgelse Rapport **2004/30**.

Andersen, S. & Pedersen, S. S. (1998) Isafsmeltning og israndlinier i Danmark. In: Andersen, S. & Pedersen, S. S. (eds) *Israndlinier i Norden.* Nordisk Ministerråd, Tema Nord **584**, 47–60.

Andrews, J. T. (1987) The Late Wisconsin glaciation and deglaciation of the Laurentide Ice Sheet. In: Ruddiman, W. F. & Wright, H. E. Jr (eds) *North American and Adjacent Oceans During the Last Deglaciation.* Boulder: Geological Society of America, The Geology of North America **K-3**, pp. 13–37.

Andrews, J. T. & Tedesco, K. (1992) Detrital carbonate-rich sediments, northwestern Labrador Sea: Implications for ice-sheet dynamics and iceberg rafting (Heinrich) events in the North Atlantic. *Geology* **20**, 1087–1090.

Andrews, J. T., Shilts, W. W. & Miller, G. H. (1983) Multiple deglaciations of the Hudson Bay Lowlands, Canada, since deposition of the Missinaibi (last-interglacial?) Formation. *Quaternary Research* **19**, 18–37.

Andrews, J. T., Erlenkeuser, H., Tedesco, K., Aksu, A.E. & Jull, A.J.T. (1994) Late Quaternary (Stage 2 and 3) Meltwater and Heinrich Events, Northwest Labrador Sea. *Quaternary Research* **41**, 26–34.

Antevs, E. (1925) *Retreat of the Last Ice Sheet in Eastern Canada*. Geological Survey of Canada, Memoir no. 146.

Antevs, E. (1928) *The Last Glaciation*. American Geographical Society, Research Series no. 17.

Arduino, G. (1760) *Sopra varie sue Osservazioni fatte in diverse parti del Territorio di Vicenza, ed altrove, appartenenti alla Teoria Terrestre, ed alla Mineralogia*. Nuova Raccolta di Opuscoli Scientifici e Filologici (Venezia), **6** (lettera al Prof. Antonio Vallisnieri, 30 marzo 1759).

Arnold, N. S., Andel, T. H. van & Valen, V. (2002) Extent and dynamics of the Scandinavialn ice sheet during Oxygen Isotope Stage 3 (65,000–25,000 yr BP). *Quaternary Research* **57**, 38–48.

Aseev, A. A. (1968) Dynamik und geomorphologische Wirkung der europäischen Eisschilde. *Petermanns Geographische Mitteilungen* **112**, 112–115.

Ashton, N., Lewis, S.J., Parfitt, S.A., Penkman, K.E.H. & Coope, G.R. (2008) New evidence for complex climate change in MIS 11 from Hoxne, Suffolk, UK. *Quaternary Science Reviews* **27**, 652–668.

Astakhov, V. (2008) Geographical extremes in the glacial history of northern Eurasia: post-QUEEN considerations. *Polar Research* **27**, 280–288.

Augustowski, B. (1965) Pattern and development of ice marginal streamways of the Kashubian coast. *Geographia Polonica* **6**, 35–42.

Baden-Powell, D.F.W. (1948) The chalky bounder clays of Norfolk and Suffolk. *Geological Magazine* **85**, 279–296.

Baker, V. R. (2008) The Spokane Flood debates: Historical background and philosophical perspective. In: Grapes, R., Oldroyd, D. & Grigelis, A. (eds) *History of Geomorphology and Quaternary Geology*. The Geological Society of London, Special Publication **301**, 33–50.

Baker, V. R., Benito, G. & Rudoi, A. N. (1993) Palaeohydrology of late Pleistocene superflooding, Altay Mountains, Siberia. *Science* **259**, 348–350.

Balco, G. (2011) Contributions and unrealized potential contributions of cosmogenic-nulcide exposure dating to glacier chronology, 1990-2010. *Quaternary Science Reviews* **30**, 3–27.

Balco, G., Briner, J., Finkel, R.C., Rayburn, J.A., Ridge, J.C. & Schäfer, J.M. (2009) Regional beryllium-10 production rate calibration for late-glacial northeastern North America. *Quaternary Geochronology* **4**, 93–107.

Ballais, J.-L. (1983) Moraines et glaciers quaternaires des Aurès (Algerie). *108ème Congrès national de Sociétés savantes*, Grenoble, 1983, Géographie, 291–303.

Ballantyne, C. K. (1984) The Late Devensian periglaciation of upland Scotland. *Quaternary Science Reviews* **3**, 311–343.

Ballantyne, C. K. (1987) The present-day periglaciation of upland Britain. In: Boardman, J. (ed.) *Periglacial Processes and Landforms in Britain and Ireland*. Cambridge: Cambridge University Press, pp. 113–126.

Ballantyne, C. K. (1998) Age and significance of mountain-top detritus. *Permafrost and Periglacial Processes* **9**, 327–345.

Ballantyne, C. K. (2002) Paraglacial geomorphology. *Quaternary Science Reviews* **21**, 1935–2017.

Ballantyne, C. K. (2007) The Loch Lomond Readvance on north Arran, Scotland: glacier reconstruction and palaeoclimatic implications. *Journal of Quaternary Science* **22**, 343–359.

Ballantyne, C. K. (2010) Extent and deglacial chronology of the last British-Irish Ice Sheet: implications of exposure dating using cosmogenic isotopes. *Journal of Quaternary Science* **25**, 515–534.

Ballantyne, C. K. & Harris, C. (1994) *The Periglaciation of Great Britain*. Cambridge: Cambridge University Press.

Ballantyne, C. K. & Hall, A. M. (2008) The altitude of the last ice sheet in Caithness and east Sutherland, northern Scotland. *Scottish Journal of Geology* **44**, 169–181.

Ballantyne, C. K. & Stone, J.O. (2013) Timing and periodicity of paraglacial rock-slope failures in the Scottish Highlands. *Geomorphology* **186**, 150–161.

Banham, P. H., Gibbard, P.L., Lunkka, J.P., Parfitt, S.A., Preece, R.C. & Turner, C. (2001) A critical assessment of 'A New Glacial Stratigraphy for Eastern England'. *Quaternary Newsletter*, 93, 5–14.

Barbier, A. & Cailleux, A. (1950) Glaciare et périglaciaire dans le Djurdjura occidental (Algérie). *Comptes Rendus des Séances de l'Académie des Sciences, Paris*, Juillet–Décembre 1950, 365–366.

Barckhausen, J. (1969) Entstehung und Entwicklung der Insel Langeoog. *Oldenburger Jahrbuch* **68**, 239–281.

Barclay, W. J., Brandon, A., Ellison, R.A. & Moorlock, B.S.P. (1992) A Middle Pleistocene palaeovalley-fill west of the Malvern Hills. *Journal of the Geological Society* **149**, 75–92.

Barendredgt, R. W. & Duk-Rodkin, A. (2011) Chronology and extent of late Cenozoic ice sheets in North America. In: Ehlers, J., Gibbard, P.L. & Hughes, P.D. (eds) *Quaternary Glaciations - Extent and Chronology - A Closer Look*. Amsterdam: Elsevier, Developments in Quaternary Science no. 15, pp. 419–426.

Barnett, P. J. (1992) Quaternary geology of Ontario. In: Thurston, P.C., Williams, H.R., Sutcliffe, R.H. & Scott, G.M. (eds) *Geology of Ontario*. Toronto, Ontario Geological Survey, Special Volume 4, pp. 1011–1088.

Barrell, D. J. A. (2011) Quaternary glaciers of New Zealand. In: Ehlers, J., Gibbard, P.L. & Hughes, P.D. (eds) *Quaternary Glaciations - Extent and Chronology - A Closer Look*. Elsevier, Developments in Quaternary Science **15**, 104–1064.

Barsch, D. (1996) *Rockglaciers: Indicators for the Present and Former Geoecology in High Mountain Environments*. Berlin: Springer.

Barton, I. G. (2008) *Calculating the peak discharge of the Glacial Lake East Fork outburst flood, Big Lost River, Idaho*. M.Sc. thesis. Lehigh University.

Bäsemann, H. (1979) *Feinkiesanalytische und morphometrische Untersuchungen an Oberflächensedimenten der Deutschen Bucht*. Dissertation, Hamburg.

Bateman, M. D., Buckland, P.C., Chase, B., Frederick, C.D. & Gaunt, G.D. (2008) The Late Devensian proglacial Lake Humber: new evidence from littoral deposits east of Ferrybridge, Yorkshire, England. *Boreas* **37**, 195–210.

Becker-Haumann, R. (1998) Das älteste Pleistozän im Illergletscher-Vorland - Neue Ergebnisse zur Stratigraphie und Paläogeographie der Biberkaltzeit im schwäbischen Alpenvorland. *Eiszeitalter und Gegenwart* **48**, 87–101.

Becker-Haumann, R. (2002) Ein neues Konzept für das drittletzte Glazial (Mindel-Glazial) im bayerischen Alpenvorland. *Zeitschrift für Geologische Wissenschaften* **30**, 173–190.

Beets, D. J., de Groot, T. A. M. & Davies, H. A. (2003) Holocene tidal back-barrier development at decelerating sea-level rise: a 5 millennia record, exposed in the western Netherlands. *Sedimentary Geology* **158**, 117–144.

Beets, D. J., Meijer, T., Beets, C. J., Cleveringa, P., Laban, C. & van der Spek, A. J. F. (2005) Evidence for a Middle Pleistocene glaciation of MIS 8 age in the southern North Sea. *Quaternary International* **133/134**, 7–19.

Behling, H. (2007) Pollen records, postglacial South America. In: Elias, S. A. (ed.) *Encyclopaedia of Quaternary Science*. Amsterdam: Elsevier, pp. 2745–2752.

Behre, K.-E. (1989) Biostratigraphy of the Last Glacial Period in Europe. *Quaternary Science Reviews* **8**, 25–44.

Behre, K.-E. (2008) *Landschaftsgeschichte Norddeutschlands. Umwelt und Siedlung von der Steinzeit bis zur Gegenwart*. Neumünster: Wachholtz.

Behre, K.-E. (2014) *Ostfriesland - Die Geschichte seiner Landschaft und ihrer Besiedelung*. Wilhelmshaven: Brune-Mettcker.

Behre, K.-E. & Lade, U. (1986) Eine Folge von Eem und 4 Weichsel-Interstadialen in Oerel/Niedersachsen und ihr Vegetationsablauf. *Eiszeitalter und Gegenwart* **36**, 11–36.

Bell, D. (1888) On the glacial phenomena of Scotland, with reference to the reports of the Boulder Committee of the Royal Society of Edinburgh. *Transactions of the Geological Society of Glasgow* **8**, 237–254.

Benn, D. I. & Lehmkuhl, F. (2000) Mass balance and equilibrium-line altitudes of glaciers in high mountain environments. *Quaternary International*, **65/66**, 15–29.

Benn, D. I. & Evans, D. J. A. (2006) Subglacial megafloods: outrageous hypothesis or just outrageous? In: Knight, P.G. (ed.) *Glacier Science and Environmental Change*. Oxford: Blackwell, pp. 42–46.

Benn, D. I. & Evans, D. J. A. (2010) *Glaciers and Glaciation*. Second edition. Hodder Education.

Bennett, M. R. (1994) Morphological evidence as a guide to deglaciation following the Loch Lomond Stadial: a review of research approaches and models. *Scottish Geographical Magazine* **110**, 24–32.

Bennett, M. R. & Glasser, N. F. (1991) The glacial landforms of Glen Geusachan, Cairngorms: a reinterpretation. *Scottish Geographical Magazine* **107**, 116–123.

Bennett, M. R. & Boulton, G. S. (1993) A reinterpretation of Scottish 'hummocky moraine' and its significance for the deglaciation of the Scottish Highlands during the Younger Dryas or Loch Lomond Stadial. *Geological Magazine* **130**, 301–318.

Bennett, M. R. & Glasser, N. F. (2009) *Glacial Geology: Ice Sheets and Landforms*. Second edition. Wiley-Blackwell.

Bennett, M. R., Hambrey, M. J., Huddart, D. & Glasser, N. F. (1998) Glacial thrusting and moraine-mound formation in Svalbard and Britain: the example of Coire a'Cheud-chnoic (Valley of the Hundred Hills), Torridon, Scotland. *Quaternary Proceedings* **6**, 17–34.

Bennett, M. R., Waller, R. I., Glasser, N. F., Hambrey, M. J. & Huddart, D. (1999) Glaci-genic clast fabrics: genetic fingerprint or wishful thinking? *Journal of Quaternary Science* **14**, 125–135.

Berendt, G. (1879) Gletschertheorie oder Drifttheorie in Norddeutschland. *Zeitschrift der Deutschen Geologischen Gesellschaft* **31**, 1–20.

Berger, A. & Loutre, M.-F. (2002) An exceptionally long interglacial ahead? *Science* **297**, 1287–1288.

Berger, A. & Loutre, M.-F. (2004) Milankovitch theory and paleoclimate. In: Elias, S. A. (ed.) *Encyclopaedia of Quaternary Science.* Amsterdam: Elsevier, pp. 1017–1022.

Bernhardi, A. (1832) Wie kamen die aus dem Norden stammenden Felsbruchstücke und Geschiebe, welche man in Norddeutschland und den benachbarten Ländern findet, an ihre gegenwärtigen Fundorte? *Jahrbuch für Mineralogie* **3**, 257–267.

Berson, A., Samoilowitsch, R. L. & Weickmann, L. (1933) *Die Arktisfahrt des Luftschiffes "Graf Zeppelin" im Juli 1931.* Petermanns Geographische Mitteilungen, Ergänzungsheft **216**.

Bettis III, E.A., Muhs, D.R., Roberts, H.M., Wintle, A.G. (2003). Last Glacial loess in the conterminous USA. *Quaternary Science Reviews* **22**: 1907–1946.

Beug, H. J. (1972) *Das Riß/Würm-Interglazial von Zeifen, Landkreis Laufen a. d. Salzach.* Bayerische Akademie der Wissenschaften, Mathematisch-Naturwissenschaftliche Klasse N. F. **151**, 46–75.

Beug, H. J. (1979) Vegetationsgeschichtlich-pollenanalytische Untersuchungen am Riß/Würm-Interglazial von Eurach am Starnberger See/Obb. *Geologica Bavarica* **80**, 91–106.

Bibus, E. & Kösel, M. (1987) *Paläoböden und periglaziale Deckschichten im Rheingletschergebiet von Oberschwaben und ihre Bedeutung für Stratigraphie, Reliefentwicklung und Standort.* Tübinger Geowissenschaftliche Arbeiten **D3**, 89.

Bibus, E., Bludau, W., Ellwanger, D., Fromm, K., Kösel, M. & Schreiner, A. (1996) On Pre-Würm glacial and interglacial deposits of The Rhine Glacier (South German Alpine Foreland, Upper Swabia, Baden-Württemberg). In: Turner, Ch. (ed.) *The Early Middle Pleistocene in Europe.* Rotterdam: Balkema, pp. 195–204.

Billard, A. & Orombelli, G. (1986) Quaternary glaciations in the French and Italian piedmonts of the Alps. *Quaternary Science Reviews* **5**, 407–411.

Björck, S. (1995) A review of the history of the Baltic Sea, 13.0-8.0 ka BP. *Quaternary International* **27**, 19–40.

Björck, S. & Digerfeldt, G. (1984) Climatic changes at Pleistocene/Holocene boundary in the Middle Swedish end moraine zone, mainly inferred from stratigraphic indications. In: Mörner, N.-A. & Karlén, W. (eds) *Climatic Changes on a Yearly to Millennial Basis.* Dordrecht, Boston, Lancaster: Reidel, pp. 37–56.

Björck, S. & Digerfeldt, G. (1989) Lake Mullsjön–a key site for understanding the final stage of the Baltic Ice Lake east of Mt. Billingen. *Boreas* **18**, 209–219.

Björck, S., Walker, M.J.C., Cwynar, L.C., Johnsen, S., Knudsen, K.-L., Lowe, J.J., Wohlfarth, B. & Members of INTIMATE (1998) An event stratigraphy for the last termination in the North Atlantic region based on the Greenland ice-core record: a proposal by the INTIMATE group. *Journal of Quaternary Science* **13**, 283–292.

Björnsson, H. (1992) Jökulhlaups in Iceland: prediction, characteristics and simulation. *Annals of Glaciology* **16**, 95–106.

Björnsson, H. (2002) Subglacial lakes and jökulhlaups in Iceland. *Global and Planetary Change* **35**, 255–271.

Björnsson, H. (2009) Jökulhlaups in Iceland: sources, release and drainage. In: Burr, D. M., Carling, P. A. & Baker, V. R. (eds) *Megaflooding on Earth and Mars*. Cambridge: Cambridge University Press, pp. 50–62.

Black, R. F. (1965) Ice wedge casts of Wisconsin. *Wisconsin Academy of Sciences, Arts and Letters* **54**, 187–222.

Black, R. F. (1976) Periglacial features indicative of permafrost. *Quaternary Research* **6**, 3–26.

Blakey, R. C. (2008) Gondwana paleogeography from assembly to breakup–a 500 million year odyssey. In: Fielding, C. R., Frank, T. D. & Isbell, J. L. (eds) *Resolving the Late Paleozoic Ice Age in Time and Space*. Geological Society of America, Special Paper **441**, 1–28.

Bloom, A. L., Broecker, W. S., Chappell, M. A., Matthews, R. K. & Mesolella, K. J. (1974) Quaternary sea level fluctuations on a tectonic coast: New 230Th/234U dates from the Huon Peninsula, New Guinea. *Quaternary Research* **4**, 185–205.

Blytt, A. (1876) *Immigration of the Norwegian Flora*. Christiania: Alb. Cammermeyer.

Bobrowsky, P. & Rutter, N.W. (1992) The Quaternary geologic history of the Canadian Rocky Mountains. *Géographie physique et Quaternaire* **46**, 5–50.

Boenigk, W. (1981) Die Gliederung der tertiären Braunkohlendeckschichten in der Ville (Niederrheinische Bucht). *Fortschritte in der Geologie im Rheinland und Westfalen* **29**, 193–263.

Boenigk, W. & Frechen, M. (2006) The Pliocene and Quaternary fluvial archives of the Rhine system. *Quaternary Science Reviews* **25**(5–6), 550–574.

Bogaard, P. van den & Schmincke, H. (1985) A widespread isochronous Late Quaternary tephra layer in central and northern Europe. *Geological Society of America Bulletin* **96**, 1554–1571.

Bondevik, S., Mangerud, J., Dawson, S., Dawson, A. & Lohne, Ø. (2003) Record-breaking Height for 8000-Year-Old Tsunami in the North Atlantic. *EOS Transactions of American Geophysical Union* **84**, 289–300.

Bondevik, S., Løvholt, F., Harbitz, C., Mangerud, J., Dawson, A. & Svendsen, J. I. (2005) The Storegga Slide tsunami–comparing field observations with numerical simulations. *Marine and Petroleum Geology* **22**, 195–208.

Boreham, S. & Gibbard, P.L. (1995) Middle Pleistocene Hoxnian Stage interglacial deposits at Hitchin, Hertfordshire, England. *Proceedings of the Geologists' Association* **106**, 259–270.

Bortenschlager, I. & Bortenschlager, S. (1978) Pollenanalytische untersuchung am bänderton von Baumkirchen (Inntal, Tirol). *Zeitschrift für Gletscherkunde und Glazialgeologie* **14**, 95–103.

Borth-Hoffmann, B. (1980) *Flachseismische Untersuchung geologischer Strukturen in der östlichen Deutschen Bucht*. Universität Kiel, Unveröffentlichte Diplornarbeit.

Böse, M. (1991) A palaeoclimatic interpretation of frost-wedge casts and aeolian sand deposits in the lowlands between Rhine and Vistula in the Upper Pleniglacial and Late Glacial. *Zeitschrift für Geomorphologie* N. F. Supplement-Band **90**, 15–28.

Böse, M. & Hebenstreit, R. (2011) Late Pleistocene and Early Holocene Glaciations in the Taiwanese High Mountain Ranges. In: Ehlers, J., Gibbard, P.L. & Hughes, P.D. (eds)

Quaternary Glaciations - Extent and Chronology - A Closer Look. Elsevier, Developments in Quaternary Science **15**, 1003–1012.

Bosinski, G. (1985) *Der Neandertaler und seine Zeit.* Kunst und Altertum am Rhein 118.

Böttger, T., Junge, F. W. & Litt, Th. (2000) Stable climatic conditions in central Germany during the last interglacial. *Journal of Quaternary Science* **15**, 469–473.

Boulton, G. S. (1968) Flow Tills and Related Deposits on some Vestspitsbergen Glaciers. *Journal of Glaciology* **7**, 391–412.

Boulton, G.S . (1978) Boulder shapes and grain-size distributions of debris as indicators of transport paths through a glacier and till genesis. *Sedimentology* **25**, 773–799.

Boulton, G. S. (1987) A theory of drumlin formation by subglacial deformation. In: Menzies, J. & Rose, J. (eds) *Drumlin Symposium.* Rotterdam: Balkema, 25–80.

Boulton, G. S. & Jones, A. S. (1979) *Stability of temperate ice caps and ice sheets resting on beds of deformable sediment.* Journal of Glaciology **24**, 29–43.

Boulton, G. S., Dent, D.L. & Morris, E.M. (1974) Subglacial shear formation and crushing in lodgement tills from south-east Iceland. *Geografiska Annaler* **56A**, 135–145.

Boulton, G. S., Smith, G. D., Jones, A.S. & Newsome, J. (1985) Glacial geology and glaciology of the last mid-latitude ice-sheets. *Journal of the Geological Society of London* **142**, 447–474.

Boulton, G. S., Dongelmans, P., Punkari, M. & Broadgate, M. (2001) Palaeoglaciology of an ice sheet through a glacial cycle: the European ice sheet through the Weichselian. *Quaternary Science Reviews* **20**, 591–625.

Bowen, D. Q. (1974) The Quaternary of Wales. In Owen, T. R. (ed.) *The Upper Palaeozoic and Post-Palaeozoic Rocks of Wales.* Cardiff: University of Wales Press, pp. 373– 426.

Bowen, D. Q. & Gibbard, P. L. (2007) The Quaternary is here to stay. *Journal of Quaternary Science* **22**, 3–8.

Bradwell, T., Stoker, M.S., Golledge, N. R., Wilson, C. K., Merritt, J. W., Long, D., Everest, J. D., Hestvik, O. B., Stevenson, A. G., Hubbard, A. L., Finlayson, A. G. & Mathers, H. E. (2008) The northern sector of the last British Ice Sheet: Maximum extent and demise. *Earth-Science Reviews* **88**, 207–226.

Braithwaite, R. J. (2008) Temperature and precipitation climate at the equilibrium-line altitude of glaciers expressed by the degree-day factor for melting snow. *Journal of Glaciology* **54**, 437–444.

Braun, D. D. (2011) The Glaciation of Pennsylvania, USA. In: Ehlers, J., Gibbard, P.L.& Hughes, P.D. (eds) *Quaternary Glaciations - Extent and Chronology - A Closer Look.* Elsevier, Developments in Quaternary Science **15**, 521–529.

Bretz, J. H. (1923) The Channeled Scabland of the Columbia Plateau. *Journal of Geology* **31**, 617–649.

Bridge, D. M. (1988) Corton Cliffs. In: Gibbard, P.L. & Zalasiewicz (eds) *Pliocene–Middle Pleistocene of East Anglia, Field Guide.* Cambridge: Quaternary Research Association, 119–125.

Bridge, D. M. & Hopson, P. M. (1985) Fine gravel, heavy mineral and grain size analysis of mid-Pleistocene deposits in the lower Waveney Valley, East Anglia. *Modern Geology* **9**, 129–144.

Bridgland, D. R. (1983) The Quaternary fluvial deposits of north Kent and eastern Essex. Ph.D. thesis. City of London Polytechnic.

Bridgland, D. R. (1988) The Pleistocene fluvial stratigraphy and palaeogeography of Essex. *Proceedings of the Geologists' Association* **99**, 291–314.

Bridgland, D. R. (1994) *The Quaternary of the Thames.* London: Chapman & Hall.

Bridgland, D. R. & D'Olier, B. (1995) The Pleistocene evolution of the Thames and Rhine drainage systems in the southern North Sea Basin. In: Preece, R. C. *Island Britain: A Quaternary Perspective.* Geological Society, London, Special Publications **96**, 27–45.

Briner, J. P., Miller, G. H., Davis, P. T., Bierman, P. R. & Caffee, M. (2003) Last Glacial Maximum ice sheet dynamics in Arctic Canada inferred from young erratics perched on ancient tors. *Quaternary Science Reviews* **22**, 437–444.

Briner, J. P., Young, N. E., Goehring, B. M. & Schäfer, J. M. (2012) Constraining Holocene ^{10}Be production rates in Greenland. *Journal of Quaternary Science* **27**, 2–6.

Bristow, C. R. & Cox, F. C. (1973) The Gipping Till: a reappraisal of East Anglian glacial stratigraphy. *Journal of Geological Society* **129**, 1–37.

Bristow, C. S. Drake, N. & Armitage, S. (2009) Deflation in the dustiest place on Earth: The Bodélé Depression, Chad. *Geomorphology* **105**, 50–58.

Broecker, W. S. (1965) Isotope geochemistry and the Pleistocene climatic record. In: Wright, H. E. Jr. & Frey, D. G. (eds) *The Quaternary of the United States.* Princeton: Princeton University Press, pp. 737–753.

Broecker, W. S. & Thurber, D. L. (1965) Uranium series dating of corals and oolites from Bahaman and Floridan Key Limestones. *Science* **149**, 58–60.

Broecker, W. S. & Denton, G. H. (1989) The role of ocean-atmosphere reorganisations in glacial cycles. *Geochimia Cosmochimia Acta* **53**, 2465–2501.

Broecker, W. S., Kennett, J. P., Flower, B. P., Teller, J. P., Trumbore, S., Bonani, G. & Wolfli, W. (1989) Routing of meltwater from the Laurentide Ice Sheet during the Younger Dryas cold episode. *Nature* **341**, 318–321.

Brookes, I. A., McAndrews, J.H. & Von Bitter, P. (1982) Quaternary interglacial and associated deposits in southwest Newfoundland. *Canadian Journal of Earth Sciences* **19**, 410–423.

Brückner-Röhling, S., Forsbach, H. & Kockel, F. (2005) The structural development of the German North Sea sector during the Tertiary and the Early Quaternary. *Zeitschrift der Deutschen Gesellschaft für Geowissenschaften* **156**, 341–355.

Brugger, K. A. (2006) Late Pleistocene climate inferred from the reconstruction of the Taylor River glacier complex, southern Sawatch Range, Colorado. *Geomorphology* **75**, 318–329.

Bruhns, K. O. (1994) *Ancient South America.* Cambridge: Cambridge University Press.

Brunnacker, K. (1957) Die Geschichte der Böden im jüngeren Pleistozän in Bayern. *Geologica Bavarica* **34**, 95.

Bruns, J. (1989) Stress indicators adjacent to buried channels of Elsterian age in North Germany. *Journal of Quaternary Science* **4**, 267–272.

Buch, L. von (1815) Über die Verbreitung großer Alpengeschiebe. *Abhandlungen der Physikalischen Classe der Akademie der Wissenschaften Berlin* **1804–1811**, 161–186.

Buckland, W. (1823) *Reliquiae Diluvianae; or, Observations on the organic Remains contained in Caves, Fissures and Diluvial Gravel, and on other Geological Phenomena, attesting the Action of an universal Deluge.* London: John Murray.

Bülow, W. von (1967) Zur Quartärbasis in Mecklenburg. *Berichte der Deutschen Gesellschaft für Geologische Wissenschaften* **A 12**, 375–404.

Bülow, W. von (ed.) (2000) *Geologische Entwicklung Südwest-Mecklenburgs seit dem Ober-Oligozän.* Schriftenreihe für Geowissenschaften **11**.

Buoncristiani, J.-F. & Campy, M. (2011) Quaternary Glaciations in the French Alps and Jura. In: Ehlers, J., Gibbard, P.L. & Hughes, P. D. (eds) *Quaternary Glaciations - Extent and Chronology - A Closer Look.* Amsterdam: Elsevier, Developments in Quaternary Science 15, pp. 117–126.

Burger, A. W. (1986) Sedimentpetrographie am Morsum Kliff, Sylt (Norddeutschland). *Mededelingen van het Werkgroep voor Tertiaire en Kwartaire Geologie* **23**, 99–109.

Burn, C. R. (1997) Cryostratigraphy, paleogeography, and climate change during the early Holocene warm interval, western Arctic coast, Canada. *Canadian Journal of Earth Sciences* **34**, 912–925.

Burn, C. R. (2007) Thermokarst topography. In: Elias, S. A. (ed.) *Encyclopaedia of Quaternary Science.* Amsterdam: Elsevier, pp. 300–309.

Burn, C. R. & Kokelj, S. V. (2009) The environment and permafrost of the Mackenzie Delta area. *Permafrost and Periglacial Processes* **20**, 83–105.

BURVAL Working Group (2009) Buried Quaternary valleys - a geophysical approach. *Zeitschrift der Deutschen Gesellschaft für Geowissenschaften* **160**, 237–247.

Busche, D. (1998) *Die Zentrale Sahara.* Gotha: Perthes.

Busche, D., Kempf, J. & Stengel, I. (2005) *Landschaftsformen der Erde. Bildatlas der Geomorphologie.* Darmstadt: Wissenschaftliche Buchgesellschaft.

Busschers, F. S., Kasse, C., Balen, R. T. van, Vandenberghe, J., Cohen, K. M., Weerts, H. J. T., Wallinga, J., Johns, C., Cleveringa, P. & Bunnik, F. P. M. (2007) Late Pleistocene evolution of the Rhine-Meuse system in the southern North Sea basin: imprints of climate change, sea-level oscillation and glacio-isostacy. *Quaternary Science Reviews* **26**, 3216–3248.

Busschers, F. S., van Balen, R. T., Cohen, K. M., Kasse, C., Weerts, H. J. T., Wallinga, J. & Bunnik, F. P. M. (2008) Response of the Rhine–Meuse fluvial system to Saalian ice-sheet dynamics. *Boreas* **37**, 377–398.

Cailleux, A. (1942) *Les actions éoliennes périglaciaires en Europe.* Societé Géologique de France **21** (Mémoir **46**), 176 S.

Caldenius, C. (1932) *Las Glaciaciones Cuaternarias en Patagonia y Tierra del Fuego.* Ministerio de Agricultura de la Nación. Dirección General de Minas y Geología, **95 (1)**, 148 S.

Calvet, M. (2004) The Quaternary glaciation of the Pyrenees. In: Ehlers, J., Gibbard, P. (eds) *Quaternary Glaciations - Extent and Chronology, Part I: Europe.* Amsterdam: Elsevier, Developments in Quaternary Science, 2a, pp. 119–128.

Calvet, M., Delmas, M., Gunnell, Y., Braucher, R. & Bourlés, D. (2011) Recent advances in research on Quaternary glaciations in the Pyrenees. In: Ehlers, J., Gibbard, P.L. & Hughes, P.D. (eds) *Quaternary Glaciations - Extent and Chronology - A Closer Look.* Amsterdam: Elsevier, Developments in Quaternary Science 15, pp. 127–140.

Cammeraat, E. & Rappol, M. (1987) On the Relationship of Bedrock Lithology and Grain Size Distribution of Till in Western Allgäu (West Germany) and Vorarlberg (Austria). *Jahrbuch der Geologischen Bundesanstalt* **130**, 383–389.

Carneiro-Filho, A., Schwartz, D., Tatumi, S. H. & Rosique, T. (2002) Amazonian paleodunes provide evidence for drier climate phases during the late Pleistocene–Holocene. *Quaternary Research* **58**, 205–209.

Carr, S. (2004) The North Sea basin. In: Ehlers, J. & Gibbard, P. L. (eds) *Quaternary Glaciations–Extent and Chronology, Part I: Europe*. Amsterdam, Elsevier: Developments in Quaternary Science **2**, 261–270.

Carr, S. J., Haflidason, H. & Sejrup, H. P. (2008) Micromorphological evidence supporting Late Weichselian glaciation of the Northern North Sea. *Boreas* **29**, 315–328.

Cepek, A. G. (1967) Stand und Probleme der Quartärstratigraphie im Nordteil der DDR. *Berichte der Deutschen Gesellschaft für Geologische Wissenschaften* **A12**, 375–404.

Cepek, A. G. (1995) Stratigraphie und Inlandeisbewegungen im Pleistozän an der Struktur Rüdersdorf bei Berlin. *Berliner Geowissenschaftliche Abhandlungen* **A168**, 103–133.

Chamberlin, T. C. (1894) Glacial phenomena of North America. In: Geikie, J. (ed.) *The Great Ice Age*, 3rd edition. New York: D. Appleton & Co, pp. 724–774.

Chamberlin, T. C. (1895) The classification of American glacial deposits. *Journal of Geology* **3**, 270–277.

Chamberlin, T. C. (1896) Nomenclature of glacial formations. *Journal of Geology* **4**, 872–876.

Charlesworth, C. K. (1927) *II*. The Readvance, Marginal Kame-moraine of the South of Scotland, and some Later Stages of Retreat. *Transactions of the Royal Society of Edinburgh* **55**, 25–50.

Charlesworth, C. K. (1929) The South Wales end-moraine. *Quarterly Journal of the Geological Society* **85**, 335–358.

Charpentier, J. de (1842) Sur l'application de l'hypothése de M. Venetz aux phénomènes erratiques du Nord. *Bibliothéque universelle de Genève* **XXXIX**, 327–346.

Chiverrell, R. C. & Thomas G. S. P. (2010) Extent and timing of the Last Glacial Maximum (LGM) in Britain and Ireland: a review. *Journal of Quaternary Science* **25**, 535–549.

Church, M. (1972) Baffin Island sandurs: a study of arctic fluvial processes. *Geological Survey of Canada Bulletin* **216**, 208.

Cimiotti, U. (1983) Zur Landschaftsentwicklung des mittleren Trave-Tales zwischen Bad Oldesloe und Schwissel, Schleswig-Holstein. *Berliner Geographische Studien* **13**, 92.

Çiner, A., Deynoux, M. & Çörekcioglu, E. (1999) Hummocky moraines in the Namaras and Susam Valleys, Central Taurids, SW Turkey. *Quaternary Science Reviews* **18**, 659–669.

Çiner, A., Sarıkaya, M.A., Yıldırım, C., 2015. Late Pleistocene piedmont glaciations in the Eastern Mediterranean; insights from cosmogenic 36Cl dating of hummocky moraines in southern Turkey. *Quaternary Science Reviews* **116**, 44–56.

Clague, J. J. & Ward, B. (2011) Pleistocene glaciation of British Columbia. In: Ehlers, J., Gibbard, P. L. & Hughes, P. D. (eds) *Quaternary Glaciations - Extent and Chronology - A Closer Look*. Amsterdam: Elsevier, Developments in Quaternary Science 15, pp. 563–574.

Clapperton, C. (1993) Nature and environmental changes in South America at the Late Glacial Maximum. *Palaeogeography, Palaeoclimatology, Palaeoecology* **101**, 189–208.

Clark, C. D., Gibbard, P.L. & Rose, J. (2004) Glacial limits in the British Isles. In: Ehlers, J. & Gibbard, P.L. (eds) *Extent and Chronology of Glaciation. Volume 1 Europe*. Elsevier Science: Amsterdam, pp. 47–82.

Clark, C. D., Hughes, A. L. C., Greenwood, S. L., Jordan, C. & Sejrup, H. P. (2012) Pattern and timing of retreat of the last British-Irish Ice Sheet. *Quaternary Science Reviews* **44**, 112–146.

Clarke, G. K. C. (1987) Fast glacier flow: ice streams, surging and tidewater glaciers. *Journal of Geophysical Research* **92**, 8835–8841.

Clark, P. U. (1992) Surface form of the Laurentide Ice Sheet and its implications to ice-sheet dynamics. *Geological Society of America Bulletin* **104**, 595–605.

Clark, P. U. (1994) Unstable behaviour of the Laurentide Ice Sheet over deforming sediment and its implications for climatic change. *Quaternary Research* **41**, 19–25.

Clark, P. U. & Lea, P. D. (eds) (1992) *The Last Interglacial-Glacial Transition in North America*. Geological Society of America, Special Paper **270**.

Clark, P. U., Dyke, A. S., Shakun, J. D., Carlson, A. E., Clark, J., Wohlfarth, B., Mitrovica, J. X., Hostetler, S. W. & McCabe, A. M. (2009) The Last Glacial Maximum. *Science* **325**(5941), 710–714.

Clayton, L. Attig, J.W., Mickelson, D.M. (2008) Effects of late Pleistocene permafrost on the landscape of Wisconsin, USA. *Boreas* **30**:173–188.

CLIMAP Project Members (1976) The Surface of the Ice-Age Earth. *Science* **191**, 1131–1137.

CLIMAP Project Members (1981) Seasonal reconstructions of the Earth's surface at the last glacial maximum. *Geological Society of America Memoirs* **145**, 1–18.

Cohen, K. M. & Gibbard, P. L. (2011) Global chronostratigraphical correlation table for the last 2.7 million years. Subcommission on Quaternary Stratigraphy (International Commission on Stratigraphy), Cambridge, England. Available at http://quaternary.stratigraphy.org/charts/ (accessed 16 May 2015).

Cohen, K. M., Gibbard, P. L. & Busschers, F. S. (2005) Stratigraphical implications of an Elsterian proglacial "North Sea" lake. Bern (Switzerland). Abstract booklet, SEQS meeting.

Cohen, K. M., MacDonald, K., Joordens, J. C. A., Roebroeks, W. & Gibbard, P. L. (2012) The earliest occupation of north-west Europe: a coastal perspective. *Quaternary International* **271**, 70–83.

Coleman, A.P. (1894) Interglacial fossils from the Don Valley. *American Geologist* **13**, 85–95.

Colgan, P. M. (2007) Evidence of glacier recession. In: Elias, S. A. (ed.) *Encyclopaedia of Quaternary Science*. Amsterdam: Elsevier, pp. 798–808.

Colgan, P. M., Mickelson, D. M. & Cutler, P. M. (2003) Ice marginal terrestrial landsystems: southern Laurentide ice sheet margin. In: Evans, D.J.A. (ed.) *Glacial Landsystems*. London: Arnold, pp. 111–142.

Colhoun, E. A. & Barrows, T. T. (2011) The glaciation of Australia. In: Ehlers, J., Gibbard, P.L. & Hughes, P.D. (eds) *Quaternary Glaciations - Extent and Chronology - A Closer Look*. Elsevier, Developments in Quaternary Science **15**, pp. 1037–1046.

Colinvaux, P. A., De Oliveira, P. E., Moreno, J. E., Miller, M. C. & Bush, M. B. (1996) A long pollen record from lowland Amazonia: Forest and cooling in glacial times. *Science* **274**, 85–88.

Colinvaux, P. A., De Oliveira, P. E. & Bush, M. B. (2000) Amazon and Neotropical plant communities on glacial time scales: the failure of the aridity and refuge hypotheses. *Quaternary Science Reviews* **19**, 141–169.

Coope, G. R. (1977) Quaternary Coleoptera as aids in the interpretation of environmental history. In: Shotton, F. W. (ed.) *British Quaternary Studies, Recent Advances*. Oxford: Clarendon Press, pp. 55–68.

Coope, G. R., Field, M. H., Gibbard, P. L., Greenwood, M. & Richards, A. E. (2002) Palaeontology and biostratigraphy of Middle Pleistocene river sediment in the Mathon Member, at Mathon, Herefordshire. *Proceedings of the Geologists Association* **113**, 237–258.

Coronato, A. & Rabassa, J. (2011) Pleistocene Glaciations in Southern Patagonia and Tierra del Fuego. In: Ehlers, J., Gibbard, P. L. & Hughes, P. D. (eds) *Quaternary Glaciations - Extent and Chronology - A Closer Look*. Amsterdam: Elsevier, Developments in Quaternary Science 15, pp. 715–728.

Coronato, A., Martínez, O. & Rabassa, J. (2004a) Glaciations in Argentine Patagonia, southern South America. In: Ehlers, J. & Gibbard, P. L. (eds): *Quaternary Glaciations–Extent and Chronology, Part III: South America, Asia, Africa, Australasia, Antarctica*. Elsevier, Developments in Quaternary Science **2**, 49–67.

Coronato, A., Meglioli, A. & Rabassa, J. (2004b) Glaciations in the Magellan Straits and Tierra del Fuego, southernmost South America. In: Ehlers, J. & Gibbard, P. L. (eds) *Quaternary Glaciations–Extent and Chronology, Part III: South America, Asia, Africa, Australasia, Antarctica*. Elsevier, Developments in Quaternary Science **2**, 45–48.

Cotta, B. (1848) *Briefe über Alexander von Humboldt's Kosmos. Ein Commentar zu diesem Werke für gebildete Laien. Erster Theil*. Leipzig: Weigel.

Cotta, B. (1867) *Die Geologie der Gegenwart*, 2. Auflage. Leipzig: J. J. Weber.

Cox, F. C. (1985) The tunnel valleys of Norfolk, East Anglia. *Proceedings of the Geologists' Association* **96**, 357–369.

Cox, F. C. & Nickless, E. F. P. (1972) Some aspects of the glacial history of central Norfolk. *Bulletin of the Geological Survey of Great Britain* **42**, 79–98.

Crossland, C. J., Kremer, H. H., Lindeboom, H. J., Marshall Crossland, J. I. & Le Tissier, M. D. A. (eds) (2005) *Coastal Fluxes in the Anthropocene*. Berlin: Springer.

Crutzen, P. J. (2002) Geology of mankind. *Nature* **415**, 23.

Crutzen, P. J. & Stoermer, E. F. (2000) The Anthropocene. *Global Change Newsletter* **41**, 17–18.

Currey, D. R. (1990) Quaternary palaeolakes in the evolution of semi-desert basins, with special emphasis on Lake Bonneville and the Great Basin, USA. *Palaeogeography, Palaeoclimatology, Palaeoecology* **76**, 189–214.

Curry, B. B. & Follmer, L. R. (1992) The last interglacial-glacial transition in Illinois: 123-25 ka. In: Clark, P.U. & Lea, P.D. (ed) *The Last Interglacial-Glacial Transition in North America*. Geological Society of America, Special Paper 270, pp. 71–88.

Curry, B. B., Grimley, D. A. & McKay III, E. D. (2011) Quaternary glaciations in Illinois. In: Ehlers, J., Gibbard, P. L. & Hughes, P.D. (eds) *Quaternary Glaciations - Extent and Chronology - A Closer Look*. Elsevier, Developments in Quaternary Science **15**, 467–487.

Cuvier, G. (1827) *Essay on the Theory of the Earth*. Fifth edition. Edinburgh: Blackwood.

Cvijić, J. (1917) L'époque glaciaire dans la péninsule des Balkanique. *Annales de Géographie* **26**, 189–218.

Damuth, J. E. & Fairbridge, R. W. (1970) Equatorial Atlantic deep-sea arkosic sands and ice-age aridity in tropical South America. *Geological Society of America Bulletin* **81**, 189–206.

Dana, J. D. (1863) *Manual of Geology*. Philadelphia: Theodore Bliss & Co.

Dansgaard, W., Johnsen, S. J., Clausen, H. B., Dahl-Jensen, N. S., Gundestrup, N. S., Hammer, C. U., Hvidberg, C. S., Steffensen, J. P., Sveinbjörnsdottir, A. E., Jouzel, J. & Bond, G. (1993) Evidence for general instability of past climate from a 250-kyr ice-core record. *Nature* **364**, 218–220.

Darwin, C. R. (1839) Observations on the parallel roads of Glen Roy, and of other parts of Lochaber in Scotland, with an attempt to prove that they are of marine origin. [Read 7 February] *Philosophical Transactions of the Royal Society* **129**, 39–81.

Darwin, C. (1842) On the distribution of erratic boulders and on the contemporaneous unstratified deposits of South America. *Transactions of the Geological Society of London* **6**, 415–431.

Darwin, C. (1859) *On the Origin of Species by Means of Natural Selection, or The Preservation of Favoured Races in the Struggle for Life.* London: John Murray.

Dawson, A. G. & Smith, D.E. (1983) *Shorelines and Isostasy.* London: Academic Press, London Institute of British Geographers, Special Publication no. 16.

Dawson, A. G., Matthews, J. A. & Shakesby, R. A. (1987) Rock platform erosion on periglacial shores: a modern analogue for Pleistocene rock platforms in Britain. In: Boardman, J. (ed.) *Periglacial Processes and Landforms in Britain and Ireland.* Cambridge: Cambridge University Press, pp. 173–182.

De Schepper, S., Gibbard, P. L., Salzmann, U. & Ehlers, J. (2014) A global synthesis of the marine and terrestrial evidence for glaciation during the Pliocene Epoch. *Earth Science Reviews* **135**, 83–102.

Delmas, M., Calvet, M., Gunnell, M., Braucher, R. & Bourlès, D. (2011) Palaeogeography and ^{10}Be exposure-age chronology of Middle and Late Pleistocene glacier systems in the northern Pyrenees: implications for reconstructing regional palaeoclimates. *Palaeogeography, Palaeoclimatology, Palaeoecology* **305**, 109–122.

Dennison, B. & Mansfield, V. N. (1976) Glaciations and dense interstellar clouds. *Nature* **261**, 32–34.

Denton, G. H. & Hughes, T. J. (eds) (1981) *The Last Great Ice Sheets.* New York: John Wiley and Sons.

Denton, G. H. & Hendy, C. H. (1994) Younger Dryas Age advance of the Franz Josef glaciers in the Southern Alps of New Zealand. *Science* **264**(5164), 1434–1437.

Denton, G. H. & Hughes, T. J. (2002) Reconstructing the Antarctic Ice Sheet at the Last Glacial Maximum. *Quaternary Sciences Reviews* **21**, 193–202.

Denton G. H., Heusser C. J., Lowell, T. V., Moreno P. I., Andersen B. G., Heusser L. E., Schlüchter, C. & Marchant, D. R. (1999) Interhemispheric linkage of paleoclimate during the last glaciation. *Geografiska Annaler* **A81**(2), 107–153.

Depéret, C. (1918) Essai de coordination chronologique générale des temps quaternaires. *Comptes rendus de l'Académie des Sciences* **167**, 418–422.

Desnoyers, J. (1829) Observations sur un ensemble de dépôts marins plus récens que les terrains tertiaires du bassin de la Seine, et constituant une Formation géologique distincte; précédées d'um Aperçu de la non simultanéité des bassins tertiaires. *Annales des Sciences Naturelles (Paris)* 171–214, 402–491.

DeVogel, S. B., Magee, J. W., Manley, W. F. & Miller, G. H. (2004) A GIS-based reconstruction of late Quaternary paleohydrology: Lake Eyre, arid central Australia. *Palaeogeography, Palaeoclimatology, Palaeoecology* **204**, 1–13.

Devoy, R. J. N. (1979) Flandrian sea level changes and vegetation history of the Lower Thames estuary. *Philosophical Transactions of the Royal Society of London* **B285**, 355–407.

Deynoux, M. (1980) Les formations glaciaires du Précambrien terminal et de la fin de l'Ordovicien en Afrique de l'ouest. Deux exemples de glaciation d'inlandsis sur une plateforme stable. *Travaux des Laboratoires des Sciences de la Terre St Jérome, Marseille*, **B17**, 554. *Diatomeenassoziationen an holozänen Sedimenten der zentralen Ostsee.* Berichte Reports, Geologisch-Paläontologisches Institut der Universität Kiel **9**, 200.

Dietrich, G. (1992) *Allgemeine Meereskunde*, 3. Auflage. Stuttgart: Borntraeger.

Dionne, J.-C. (1981) A boulder-strewn tidal flat, north shore of the Gulf of St. Lawrence, Québec. *Géographie Physique et Quaternaire* **XXXV**, 261–267.

Dolgushin, L. D. (1961) Main features of the modern glaciation in the Urals. *International Association of Hydrological Sciences* **54**, 335–347.

Donner, J. (1995) *The Quaternary History of Scandinavia*. Cambridge: Cambridge University Press, World and Regional Geology no. **7**.

Donner, J. J. & West, R. G. (1955) Ett drumlinsfält pa ön Skye, Skottland. *Eripainos Terrasta* **2**, 45–48.

Doppler, G. (1980) *Das Quartär im Raum Trostberg an der Alz im Vergleich mit dem nordwestlichen Altmoränengebiet des Salzachvorlandgletschers (Südostbayern)* Dissertation, München.

Doppler, G. & Geiss, E. (2005) Der Tüttensee im Chiemgau–Toteiskessel statt Impaktkrater. Available at: http://www.lfu.bayern.de/geologie/meteorite/bayern/doc/tuettensee.pdf (accessed 13 May 2015).

Dowdeswell, J. A. & Ottesen, D. (2013) Buried iceberg ploughmarks in the early Quaternary sediments of the central North Sea: A two-million year record of glacial influence from 3D seismic data. *Marine Geology* **344**, 1–9.

Drake, L. D. (1983) Ore plumes in till. *The Journal of Geology* **91**, 707–713.

Draxler, I. & Husen, D. van (1989) Ein ^{14}C-datiertes Profil in der Niederterrasse bei Neurath (Stainz, Stmk). *Zeitschrift für Gletscherkunde und Glazialgeologie* **25**, 123–130.

Dredge, L. A. & Thorleifson, L.H. (1987) The Middle Wisconsinan history of the Laurentide Ice Sheet. *Géographie physique et Quaternaire* **XLI**, 215–235.

Dredge, L. A. & Cowan, W.R. (1989) Quaternary geology of the southwestern Canadian Shield. In: Fulton, R.T. (ed.) *Quaternary Geology of Canada and Greenland*. Geological Survey of Canada, Geology of Canada, 1 (also: Geological Society of America, The Geology of North America, K-1), pp. 214–235.

Dreesbach, R. (1985) *Sedimentpetrographische Untersuchungen zur Stratigraphie des Würmglazials im Bereich des Isar-Loisachgletschers*. Dissertation, München.

Dreimanis, A. (1976) Tills, their origin and properties. In Legget, R. F. (ed.) *Glacial Till*. The Royal Society of Canada, Special Publication **12**, 11–49.

Dreimanis, A. (1988) Tills: Their genetic terminology and classification. In: Goldthwait, R. P. & Matsch, C. L. (eds) *Genetic Classification of Glacigenic Deposits*. Rotterdam Brookfield: Balkema, pp. 17–83.

Dreimanis, A. (1990) Formation, deposition and identification of subglacial and supraglacial till. In: Kujansuu, R. & Saarnisto, M. (eds) *Glacial Indicator Tracing*. Rotterdam, Brookfield: Balkema, pp. 35–59.

Dreimanis, A., Liivrand, E. & Raukas, A. (1989) Glacially redeposited pollen in tills of southern Ontario, Canada. *Canadian Journal of Earth Sciences* **26**, 1667–1676.

Drescher-Schneider, R. (2000) Die Vegetations- und Klimaentwicklung im Riß/Würm-Interglazial und im Früh- und Mittelwürm in der Umgebung von Mondsee. *Mitteilungen der Kommission für Quartärforschung der Österreichischen Akademie der Wissenschaften* **12**, 37–92.

Drygalski, E. von (1897) *Grönlandexpedition der Gesellschaft für Erdkunde zu Berlin 1891–1893.* Berlin: Kühl.

Duk-Rodkin, A. & Barendregt, R.W. (2011) Stratigraphical record of glacials/interglacials in Northwest Canada. In: Ehlers, J., Gibbard, P.L. & Hughes, P.D. (eds) *Quaternary Glaciations - Extent and Chronology - A Closer Look.* Elsevier, Developments in Quaternary Science **15**, 661–698.

Duller, G. A. T. (2008) Single grain optical dating of Quaternary sediments: why aliquots size matters in luminescence dating. *Boreas* **37**, 589–612.

Dyke, A. S. (2004) An outline of North American deglaciation with emphasis on central and northern Canada. In: Ehlers, J. & Gibbard, P. L. (eds) *Quaternary Glaciations–Extent and Chronology Part II: North America.* Amsterdam: Elsevier, Developments in Quaternary Science **2**, 373–424.

Dyke, A. S. & Prest, V.K. (1987) The Late Wisconsinan and Holocene history of the Laurentide Ice Sheet. *Géographie Physique et Quaternaire* **41**, 237–263.

Dyke, A. S., Vincent, J.-S., Andrews, J.T., Dredge L.A. & Cowan, W.R. (1989) The Laurentide Ice Sheet and an introduction to the Quaternary geology of the Canadian Shield. In: Fulton, R.J. (ed.) *Quaternary Geology of Canada and Greenland.* Geological Survey of Canada, Geology of Canada, v. 1 (also Geological Society of America, The Geology of North America, v. K-1), pp. 178–189.

Easterbrook, D. J. (1986) Stratigraphy and chronology of Quaternary deposits of the Puget Lowland and Olympic Mountains of Washington and the Cascade Mountains of Washington and Oregon. *Quaternary Science Reviews* **5**, 135–159.

Easterbrook, D. J. (1992) Advance and retreat of Cordilleran Ice Sheets in Washington, USA. *Géographie Physique et Quaternaire* **46**, 51–68.

Easterbrook, D. J., Briggs, N. D., Westgate, J. A. & Gorton, M. P. (1981) Age of the Salmon Springs glaciation in Washington. *Geology* **9**, 87–93.

Eberl, B. (1930) *Die Eiszeitenfolge im nördlichen Alpenvorlande. Ihr Ablauf, ihre Chronologie auf Grund der Aufnahmen des Lech- und Illergletschers.* Augsburg: Benno Filser.

Edelman, C. H., Florschütz, F. & Jeswiet, J. (1936) *Über spätpleistozäne und frühholozäne kryoturbate Ablagerungen in den östlichen Niederlanden.* Verhandelingen van het Geologisch-Mijnbouwkundig Genootschap voor Nederland en Koloniën, Geologische Serie **11**, 301–336.

Ehlers, J. (1978) Die quartäre Morphogenese der Harburger Berge und ihrer Umgebung. *Mitteilungen der Geographischen Gesellschaft in Hamburg* **68**, 181.

Ehlers, J. (1990) Untersuchungen zur Morphodynamik der Vereisungen Norddeutschlands unter Berücksichtigung benachbarter Gebiete. *Bremer Beiträge zur Geographie und Raumplanung* **19**, 166.

Ehlers, J. (1996) *Quaternary and Glacial Geology.* Chichester: John Wiley and Sons.

Ehlers, J. (2008) *Die Nordsee. Vom Wattenmeer zum Nordatlantik.* Darmstadt: Primus.

Ehlers, J. (2011) *Das Eiszeitalter.* Spektrum Akademischer Verlag.

Ehlers, J. & Stephan, H.-J. (1983) Till fabric and ice movement. In: Ehlers, J. (ed.) *Glacial Deposits in North-West Europe*. Rotterdam: Balkema, pp. 267–274.

Ehlers, J. & Gibbard, P.L. (1991) Anglian glacial deposits in Britain and the adjoining off-shore regions. In: Ehlers, J., Gibbard, P.L. & Rose, J. (eds) *Glacial Deposits in Great Britain and Ireland*. Rotterdam: Balkema, pp. 17–24.

Ehlers, J. & Gibbard, P.L. (eds) (2004a) *Quaternary Glaciations: Extent and Chronology. Part I: Europe*. Amsterdam: Elsevier. Developments in Quaternary Science, **2a**.

Ehlers, J. & Gibbard, P.L. (eds) (2004b) *Quaternary Glaciations: Extent and Chronology. Part II: North America*. Amsterdam: Elsevier, Developments in Quaternary Science, **2b**.

Ehlers, J. & Gibbard, P.L. (eds) (2004c) *Quaternary Glaciations - Extent and Chronology. Part III: South America, Asia, Africa, Australasia, Antarctica*. Amsterdam: Elsevier, Developments in Quaternary Science, **2c**.

Ehlers, J. & Gibbard, P. (2008) Extent and chronology of Quaternary glaciation. *Episodes* **31**, 211–218.

Ehlers, J., Meyer, K.-D. & Stephan, H.-J. (1984) The pre-Weichselian glaciations of North-West Europe. *Quaternary Science Reviews* **3**, 1–40.

Ehlers, J., Gibbard, P.L. & Whiteman, C.A. (1987) Recent investigations of the Marly Drift of northwest Norfolk, England. In: Van der Meer, J.J.M. (ed.) *Tills and Glaciotectonics*. Rotterdam: Balkema, pp. 39–54.

Ehlers, J., Gibbard, P.L. & Hughes, P.D. (eds) (2011a) *Quaternary Glaciations - Extent and Chronology - A Closer Look*. Elsevier, Developments in Quaternary Science **15**.

Ehlers, J., Gibbard, P.L. & Hughes, P.D. (2011b) Introduction. In: Ehlers, J., Gibbard, P.L. & Hughes, P.D. (eds) *Quaternary Glaciations - Extent and Chronology - A Closer Look*. Elsevier, Developments in Quaternary Science **15**, 1–14.

Ehlers, J., Grube, A., Stephan, H.-J. & Wansa, S. (2011c) Pleistocene glaciations of North Germany: new results. In: Ehlers, J., Gibbard, P.L. & Hughes, P.D. (eds) *Quaternary Glaciations - Extent and Chronology - A Closer Look*. Elsevier, Developments in Quaternary Science **15**.

Eichler, H. & Sinn, P. (1974) Zur Gliederung der Altmoränen im westlichen Salzachgletschergebiet. *Zeitschrift für Geomorphologie NF* **18**, 132–158.

Eissmann, L. (1967) Glaziäre Destruktionszonen (Rinnen, Becken) im Altmoränengebiet des Norddeutschen Tief-landes *Geologie* **16**, 804–833.

Eissmann, L. (1975) Das Quartär der Leipziger Tieflandsbucht und angrenzender Gebiete um Saale und Elbe. Modell einer Landschaftsentwicklung am Rand der europäischen Kontinentalvereisung. Schriftenreihe für Geologische Wissenschaften **2**, 263.

Eissmann, L. (1978) Mollisoldiapirismus. *Zeitschrift für Angewandte Geologie* **24**, 130–138.

Eissmann, L. (1981) *Periglaziäre Prozesse und Permafroststrukturen aus sechs Kaltzeiten des Quartärs. Ein Beitrag zur Periglazialgeologie aus der Sicht des Saale-Elbe-Gebietes*. Altenburger Naturwissenschaftliche Forschungen 1, 171.

Eissmann, L. (1997) *Das quartäre Eiszeitalter in Sachsen und Nordostthüringen*. Altenburger Naturwissenschaftliche Forschungen **8**, 98.

Elias, S. A. & Brigham-Grette, J. (2013) Late Pleistocene glacial events in Beringia. In: Elias, S. A. & Mock, C. J. (eds) *Encyclopedia of Quaternary Science*. Amsterdam, Elsevier, 191–201.

Ellwanger, D. (1988) Würmeiszeitliche Rinnen und Schotter bei Leutkirch/Memmingen. *Jahreshefte des Geologischen Landesamtes Baden-Württemberg* **30**, 207–229.

Ellwanger, D. (1990) Zur Riß-Stratigraphie im Andelsbach-Gebiet (Baden-Württemberg). *Jahreshefte des Geologischen Landesamtes Baden-Württemberg* **32**, 235–245.

Elverhøi, A., Siegert, M., Dowdeswell, J. & Svendsen, J.-I. (2002) The Eurasian Arctic during the last Ice Age. *American Scientist* **90**, 32.

Erd, K. (1965) Pollenanalytische Gliederung des mittelpleistozänen Richtprofils Pritzwalk-Prignitz. *Eiszeitalter und Gegenwart* **16**, 252–253.

Ergenzinger, P. J. (1978) Das Gebiet Enneri Misky im Tibesti Gebirge, Republique du Tchad Erläuterungen zu einer Geomorphologischen Karte 1:200,000. Berliner Geographische Abhandlungen **23**, 71.

Eriksson, K. (1983) *Till investigations and mineral prospecting.* In: Ehlers, J. (ed.) *Glacial Deposits in North-West Europe.* Rotterdam: Balkema, pp. 107–113.

Eronen, M. (1974) The history of the Litorina Sea and associated Holocene events. *Societas Scientiarum Fennica, Commentationes Physico-Mathematicae* **44**, 79–195.

Eschman, D. F. & Mickelson, D.M. (1986) Correlation of glacial deposits in the Huron, Lake Michigan and Green Bay Lobes in Michigan and Wisconsin. *Quaternary Science Reviews* **5**, 53–57.

Evans, D. J. A. (2010) Defending and testing hypotheses: a response to John Shaw's paper 'In defence of the meltwater (megaflood) hypothesis for the formation of subglacial bedform fields'. *Journal of Quaternary Science* **25**, 822–823.

Evans, D.J.A. & Twigg, D.R. (2002) The active temperate glacial landsystem:a model based on Breiðamerkurjökull and Fjallsjökull, Iceland. *Quaternary Science Reviews* **21**, 2143–2177.

Eyles, N. (1983) Modern Icelandic glaciers as depositional models for "hummocky moraine" in the Scottish Highlands. In: Evenson, E.B., Schlüchter, C. & Rabassa, J. (eds) *Tills and Related Deposits.* Rotterdam: Balkema, 47–60.

Eyles, N. & McCabe, A. M. (1989) The Late Devensian (22,000 BP) Irish Sea Basin: The sedimentary record of a collapsed ice sheet margin. *Quaternary Science Reviews* **8**, 304–351.

Ezcurra, E. (2006) *Global Deserts Outlook.* Nairobi, United Nations Environment Programme.

Falsan, A. & Chantre, E. (1877/78) Catalogue des blocs erratiques et des surfaces de roches rayées observés dans la partie moyenne du Bassin du Rhone et classés par régions géographiques. *Annales de la Societé d'Agriculture, Histoire Naturelle et Arts Utiles de Lyon* 4e Série T **10**, 117–441, 5e Série, T. **1**, 510–572.

Fässler, H. (2005) *Reise in Schwarz-Weiß. Schweizer Ortstermine in Sachen Sklaverei.* Zürich: Rotpunktverlag.

Federici, P. R., Granger, D.E., Ribolini, A., Spagnolo, M., Pappalardo, M. & Cyr, A.J. (2012) Last Glacial Maximum and the Gschnitz stadial in the Maritime Alps according to 10Be cosmogenic dating. *Boreas* **41**, 277–291.

Felix-Henningsen, P. (1979) *Merkmale, Genese und Stratigraphie fossiler und reliktischer Bodenbildungen in saalezeitlichen Geschiebelehmen Schleswig-Holsteins und Süd-Dänemarks.* Dissertation, Universität Kiel.

Felix-Henningsen, P. (1983) Palaeosols and their stratigraphhical interpretation. In: Ehlers, J. (ed.) *Glacial Deposits in North-West Europe.* Rotterdam: Balkema, 289–295.

Felix-Henningsen, P. (1990) *Die mesozoisch-tertiäre Verwitterungsdecke (MTV) im Rheinischen Schiefergebirge. Aufbau, Genese und quartäre Überprägung.* Relief, Boden: Paläoklima **6**.

Felsch, Ph. (2010) *Wie August Petermann den Nordpol erfand.* München: Sammlung Luchterhand.

Fenton, C. R., Hermanns, R.L., Blikra, L.H., Kubik, P.W., Bryant, C., Niedermann, S. & Meixner, A. (2011) Regional ^{10}Be production rate calibration for the past 12 ka deduced from the radiocarbon-dated Grøtlandsura and Russenes rock avalanches at 69°N, Norway. *Quaternary Geochronology* **6**, 437–452.

Feruglio, E. (1944) Estudios geológicos y glaciológicos en la región del Lago Argentino (Patagonia). *Boletín de la Academia Nacional de Ciencias* **37**, 1–208.

Fiebig, M. & Preusser, F. (2003) Das Alter fluvialer Ablagerungen aus der Region Ingolstadt (Bayern) und ihre Bedeutung für die Eiszeitenchronologie des Alpenvorlandes. *Zeitschrift für Geomorphologie, Neue Folge* **47**, 449–467.

Fiebig, M. & Preusser, F. (2008) Pleistocene glaciations of the northern Alpine Foreland. *Geographica Helvetica* **63**, 145–150.

Fiebig, M., Ellwanger, D. & Doppler, G. (2011) Pleistocene glaciations of southern Germany. In: Ehlers, J., Gibbard, P.L. & Hughes, P.D. (eds) *Quaternary Glaciations - Extent and Chronology - A Closer Look.* Elsevier, Developments in Quaternary Science **15**, 163–174.

Figge, K. (1980) Das Elbe-Urstromtal im Bereich der Deutschen Bucht (Nordsee). *Eiszeitalter und Gegenwart* **30**, 203–211.

Fink, D., Hughes, P. & Fenton, C. (2012) Extent, timing and palaeoclimatic significance of glaciation in the High Atlas, Morocco, over the past 20 ka. *Proceedings of the 21st International Radiocarbon Conference*, Abstract Booklet, Abstract S18-P-348, p. 558.

Firbas, F. (1927) Beiträge zur Kenntnis der Schieferkohlen des Inntals und der interglazialen Waldgeschichte der Ostalpen. *Zeitschrift für Gletscherkunde* **15**, 261–277.

Firbas, F. (1949) *Spät- und Nacheiszeitliche Waldgeschichte Mitteleuropas Nördlich der Alpen* (2 vols). Jena: Gustav Fischer.

Firestone, R. B., West, A., Kennett, J. P., Becker, L., Bunch, T. E., Revay, Z. S., Schultz, P. H., Belgya, T., Kennett, D. J., Erlandson, J. M., Dickenson, O. J., Goodtear, A. C., Harris, R. S., Howard, G. A., Kloosterman, J. B., Lechler, P., Mayewski, P. A., Montgomery, J., Poreda, R., Darrah, T., Que Hee, S. S., Smith, A. R., Stich, A., Topping, W., Wittke, J. H. & Wolbach, W. S. (2007) Evidence for an extraterrestrial impact 12,900 years ago that contributed to the megafaunal extinctions and the Younger Dryas cooling. *PNAS* **104**(41), 16016–16021.

Fischer, K. (2008) Die Säugetierfunde aus dem Eem-Interglazial von Klinge bei Cottbus (Brandenburg). *Natur und Landschaft in der Niederlausitz* **27**, 140–166.

Fish, P. R. & Whiteman, C.A. (2001) Chalk micropalaeontology and the provenancing of Middle Pleistocene Lowestoft Formation till in eastern England. *Earth Surface Processes and Landforms* **26**, 953–970.

Flammarion, N. C. (1886–87) *De Wereld vóór de schepping van den mensch.* Zutphen: W. J. Thieme & Cie.

Fletcher, W., Sanchez-Goñi, M.F., Allen, J.R.M., Cheddadi, R., Combourieu Nebout, N., Huntley, B., Lawson, I., Londeix, L., Magri, D., Margari, V., Müller, U.C., Naughton, F., Novenko, E., Roucoux, K. & Tzedakis, P.C. (2010) Millennial-scale variability during the last glacial in vegetation records from Europe. *Quaternary Science Reviews* **29**, 2839–2864.

Flint, R. F. (1971) *Glacial and Quaternary Geology*. New York: Wiley.

Flint, R. & Fidalgo, F. (1964) Glacial geology of the East Flank of the Argentine Andes between latitude 39°10' S and latitude 41°20' S. *Geological Society of America Bulletin* **75**, 335–352.

Fliri, F. (1973) *Beiträge zur Geschichte der alpinen Würmvereisung: Forschungen am Bänderton von Baumkirchen (Inntal, Nordtirol)* Zeitschrift für Geomorphologie N. F., Supplement-Band **16**, 1–14.

Florin, M.-B. (1977) Late-Glacial and Pre-Boreal vegetation in southern central Sweden.II. Pollen and diatom diagrams. *Striae* **5**, 60.

Follmer, L. R. (1978) The Sangamon Soil in its type area - A review. In: Mahaney, W.C. (ed.) *Quaternary Soils*. Toronto: York University, pp. 125–165.

Follmer, L. R. (1982) The geomorphology of the Sangamon surface: its spatial and temporal attributes. In: Thorn, C.E. (ed.) *Space and Time in Geomorphology. The Binghampton Symposia in Geomorphology*. London: Allen & Unwin, International Series no. 12, 117–146.

Forchhammer, J. G. (1847) *Geognostische Karte der Herzogthümer Schleswig und Holstein*. Berlin: Lithographie bei Delius.

Forsström, L., Aalto, M., Eronen, M. & Grönlund, T. (1988) Stratigraphic evidence for Eemian crustal movements and relative sea-level changes in Eastern Fennoscandia. *Palaeogeography, Palaeoclimatology, Palaeoecology* **68**, 317–335.

Franke, J., Paul, A. & Schulz, M. (2008) Modeling variations of marine reservoir ages during the last 45,000 years. *Climate of the Past* **4**, 125–136.

Frechen, M., Sierralta, M., Stephan, H.-J. & Techmer, A. (2007) *Die zeitliche Stellung saale- und weichselzeitlicher Eisrandlagen in Schleswig-Holstein*. 74. Tagung der Arbeitsgemeinschaft Norddeutscher Geologen, Tagungsband und Exkursionsführer, 16.

Fredén, C. (1979) The Quaternary history of the Baltic: the western part. In: Gudelis, V. & Königsson, L. -K. (eds) *The Quaternary History of the Baltic*. Acta Universitatis Upsaliensis, Symposia Universitatis Upsaliensis Annum Quingentesimum Celebrantis **1**, 59–74.

French, H. M. (1986) Periglacial involutions and mass displacement structures, Banks Island, Canada. Geografiska Annaler **A68**, 167–174.

French, H. M. (2007) *The Periglacial Environment*. Third edition. London: Wiley.

French, H. M. & Harry, D. G. (1988) Nature and origin of ground ice, Sandhills Moraine, southwest Banks Island, Western Canadian Arctic. *Journal of Quaternary Science* **3**, 19–30.

French, H. M., Demitroff, M. & Newell, W. L. (2009) Past permafrost on the Mid-Atlantic Coastal Plain, eastern United States. *Permafrost and Periglacial Processes* **20**, 285–294.

Fricker, H. A. & Scambos, T. (2009) Connected subglacial lake activity on lower Mercer and Whillans Ice Streams, West Antarctica, 2003–2008. *Journal of Glaciology* **55**, 303–315.

Friis, H. & Larsen, G. (1975) Tungmineralanalytisk bidrag til forståelsen af dannelsesforholdene for det sydfynske hvide sand (Kvartær). *Dansk Geologisk Forening, Årsskrift* **1974**, 25–31.

Frye, J. C. & Willman, H.B. (1960) Classification of the Wisconsinan Stage in the Lake Michigan glacial lobe. *Illinois State Geological Survey Circular* **285**, 16 pp.

Frye, J. C. & Willman, H.B. (1975) Quaternary system. In: Willman, H.B., Atherton, E., Buschbach, T.C., Collinson, C., Frye, J.C., Hopkins, M.E., Lineback, J.A. & Simon, J.A. (eds) *Handbook of Illinois Stratigraphy*. Illinois State Geological Survey Bulletin no. 95, pp. 211–239.

Frye, J. C., Willman, H.B. & Black, R.F. (1965) Outline of glacial geology of Illinois and Wisconsin. In: Wright, H.E. Jr. & Frey, D.G. (eds) *The Quaternary of the United States*. Princeton, New Jersey: Princeton University Press, pp. 43–61.

Fuhrmann, R. (1990) Die Molluskenfauna des Interglazials von Gröbern. *Altenburger Naturwissenschaftliche Forschungen* **5**, 148–167.

Fullerton, D. S. (1986) Stratigraphy and correlation of glacial deposits from Indiana to New York and New Jersey. *Quaternary Science Reviews* **5**, 23–52.

Fullerton, D. S. & Colton, R.B. (1986) Stratigraphy and correlation of the glacial deposits on the Montana plains. In: Richmond, G.M. & Fullerton, D.S. (eds) Quaternary Glaciations in the United States of America. *Quaternary Science Reviews* **5**, 69–82.

Fulton, R. J. (1984) Summary: Quaternary stratigraphy of Canada. In: R.J. Fulton (ed.) *Quaternary Stratigraphy of Canada*. A Canadian Contribution to IGCP project 24, Geological Survey of Canada, Paper no. 84(10), 1–5.

Fulton, R. J. (1989) *Quaternary Geology of Canada and Greenland*. Geological Survey of Canada, Geology of Canada no. 1.

Funder, S., Hjort, Ch. & Kelly, M. (1991) Isotope Stage 5 (130-74 ka) in Greenland, a review. *Quaternary International* **10–12**, 107–122.

Funder, S., Hjort, Ch. & Landvik, J.Y. (1994) The last glacial cycles in East Greenland, an overview. *Boreas* **23**, 283–293.

Funder, S., Demidov, I. & Yelovicheva, Y. (2002) Hydrography and mollusc faunas of the Baltic and the White Sea - North Sea seaway in the Eemian. *Palaeogeography, Palaeoclimatology, Palaeoecology* **184**, 275–304.

Funder, S., Kjeldsen, K. K., Kjær, K. H., Ó Cofaigh, C. (2011) The Greenland Ice Sheet during the past 300,000 years: A review. In: Ehlers, J., Gibbard, P.L. & Hughes, P.D. (eds) *Quaternary Glaciations - Extent and Chronology - A Closer Look*. Elsevier, Developments in Quaternary Science **15**, 699–714.

Galon, R. (1965) Some new problems concerning subglacial channels. *Geographia Polonica* **6**, 19–28.

Galon, R., Lankauf, K. & Noryskiewicz, B. (1983) Zur Entstehung der subglaziären Rinnen im nordischen Vereisungsgebiet an einem Beispiel aus der Tuchola-Heide. *Petermanns Geographische Mitteilungen, Ergänzungsheft* **282**, 176–183.

Gao, C., McAndrews, J. H., Wang, X., Menzies, J., Turton, C. L., Wood, B. D., Pei, J. & Kodors, C. (2012) Glaciation of North America in the James Bay Lowland, Canada, 3.5 Ma. *Geology* **40**, 975–978.

Garleff, K. (1968) *Geomorphologische Untersuchungen an geschlossenen Hohlformen ("Kaven") des Niedersächsischen Tieflandes*. Göttinger Geographische Abhandlungen 44.

Garry, C. E., Schwert, D.P., Baker, R.G., Kemnis, T.J., Horton, D.G. & Sullivan, A.E. (1990) Plant and insect remains from the Wisconsinan interstadial/stadial transition at Wedron, north-central Illinois. *Quaternary Research* **33**, 387–399.

Gassert, D. (1975) Stausee- und Rinnenbildung an den südlichsten Eisrandlagen in Norddeutschland. *Würzburger Geographische Arbeiten* **43**, 55–65.

Gaunt, G. D. (1974) A radiocarbon date relating to Lake Humber. *Proceedings of the Yorkshire Geological Society* **40**, 195–197.

Gaunt, G. D. (1976) The Devensian maximum ice limit in the Vale of York. *Proceedings of the Yorkshire Geological Society* **40**, 631–637.

Geer, G. de (1888–90) Om Skandinaviens nivåförändringar under Kvartärperioden. *Geologiska Föreningens i Stockholm Förhandlingar* **10**, 366–379; **12**, 61–110.

Geer, G. de (1896) *Om Skandinaviens geografiska utveckling efter istiden.* Stockholm: P. A. Norstedt & Söner.

Geer, G. de (1940) Geochronologia Suecica Principles. *Kungliga Svenska Vetenskabs-Akademiens Handlingar* **3**, **18:6**, 367.

Gehrels, W. R. (2010) Late Holocene land- and sea-level changes in the British Isles: implications for future sea-level predictions. *Quaternary Science Reviews* **29**, 1648–1660.

Geikie, A. (1863) On the phenomena of the glacial drift of Scotland. *Transactions of the Geological Society of Glasgow* **1**, 190.

Geikie, J. (1874) *The Great Ice Age and its Relationship to the Antiquity of Man.* London: Isbister.

Geikie, J. (1894) *The Great Ice Age and its Relationship to the Antiquity of Man.* Third edition. London: Stanford.

Geirsdóttir, Á. (2011) Pliocene and Pleistocene glaciations of Iceland: A brief overview of the glacial history. In: Ehlers, J., Gibbard, P.L. & Hughes, P.D. (eds) *Quaternary Glaciations - Extent and Chronology - A Closer Look.* Elsevier, Developments in Quaternary Science **15**, 199–210.

Gemmell A. M. D. (1973) The deglaciation of the Isle of Arran, Scotland. *Transactions of the Institute of British Geographers* **59**, 25–39.

Geyh, M. A. & Schleicher, H. (1990) *Absolute Age Determination. Physical und Chemical Dating Methods and Their Application.* Berlin: Springer.

Gibbard, P. L. (1977) Pleistocene history of the Vale of St Albans. *Philosophical Transactions of the Royal Society of London* **B280**, 445–483.

Gibbard, P. L. (1979) Middle Pleistocene drainage in the Thames Valley. *Geological Magazine* **116**, 35–44.

Gibbard, P. L. (1985) *The Pleistocene History of the Middle Thames Valley.* Cambridge: Cambridge University Press.

Gibbard, P. L. (1988) The history of the great northwest European rivers during the past three million years. *Philosophical Transactions of the Royal Society of London B* **318**, 559–602.

Gibbard, P. L. (1989) The Geomorphology of a part of the Middle Thames: forty years on: a reappraisal of the work of F. Kenneth Hare. *Proceedings of the Geologists' Association* **100**, 481–503.

Gibbard, P. L. (1994) *Pleistocene History of the Lower Thames Valley.* Cambridge: Cambridge University Press.

Gibbard, P. L. (1995) Formation of the Strait of Dover. In: Preece, R.C. (ed.) *Island Britain – a Quaternary Perspective.* Geological Society of London, Special Publication no.96, 15–26.

Gibbard, P. L. (2007) Palaeogeography: Europe cut adrift. *Nature* **448**, 259–260.

Gibbard, P. L. & Allen, L. G. (1994) Drainage evolution in south and east England during the Pleistocene. *Terra Nova* **6**, 444–452.

Gibbard, P. L. & West, R.G. (2000) Quaternary chronostratigraphy: the nomenclature of terrestrial sequences. *Boreas* **29**, 329–336.

Gibbard, P. L. & Clark, C.D. (2011) Pleistocene glaciation limits in Great Britain. In: Ehlers, J., Gibbard, P.L. & Hughes, P.D. (eds) *Quaternary Glaciations–Extent and Chronology–A Closer Look*. Amsterdam: Elsevier, Developments in Quaternary Science 15, pp. 75–93.

Gibbard, P. L. & van der Vegt, P. (2012) The genesis and significance of the Middle Pleistocene glacial meltwater and associated deposits in East Anglia. In: Dixon, R. & Markham, C. (eds) *A Celebration of Suffolk Geology. The GeoSuffolk 10th Anniversary Volume*. Ipswich: Geoscience Suffolk, 303–326.

Gibbard, P. L., Aalto, M. M. & Beales, P.W. (1977) A Hoxnian interglacial site at Fishers Green, Stevenage. *New Phytologist* **78**, 505–520.

Gibbard, P. L., Whiteman, C.A. & Bridgland, D.R. (1988) A preliminary report on the stratigraphy of the Lower Thames valley. *Quaternary Newsletter* **56**, 1–8.

Gibbard, P. L., West, R.G., Andrew, R. & Pettit, M. (1992) The margin of a Middle Pleistocene ice advance at Tottenhill, Norfolk, England. *Geological Magazine* **129**, 59–76.

Gibbard, P. L., Boreham, S., Burger, A.W. & Roe, H.M. (1996) Middle Pleistocene lacustrine deposits in eastern Essex and their palaeogeographical implications. *Journal of Quaternary Science* **11**, 281–298.

Gibbard, P. L., Moscariello, A., Bailey, H. W., Boreham, S., Koch, C., Lord, A. R., Whittaker, J. E. & Whiteman, C. A. (2008) Reply: Middle Pleistocene sedimentation at Pakefield, Suffolk, England: a response to Lee et al. (2006). *Journal of Quaternary Science* **23**, 85–92.

Gibbard, P. L., Pasanen, A., West, R. G., Lunkka, J. P., Boreham, S., Cohen, K. M. & Rolfe, C. (2009) Late Middle Pleistocene glaciation in eastern England. *Boreas* **38**, 504–528.

Gibbard, P., Head, M. J., Walker, M. J. C. & The Subcommission on Quaternary Stratigraphy (2010) Formal ratification of the Quaternary System/Period and the Pleistocene Series/Epoch with a base at 2.58 Ma. *Journal of Quaternary Science* **25**, 96–102.

Gibbard, P. L., Turner, C. & West, R.G. (2013) The Bytham river reconsidered. *Quaternary International* **292**, 15–32.

Gilbert, G.K. (1890) *Lake Bonneville*. US Geological Survey Monograph 1: 438 pp.

Gillespie, A. R. & Clark, D. H. (2011) Glaciations of the Sierra Nevada, California, USA. In: Ehlers, J., Gibbard, P.L. & Hughes, P.D. (eds) *Quaternary Glaciations - Extent and Chronology - A Closer Look*. Elsevier, Developments in Quaternary Science **15**, 447–462.

Girard, H. (1855) *Die norddeutsche Ebene, insbesondere zwischen Elbe und Weichsel, geologisch dargestellt*. Berlin: Reimer.

Giraudi, C. (2011) Middle Pleistocene to Holocene Glaciations in the Italian Apennines. In: Ehlers, J., Gibbard, P.L. & Hughes, P.D. (eds) *Quaternary Glaciations - Extent and Chronology - A Closer Look*. Elsevier, Developments in Quaternary Science **15**, 211–220.

Giraudi, C., Bodrato, G., Ricci Lucchi, M., Cipriani, N., Villa, I.M., Giaccio, B. & Zuppi, G.M. (2011) Middle and Late Pleistocene glaciations in the Campo Felice basin (Central Apennines – Italy). *Quaternary Research* **75**, 219–230.

Glasser, N. F., Hughes, P.D., Fenton. C.R., Schnabel, C. & Rother, H. (2012) ^{10}Be and ^{26}Al exposure-age dating of bedrock surfaces on the Aran Ridge, Wales: Evidence for a thick Welsh Ice Cap at the LGM. *Journal of Quaternary Science* **27**, 97–104.

Golding, W. (1955) *The Inheritors*. London: Faber & Faber.

Goldthwait, R. P. (1959) Scenes in Ohio during the last ice age. *The Ohio Journal of Science* **59**, 193–216.

Gordon, J. E. (1993) Glen Roy and the parallel roads of Lochaber. In: Gordon, J. E. & Sutherland, D. G. (eds) *Quaternary of Scotland*. London: Geological Conservation Review Series no. **6**.

Göttlich, K. H. & Werner, J. (1974) Vorrißzeitliche Interglazialvorkommen in der Altmoräne des östlichen Rheingletschers. *Geologisches Jahrbuch A* **18**, 49–79.

Gottsche, C. (1897a) Die tiefsten Glacialablagerungen der Gegend von Hamburg. Vorläufige Mittheilung. *Mittheilungen der Geographischen Gesellschaft in Hamburg* **XIII**, 131–140.

Gottsche, C. (1897b) Die Endmoränen und das marine Diluvium Schleswig-Holstein's, im Auftrage der Geographischen Gesellschaft in Hamburg untersucht. Theil I: Die Endmoränen. *Mittheilungen der Geographischen Gesellschaft in Hamburg* **XIII**, 57.

Gottsche, C. (1898) Die Endmoränen und das marine Diluvium Schleswig-Holstein's, im Auftrage der Geographischen Gesellschaft in Hamburg untersucht. Theil II: Das marine Diluvium. *Mittheilungen der Geographischen Gesellschaft in Hamburg* **XIV**, 74 S.

Götzinger, G. (1938) *Verhandlungen der III.* Internationalen Quartär-Konferenz, Wien, September 1936. Wien: Geologische Landesanstalt.

Gould, S. J. (1980) *The Panda's Thumb*. W. W. Norton & Co.

Gradstein, F. Ogg, J. & Smith, A. (eds) (2005) *A Geologic Time Scale (2004)*. Cambridge: Cambridge University Press.

Graham, A. G. C. (2007) *Reconstructing Pleistocene Glacial Environments in the Central North Sea Using 3D Seismic and Borehole Data*. PhD thesis, University of London.

Graham, A. G. C., Lonergan, L. & Stoker, M. S. (2007) Evidence for Late Pleistocene ice stream activity in the Witch Ground basin, central North Sea, from 3D seismic reflection data. *Quaternary Science Reviews* **26**, 627–643.

Graham, A. G. C., Lonergan, L. & Stoker, M. S. (2009) Seafloor glacial features reveal the extent and decay of the last British Ice Sheet, east of Scotland. *Journal of Quaternary Science* **24**, 117–138.

Graham, A. C. C., Stoker, M.S., Lonergan, L., Bradwell, T. & Stewart, M.A. (2011) The Pleistocene glaciations of the North Sea Basin. In: Ehlers, J., Gibbard, P.L. & Hughes, P.D. (eds) *Quaternary Glaciations - Extent and Chronology - A Closer Look*. Elsevier, Developments in Quaternary Science **15**, 261–278.

Grahmann, R. (1925) *Diluvium und Pliozän in Nordwestsachsen*. Abhandlungen der Mathematisch-Physikalischen Klasse der Sächsischen Akademie der Wissenschaften, Leipzig **39**(4), 82.

GRASP (2009) Stratigraphic architecture of the southern North Sea tunnel valleys. *Glaciogenic Reservoirs and Hydrocarbon Systems*. Petroleum Group Geological Society, Abstracts 1–2.

Gravenor, C. P. & Kupsch, W. O. (1959) Ice-disintegration features in western Canada. *Journal of Geology* **67**, 48–64.

Gray, J. M. (1991) Glaciofluvial landforms. In: Ehlers, J., Gibbard, P. L. & Rose, J. (eds) *Glacial Deposits in Great Britain and Ireland*. Rotterdam, Brookfield: Balkema, pp. 443–454.

Gray, J. M. & Brooks, C. L. (1972) The Loch Lomond Readvance moraines of Mull and Meinteith. *Scottish Journal of Geology* **8**, 95–103.

Gray, J. M. & Coxon, P. (1991) The Loch Lomond Stadial glaciation in Britain and Ireland. In: Ehlers, J., Gibbard, P.L. & Rose, J. (eds) *Glacial Deposits in Great Britain and Ireland*. Rotterdam: Balkema, pp. 89–105.

Green, H., Woodhead, J., Hellstrom, J. & Drysdale, R. (2013) Re-analysis of key evidence in the case for a hemispherically synchronous response to the Younger Dryas event. *Journal of Quaternary Science* **28**, 8–12.

Green, R. E., Krause, J., Briggs, A. W., Maricic, T., Stenzel, U., Kircher, M., Patterson, N., Li, H., Zhai, W., Fritz, M. H-Y., Hansen, N. F., Durand, E. Y., Malaspinas, A. -S., Jensen, J. D., Marques-Bonet, T., Alkan, C., Prüfer, K., Meyer, M., Burbano, H. A., Good, J. M., Schultz, R., Aximu-Petri, A., Butthof, A., Höber, B., Höffner, B., Siegemund, M., Weihmann, A., Nusbaum, C., Lander, E. S., Russ, C., Novod, N., Affourtit, J., Egholm, M., Verna, C., Rudan, P., Brajkovic, D., Kucan, Ž., Gušic, I., Doronichev, V. B., Golovanova, L. V., Lalueza-Fox, C., de la Rasilla, M., Fortea, F., Rosas, A., Schmitz, R. W., Johnson, P. L. F., Eichler, E. E., Falush, D., Birney, E., Mullikin, J. C., Slatkin, M., Nielsen, R., Lachmann, M., Reich, D. & Pääbo, S. (2010) A draft sequence of the Neandertal genome. *Science* **328**, 710–722.

Grimm, W.-D., Bläsig, H., Doppler, G., Fakhrai, M., Goroncek, K., Hintermaier, G., Just, J., Kiechle, W., Lobinger, W. H., Ludwig, H., Muzavor, S., Pakzad, M., Schwarz, U. & Sidiropoulos, T. (1979) Quartärgeologische Untersuchungen im Nordwestteil des Salzach-Vorlandgletschers (Oberbayern). In: Schlüchter, Ch. (ed.) *Moraines and Varves - Origin, Genesis, Classification*. Rotterdam: Balkema, pp. 101–119.

Gripp, K. (1924) Über die äußerste Grenze der letzten Vereisung in Nordwest-Deutschland. *Mitteilungen der Geographischen Gesellschaft in Hamburg* **36**, 159–245.

Gripp, K. (1929) Glaciologische und geologische Ergebnisse der Hamburgischen Spitzbergen-Expedition 1927. *Abhandlungen aus dem Gebiete der Naturwissenschaften, herausgegeben vom Naturwissenschaftlichen Verein in Hamburg* **XXII (3/4)**, 145–249.

Gripp, K. (1938) Endmoränen. *Comptes Rendus du Congrès International de Géographie*, Amsterdam, T. II, Sect. IIa Géographie Physique, 215–228.

Gripp, K. (1954) Die Entstehung der Landschaft Ost-Schleswigs vom Dänischen Wohld bis Alsen. *Meyniana* **2**, 81–123.

Gripp, K. (1964) *Erdgeschichte von Schleswig-Holstein*. Neumünster: Wachholtz.

Gripp, K. (1974) Untermoräne - Grundmoräne - Grundmoränenlandschaft. *Eiszeitalter und Gegenwart* **25**, 5–9.

Gripp, K. & Todtmann, E. M. (1926) Die Endmoränen des Green Bay Gletschers auf Spitzbergen; eine Studie zum Verständnis norddeutscher Diluvialgebilde. *Mitteilungen der Geographischen Gesellschaft in Hamburg* **37**, 43–75.

Grosswald, M. G. & Kuhle, M. (1994) Impact of glaciations on Lake Baikal. In: Horie, S. & Toyoda, K. (eds) *International Project on Paleolimnology and Late Cenozoic Climate* **8**. Innsbruck: Universitätsverlag Wagner, pp. 48–60.

Grosswald, M. G. & Rudoy, A. N. (1996) Quaternary glacier-dammed lakes in the mountains of Siberia. *Polar Geography* **20**, 180–198.

Grosswald, M. G. & Hughes, T. J. (2002) The Russian component of an Arctic Ice Sheet during the Last Glacial Maximum. *Quaternary Science Reviews* **21**, 121–146.

Grottenthaler, W. (1985) *Geologische Karte von Bayern 1:25 000, Erläuterungen zum Blatt Nr. 8036 Otterfing und zum Blatt Nr. 8136 Holzkirchen.* München: Bayerisches Geologisches Landesamt.

Grottenthaler, W. (1989) Lithofazielle Untersuchungen von Moränen und Schottern in der Typusregion des Würm. In: Rose, J. & Schlüchter, Ch. (eds) *Quaternary Type Sections: Imagination or Reality?* Rotterdam, Brookfield: Balkema, pp. 101–112.

Grube, F. (1967) Die Gliederung der Saale-(Riß-)Kaltzeit im Hamburger Raum. *Fundamenta* **B2**, 168–195.

Grube, F. (1968) Zur Geologie der weichsel-eiszeitlichen Gletscherrandzone von Rahlstedt-Meiendorf. Ein Beitrag zur regionalen Geologie von Hamburg. *Abhandlungen und Verhandlungen des Naturwissenschaftlichen Vereins in Hamburg NF* **XIII**, 141–194.

Grube, F. (1979) Zur Morphogenese und Sedimentation im quartären Vereisungsgebiet Norddeutschlands. *Verhandlungen des Naturwissenschaftlichen Vereins in Hamburg (NF)* **23**, 69–80.

Grube, F. (1981) The subdivision of the Saalian in the Hamburg Region. *Mededelingen Rijks Geologische Dienst* **34**(4), 15–25.

Grüger, E. (1972) Late Quaternary vegetation development in south-central Illinois. *Quaternary Research* **2**, 217–231.

Grüger, E. (1979) Spätriß, Riß/Würm und Frühwürm am Samerberg in Oberbayern - ein vegetationsgeschichtlicher Beitrag zur Gliederung des Jungpleistozäns. *Geologica Bavarica* **80**, 5–64.

Grüger, E. (1983) Untersuchungen zur Gliederung und Vegetationsgeschichte des Mittelpleistozäns am Samerberg in Oberbayern. *Geologica Bavarica* **84**, 21–40.

Grüger, E. (1989) Palynostratigraphy of the last interglacial/glacial cycle in Germany. *Quaternary International* **3/4**, 69–70.

Gualtieri, L., Vartanyan, S. L., Brigham-Grette, J. & Anderson, P. M. (2005) Evidence for an ice-free Wrangel Island, northeast Siberia during the Last Glacial Maximum. *Boreas* **34**, 264–273.

Guobytė, R. & Satkūnas, J. (2011) Pleistocene glaciations in Lithuania. In: Ehlers, J., Gibbard, P.L. & Hughes, P.D. (eds) *Quaternary Glaciations - Extent and Chronology - A Closer Look.* Elsevier, Developments in Quaternary Science **15**, 231–246.

Habbe, K. A. (1986a) Bemerkungen zum Altpleistozän des Illergletscher-Gebietes. *Eiszeitalter und Gegenwart* **36**, 121–134.

Habbe, K. A. (1986b) Zur geomorphologischen Kartierung von Blatt Grönenbach (I)–Probleme, Beobachtungen, Schlußfolgerungen. *Erlanger Geographische Arbeiten* **47**, 119.

Habbe, K. A. (2007) Stratigraphische Begriffe für das Quartär des süddeutschen Alpenvorlandes. *Eiszeitalter und Gegenwart* **56**(1/2), 66–83.

Hacht, U. von (1987) Spuren früher Kaltzeiten im Kaolinsand von Braderup/Sylt. In: Hacht, U. von (ed.) *Fossilien von Sylt* II. Verlag und Verlagsbuchhandlung Hacht, pp. 269–301.

Hachtmann, R. (2007) *Tourismus-Geschichte.* Göttingen: UTB.

Hagedorn, E.-M. & Boenigk, W. (2008) The Pliocene and Quaternary sedimentary and fluvial history in the Upper Rhine Graben based on heavy mineral analyses. *Netherlands Journal of Geosciences–Geologie en Mijnbouw* **87**, 21–32.

Hahne, J., Mengeling, H., Merkt, J. & Grahmann, F. (1994) Die Hunteburg-Warmzeit ("Cromer-Komplex") und Ablagerungen der Elster-, Saale- und Weichsel-Kaltzeit in der Forschungsbohrung Hunteburg GE 58 bei Osnabrück. *Geologisches Jahrbuch* **A134**, 117–166.

Hall, A. (1978) Some new palaeobotanical records for the British Ipswichian interglacial. *New Phytologist* **81**, 805–812.

Hall, K. (2004) Glaciation in southern Africa. In: Ehlers J. & Gibbard P. L. (eds) *Quaternary Glaciations–Extent and Chronology, Part III: South America, Asia, Africa, Australasia, Antarctica.* Amsterdam: Elsevier, Developments in Quaternary Science **2**, 337–338.

Hall, K. & Meiklejohn, I. (2011) Quaternary glaciation of the sub-Antarctic Islands. In: Ehlers, J., Gibbard, P. L. & Hughes, P. D. (eds) *Quaternary Glaciations - Extent and Chronology - A Closer Look.* Amsterdam: Elsevier, Developments in Quaternary Science 15, pp. 1081–1085.

Hallberg, G. R. (1986) Pre-Wisconsin glacial stratigraphy of the Central Plains Region in Iowa, Nebraska, Kansas and Missouri. *Quaternary Science Reviews* **5**, 11–15.

Hallberg, G.R. & Kemnis, T.J. (1986) Stratigraphy and correlation of the glacial deposits of the Des Moines and James Lobes and adjacent areas in North Dakota, South Dakota, Minnesota, and Iowa. *Quaternary Science Reviews* **5**, 65–68.

Hallik, R. (1975) Moortypen Nordeuropas, unter besonderer Berücksichtigung der Verhältnisse in Schweden. *Abhandlungen und Verhandlungen des Naturwissenschaftlichen Vereins in Hamburg NF* **18/19**, 33–41.

Hambach, U., Rolf, C. & Schnepp, E. (2008) Magnetic dating of Quaternary sediments, volcanites and archaeological materials: an overview. *Eiszeitalter und Gegenwart* **57**, 25–51.

Hambrey, M. J. & Harland, W. B. (eds) (1981) *Earth's Pre-Pleistocene Glacial Record.* Cambridge: Cambridge University Press.

Hammen, T. van der (1974) The Pleistocene changes of vegetation and climate in Tropical South America. *Journal of Biogeography* **1**, 3–26.

Hammen, T. van der & Hooghiemstra, H. (2000) Neogene and Quaternary history of vegetation, climate and plant diversity in Amazonia. *Quaternary Science Reviews* **19**, 725–742.

Hannemann, M. (1964) Quartärbasis und älteres Quartär in Ostbrandenburg. *Zeitschrift für Angewandte Geologie* **10**, 370–376.

Hantke, R. (1974) Zur *Vergletscherung* der Schwäbischen *Alb. Eiszeitalter und Gegenwart* **25**, 214.

Hantke, R. (1978) *Eiszeitalter. Die jüngste Erdgeschichte der Schweiz und ihrer Nachbargebiete, Band 1.* Thun: Ott.

Harington, C.R. (1990) Vertebrates of the last interglaciation in Canada: a review, with new data. *Géographie physique et Quaternaire* **44**, 375–387.

Harkness, R. (1870) On the distribution of Wastdale-Crag blocks, "Shap-Fell Granite Boulders," in Westmoreland. *Quarterly Journal of the Geological Society* **26**, 517–528.

Harland, W. B. (2007) Origins and assessment of snowball Earth hypotheses. *Geological Magazine* **144**, 633–642.

Harland, W. B., Armstrong, R. L., Cox, A. V., Craig, L. E., Smith, A. G. & Smith, D. G. (1990) *A Geologic Time Scale.* Cambridge: Cambridge University Press.

Harris, C. & Murton, J. B. (2005) Interactions between glaciers and permafrost: an introduction. In: Harris, C. & Murton, J. B. (eds) *Cryospheric Systems: Glaciers and Permafrost.* Geological Society of London, Special Publication no. **242**, 1–9.

Harris, C. & Ross, N. (2007) *Pingos and pingo scars.* In: Elias, S. A. (Ed.): *Encyclopaedia of Quaternary Science.* Amsterdam: Elsevier, pp. 2200–2207.

Harris, C., Vonder Mühll, D., Isaksen, K., Haeberli, W., Sollid, J. L., King, L., Holmlund, P., Dramis, F., Guglielmin, M. & Palacios, D. (2003) Warming permafrost in European mountains. *Global and Planetary Change* **39**, 215–225.

Harris, C., Arenson, L. U., Christiansen, H. H., Etzelmüller, B., Frauenfelder, R., Gruber, S., Haeberli, W., Hauck, C., Hölzle, M., Humlum, O., Isaksen, K., Kääb, A., Kern-Lütschg, M. A., Lehning, M., Matsuoka, N., Murton, J. B., Nötzli, J., Phillips, M., Ross, N., Seppälä, M., Springman, S. M. & Mühll, D. von der (2009) Permafrost and climate in Europe: Monitoring and modelling thermal, geomorphological and geotechnical responses. *Earth Science Reviews* **92**, 117–171.

Harrison, S. & Glasser, N.F. (2011) The Pleistocene glaciations of Chile. In: Ehlers, J., Gibbard, P.L. & Hughes, P.D. (eds) *Quaternary Glaciations - Extent and Chronology - A Closer Look.* Elsevier, Developments in Quaternary Science **15**, 739–756.

Hart, J. K. (1987) *The genesis of north-east Norfolk Drift.* Unpublished PhD thesis, University of East Anglia.

Hart, J. K. (1997) The relationship between drumlins and other forms of subglacial glaciotectonic deformation. *Quaternary Science Reviews* **16**, 93–107.

Hart, J. K. & Boulton, G. S. (1991) The interelation of glaciotectonic and glaciodepositional processes within the glacial environment. *Quaternary Science Reviews* **10**, 335–350.

Harting, P. (1874) De bodem van het Eemdal. *Verslag Koninklijke Akademie van Wetenschappen*, Afdeling **N, II**, Deel **VIII**, 282–290.

Hättestrand, C. & Stroeven, A. (2002) A relict landscape in the centre of Fennoscandian glaciation: Geomorphological evidence of minimal Quaternary glacial erosion. *Geomorphology* **44**, 127–143.

Haug, H. H., Ganopolski, A., Sigman, D.M., Rosell-Mele, A., Swann, G.E.A., Tiedemann, R., Jaccard, S.L., Bollman, J., Maslin, M.A., Leng, M.J. & Eglinton, G. (2005) North Pacific seasonality and the glaciation of North America 2.7 million years ago. *Nature* **433**, 821–825.

Häuselmann, Ph., Fiebig, M., Kubik, P. W. & Adrian, H. (2007) A first attempt to date the original "Deckenschotter" of Penck and Brückner with cosmogenic nuclides. *Quaternary International* **164–165**, 33–42.

Hays, J. D., Imbrie, J. & Shackleton, N. J. (1976) Variations in the earth's orbit: pacemaker of the ice ages. *Science* **194**, 1121–1132.

Hedberg, H. D. (ed.) (1976) *International Stratigraphic Guide.* New York, Chichester, Brisbane, Toronto, Singapore: Wiley.

Heinrich, H. (1988) Origin and consequences of cyclic ice rafting in the Northeast Atlantic Ocean during the past 130,000 years. *Quaternary Research* **29**, 142–152.

Heiri, O., Brooks, S.J., Renssen, H. et al. (2014) Validation of climate model-inferred regional temperature change for late-glacial Europe. *Nature Communications* **5**, doi: 10.1038/ncomms5914.

Helmens, K. (2011) Quaternary glaciations of Colombia. In: Ehlers, J., Gibbard, P.L. & Hughes, P.D. (eds) *Quaternary Glaciations - Extent and Chronology - A Closer Look*. Elsevier, Developments in Quaternary Science **15**, 815–834.

Hermans, W. F. (1966) *Nooit meer Slapen*. Amsterdam: De Bezige Bij.

Hermsdorf, N. & Strahl, J. (2006) Zum Problem der so genannten Uecker-Warmzeit (Intra-Saale)–Untersuchungen an neuen Bohrkernen aus dem Raum Prenzlau. *Brandenburger Geowissenschaftliche Beiträge* **13**, 49–61.

Hesemann, J. (1975) *Kristalline Geschiebe der Nordischen Vereisungen*. Krefeld: Geologisches Landesamt Nordrhein-Westfalen.

Heuberger, H. & Weingartner, H. (1985) Die Ausdehnung der letzteiszeitlichen Vergletscherung an der Mount-Everest-Südflanke, Nepal. *Mitteilungen der Österreichischen Geographischen Gesellschaft* **127**, 71–80.

Heuberger, H. & Sgibnev, V. V. (1998) Paleoglaciological studies in the Ala-Archa National Park, Kyrgystan, NW Tian-Shan mountains, and using multitextural analysis as a sedimentological tool for solving stratigraphic problems. *Zeitschrift für Gletscherkunde und Glazialgeologie* **34**, 95–123.

Heusser, C. J. (2003) *Ice Age Southern Andes - a Chronicle of Paleoecological Events*. Amsterdam: Elsevier, Developments in Quaternary Science **3**,

Heusser, L. E. & King, J.E. (1988) North America, with special emphasis on the development of the Pacific coastal forest and prairie/forest boundary prior to the last glacial maximum. In: Huntley, B. & Webb, T. III (eds) *Vegetational History*. Dordrecht: Kluwer, pp. 193–236.

Hey, R. W. (1976) Provenance of far-travelled pebbles in the pre-Anglian Pleistocene of East Anglia. *Proceedings of the Geologists' Association* **87**, 69–82.

Hey, R. W. (1980) Equivalents of the Westland Green Gravels in Essex and East Anglia. *Proceedings of the Geologists' Association* **91**, 279–290.

Hey, R. W. (1991) Pre-Anglian glacial and glaciations in Britain. In: Ehlers, J., Gibbard, P.L. & Rose, J. (eds) *Glacial Deposits in Great Britain and Ireland*. Balkema: Rotterdam, pp. 13–16.

Heyman, J., Stroeven, A., Caffee, M.W., Hättestrand, C., Harbor, J.M., Li, Y., Alexanderson, H., Zhou, L.& Hubbard, A. (2011) Palaeoglaciology of Bayan Har Shan, NE Tibetan Plateau, exposure ages reveal a missing LGM expansion. *Quaternary Science Reviews* **30**, 1988–2001.

Hicock, S. R. & Dreimanis, A. (1992) Sunnybrook Drift in the Toronto Area, Canada: reinvestigation and reinterpretation. In: Clark, P.U. & Lea, P.D. (eds) *The Last Interglacial-Glacial Transition in North America*. Geological Society of America, Special Paper no. 270, pp. 139–161.

Hinsch, W. (1993) Marine Molluskenfaunen in Typusprofilen des Elster-Saale-Interglazials und des Elster-Spätglazials. *Geologisches Jahrbuch A* **138**, 9–34.

Hinze, C. (1982) *Geologische Karte von Niedersachsen 1:25. 000, Erläuterungen zu Blatt Nr. 3615 Bohmte*. Hannover: Niedersächsisches Landesamt für Bodenforschung.

Hoare P.G. & Connell, E.R. (2004) The first appearance of Scandinavian indicators in East Anglia's glacial record. *Bulletin of the Geological Society of Norfolk* **54**, 3–12.

Hoare, P. G., Gale, S. J., Robinson, R. A. J., Connell, E. R. & Larkin, N. R. (2009) Marine Isotope Stage 7-6 transition age for beach sediments at Morston, north Norfolk, UK: implications for Pleistocene chronology, stratigraphy and tectonics. *Journal of Quaternary Science* **24**, 311–316.

Hodgson, D. M. (1982) *Hummocky and fluted moraines in parts of northwest Scotland*. Ph.D. Thesis, University of Edinburgh.

Hoek, W. Z. (2008) The Last Glacial-Interglacial Transition. *Episodes* **31**, 226–229.

Hoelzmann, P., Keding, B., Berke, H., Kruse, A. & Kröpelin, S. (2001) Environmental change and archaeology: Lake evolution and human occupation in the Eastern Sahara during the Holocene. *Palaeogeography, Palaeoclimatology, Palaeoecology* **169**, 193–217.

Hoff, K. E. A. von (1834) *Geschichte der durch Ueberlieferung nachgewiesenen natürlichen Veränderungen der Erdoberfläche*, 1–4. Gotha: Perthes.

Hoffmann, G., Lampe, R. & Barnasch, J. (2005) Postglacial evolution of coastal barriers along the West Pomeranian coast, NE Germany. *Quaternary International* **133–134**, 47–59.

Höfle, H.-C. (1980) Klassifikation von Grundmoränen in Niedersachsen. *Verhandlungen des Naturwissenschaftlichen Vereins in Hamburg (NF)* **23**, 81–91.

Höfle, H.-C. (1991) Über die innere Struktur und die stratigraphische Stellung mehrerer Endmoränenwälle im Bereich der Nordheide bis östlich Lüneburg. *Geologisches Jahrbuch A* **126**, 151–169.

Hofmann, F. (1987) Geologie und Entstehungsgeschichte des Rheinfalls. *Neujahrsblatt der Naturforschenden Gesellschaft Schaffhausen* **39**, 10–20.

Högbom, A. G. (1912) Summary of lecture. *Bulletin of the Geological Institutions of the University of Uppsala* **11**, 302.

Hoinkes, H. (1970) Methoden und Möglichkeiten von Massenhaushaltsstudien auf Gletschern. Ergebnisse der Meßreihe Hintereisferner (Ötztaler Alpen) 1953–1968. *Zeitschrift für Gletscherkunde und Glazialgeologie* **VI**, 37–90.

Hollin, J. T. & Schilling, D. H. (1981) Late Wisconsin-Weichselian mountain glaciers and small ice caps. In: Denton, G. H. & Hughes, T. J. (eds) *The Last Great Ice Sheets*. New York: John Wiley & Sons, pp. 179–206.

Holmes, C. D. (1941) Till fabric. *Bulletin of the Geological Society of America* **52**, 1299–1354.

Hölting, B. (1958) Die Entwässerung des würmzeitlichen Eisrandes in Mittelholstein. *Meyniana* **7**, 61–98.

Hoppe, G. (1952) Hummocky moraine regions with special reference to the interior of Norbotten. *Geografiska Annaler* **34**, 1–26.

Hopson, P. M. & Bridge, D.M.C. (1987) Middle Pleistocene stratigraphy in the lower Waveney valley, East Anglia. *Proceedings of the Geologists' Association* **98**, 171–185.

Hoselmann, C. (1996) Der Hauptterrassen-Komplex am unteren Mittelrhein. *Zeitschrift der Deutschen Geologischen Gesellschaft* **147**, 481–497.

Hoselmann, C. (2008) The Pliocene and Pleistocene fluvial evolution in the northern Upper Rhine Graben based on results of the research borehole at Viernheim (Hessen, Germany). *Eiszeitalter und Gegenwart* **57**(3/4), 286–315.

Houmark-Nielsen, M. (1987) Pleistocene stratigraphy and glacial history of the central part of Denmark. *Bulletin of the Geological Society of Denmark* **36**, 1–189.

Houmark-Nielsen, M. (2003) Signature and timing of the Kattegat Ice Stream: onset of the LGM -sequence in the southwestern part of the Scandinavian Ice Sheet. *Boreas* **32**, 227–241.

Houmark-Nielsen, M. (2004) The Pleistocene of Denmark: a review of stratigraphy and glaciation history. In: Ehlers, J. & Gibbard, P. L. (eds) *Quaternary Glaciations–Extent and Chronology. Part I, Europe.* Amsterdam: Elsevier. Developments in Quaternary Science **2**, 35–46.

Houmark-Nielsen, M. (2007) Extent and age of Middle and Late Pleistocene glaciations and periglacial episodes in southern Jylland, Denmark. *Bulletin of the Geological Society of Denmark* **55**, 9–35.

Houmark-Nielsen, M. (2010) Extent, age and dynamics of Marine Isotope Stage 3 glaciations in the southwestern Baltic Basin. *Boreas* **39**, 343–359.

Houmark-Nielsen, M. (2011) Pleistocene Glaciations in Denmark: A Closer Look at Chronology, Ice Dynamics and Landforms. In: Ehlers, J., Gibbard, P.L. & Hughes, P.D. (eds) *Quaternary Glaciations - Extent and Chronology - A Closer Look.* Elsevier, Developments in Quaternary Science **15**, 47–58.

Howard, A. D., Fairbridge, R. W. & Quinn, J. H. (1968) Terraces - fluvial. In: Fairbridge, R.W. (ed.) *The Encyclopedia of Geomorphology.* New York: Reinhold Book Corporation, pp. 1117–1123.

Hudjashov, G., Kivisild, T., Underhill, P. A., Endicott, P., Sanchez, J. J., Lin, A. A., Shen, P., Oefner, C. R., Villems, R. & Forster, P. (2007) Revealing the prehistoric settlement of Australia by Y chromosome and mtDNA analysis. *Proceedings of the National Academy of Sciences of the United States of America* **104**(21), 8726–8730.

Hughes, P. D. (2002) Loch Lomond Stadial glaciers in the Aran and Arenig Mountains, North Wales, Great Britain. *Geological Journal* **37**, 9–15.

Hughes, P. D. (2008) Response of a Montenegro glacier to extreme summer heatwaves in 2003 and 2007. *Geografiska Annaler* **90A**, 259–267.

Hughes, P. D. (2009a) Twenty-first Century Glaciers in the Prokletije Mountains, Albania. *Arctic, Antarctic and Alpine Research* **41**, 455–459.

Hughes, P. D. (2009b) Loch Lomond Stadial (Younger Dryas) glaciers and climate in Wales. *Geological Journal* **44**, 375–391.

Hughes, P. D. (2011) Mediterranean glaciers and glaciation. In: In: Singh, V.P., Singh, P. & Haritashya, U.K. (eds) *Encyclopedia of Snow, Ice and Glaciers.* Springer, pp. 726–730.

Hughes, P. D. (2012) Glacial history. In: Vogiatzakis, I.N. & Tzanopoulos, J. (eds) *Mediterranean Mountain Environments.* Wiley-Blackwell, 35–63.

Hughes, P. D. (2014) Little Ice Age glaciers in the Mediterranean mountains. In: Carozza, J.-M., Devillers, B. & Morhange, C. (eds) *Little Ice Age in the Mediterranean,* Méditerranée **122**, 62–80.

Hughes, P. D. & Braithwaite, R. J. (2008) Application of a degree-day model to reconstruct Pleistocene glacial climates. *Quaternary Research* **69**, 110–116.

Hughes, P. D. & Woodward, J.C. (2008) Timing of glaciation in the Mediterranean mountains during the Last Cold Stage. *Journal of Quaternary Science* **23**, 575–588.

Hughes, P.D. & Woodward, J.C. (2009) Glacial and Periglacial Environments. In: Woodward, J.C. (Ed) *The Physical Geography of the Mediterranean*. Oxford University Press. p. 353–383.

Hughes, P. D. & Gibbard, P.L. (2014) A stratigraphical basis for the Last Glacial Maximum (LGM). *Quaternary International*, doi: 10.1016/j.quaint.2014.06.006.

Hughes, P. D., Gibbard, P. L. & Woodward, J. C. (2003) Relict rock glaciers as indicators of Mediterranean palaeoclimate during the Last Glacial Maximum (Late Würmian) of northwest Greece. *Journal of Quaternary Science* **18**, 431–440.

Hughes, P.D., Gibbard, P.L. & Woodward, J.C. (2005) Quaternary glacial records in mountain regions: A formal stratigraphical approach. *Episodes* **28**, 85–92.

Hughes, P. D., Woodward, J. C. & Gibbard, P. L. (2006a) Middle Pleistocene glacier-sediment dynamics in the Pindus Mountains, Greece. *Journal of the Geological Society of London* **163**, 857–867.

Hughes, P. D., Woodward, J. C., Gibbard, P. L., Macklin, M. G., Gilmour, M. A. & Smith G. R. (2006b) The glacial history of the Pindus Mountains, Greece. *Journal of Geology* **114**, 413–434.

Hughes, P. D., Gibbard, P.L. & Woodward, J.C. (2006c) Quaternary glacial history of the Mediterranean Mountains. *Progress in Physical Geography* **30**, 334–364.

Hughes, P. D., Woodward, J.C., van Calsteren, P.C., Thomas, L.E. & Adamson, K. (2010) Pleistocene ice caps on the coastal mountains of the Adriatic Sea: palaeoclimatic and wider palaeoenvironmental implications. *Quaternary Science Reviews* **29**, 3690–3708.

Hughes, P. D., Woodward, J. C., van Calsteren, P. C. & Thomas, L. E. (2011) The Glacial History of The Dinaric Alps, Montenegro. *Quaternary Science Reviews* **30**, 3393–3412.

Hughes, P. D., Braithwaite, R. J., Fenton, C. R. & Schnabel, C. (2012) Two Younger Dryas glacier phases in the English Lake District: geomorphological evidence and preliminary [10]Be exposure ages. *North West Geography* **12**, 10–19.

Hughes, P. D., Gibbard, P. L. & Ehlers, J. (2013) Timing of glaciation during the last glacial cycle: evaluating the meaning and significance of the 'Last Glacial Maximum' (LGM). *Earth Science Reviews* **125**, 171–198.

Hughes, P. D., Fink, D., Fletcher, W. J. & Hannah, G. (2014) Catastrophic rock avalanches in a glaciated valley of the High Atlas, Morocco: [10]Be exposure ages reveal a 4 ka seismic event. *Geological Society of America–Bulletin* **126**,1093–1104.

Hughes, T. J. (1987) Deluge II and the continent of doom: rising sea level and collapsing Antarctic ice. *Boreas* **16**, 89–100.

Hughes, T. (1992) Abrupt climatic change related to unstable ice-sheet dynamics: toward a new paradigm. *Palaeogeography, Palaeoclimatology, Palaeoecology* **97**, 203–234.

Hughes, T. J. (2006) Topsy-Turvy Science: A Personal Narrative of a Half-Century in Science. Available at: http://nia.ecsu.edu/ureoms2006/hughes/tts_printable. htm (accessed 22 April 2015).

Humboldt, A. von (1845) *Kosmos - Entwurf einer physischen Weltbeschreibung, Bd. 1.* Stuttgart, Augsburg: Cotta.

Humlum, O. (1998) The Climatic Significance of Rock Glaciers. *Permafrost and Periglacial Processes* **9**, 375–395.

Huntley, D. J., Godfrey-Smith, D. I. & Thewalt, M. L. W. (1985) Optical dating of sediments. *Nature* **313**, 105–107.

Husen, D. van (1968) Ein Beitrag zur Talgeschichte des Ennstales im Quartär. *Mitteilungen der Gesellschaft der Geologie- und Bergbaustudenten* **18** (1967), 249–286.

Husen, D. van (1977) Zur Fazies und Stratigraphie der jungpleistozänen Ablagerungen im Trauntal. *Jahrbuch, Geologische Bundesanstalt* **120**, 1–130.

Husen, D. van (1980) *Erläuterungen zu Blatt 160, Neumarkt in Steiermark*. Wien: Geologische Bundesanstalt.

Husen, D. van (1981) Geologisch-sedimentologische Aspekte im Quartär von Österreich. *Mitteilungen der Österreichischen Geologischen Gesellschaft* **74/75**, 197–230.

Husen, D. van (1983) General sediment development in relation to the climatic changes during Würm in the eastern Alps. In: Evenson, E. B., Schlüchter, Ch. & Rabassa, J. (eds) *Tills & Related Deposits*. Rotterdam: Balkema, pp. 345–349.

Husen, D. van (1987) *Die Ostalpen in den Eiszeiten*. Wien: Geologische Bundesanstalt.

Husen, D. van (2004) Quaternary glaciations in Austria. In: Ehlers, J. & Gibbard, P. L. (eds) *Quaternary Glaciations–Extent and Chronology, Part I, Europe*. Amsterdam: Elsevier, Developments in Quaternary Science **2**, 1–13.

Husen, D. van (2011) Quaternary Glaciations in Austria. In: Ehlers, J., Gibbard, P.L. & Hughes, P.D. (eds) *Quaternary Glaciations - Extent and Chronology - A Closer Look*. Elsevier, Developments in Quaternary Science **15**, 15–29.

Hütt, G., Jaek, I. & Tchonka, J. (1988) Optical dating: K-feldspars optical response stimulation spectra. *Quaternary Science Reviews* **7**, 381–385.

Huusc, M. & Lykke-Andersen, H. (2000) Overdeepened Quaternary valleys in the eastern Danish North Sea: morphology and origin. *Quaternary Science Reviews* **19**, 1233–1253.

Hyvärinen, H. & Eronen, M. (1979) The Quaternary History of the Baltic. The Northern Part. In Gudelis, V. & Königsson, L.-K. (eds) *The Quaternary History of the Baltic*. Acta Universitatis Upsaliensis, Symposia Universitatis Upsaliensis Annum Quingentesimum Celebrantis **1**, 7–27.

Illies, H. (1952a) Die eiszeitliche Fluß- und Formengeschichte des Unterelbe-Gebietes. *Geologisches Jahrbuch* **66**, 525–558.

Illies, H. (1952b) Eisrandlagen und eiszeitliche Entwässerung in der Umgebung von Bremen. *Abhandlungen des Naturwissenschaftlichen Vereins zu Bremen* **33**, 19–56.

Imbrie, J. & Imbrie, K.P. (1979) *Ice Ages: Solving the Mystery*. Cambridge (Mass.), London: Harvard University Press.

Imbrie, J. & Imbrie, K. P. (1986) *Ice Ages: Solving the Mystery*. Cambridge (Mass.), London: Harvard University Press.

Imbrie, J., Hays, J.D., Martinson, D.G., MacIntyre, A., Mix, A.C., Morley, J.J., Pisias, N.G., Prell, W.L. & Shackleton, N.J. (1984) The orbital theory of Pleistocene climate: support from a revised chronology of the marine ‰18O record. In: Berger, A., Imbrie, J., Hays, J., Kukla, G. & Saltzman, B. (eds) *Milankovitch and Climate*. Dordrecht: Reidel, pp. 269–305.

Ingólfsson, Ó. (2004) Quaternary glacial and climate history of Antarctica. In: Ehlers, J. & Gibbard, P.L. (eds) *Quaternary Glaciations–Extent and Chronology, Part III*. Amsterdam: Elsevier, Developments in Quaternary Science **2c**, 3–43.

IPCC (2007) *Climate Change 2007: Contribution of Working Group I to the Fourth Assessment Report of the Intergovernmental Panel on Climate Change*. (S. Solomon, D. Qin, M. Manning, Z. Chen, M. Marquis, K. B. Averyt, M. Tignor & H. L. Miller, eds) Cambridge, New York: Cambridge University Press.

IPCC (2013) *Climate Change 2013: The Physical Science Basis. Contribution of Working Group I to the Fifth Assessment Report of the Intergovernmental Panel on Climate Change*. Stocker, T.F., D. Qin, G.-K. Plattner, M. Tignor, S.K. Allen, J. Boschung, A. Nauels, Y. Xia, V. Bex & P.M. Midgley (eds) Cambrudge: Cambridge University Press.

Isarin, R. F. B. (1997) Permafrost distribution and temperatures in Europe during the Younger Dryas. *Permafrost and Periglacial Processes* **8**, 313–333.

Israelson, C., Funder, S. & Kelly, M. (1994) The Aucellaelv stade at Aucellaelv, the first Weichselian glacier advance in Scoresby Sund, East Greenland. *Boreas* **23**, 424–431.

Ives, J. D. & Andrews, J. T. (1963) Studies in the physical geography of north-central Baffin Island. *Geographical Bulletin* **19**, 5–48.

Ives, J. D., Andrews, J. T. & Barry, R. G. (1975) Growth and Decay of the Laurentide Ice Sheet and Comparisons with Fenno-Scandinavia. *Naturwissenschaften* **62**, 118–125.

Ivy-Ochs, S., Schlüchter, C., Kubik, P. W. & Denton, G. H. (1999) Moraine exposure dates imply synchronous Younger Dryas glacier advnaces in the European Alps and in the Souhtern Alps of New Zealand. *Geografiska Annaler* **81**, 313–323.

Ivy-Ochs, S., Schäfer, J., Kubik, P. W., Synal, H.-A. & Schlüchter, C. (2004) Timing of deglaciation on the northern alpine foreland (Switzerland). *Eclogae geologicae Helvetiae* **97**, 47–55.

Jamieson T. F. (1865) On the history of the last geological changes in Scotland. *Quarterly Journal of the Geological Society of London* **21**, 161–203.

Jansen, F. (2004) Geologische Karte von Nordrhein-Westfalen 1:25 000, Erläuterungen zu Blatt 4205 Hamminkeln. Geologischer Dienst Nordrhein-Westfalen, Krefeld.

Jerusalem, J. F. W. (1774) *Betrachtungen über die vornehmsten Wahrheiten der Religion. Zweyter Theil*. Braunschweig: Verlag der Fürstlichen Waisenhaus-Buchhandlung.

Jerz, H. (1982) Paläoböden in Südbayern (Alpenvorland und Alpen). *Geologisches Jahrbuch* **F 14**, 27–43.

Jerz, H. (1987) *Geologische Karte von Bayern 1:25 000, Erläuterungen zum Blatt Nr. 8034 Starnberg-Süd*. München: Bayerisches Geologisches Landesamt.

Jessen, K. & Milthers, V. (1928) *Stratigraphical and Paleontological Studies of Interglacial Fresh-Water Deposits in Jutland and Northwest Germany*. Danmarks Geologiske Undersøgelse, 2. Række **48**.

Johansson, P., Lunkka, J.P. & Sarala, P. (2011) The Glaciation of Finland. In: Ehlers, J., Gibbard, P.L. & Hughes, P.D. (eds) *Quaternary Glaciations - Extent and Chronology - A Closer Look*. Elsevier, Developments in Quaternary Science **15**, 105–116.

Johnsen, S. J. & Vinther, B. M. (2007) *Greenland Stable Isotopes*. In: Elias, S. A. (ed.) *Encyclopaedia of Quaternary Science*. Amsterdam: Elsevier, pp. 1251–1258.

Johnson, W.H. (1990) Ice-wedge casts and relict patterned ground in central Illinois and their environmental significance. *Quaternary Research* **33**, 51–72.

Johnstrup, F. (1874) *Om hævningsfænomenerne i Møens Klint*. København: Schultz.

Jolin, L. (2010) Heroes frozen in time. *Cambridge Alumni Magazine* **59**, 18–21.

Jones, R.L. & Keen, D.H. (1993) *Pleistocene Environments in the British Isles*. Chapman & Hall.

Joon, B., Laban, C. & van der Meer, J. J. M. (1990) The Saalian glaciation in the Dutch part of the North Sea. *Geologie en Mijnbouw* **69**, 151–158.

Jørgensen, F. & Sandersen, P. B. E. (2009) Buried valley mapping in Denmark: evaluating mapping method constraints and the importance of data density. *Zeitschrift der Deutschen Gesellschaft für Geowissenschaften* **160**, 211–223.

Jouzel, J., Masson-Delmotte, V., Cattani, O., Dreyfus, G., Falourd, S., Hoffmann, G., Minster, B., Nouet, J., Barnola, J. M., Chappellaz J., Fischer, H., Gallet, J. C., Johnsen S., Leuenberger, M., Loulergue, L., Luethi, D., Oerter, H., Parrenin, F., Raisbeck, G., Raynaud, D., Schilt, A., Schwander, J., Selmo, E., Souchez, R., Spahni, R., Stauffer, B., Steffensen, J. P., Stenni, B., Stocker, T. F., Tison, J. L., Werner, M. & Wolff, E. W. (2007) Orbital and Millennial Antarctic Climate Variability over the Past 800,000 Years. *Science* **317**, 793–797.

Junge, F. W. (1998) Die Bändertone Mitteldeutschlands und angrenzender Gebiete. Ein regionaler Beitrag zur quartären Stausee-Entwicklung im Randbereich des elsterglazialen skandinavischen Inlandeises. *Altenburger Naturwissenschaftliche Forschungen* **9**, 210.

Juschus O., Melles M., Gebhardt A. C. & Niessen F. (2009) Late Quaternary mass movement events in Lake El´gygytgyn, north-eastern Siberia. *Sedimentology* **56**, 2155–2174.

Kääb, A. (2007) Rock Glaciers and Protalus Forms. In: Elias, S. A. (ed.) *Encyclopaedia of Quaternary Science*. Amsterdam: Elsevier, pp. 2236–2242.

Kabel, C. (1982) *Geschiebestratigraphische Untersuchungen im Pleistozän Schleswig Holsteins und angrenzender Gebiete*. Dissertation, Universität Kiel.

Kadomura, H. (1995) Palaeoecological and Palaeohydrological Changes in the Humid Tropis during the Last 20. 000 Years, with Reference to Equatorial Africa. In: Gregory, K. J., Starkel, L. & Baker, V. R. (eds) *Global Continental Palaeohydrology*. Chichester: Wiley.

Kahlke, R. D. (1997–2001) *Das Pleistozän von Untermaßfeld bei Meiningen (Thüringen)*. Monographien der Römisch-Germanischen Zentralmuseums **40/1** (1997): 426 S. Herausgegeben in Verbindung mit dem Bereich Quartärpaläontologie Weimar, Institut für Geowissenschaften der Friedrich-Schiller-Universität Jena. **40/2** (2001): 288 S. **40/3** (2001): 340 S. Herausgegeben in Verbindung mit der Senckenbergischen Naturforschenden Gesellschaft, Forschungsstation für Quartärpaläontologie Weimar.

Kalm, V. & Mahaney, W.C. (2011) Late Quaternary Glaciations in the Venezuelan (Merida) Andes. In: Ehlers, J., Gibbard, P.L. & Hughes, P.D. (eds) *Quaternary Glaciations - Extent and Chronology - A Closer Look*. Elsevier, Developments in Quaternary Science **15**, 835–842.

Kalm, V., Raukas, A., Rattas, M. & Lasberg, K. (2011) Pleistocene Glaciations in Estonia. In: Ehlers, J., Gibbard, P.L. & Hughes, P.D. (eds) *Quaternary Glaciations - Extent and Chronology - A Closer Look*. Elsevier, Developments in Quaternary Science **15**, 95–104.

Kaltwang, J. (1992) Die pleistozäne Vereisungsgrenze im südlichen Niedersachsen und im östlichen Westfalen. *Mitteilungen aus dem Geologischen Institut der Universität Hannover* **33**, 161.

Kaplan, M. R., Schaefer, J., Denton, G. H., Barrell, D. J. A., Chinn, T. J. H., Putnam, A. E., Andersen, B. G., Finkel, R. C., Schwatrtz, R. & Doughty, A. M. (2010) Glacier retreat in New Zealand during the Younger Dryas stadial. *Nature* **467**, 194–197.

Karabanov, A. K. & Matveyev, A.V. (2011) The Pleistocene Glaciations in Belarus. In: Ehlers, J., Gibbard, P.L. & Hughes, P.D. (eds) *Quaternary Glaciations - Extent and Chronology - A Closer Look*. Elsevier, Developments in Quaternary Science **15**, 29–37.

Karrow, P. F. (1969) Stratigraphic studies in the Toronto Pleistocene. *Proceedings of the Geological Association of Canada* **20**, 4–16.

Karrow, P. F. 1990. Interglacial beds at Toronto, Ontario. *Geographie Physique et Quaternaire* **44**, 289–297.

Karte, J. (1979) Räumliche Abgrenzung und regionale Differenzierung des Periglaziärs. *Bochumer Geographische Arbeiten* **35**, 211.

Kasse, C. (1988) *Early-Pleistocene tidal and fluviatile environments in the southern Netherlands and northern Belgium*. Thesis, Free University Press, Amsterdam.

Kaufman, D. S., Miller, G.H., Stravers, J.A. & Andrews, J.T. (1993) Abrupt early Holocene (9.9-9.6 ka) ice-stream advance at the mouth of Hudson Strait, Arctic Canada. *Geology* **21**, 1063–1066.

Keilhack, K. (1883) Beobachtungen an isländischen Gletscher- und norddeutschen Diluvialablagerungen. *Jahrbuch der Königlich Preußischen Geologischen Landesanstalt und Bergakademie zu Berlin für das Jahr 1882*, 159–176.

Keilhack, K. (1896) Die Geikiessche Gliederung der nordeuropäischen Glazialablagerungen. *Jahrbuch der Königlich Preußischen Geologischen Landesanstalt für 1895*, **XV**, 111–124.

Keilhack, K. (1898) Glaciale hydrographie. In: Berendt, G., Keilhack, K., Schröder, H. & Wahnschaffe, F. (eds) Neuere Forschungsergebnisse auf dem Gebiete der Glacialgeologie in Norddeutschland erläutert an einigen Beispielen. *Jahrbuch der Königlich Preußischen Geologischen Landesanstalt und Bergakademie zu Berlin für das Jahr 1897*, **XVIII**, 113–129.

Keilhack, K. (1899) Die Stillstandslagen des letzten Inlandeises und die hydrographische Entwicklung des pommerschen Küstengebietes. *Jahrbuch der Königlich Preußischen Geologischen Landesanstalt und Bergakademie zu Berlin für das Jahr 1898*, **XIX**, 90–152.

Keilhack, K. (1904) Die große baltische Endmoräne und das Thorn-Eberswalder Haupttal. *Zeitschrift der Deutschen Geologischen Gesellschaft* **56**, Monatsberichte, 132–141.

Keilhack, K. (1909) *Erdgeschichtliche Entwicklung und geologische Verhältnisse der Gegend von Magdeburg*. Magdeburg: Faber.

Keilhack, K. (1910) *Geologische Karte von Preußen 1:25.000, Erläuterungen zu Blatt Charlottenburg, 2. Auflage*, Berlin: Königlich Preußische Geologische Landesanstalt.

Keller, O. & Krayss, E. (1993) The Rhine-Linth Glacier in the Upper Wurm: A Model of the Last Alpine Glaciation. *Quaternary International* **18**, 15–27.

Kelts, K. R. (1978) *Geological and Sedimentological Evolution of Lake Zurich and Lake Zug*. Dissertation, ETH Zürich.

Kemna, H. A. (2005) Pliocene and Lower Pleistocene Stratigraphy in the Lower Rhine Embayment, Germany. *Kölner Forum für Geologie und Paläontologie* **14**, 121.

Kjær, K. H. & Krüger, J. (2001) The final phase of dead-ice moraine development: processes and sediment architecture, Kötlujökull, Iceland. *Sedimentology* **48**, 935–952.

Kjær, K. H., Larsen, E. & Funder, S. (2006a) Late Quaternary in northwestern Russia–Introduction. *Boreas* **35**, 391–393.

Kjær, K. H., Larsen, E., Funder, S., Demidov, I., Jensen, M., Håkansson, L. & Murray, A. (2006b) Eurasian ice sheet interaction in northwestern Russia throughout the late Quaternary. *Boreas* **35**, 444–475.

Klassen, R. A. & Bolduc, A.M. (1984) Ice flow directions and drift composition, Churchill Falls, Labrador. Current Research. Geological Survey of Canada, Part A, Paper 84-1A, pp. 255–258.

Klassen, R. W. (1983) Lake Agassiz and the glacial history of northern Manitoba. In: Teller, J.T. & Clayton, L. (eds) *Glacial Lake Agassiz.* Geological Association of Canada, Special Paper no. 26, pp. 97–115.

Klebelsberg, R. von (1948–49) *Handbuch der Gletscherkunde und Glazialgeologie.* Wien: Springer.

Kleman, J. (1990) On the use of glacial striae for reconstruction of paleo-ice sheet flow patterns–With application to the Scandinavian ice sheet. *Geografiska Annaler* **72A**, 217–236.

Kliewe, H. & Janke, W. (1982) Der holozäne Wasserspiegelanstieg der Ostsee im nordöstlichen Küstengebiet der DDR. *Petermanns Geographische Mitteilungen* **126**, 65–74.

Klostermann, J. (1985) Versuch einer Neugliederung des späten Elster- und des Saale-Glazials der Niederrheinischen Bucht. *Geologisches Jahrbuch A* **83**, 46.

Klostermann, J. (1992) *Das Quartär der Niederrrheinischen Bucht. Ablagerungen der letzten Eiszeit am Niederrhein.* Krefeld: Geologisches Landesamt Nordrhein-Westfalen.

Knies, J., Matthiessen, J., Vogt, C., Laberg, J. S., Hjelstuen, B. O., Smelror, M. et al. (2009) The Plio-Pleistocene glaciations of the Barents Sea-Svalbard region: A new model based on revised chronostratigraphy. *Quaternary Science Reviews* **28**, 812–829.

Knipping, M. (2008) Early and Middle Pleistocene pollen assemblages of deep core drillings in the northern Upper Rhine Graben, Germany. *Netherlands Journal of Geosciences–Geologie en Mijnbouw* **87**, 51–65.

Knox, A.S. 1962. Pollen from the Pleistocene terrace deposits of Washington, DC. *Pollen and Spores* **4**, 356–358.

Knudsen, K. L. (1987) Foraminifera in Late Elsterian-Holsteinian deposits of the Tornskov area in South Jutland, Denmark. *Danmarks Geologiske Undersøgelse* **B 10**, 7–31.

Knudsen, K. L. & Lykke-Andersen, A.-L. (1982) Foraminifera in Late Saalian, Eemian, Early and Middle Weichselian of the Skærumhede I boring. *Bulletin of the Geological Society of Denmark* **30**, 97–109.

Knudsen, K. L. & Gibbard, P. L. (2006) Eemian and Weichselian environmental development in the western Baltic area, NW Europe–Introduction. *Boreas* **35**, 317–319

Knudsen, K. L., Jiang, H., Gibbard, P. L., Kristensen, P., Seidenkrantz, M.-S., Janczyk-Kopinowa, Z. & Marks, L. (2012) Environmental reconstructions of Last Interglacial (Eemian) marine records in the Lower Vistula area, southern Baltic. *Boreas* **41**, 209–234.

Koch, E. (1924) Die prädiluviale Auflagerungsfläche unter Hamburg und Umgebung. *Mitteilungen aus dem Mineralogisch-Geologischen Staatsinstitut* **VI**, 31–95.

Kocureka, G., Carra, M., Ewinga, R., Havholma, K. G., Nagarb, Y. C. & Singhvi, A. K. (2006) White Sands Dune Field, New Mexico: Age, dune dynamics and recent accumulations. *Sedimentary Geology* **197**, 313–331.

Koenigswald, W. von (2002) *Lebendige Eiszeit.* Stuttgart: Theiss.

Koenigswald, W. von (2007) Mammalian Faunas from the interglacial periods in Central Europe and their stratigraphic correlation. In: Sirocko, F., Claussen, M., Sanchez Goñi, M. F. & Litt, T. (eds) *The Climate of Past Interglacials*. Amsterdam: Elsevier, Developments in Quaternary Science **7**, 445–454.

Koenigswald, W. von & Löscher, M. (1982) Jungpleistozäne HippopotamusFunde aus der Oberrheinebene und ihre biogeographische Bedeutung. *Neues Jahrbuch für Geologie und Paläontologie Abhandlungen* **163**, 331–348.

Koenigswald, W. von & Sander, M. (1995) Eiszeitliche Tierfährten aus Bottrop Welheim. *Münchner Geowissenschaftliche Abhandlungen A* **27**, 1–80.

Koenigswald, W. von & Heinrich, W.-D. (2007) Biostratigraphische Begriffe in der Säugetier-stratigraphie. *Eiszeitalter und Gegenwart* **56**, 96–115.

Kohl, H. (1958) Unbekannte Altmoränen in der südwestlichen Traun-Enns-Platte. *Mitteilungen der Geographischen Gesellschaft Wien* **100**, 131–143.

Kohl, H. (1968) *Beiträge über Aufbau und Alter der Donausohle bei Linz*. Naturkundliches Jahrbuch der Stadt Linz: 7–60.

Kolp, O. (1982) Entwicklung und Chronologie des Vor- und Neudarßes. *Petermanns Geographische Mitteilungen* **126**, 85–94.

Kolp, O. (1986) Entwicklungsphasen des Ancylus-Sees. *Petermanns Geographische Mitteilungen* **130**, 79–94.

Komatsu, G., Arzhannikov, S. G., Gillespie, A. R., Burke, R. M., Miyamoto, H. & Baker, V. R. (2008) Quaternary paleolake formation and cataclysmic flooding along the upper Yenisei River. *Geomorphology* **104**, 143–164.

Konradi, P. B., Larsen, B. & Sørensen, A. B. (2005) Marine Eemian in the Danish eastern North Sea. *Quaternary International* **133–134**, 21–31.

Kopp, G. (2000) *Evolution und Lücke. Potenziale der historischen Geo- und Biowissenschaften für die Umweltbildung*. Dissertation, Kiel.

Kosack, B. & Lange, W. (1985) Das Eem-Vorkommen von Offenbüttel/Schnittlohe und die Ausbreitung des Eem-Meeres zwischen Nord- und Ostsee. *Geologisches Jahrbuch* **A86**, 3–17.

Koster, E. A. (1982) Terminology and lithostratigraphic division of (surficial) sandy eolian deposits in The Netherlands: an evaluation. *Geologie en Mijnbouw* **61**, 121–129.

Koster, E. A. (1988) Ancient and modern cold-climate aeolian sand deposition: a review. *Journal of Quaternary Science* **3**, 69–83.

Kotlyakov, V. & Khromova, T. (2002) Maps of permafrost and ground ice. In: Stolbovoi V. & McCallum, I. (eds) CD-ROM *Land Resources of Russia*. Laxenburg, Austria: International Institute for Applied Systems Analysis and the Russian Academy of Science. CD-ROM. Distributed by the National Snow and Ice Data Center/World Data Center for Glaciology, Boulder.

Koutaniemi, L. (1999) Twenty-one years of string movements Liippasuo aapa mire, Finland. *Boreas* **28**, 521–530.

Kozarski, S. (1975) Oriented kettle holes in outwash plains. *Quaestiones Geographicae* **2**, 99–112.

Krause. E., Huber, L. & Fischer, H. (eds) (1991) *Hochschulalltag im Dritten Reich. Die Hamburger Universität 1933–1945*. 3 Bände. Berlin: Reimer.

Krigström, A. (1962) Geomorphological studies of sandur plains and their braided rivers in Iceland. *Geografiska Annaler* **44**, 328–346.

Kristensen, P. H. & Knudsen, K. L. (2006) Palaeoenvironments of a complete Eemian sequence at Mommark, South Denmark: foraminifera, ostracods and stable isotopes. *Boreas* **35**, 349–366.

Kristensen, P., Gibbard, P., Knudsen, K. L. & Ehlers, J. (2000) Last Interglacial stratigraphy at Ristinge Klint, South Denmark. *Boreas* **29**, 103–116.

Kristensen, T. B., Huuse, M., Piotrowski, J. A. & Clausen, O. R. (2007) A morphometric analysis of tunnel valleys in the eastern North Sea based on 3D seismic data. *Journal of Quaternary Science* **22**, 801–815.

Kristensen, T. B., Piotrowski, J. A., Huuse, M., Clausen, O. R. & Hamberg, I. (2008) Time-transgressive tunnel valley formation indicated by infill sediment structure, North Sea; the role of glaciohydraulic supercooling. *Earth Surface Processes and Landforms* **33**, 546–559.

Kromer, B. (2009) Radiocarbon and dendrochronology. *Dendrochronologia* **27**, 15–19.

Krüger, J. (1970) Till Fabric in Relation to Direction of Ice Movement–A study from the Fakse Banke, Denmark. *Geografisk Tidsskrift* **69**, 133–170.

Krüger, J. (1987) Træk af et glaciallandskabs udvikling ved nordranden af Mýrdalsjökull, Island. *Dansk Geologisk Forening, Årsskrift for 1986*, 49–65.

Kühl, N. & Litt, T. (2007) Quantitative time-series reconstructions of Holsteinian and Eemian *Temperatures Using Botanical Data*. In: Sirocko, F., Claussen, M., Sanchez-Goñi, M. F. & Litt, T. (eds) *The Climate of Past Interglacials*. Amsterdam: Elsevier, Developments in Quaternary Science 7, 239–259.

Kuhle, M. (1985) Ein subtropisches Inlandeis als Eiszeitauslöser. Südtibet- und Mt. Everest-Expedition 1984. *Georgia Augusta, Nachrichten aus der Universität Göttingen* **42**, 35–51.

Kuhle, M. (1986) Die Vergletscherung Tibets und die Entstehung von Eiszeiten. *Spektrum der Wissenschaften* **9/86**, 42–54.

Kuhle, M. (1989) Die Inlandvereisung Tibets als Basis einer in der Globalstrahlungs geometrie fußenden, reliefspezifischen Eiszeittheorie. *Petermanns Geographische Mitteilungen* **113**, 265–285.

Kuhle, M. (2004) The High Glacial (Last Ice Age and LGM) ice cover in High and Central Asia. In: Ehlers, J. & Gibbard, P. L. (eds) *Quaternary Glaciations–Extent and Chronology, Part III: South America, Asia, Africa, Australasia, Antarctica*, Amsterdam: Elsevier, Developments in Quaternary Science **2**, 175–199.

Kuhle, M. (2011) The High Glacial (Last Ice Age and Last Glacial Maximum) Ice Cover of High and Central Asia, with a Critical Review of Some Recent OSL and TCN dates. In: Ehlers, J., Gibbard, P.L. & Hughes, P.D. (eds) *Quaternary Glaciations - Extent and Chronology - A Closer Look*. Elsevier, Developments in Quaternary Science **15**, 943–967.

Kuhlemann, J., Rohling, E. J., Krumrei, I., Kubik, P., Ivy-Ochs, S. & Kucera, M. (2008) Regional synthesis of Mediterranean atmospheric circulation during the Last Glacial Maximum. *Science* **321**, 1338–1340.

Kuster, H. & Meyer, K.-D. (1979) Glaziäre Rinnen im mittleren und nordöstlichen Niedersachsen. *Eiszeitalter und Gegenwart* **29**, 135–156.

Kvasov, D. D. (1978) *The Late-Quaternary history of large lakes and inland seas of Eastern Europe*. Annales Academiæ Scientiarum Fennicæ, Series A, III. Geologica–Geographica **127**, 71.

La Frenierre, J., In Huh, K. & Mark, B. G. (2011) Ecuador, Peru and Bolivia. In: Ehlers, J., Gibbard, P.L. & Hughes, P.D. (eds) *Quaternary Glaciations - Extent and Chronology - A Closer Look*. Elsevier, Developments in Quaternary Science **15** 773–802.

Laban, C. & van der Meer. J. J. M. (2011) Pleistocene Glaciation in the Netherlands. In: Ehlers, J., Gibbard, P.L. & Hughes, P.D. (eds) *Quaternary Glaciations - Extent and Chronology - A Closer Look*. Elsevier, Developments in Quaternary Science **15**, 199–210.

Lade, U. (1980) Quartärmorphologische und -geologische Untersuchungen in der Bremervörder-Wesermünder Geest. *Würzburger Geographische Arbeiten* **50**, 173.

Lambeck, K., Smither, C. & Johnston, P. (1998) Sea-level change, glacial rebound and mantle viscosity for northern Europe. *Geophysical Journal International* **134**, 102–144.

Lambeck, K., Purcell, A., Funder, S., Kjær, K., Larsen, E. & Möller, P. (2006) Constraints on the Late Saalian to early Middle Weichselian ice sheet of Eurasia from field data and rebound modelling. *Boreas* **35**, 539–575.

Lambeck, K., Purcell, A., Zhao, J. & Svensson, N.-O. (2010) The Scandinavian Ice Sheet: from MIS 4 to the end of the Last Glacial Maximum. *Boreas* **39**, 410–435.

Lambert, A. M. & Hsü, K. J. (1979) Varve-like sediments of the Walensee, Switzerland. In: Schlüchter, C. (ed.) *Moraines and Varves - Origin, Genesis, Classification*. Rotterdam: Balkema, pp. 287–294.

Lambert, F., Bigler, M., Steffensen, J.P., Hutterli, M. & Fischer, H. (2012) Centennial mineral dust variability in high-resolution ice core data from Dome C, Antarctica. *Climate of the Past* **8**, 609–623.

Lancaster, N. (2007) Dune Fields–Low Latitudes. In: Elias, S. A. (ed.) *Encyclopaedia of Quaternary Science*. Amsterdam: Elsevier, pp. 626–642.

Lancaster, N., Kocurek, G., Singhvi, A., Pandey, V., Deynoux, M., Ghienne, J.-F. & Lô, K. (2002) Late Pleistocene and Holocene dune activity and wind regimes in the western Sahara of Mauretania. *Geology* **30**, 991–994.

Lang, H. O. (1879) *Erratische Gesteine aus dem Herzogthum Bremen*. Göttingen: Peppmüller.

Larsen, E., Kjær, K. H., Demidov, I., Funder, S., Grøsfjeld, K., Houmark-Nielsen, M., Jensen, M., Linge, H. & Lyså, A. (2006) Late Pleistocene glacial and lake history of northwestern Russia. *Boreas* **35**, 394–424.

Larson, G. J., Ehlers, J. & Gibbard, P. L. (2003) Large-scale glaciotectonic deformation in the Great Lakes basin, USA-Canada. *Boreas* **32**, 370–385.

Larter, R. D., Graham, A. G. C., Gohl, K., Kuhn, G., Hillenbrand, C.-D., Smith, J. A., Deen, T. J., Livermore, R. A. & Schenke, H.-W. (2009) Subglacial bedforms reveal complex basal regime in a zone of paleo-ice stream convergence, Amundsen Sea embayment, West Antarctica. *Geology* **37**, 411–414.

Lawson, D. (1979) Sedimentological analysis of the western terminus region of the Matanuska Glacier, Alaska. CRREL Report **79-9**, 112.

Laymon, C. A. (1992) Glacial geology of western Hudson Strait, Canada, with reference to Laurentide Ice Sheet dynamics. *Geological Society of America Bulletin* **104**, 1169–1177.

Lebrouc, V., Schwartz, S., Baillet, L., Jongmans, D. & Gamond, J.F. (2013) Modeling permafrost extension in a rock slope since the Last Glacial maximum: Application to the large Séchilienne landslide (French Alps). *Geomorphology* **198**, 189–200.

Lee, J.R., Rose, J., Hamblin, R.J.O., Moorlock, B.S.P., Riding, J.B., Phillips, E., Barendregt, R. & Candy, I. (2011) The Glacial History of the British Isles during the Early, Middle and Late Pleistocene: Implications for the long-term development of the British Ice Sheet. In: Ehlers, J., Gibbard, P.L. & Hughes, P.D. (eds) *Quaternary Glaciations - Extent and Chronology - A Closer Look*. Elsevier, Developments in Quaternary Science **15**, 59–74.

Lehmkuhl, F. & Owen, L. A. (2005) Late Quaternary glaciation of Tibet and the bordering mountains: synthesis and new research. *Boreas* **34**, 87–100.

Leighton, M. M. (1960) The classification of the Wisconsin glacial stage of the north-central United States. *Journal of Geology* **68**, 529–552.

Lemdahl, G. (1988) *Palaeoclimatic and palaeoecological studies based on subfossil insects from Late Weichselian sediments in southern Sweden*. LUNDQUA Thesis **22**. 11.

Leszczynska, K. (2009) Pleistocene glacigenic deposits of the Danbury-Tiptree Ridge, Essex, England. In Proceedings of Conference on *Glacigenic Reservoirs and Hydrocarbon Systems*. London: Geological Society.

Leverett, F. (1898a) The weathered zone (Sangamon) between the Iowan loess and Illinoian till sheet. *Journal of Geology* **6**, 171–181.

Leverett, F. (1898b) The Peorian soil and weathered zone (Toronto Formation?). *Journal of Geology* **6**, 244–249.

Leverett, F. (1898c) The Peorian soil and weathered zone (Toronto Formation?) *Journal of Geology* **6**, 244–249.

Leverett, F. (1929) Moraines and shorelines of the Lake Superior region. US Geological Survey Professional Paper no. 154.

Levine, M. A. (2007) Determining the provenance of native copper artifacts from Northeastern North America: evidence from instrumental neutron activation analysis. *Journal of Archaeological Science* **34**, 572–587.

Liedtke, H. (1975) Die nordischen Vereisungen in Mitteleuropa. *Forschungen zur Deutschen Landeskunde* **204**, 160.

Liedtke, H. (1981) Die nordischen Vereisungen in Mitteleuropa, 2. Auflage. *Forschungen zur Deutschen Landeskunde* **204**, 307.

Liestøl, O. (2000) *Glaciology*. Oslo: Unipub.

Lilly, K., Fink, D., Fabel, D. & Lambeck, K. (2010) Pleistocene dynamics of the interior East Antarctic ice sheet. *Geology* **38**, 703–706.

Lindén, A. (1975) Till petrographic studies in an Archaean bedrock area in southern central Sweden. *Striae* **1**, 57.

Linke, G. & Hallik, R. (1993) Die pollenanalytischen Ergebnisse der Bohrungen Hamburg-Dockenhuden (qho4), Wedel (qho2) und Hamburg-Billbrook. *Geologisches Jahrbuch* **A138**, 169–184.

Linné, C. von (1745) *Ölandska och Gothländska Resa, på Riksens Högloflige Ständers befallning förrättad Åhr 1741*. Stockholm und Uppsala: Kiesewetter.

Linton, D. L. (1955) The problem of tors. *The Geographical Journal* **121**(4), 470–487.

Lisiecki, L. E. & Raymo, M. E. (2005) A Pliocene–Pleistocene stack of 57 globally distributed benthic $\delta^{18}O$ records. *Paleoceanography* **20**, PA1003.

Litt, T. (1990) Stratigraphie und Ökologie des eemzeitlichen Waldelefanten-Schlachtplatzes von Gröbern, Kreis Gräfenhainichen. In: Mania, D., Thomae, M., Litt, T. & Weber, T. (eds)

Neumark-Gröbern. Beiträge zur Jagd des mittelpaläolithischen Menschen. Berlin: Deutscher Verlag der Wissenschaften, pp. 193–208.

Litt, T. (1994) Paläoökologie, Paläobotanik und Stratigraphie des Jungquartärs im nordmitteleuropäischen Tiefland unter besonderer Berücksichtigung des Elbe-Saale-Gebietes. *Dissertationes Botanicae* **227**, 185.

Litt, T. & Wansa, S. (2008) Quartär. In: Bachmann, G. H., Ehling, B.-C., Eichner, R. & Schwab, M. (eds) *Geologie von Sachsen-Anhalt.* Stuttgart: Schweizerbart, pp. 293–325.

Litt, T., Brauer, A., Goslar, T., Merkt, J., Balaga, K., Müller, H., Ralska-Jasiewiczowa, M., Stebich, M. & Negendank, J. F. W. (2001) Correlation and synchronisation of Lateglacial continental sequences in northern central Europe based on annually laminated lacustrine sediments. *Quaternary Science Reviews* **20**, 1233–1249.

Litt, T., Schmincke, H.-U. & Kromer, B. (2003) Environmental response to climate and volcanic events in central Europe during the Weichselian Lateglacial. *Quaternary Science Reviews* **22**, 7–32.

Litt, T., Behre, K.-E., Meyer, K.-D., Stephan, H.-J. & Wansa, S. (2007) Stratigraphische Begriffe für das Quartär des norddeutschen Vereisungsgebietes. *Eiszeitalter und Gegenwart* **56**(1/2), 7–65.

Litt, T., Schmincke, H.-U., Frechen, M. & Schlüchter, C. (2008) Quaternary. In: McCann, T. (ed.) *The Geology of Central Europe.* The Geological Society of London, pp. 1287–1340.

Liverman, D. G. E., Catto, N.R. & Rutter, N.W. (1988) Laurentide glaciation in west-central Alberta: a single (Late Wisconsinan) event. *Canadian Journal of Earth Sciences* **26**, 266–274.

Lliboutry, L. (1968) General theory of subglacial cavitation and sliding of temperate glaciers. *Journal of Glaciology* **7**, 21–58.

Long, D. & Stoker, M. S. (1986) *Channels in the North Sea: The Nature of a Hazard.* Springer, Advances in Underwater Technology. Ocean Science and Offshore Engineering **6**, pp. 339–351.

Long, D., Laban, C., Streif, H., Cameron, T. D. J. & Schüttenhelm, R. T. E. (1988) The sedimentary record of climatic variation in the southern North Sea. *Philosophical Transactions of the Royal Society of London B* **318**, 523–537.

Lordkipanidze, D., Ponce de Leòn, M. S., Margvelashvili, A., Rak, Y., Rightmire, G. P., Vekua, A. & Zollikofer, C. P. E. (2013) A complete skull from Dmanisi, Georgia, and the evolutionary biology of early Homo. *Science* **342**(6156), 326–331.

Lourens, L. J., Sluijs, A., Kroon, D., Zachos, J.C., Thomas, E., Röhl, U., Bowles, J. & Raffi, I. (2005) Astronomical pacing of late Palaeocene to early Eocene global warming events. *Nature* **435**, 1083–1087.

Lowe, J. J., Rasmussen, S. O., Björck, S., Hoek, W. Z., Steffensen, J. P., Walker, M. J. C., Yu, Z. C. & the INTIMATE group (2008) Synchronisation of palaeoenvironmental events in the North Atlantic region during the Last Termination: a revised protocol recommended by the INTIMATE group. *Quaternary Science Reviews* **27**, 6–17.

Ložek, V. (1964) *Quartärmollusken der Tschechoslowakei.* Rozpravy ústředního ústavu geologického **31**, 374.

Lukas, S., Graf, A., Coray, S. & Schlüchter, C. (2012) Genesis, stability and preservation potential of large lateral moraines of Alpine valley glaciers – towards a unifying theory based on Findelengletscher, Switzerland. *Quaternary Science Reviews* **38**, 27–48.

Lundqvist, J. (1990) Glacial morphology as an indicator of the direction of glacial transport. In: Kujansuu, R. & Saarnisto, M. (eds) *Glacial Indicator Tracing*. Rotterdam, Brookfield: Balkema, pp. 61–70.

Lundqvist, J. (1991) Carl C:zon Caldenius–geologist, geotechnician, predecessor of IGCP. *Boreas* **20**, 183–189.

Lunkka, J. P. (1994) Sedimentology and lithostratigraphy of the North Sea Drift and Lowestoft Till Formations in the coastal cliffs of NE Norfolk. *Journal of Quaternary Science* **9**, 209–233.

Lüthgens, C. & Böse, M. (2010) Morphostratigraphy to geochronology–on the dating of ice marginal positions. *Quaternary Science Reviews*, doi: 10.1016/jquascirev.2010.10.009.

Lüttig, G. (1958) Methodische Fragen der Geschiebeforschung. *Geologisches Jahrbuch* **75**, 361–418.

Lutz, R., Kalka, S., Gaedicke, C., Reinhardt, L. & Winsemann, J. (2009) Pleistocene tunnel valleys in the German North Sea: spatial distribution and morphology. *Zeitschrift der Deutschen Gesellschaft für Geowissenschaften* **160**, 225–235.

Lyell, C. (1830–33) *Principles of Geology*. London: John Murray. (1830) **1**, 511; (1832) **2**, 330; (1833) **3**, 109.

Lyell, C. (1834) *Observations on the Loamy Deposit Called 'Loess' in the Valley of the Rhine*. The Geological Society of London, Proceedings **2**(36), 83–85.

Lyell, C. (1840) On the Boulder Formation, or drift and associated freshwater deposits composing the mud-cliffs of Eastern Norfolk. *The London and Edinburgh Philosophical Magazine and Journal of Science*, Third Series **16**(104), 345–380.

Lyell, C. (1863) *The Geological Evidences of the Antiquity of Man*. London: John Murray.

Maarleveld, G. C. & van den Toorn, J. C. (1955) Pseude-sölle in Noord-Nederland. *Tijdschrift van het Koninklijk Nederlandsch Aardrijkskundig Genootschap* **LXXII**, 347–360.

MacClintock, P. & Dreimanis, A. (1964) Reorientation of till fabric by overriding glacier in the St. Lawrence valley. *American Journal of Science* **262**, 133–142.

Mackay, J. R. (1974) Ice-wedge cracks, Garry Island, Northwest Territories. *Canadian Journal of Earth Sciences* **11**, 1366–1383.

Mackay, J. R. (1979) An equilibrium model for hummocks (non-sorted circles), Garry Island, Northwest Territories. Geological Survey of Canada, Paper **79-1A**, 165–167.

Mackay, J. R. (1980) The origin of hummocks, western Arctic coast, Canada. *Canadian Journal of Earth Sciences* **17**, 996–1006.

Mackay, J. R. (1990) Some observations on the growth and deformation of epigenetic, syngenetic and anti-syngenetic ice wedges. *Permafrost and Periglacial Processes* **1**, 15–29.

Mackay, J. R. (1999) Cold-climate shattering (1974 to 1993) of 200 glacial erratics on the exposed bottom of a recently drained Arctic Lake, Western Arctic Coast, Canada. *Permafrost and Periglacial Processes* **10**, 125–136.

Mackay, J. R. & MacKay, D. K. (1976) Cryostatic pressures in nonsorted circles (mud hummocks), Inuvik, Northwest Territories. *Canadian Journal of Earth Sciences* **13**, 889–897.

Mackintosh, A., White, D., Fink, D., Gore, D.B., Pickard, J. & Fanning, P.C. (2007) Exposure ages from mountain dipsticks in Mac. Robertson Land, East Antarctica, indicate little change in ice sheet thickness since the Last Glacial Maximum. *Geology* **35**, 551–554.

Mackintosh, D. (1873) Observations on the more remarkable Boulders of the Northwest of England and the Welsh Borders. *Quarterly Journal of the Geological Society* **29**, 351–360.

Madsen, V., Nordmann, V. & Hartz, N. (1908) *Eem-Zonerne. Studier over Cyrinaleret og andre Eem-Aflejringer i Danmark, Nord-Tyskland og Holland.* Danmarks Geologiske Undersøgelse, II. Række **17**, 302.

Makaske, B., Maas, G. J. & Smeerdijk, D. G. van (2008) The age and origin of the Gelderse Ijssel. *Netherlands Journal of Geosciences–Geologie en Mijnbouw* **87**, 323–337.

Makowska, A. (1979) Interglacjal eemski w dolinie dolnej Wisły. *Studia Geologica Polonica* **63**, 1–90.

Mangerud, J. (1982) The Chronostratigraphical Subdivision of the Holocene in Norden: a review. *Striae* **16**, 65–70.

Mangerud, J. (1987) The Alleröd/Younger Dryas Boundary. In: Berger, W. H. & Labeyrie, L. D. (eds) *Abrupt Climatic Change.* Dordrecht: Reidel, pp. 163–171.

Mangerud, J., Andersen, S. T., Berglund, B. E. & Donner, J. J. (1974) Quaternary stratigraphy of Norden, a proposal for terminology and classification. *Boreas* **3**, 109–128.

Mangerud, J., Sønstegaard, E. & Sejrup, H.-P. (1977) Saalian–Eemian–Weichselian stratigraphy at Fjøsanger, Western Norway. Abstracts. *X INQUA Congress.* Birmingham 1977, 286.

Mangerud, J., Sønstegaard, E., Sejrup, H. P. & Haldorsen, S. (1981) A continuous Eemian–Early Weichselian sequence containing pollen and marine fossils at Fjøsanger, western Norway. *Boreas* **10**, 137–208.

Mangerud, J., Astakhov, V. I., Murray, A. & Svendsen, J. I. (2001) The chronology of a large ice-dammed lake and the Barents-Kara Ice Sheet advance, Northern Russia. *Global and Planetary Change* **31**, 319–334.

Mangerud, J., Jakobsson, M., Alexanderson, H., Asthakov, V., Clarke, G. K. C., Henriksen, M., Hjort, C., Krinner, G., Lunkka, P. J., Müller, P., Murray, A., Nikolskaya, O., Saarnisto, M. & Svendsen, J. I. (2004) Ice-dammed lakes and rerouting of the drainage of northern Eurasia during the last glaciation. *Quaternary Science Reviews* **23**, 1313–1332.

Mangerud, J., Kaufman, D., Hansen, J. & Svendsen, J. I. (2008) Ice-free conditions in Nowaja Semlja 35 000–30 000 cal years B. P., as indicated by radiocarbon ages and amino acid racemization evidence from marine molluscs. *Polar Research* **27**, 187–208.

Mangerud, J., Gyllencreutz, R., Lohne, Ø. & Svendsen, J.I. (2011) In: Ehlers, J., Gibbard, P.L. & Hughes, P.D. (eds) *Quaternary Glaciations - Extent and Chronology - A Closer Look.* Elsevier, Developments in Quaternary Science **15**, 279–298.

Mania, D. (1973) Paläoökologie, Faunenentwicklung und Stratigraphie des Eiszeitalters im mittleren Elbe-Saalegebiet auf Grund von Molluskengesellschaften. *Geologie* **21**, Beiheft **78/79**, 175.

Mania, D. (1975) Zur Stellung der Travertinablagerungen von Weimar-Ehringsdorf im Jungpleistozän des nördlichen Mittelgebirgsraumes. In: III. Internationales Paläontologisches Kolloquium (1968) *Das Pleistozän von Weimar-Ehringsdorf, Teil 2.* Abhandlungen des Zentralen Geologischen Instituts **23**, 571–589.

Mania, D. (1990) Stratigraphie, Ökologie und mittelpaläolithische Jagdbefunde des Interglazials von Neumark-Nord (Geiseltal). *Veröffentlichungen des Landesmuseums für Vorgeschichte in Halle* **43**, 9–130.

Mania, D. & Mai, D. H. (2001) Molluskenfaunen und Floren im Elbe-Saalegebiet während des mittleren Eiszeitalters. *Praehistoria Thuringica* **6/7**, 46–91.

Mannerfelt, G. M. (1945) Några glacialmorfologiska formelement. *Geografiska Annaler* **27**, 1–239.

Marchant, R., Harrison, S. P., Hooghiemstra, H., Markgraf, V., van Boxel, J. H., Ager, T., Almeida, L., Anderson, R., Baied, C., Behling, H., Berrio, J. C., Burbridge, R., Björck, S., Byrne, R., Bush, M. B., Cleef, A. M., Duivenvoorden, J. F., Flenley, J. R., De Oliveira, P., van Geel, B., Graf, K. J., Gosling, W. D., Harbele, S., van der Hammen, T., Hansen, B. C. S., Horn, S. P., Islebe G. A., Kuhry, P., Ledru, M.-P., Mayle, F. E., Leyden, B. W., Lozano-García, S., Melief, A. B. M., Moreno, P., Moar, N. T., Prieto, A., van Reenen, G. B., Salgado-Labouriau, M. L., Schäbitz, F., Schreve-Brinkman, E. J. & Wille, M. (2009) Pollen-based biome reconstructions for Latin America at 0, 6000 and 18 000 radiocarbon years. *Climate of the Past Discussions* **5**, 369–461.

Marcinek, J., Präger, F. & Steinmüller, A. (1970) Periglaziäre Gestaltung der Täler. In: Richter, H., Haase, G., Lieberoth, I. & Ruske, R. (eds) *Periglazial - Löß - Paläolithikum im Jungpleistozän der Deutschen Demokratischen Republik*. Petermanns Geographische Mitteilungen, Ergänzungsheft no. 274, pp. 281–328.

Marcus, L. F. & Berger, R. (1984) The Significance of Radiocarbon Dates for Rancho La Brea. In: Martin, P. S. & Klein, R. G. (eds) *Quaternary Extinctions–A Prehistoric Revolution*. Tucson: University of Arizona Press, pp. 159–183.

MARGO Project Members (2009) Constraints on the magnitude and patterns of ocean cooling at the Last Glacial Maximum. *Nature Geoscience* **2**, 127–132.

Mark, B.G. & Osmaston, H.A. (2008) Quaternary glaciation in Africa: key chronologies and climatic implications. *Journal of Quaternary Science* **23**, 589–608.

Marks, L. (2004) Pleistocene glacial limits in Poland. In: Ehlers, J. & Gibbard, P. L. (eds) *Quaternary Glaciations–Extent and Chronology Part I, Europe*. Amsterdam: Elsevier, Developments in Quaternary Science **2**, 295–300.

Marks, L. (2011) Quaternary Glaciations in Poland. In: Ehlers, J., Gibbard, P.L. & Hughes, P.D. (eds) *Quaternary Glaciations - Extent and Chronology - A Closer Look*. Elsevier, Developments in Quaternary Science **15**, 299–304.

Marks, L. & Pavlovskaya, I. E. (2003) The Holsteinian Interglacial river network of mid-eastern Poland and western Belarus. *Boreas* **32**, 337–346.

Marks, L., Piotrowski, J. A., Stephan, H.-J., Fedorowicz, S. & Butrym, J. (1995) Thermoluminescence indications of the Middle Weichselian (Vistulian) Glaciation in Northwest Germany. *Meyniana* **47**, 69–82.

Marquette, G. C., Gray, J. T., Gosse, J. C., Courchesne, F., Stockli, L., Macpherson, G. & Finkel, R. (2004) Felsenmeer persistence under non-erosive ice in the Torngat and Kaumajet mountains, Quebec and Labrador, as determined by soil weathering and cosmogenic nuclide exposure dating. *Canadian Journal of Earth Sciences* **41**, 19–38.

Martínez, O., Coronato, A. & Rabassa, J. (2011) Pleistocene Glaciations in Northern Patagonia, Argentina: An Updated Review. In: Ehlers, J., Gibbard, P.L. & Hughes, P.D. (eds) *Quaternary Glaciations - Extent and Chronology - A Closer Look*. Elsevier, Developments in Quaternary Science **15**, 729–734.

Martonne, E. de (1924) Les formes glaciaires sur le versant Nord du Haut Atlas. *Annales de Géographie* (Paris, 15 Mai 1924) **183**, 296–302.

Mathers, S. J. & Zalasiewicz, J. A. (1986) A sedimentation pattern in Anglian meltwater channels from Suffolk, England. *Sedimentology* **33**, 559–573.

Mathers, S. J. & Zalasiewicz, J. A. (1988) The Red Crag and Norwich Crag formations of southern East Anglia. *Procceedings of the Geologists' Association* **99**, 261–278.

Mathers, S. J., Zalasiewicz, J.A., Bloodworth, A.J. & Morton, A.C. (1987) The Banham Beds: a petrologically distinct suite of Anglian glacigenic deposits from central East Anglia. *Proceedings of the Geologists' Association* **98**, 229–240.

Mathewes, J. A. (2007) Neoglaciation in Europe. In: Elias, S. A. (ed.) *Encyclopaedia of Quaternary Science*. Amsterdam: Elsevier, pp. 1122–1133.

Matsch, C. L. & Schneider, A.F. (1986) Stratigraphy and correlation of the glacial deposits of the glacial lobe complex in Minnesota and northwestern Wisconsin. *In*: Sibrava, V., Bowen, D.Q. & Richmond, G.M. (eds) *Glaciations in the Northern Hemisphere*. Quaternary Science Reviews, Special Volume no. 5, 59–64.

McCabe, A. M. (1991) The distribution and stratigraphy of drumlins in Ireland. In: Ehlers, J., Gibbard, P. L. & Rose, J. (eds) *Glacial Deposits in Great Britain and Ireland*. Rotterdam, Brookfield: Balkema, pp. 421–435.

McCarroll, D. & Ballantyne, C. K. (2000) The last ice sheet in Snowdonia. *Journal of Quaternary Science* **15**, 765–778.

McCarthy, D. P. (2007) Lichenometry. In: Elias, S. A. (ed.) *Encyclopaedia of Quaternary Science*. Amsterdam: Elsevier, pp. 1399–1405.

McCarthy, A., Mackintosh, A., Rieser, U. & Fink, D. (2008) Mountain glacier chronology from Boulder Lake, New Zealand, indicates MIS 4 and MIS 2 ice advances of similar extent. *Arctic, Antarctic, and Alpine Research* **40**, 695–708.

McCave, I. N., Holligan, P.M., Street-Perrott, F.A., Labeyrie, L. & Sarnthein, M. (1995) Sedimentary processes and the creation of the stratigraphic record in the Late Quaternary North Atlantic Ocean [and discussion]. *Philosophical Transactions of the Royal Society B* **348**, 1324, 229–241.

McCrea, W. H. (1975) Ice ages and the galaxy. *Nature* **255**, 607–609.

McIntyre, A., Ruddiman, W. F. & Jantzen, R. (1972) Southward penetrations of the North Atlantic polar front: faunal and floral evidence of large-scale surface water mass movements over the last 225,000 years. *Deep Sea Research and Oceanographic Abstracts* **19**, 61–77.

Meadows, M. E. & Chase, B. M. (2007) Pollen Records, Late Pleistocene–Africa. In: Elias, S. A. (ed.) *Encyclopaedia of Quaternary Science*. Amsterdam: Elsevier, pp. 2606–2613.

Meer, J. J. M. van der (1993) Microscopic evidence of subglacial deformation. *Quaternary Science Reviews* **12**, 553–587.

Meier, M. F. & Post, A. (1969) What are glacier surges? *Canadian Journal of Earth Sciences* **6**, 807–817.

Meltzer, D. J., Holliday, V. T., Cannon, M. D. & Miller, D. S. (2014) Chronological evidence fails to support claim of an isochronous widespread layer of cosmic impact indicators dated to 12,800 years ago. *Proceedings of the National Academy of Science U.S.A.* **111** (**21**), E2162–71.

Menke, B. (1968) Beiträge zur Biostratigraphie des Mittelpleistozäns in Norddeutschland. *Meyniana* **18**, 35–42.

Menke, B. (1969) Vegetationsgeschichtliche Untersuchungen an altpleistozänen Ablagerungen aus Lieth bei Elmshorn. *Eiszeitalter und Gegenwart* **20**, 76–83.

Menke, B. (1975) Vegetationsgeschichte und Florenstratigraphie Nordwestdeutschlands im Pliozän und Frühquartär. Mit einem Beitrag zur Biostratigraphie des Weichsel-Frühglazials. *Geologisches Jahrbuch A* **26**, 3–151.

Menke, B. (1985) Eem-Interglazial und "Treene-Warmzeit" in Husum/Nordfriesland. Mit einem Beitrag von Risto Tynni, unter Mitarbeit von Holger Ziemus. *Geologisches Jahrbuch A* **86**, 63–99.

Menke, B. (1992) Eeminterglaziale und nacheiszeitliche Wälder in Schleswig-Holstein. *Berichte des Geologischen Landesamtes Schleswig-Holstein* **1**, 28–101.

Menke, B. & Tynni, R. (1984) Das Eeminterglazial und das Weichselfrühglazial von Rederstall/Dithmarschen und ihre Bedeutung für die mitteleuropäische Jungpleistozän-Gliederung. *Geologisches Jahrbuch A* **76**, 120.

Menning, M. & Hendrich, A. (2005) Erläuterungen zur Stratigraphischen Tabelle von Deutschland 2005 (ESTD 2005). *Newsletters on Stratigraphy* **41**, 1–405.

Menzies, J., van der Meer, J. J. M. & Rose, J. (2006) Till–as a glacial "tectomict", its internal architecture, and the development of a "typing" method for till differentiation. Binghamton Symposium 2003, *Geomorphology* **75**, 172–200.

Mercer, J., Fleck, R., Mankinen, E. & Sander, W. (1975) *Southern Patagonia: Glacial Events Between 4 m. y. and 1 m. y. Ago.* Wellington: The Royal Society of New Zealand, Quaternary Studies pp. 223–230.

Messerli, B. (1967) Die eiszeitliche und die gegenwartige Vertgletscherung im Mittelemeeraum. *Geographica Helvetica* **22**, 105–228.

Messerli, B. (1972) Formen und Formungsprozesse in der Hochgebirgsregion des Tibesti. Hochgebirgsforschung H. 2, Universitätsverlag Wagner Innsbruck-München, pp. 23–86.

Messerli, B. & Winiger M. (1992) Climate, environmental changeand resources of the African mountains from the Mediterranean to the Equator. *Mountain Research and Development* **12**, 315–336.

Meyer, K.-D. (1970) Zur Geschiebeführung des Oldenburgisch-Ostfriesischen Geestrückens. *Abhandlungen des Naturwissenschaftlichen Vereins zu Bremen* **37**, 227–246.

Meyer, K.-D. (1973) Zur Entstehung der abflußlosen Hohlformen auf der Neuenwalder Geest. *Jahrbuch der Männer vom Morgenstern* **53**, 23–29.

Meyer, K.-D. (1980) Zur Geologie der Dammer und Fürstenauer Stauchendmoränen (Rehburger Phase des Drenthe-Stadiums). In: *Festschrift G. Keller.* Osnabrück: Wenner, pp. 83–104.

Meyer, K.-D. (1983a) Indicator pebbles and stone count methods. In: Ehlers, J. (ed.) *Glacial Deposits in North-West Europe.* Rotterdam: Balkema, pp. 275–287.

Meyer, K.-D. (1983b) Zur Anlage der Urstromtäler in Niedersachsen. *Zeitschrift für Geomorphologie NF* **27**, 147–160.

Meyer, K.-D. (1983c) Zur Anlage der Urstromtäler in Niedersachsen. *Zeitschrift für Geomorphologie NF* **27**, 147–160.

Meyer, K.-D. (1991) Zur Entstehung der westlichen Ostsee. *Geologisches Jahrbuch A* **127**, 429–446.

Meyer, R. K. F. & Schmidt-Kaler, H. (2002) *Auf den Spuren der Eiszeit südlich von München, westlicher Teil.* München: Pfeil, Wanderungen in die Erdgeschichte **9**.

Miara, S. (1995) Gliederung der rißzeitlichen Schotter und ihrer Deckschichten beiderseits der unteren Iller nördlich der Würmendmoränen. *Münchner Geographische Abhandlungen* **22**, 185.

Mickelson, D. M. & Evenson, E.B. (1975) Pre-Twocreekan age of the type Valders till, Wisconsin. *Geology* **3**, 587–590.

Mickelson, D. M., Clayton, L., Fullerton, D.S. & Borns, H.W. (1983) The Late Wisconsin glacial record of the Laurentide Ice Sheet in the United States. In: Wright, H.E., Jr. (ed.) *Late-Quaternary Environments of the United States*. Minneapolis: University of Minnesota Press, pp. 3–37.

Milankovitch, M. (1941) Kanon der Erdbestrahlung und seine Anwendung auf das Eiszeitenproblem. *Académie Royale Serbe. Éditions Speciales* **133, XX**, 633.

Miller, B. B., McCoy, W.D., Wayne, W.J. & Brockman, C.S. (1992) Ages of the Whitewater and Fairhaven tills in southwestern Ohio and southeastern Indiana. In: Clark, P.U. & Lea, P.D. (eds) *The Last Interglacial-Glacial Transition in North America*. Geological Society of America, Special Paper no. 270, pp. 89–98.

Miller, H. (1850) On peculiar scratched pebbles, etc., in the boulder clay in Caithness. *Report of the British Association for the Advancement of Science* 1850 (Edinburgh) 93–96.

Miller, K. G., Fairbanks, R.G. & Mountain, G.S. (1987) Tertiary oxygen isotope synthesis, sea level history, and continental margin erosion. *Paleoceanography* **2**, 1–19.

Mills, S. C., Grab, S. W. & Carr, S. J. (2009) Recognition and palaeoclimatic implications of late Quaternary niche glaciation in eastern Lesotho. *Journal of Quaternary Science* **24**, 647–663.

Milthers, K. (1959) Beskrivelse til Geologisk Kort over Danmark. Kortbladene Fåborg, Svendborg og Gulstav. *Danmarks Geologiske Undersøgelse* **I, 21A**, 112.

Mitchell, G. F., Penny, L.F., Shotton, F.W. & West, R.G. (1973) *A Correlation of Quaternary Deposits in the British Isles*. Geological Society of London, Special Report no. 4.

Mix, A. C., Bard, E. & Schneider, R. (2001) Environmental processes of the ice age: land, oceans, glaciers (EPILOG). *Quaternary Science Reviews* **20**, 627–657.

Mohn, H. & Nansen, F. (1893) Wissenschaftliche Ergebnisse von Dr. F. Nansens Durchquerung von Grönland 1888. *Petermanns Geographische Mitteilungen, Ergänzungsband* **XXIII (105)**, 111.

Moorlock, B. S., Hamblin, R.J.O., Morigi, A.N., Booth, S.J. & Jeffery, D.H. (2000) The geology of the country around Lowestoft and Saxmundham. Memoir of the British Geological Survey, sheets 176 and 191 (England and Wales).

Morgan, A. V. & Morgan, A. (1990) Beetles. In: Warner, B. G. (ed.) *Methods in Quaternary Ecology*. Geoscience Canada, Reprint Series **5**, 113–126.

Morgan, L. A., Cathey, H. E. & Pierce, K. L. (eds) (2009) The Track of the Yellowstone Hot Spot: Multi-disciplinary Perspectives on the Origin of the Yellowstone-Snake River Plain Volcanic Province. *Journal of Volcanology and Geothermal Research* **188**(1–3), 304.

Morlot, A. von (1844) *Ueber die Gletscher der Vorwelt und ihre Bedeutung*. Bern: Rätzer.

Morlot, A. von (1847) *Erläuterungen zur Geologischen Übersichtskarte der nordöstlichen Alpen*. Wien: Braumüller und Seidel.

Morrison, R. B. (ed.) (1991) *Quaternary Nonglacial Geology: Conterminous US*. Geological Society of America, The Geology of North America no. K-2.

Moscariello, A. (1996) Quaternary Geology of the Geneva Bay (Lake Geneva, Switzerland): Sedimentary Record, Palaeoenvironmental and Palaeoclimatic Reconstruction since the Last Glacial Cycle. *Terre und Environement* **4**, 230.

Mott, R. J. & Grant, D.R.G. (1985) Pre -Late Wisconsinan paleoenvironments in Atlantic Canada. *Géographie Physique et Quaternaire* **39**, 239–254.

Mott, R. J. (1990) Sangamonian forest history and climate in Atlantic Canada. *Géographie Physique et Quaternaire* **44**, 257–270.

Müller, H. (1974a) Pollenanalytische Untersuchungen und Jahresschichtenzählungen an der eemzeitlichen Kieselgur von Bispingen/Luhe. *Geologisches Jahrbuch A* **21**, 149–169.

Müller, H. (1974b) Pollenanalytische Untersuchungen und Jahresschichtenzählungen an der holsteinzeitlichen Kieselgur von Munster-Breloh. *Geologisches Jahrbuch A* **21**, 107–140.

Müller, H. (1992) Climate changes during and at the end of the interglacials of the Cromerian Complex. In: Kukla, G. & Went, E. (eds) *Start of a Glacial*. NATO ASI Series **I (3)**, 51–69.

Müller, U. (2001) Die Vegetations- und Klimaentwicklung im jüngeren Quartär anhand ausgewählter Profile aus dem südwestdeutschen Alpenvorland. *Tübinger Geowissenschaftliche Arbeiten* **D7**, 118.

Müller, U. (2004a) Alt- und Mittel-Pleistozän. In: Katzung, G. (ed.) *Geologie von Mecklenburg-Vorpommern*. Stuttgart: Schweizerbart, pp. 226–233.

Müller, U. (2004b) Jung-Pleistozän–Eem-Warmzeit bis Weichsel-Hochglazial. In: Katzung, G. (ed.) *Geologie von Mecklenburg-Vorpommern*. Stuttgart: Schweizerbart, pp. 234–242.

Munthe, H. (1902) *Beskrivning til kartbladet Kalmar*. Sveriges Geologiska Undersökning **C 4**, 213.

Munthe, H. (1927) *Studier över Ancylussjöns avlopp*. Sveriges Geologiska Undersökning **C 346**, 107.

Murton, D. K., Pawley, S.W. and Murton, J.B. (2009) Sedimentology and luminescence ages of Glacial Lake Humber deposits in the central Vale of York. *Proceedings of the Geologists' Association* **120**, 209–222.

Muttoni, G., Carcano, C., Garzanti, E., Ghielmi, M., Piccin, A., Pini, R., Rogledi, S. & Sciunnach, D. (2003) Onset of Pleistocene glaciations in the Alps. *Geology* **31**, 989–992.

Muttoni, G., Ravazzi, C., Breda, M., Pini, R., Laj, C., Kissel, C., Mazaud, A. & Garzanti, E. (2007) Magnetostratigraphic dating of an intensification of glacial activity in the southern Italian Alps during Marine Isotope Stage 22. *Quaternary Research* **67**, 161–173.

Nansen, F. (1922) *The Strandflat and Isostasy*. Videnskabsselskabets i Kristiania skrifter 1921, I. Matematisk- Naturvidenskabelig Klasse **11**, 313.

Nelson, F. E., Shiklomanov, N. I., Christiansen H. H., Hinkel, K. M. (2004) The Circumpolar-Active-Layer-Monitoring (CALM) Workshop: introduction. *Permafrost and Periglacial Processes* **15**, 99–188.

Nesje, A. & Sejrup, H. P. (1988) Late Weichselian/Devensian ice sheets in the North Sea and adjacent land areas. *Boreas* **17**, 371–384.

Nesje, A. & Dahl, S. V. (2000) *Glaciers and Environmental Change*. London: Arnold.

Nesje, A., Kvamme, M., Rye, N. & Løvlie, R. (1991) Holocene glacial and climate history of the Jostedalsbreen region, western Norway; evidence from lake sediments and terrestrial deposits. *Quaternary Science Reviews* **10**, 87–114.

Nicolas, A. (1995) *Die ozeanischen Rücken–Gebirge unter dem Meer*. Heidelberg: Springer.

Noormets, R. & Flodén, T. (2002) Glacial deposits and Late Weichselian ice-sheet dynamics in the northeastern Baltic Sea. *Boreas* **31**, 36–56.

Nye, J. F. (1952) The mechanics of glacier flow. *Journal of Glaciology* **2**, 82–93.

Nývlt, D., Engel, Z. & Tyráček, J. (2011) Pleistocene Glaciations of Czechia. In: Ehlers, J., Gibbard, P.L. & Hughes, P.D. (eds) *Quaternary Glaciations - Extent and Chronology - A Closer Look*. Elsevier, Developments in Quaternary Science **15**, 37–46.

Ó Cofaigh, C. & Evans, D.J.A. (2007) Radiocarbon constraints on the age of the maximum advance of the British-Irish Ice Sheet in the Celtic Sea. *Quaternary Science Reviews* **26**, 1197–1203.

Ó Cofaigh, C., Dunlop, P. & Benetti, S. (2010) Marine geophysical evidence for Late Pleistocene ice sheet extent and recession off northwest Ireland. *Quaternary Science Reviews* **44**, 147–169.

O'Connor, J.E. (1993) *Hydrology, Hydraulics, and Geomorphology of the Bonneville Flood*. Geological Society of America, Special Paper no. 274.

Ohl, C., Frew, P., Sayers, P., Watson, G., Lawton, P., Farrow, B., Walkden, M. & Hall, J. (2003) North Norfolk - a regional approach to coastal erosion management and sustainability practice. In: McInnes, R.G. (ed.) *International Conference on Coastal Management 2003*. Thomas Telford: Brighton, 226–240.

Ohmura, A., Kasser, P. & Funk, M. (1992) Climate at the equilibrium line of glaciers. *Journal of Glaciology* **38**, 397–411.

Ohse, W. (1983) *Lösungs- und Fällungserscheinungen im System oberflächennahes unterirdisches Wasser/gesteinsbildende Minerale–eine Untersuchung auf der Grundlage der chemischen Gleichgewichts-Thermodynamik*. Kiel, Dissertation.

Olcott, A. N., Sessions, A. L., Corsetti, F. A., Kaufman, A. J. & de Oliviera, T. F. (2005) Biomarker evidence for photosynthesis during Neoproterozoic glaciation. *Science* **310**, 471–474.

Oldale, R.N. & Colman, S.M. (1992) On the age of the penultimate full glaciation of New England. In: Clark, P.U. & Lea, P.D. (eds) *The Last Interglacial-Glacial Transition in North America*. Geological Society of America, Special Paper no. 270, pp. 163–170.

Oppenheimer, S. (2009) The great arc of dispersal of modern humans: Africa to Australia. *Quaternary International* **202**, 2–13.

Østrem, G., Haakensen, N. & Melander, O. (1973) *Atlas over Breer i Nord-Skandinavia*. Norges Vassdrags og Elektrisitetsvesen og Stockholm Universitet.

Ostry, R. C. & Deane, R. E. (1963) Microfabric analyses of till. *Bulletin of the Geomorphological Society of America* **74**, 165–168.

Otvos, E. G. (2014) The Last Interglacial Stage: Definitions and marine highstand, North America and Eurasia. *Quaternary International*, doi:10.1016/j.quaint.2014.05.010.

Overbeck, F. (1975) *Botanisch-geologische Moorkunde unter besonderer Berücksichtigung der Moore Nordwestdeutschlands als Quellen zur Vegetations-, Klima- und Siedlungsgeschichte*. Neumünster: Wachholtz.

Pachur, H.-J. (1997) Der Ptolemäus-See in Westnubien als Paläoklimaindikator. *Petermanns Geographische Mitteilungen* **141**, 227–250.

Pachur, H.-J. & Röper, H.-P. (1987) Zur Paläolimnologie Berliner Seen. *Berliner Geographische Abhandlungen* **44**, 150.

Palacios, D., Andrés, N. de, Marcos, J. De & Vázquez-Selem, L. (2012) Glacial landforms and their palaeoclimatic significance in Sierra de Guadarrama, Central Iberian Peninsula. *Geomorphology* **139–140**, 67–78.

Pallàs, R., Rodés, Á., Braucher, R., Bourlès, D., Delmas, M., Calvet, M. & Gunnell, Y. (2010) Small, isolated glacial catchments as priority targets for cosmogenic surface exposure dating of Pleistocene climate fluctuations, southeastern Pyrenees. *Geology* **38**, 891–894.

Palmer, J. & Radley J. (1961) Gritstone tors of the English Pennines. *Zeitschrift für Geomorphologie* **5**, 37–52.

Palmer, J. & Neilson, R. A. (1962) The origin of granite tors on Dartmoor, Devonshire. *Proceedings of the Yorkshire Geological Society* **33**(3/15), 315–340.

Parfitt, S. A., Barendregt, R. W., Breda, M., Candy, I., Collins, M. J., Coope, G. R., Durbidge, P., Field. M. H., Lee, J. R., Lister, A. M., Mutch, R., Penkman, K. E. H., Preece, R. C., Rose, J., Stringer, C. B., Symmons, R., Whittaker, J. E.,Wymer, J. J.& Stuart, A. J. (2005) The earliest record of human activity in northern Europe. *Nature* **438**, 1008–1012.

Parfitt, S. A., Ashton, N. M., Lewis, S. G., Abel, R. L., Coope, G. R., Field, M. H., Gale, R., Hoare, P. G., Larkin, N. R., Lewis, M. D., Karloukovski, V., Maher, B. A., Peglar, S. M., Preece, R. C., Whittaker, J. E. & Stringer, C. B. (2010) Early Pleistocene human occupation at the edge of the boreal zone in northwest Europe. *Nature* **466**, 229–233.

Park, C. & Schmincke, H.-U. (2009) Apokalypse im Rheintal. *Spektrum der Wissenschaft Heft* **02/09**, 78–87.

Patzelt, G. (1983) Die spätglazialen Gletscherstände im Bereich des Mieslkopfes und im Arztal, Tuxer Voralpen, Tirol. *Innsbrucker Geographische Studien* **8**, 35–58.

Pawley, S. M., Rose, J., Lee, J.R., Moorlock, B.S.P. & Hamblin, R.J.O. (2004) Middle Pleistocene sedimentology and lithostratigraphy of Weybourne northeast Norfolk, England. *Proceedings of the Geologists' Association* **115**, 25–42.

Pawley S. M., Bailey, R.M., Rose, J., Moorlock, B.S.P., Hamblin, R.J.O., Booth, S.J. & Lee, J.R. (2008) Age limits on Middle Pleistocene glacial sediments from OSL dating, north Norfolk, UK. *Quaternary Science Reviews* **27**, 1363–1377.

Peacock, J.D. (1967) West Highland morainic features aligned in the direction of ice flow. *Scottish Journal of Geology* **3**, 372–373.

Peck, V.L., Hall, I.R., Zahn, R., Elderfield, H., Grousset, F., Hemming, S.R. & Scourse, J.D. (2006) High resolution evidence for linkages between NW European ice sheet instability and Atlantic meridional overturning circulation. *Earth and Planetary Science Letters* **243**, 475–488.

Pedersen, S. A. S. (2005) Structural analysis of the Rubjerg Knude Glaciotectonic Complex, Vendsyssel, northern Denmark. *Geological Survey of Denmark and Greenland Bulletin* **8**, 192.

Pedersen, S. A. S. (2006) Structurer og dynamisk udvikling af Rubjerg Knude Glaciotectonic Complex, Vendsyssel, Denmark. *Geologisk Tidsskrift* **2006**(1), 46.

Pellitero, R., Rea, B., Spagnolo, M., Bakke, J., Hughes, P.D., Ivy-Ochs, S., Lukas, S., Ribolini, A., 2015. A GIS tool for automatic calculation of glacier Equilibrium Line Altitudes. *Computers and Geosciences* **82**, 55–62.

Peltier, W.R. & Fairbanks, R.G. (2006) Global glacial ice volume and Last Glacial Maximum duration from an extended Barbados sea level record. *Quaternary Science Reviews* **26**, 862–875.

Penck, A. (1879) Die Geschiebeformation Norddeutschlands. *Zeitschrift der Deutschen Geologischen Gesellschaft* **31**, 117–203.

Penck, A. (1882) *Die Vergletscherung der deutschen Alpen, ihre Ursachen, periodische Wiederkehr und ihr Einfluß auf die Bodengestaltung.* Leipzig: Barth.

Penck, A. (1899) Die vierte Eiszeit im Bereich der Alpen. *Schriften der Vereinigung zur Verbreitung naturwissenschaftlicher Kenntnisse* **39**, 1–20.

Penck, A. (1900) Die Eiszeit auf der Balkanhalbinsel. *Globus* **78**, 133–178.

Penck, A. & Brückner, E. (1901–09) *Die Alpen im Eiszeitalter*, 3 Bände. Leipzig: Alpine Lakes Protection Society.

Penny, L. F. & Rawson, P.F. (1969) Field Meeting in East Yorkshire and North Lincolnshire. *Proceedings of the Geologists' Association* **80**, 193–218.

Perrin, R.M.S., Rose, J. & Davies, H. (1979) The distribution, variations and origins of pre-Devensian tills in eastern England. *Philosophical Transactions of the Royal Society of London B* **287**, 536–570.

Petersen, K. S. (1983) Redeposited biological material. In: Ehlers, J. (ed.) *Glacial Deposits in North-West Europe*. Rotterdam: Balkema, pp. 203–206.

Petersen, K. S. & Konradi, P. B. (1974) Lithologiske og palæontologiske beskrivelser af profiler i kvartæret pa Sjælland. *Dansk Geologisk Forening*, Årsskrift for **1973**, 47–56.

Petit-Maire, N. (ed.) (1991) *Paléoenvironnements du Sahara. Lacs Holocenes a Taoudenni (Mali)*. Editons du CNRS, Paris, France.

Petroleum Exploration Society of Great Britain (2007) *Structural Framework of the North Sea and Atlantic Margin*. London: Petroleum Exploration Society of Great Britain.

Phillips, F.M., Bowen, D.Q. & Elmore, D. (1994) Surface exposure dating of glacial features in Great Britain using cosmogenic chlorine-36: preliminary results. *Mineralogical Magazine* **58**A, 722–723.

Phillips, L. (1974) Vegetational history of the Ipswichian/Eemian Interglacial in Britain and Continental Europe. *New Phytologist* **73**, 589–604.

Piehler, H. (1974) Die Entwicklung der Nahtstelle von Lech-, Loisach- und Ammergletscher vom Hoch- bis Spätglazial der letzten Vereisung. *Münchener Geographische Abhandlungen* **13**, 105.

Pillans, B. (2007) Quaternary Stratigraphy - Overview. In: Elias, S. A. (ed.) *Encyclopaedia of Quaternary Science*. Amsterdam: Elsevier, pp. 2785–2802.

Piotrowski, J. A. (1992) Was ist ein Till?–Faziesstudien an glazialen Sedimenten. *Die Geowissenschaften* **10**(4), 100–108.

Piotrowski, J. A. (1997) Subglacial hydrology in north-western Germany during the last glaciation; groundwater flow, tunnel valleys and hydrological cycles. *Quaternary Science Reviews* **16**, 169–185.

Pissart, A. (1987) Weichselian periglacial structures and their environmental significance: Belgium, the Netherlands, and northern France. In: Boardman, J. (ed.) *Periglacial Processes and Landforms in Britain and Ireland*. Cambridge: Cambridge University Press, pp. 77–85.

Pissart, A. (2003) The remnants of Younger Dryas lithalsas on the Hautes Fagnes Plateau in Belgium and elsewhere in the world. *Geomorphology* **52**, 5–38.

Pitts, M. & Roberts, M. (1998) *Fairweather Eden: Life in Britain half a million years ago as revealed by the excavations at Boxgrove*. London: Arrow Books.

Plummer, M. & Phillips, F.M. (2003) A 2-D numerical model of snow/ice energy balance and ice flow for paleoclimatic interpretation of glacial geomorphic features. *Quaternary Science Reviews* **14**, 1389–1406.

Pollard, D. & DeConto, R.M. (2009) Modelling West Antarctic ice sheet growth and collapse through the past five million years. *Nature* **458**, 329–332.

Polyak, L., Edwards, M.H., Coakley, B.J. & Jakobsson, M. (2001) Ice shelves in the Pleistocene Arctic Ocean inferred from glaciogenic deep-sea bedforms. *Nature* **410**, 453–457.

Polyak, L., Darby, D.A., Bischof, J. & Jakobsson, M. (2007) Stratigraphic constraints on late Pleistocene glacial erosion and deglaciation of the Chukchi margin, Arctic Ocean. *Quaternary Research* **67**, 234–245.

Polyak, L., Alley, R. B., Andrews, J. T., Brigham-Grette, J., Cronin, T. M., Darby, D. A., Dyke, A. S., Fitzpatrick, J. J., Funder, S., Holland, M., Jennings, A. E., Miller, G. H., O'Regan, M., Savelle, J., Serreze, M., John, K. S., White, J. W. C. & Wolff, E. (2010) History of sea ice in the Arctic. *Quaternary Science Reviews* **29**, 1757–1778.

Porter, S. C. (2004) Late Pleistocene glaciation of the Hindu Kush, Afghanistan. In: Ehlers, J. & Gibbard, P. L. (eds) *Quaternary Glaciations–Extent and Chronology Part III: South America, Asia, Africa, Australasia, Antarctica*. Amsterdam: Elsevier, Developments in Quaternary Science **2**, 1–2.

Porter, S. C. (2007) Loess Records/China. In: Elias, S. A. (ed.) *Encyclopaedia of Quaternary Science*. Amsterdam: Elsevier, pp. 1429–1440.

Porter, S. C. & Zhisheng, A. (1995) Correlation between climate events in the North Atlantic and China during the last glaciation. *Nature* **375**, 305–308.

Post, L. von (1916) Skogsträdpollen i sydsvenska torvmosselagerföldjcr. *Geologiska Foreningens i Stockholm Förhandlingar* **38**, 384–394.

Post, L. von (1928) Svea älvs geologiska tidsställning. *Sveriges Geologiska Undersökning C* **347**, 132.

Poulton, C.V.L., Lee, J.R., Jones, L.D., Hobbs, P.R.N. & Hall, M. (2006) Preliminary investigation into monitoring coastal erosion using terrestrial laser scanning: case study at Happisburgh, Norfolk, UK. *Bulletin of the Geological Society of Norfolk* **56**, 45–65.

Praeg, D. (2003) Seismic imaging of mid-Pleistocene tunnel valleys in the North Sea Basin - high resolution from low frequencies. *Journal of Applied Geophysics* **53**, 273–298.

Prange, W. (1979) Geologie der Steilufer von Schwansen, Schleswig-Holstein. *Schriften des Naturwissenschaftlichen Vereins in Schleswig-Holstein* **49**, 1–24.

Preece, R. C. & Parfitt, S.A. (2000) The Cromer Forest-bed Formation: new thoughts on an old problem. In: Lewis, S.G., Whiteman, C.A. & Preece, R.C. (eds) *The Quaternary of Norfolk and Suffolk, Field Guide*. London: Quaternary Research Association, pp. 1–27.

Preece, R. C., Parfitt, S. A., Coope, G. R., Penkman, K. E. H., Ponel, P. & Whittaker, J. E. (2009) Biostratigraphic and aminostratigraphic constraints on the age of the Middle Pleistocene glacial succession in north Norfolk, UK. *Journal of Quaternary Science* **24**, 557–580.

Prentice, M. L., Hope, G. S., Peterson, J. A. & Barrows, T. T. (2011) The glaciation of the south-east Asian equatorial region. In: Ehlers, J., Gibbard, P.L. & Hughes, P.D. (eds) *Quaternary Glaciations - Extent and Chronology - A Closer Look*. Elsevier, Developments in Quaternary Science **15**, 1023–1036.

Prest, V. K. (1983) *Canada's Heritage of Glacial Features*. Geological Survey of Canada, Miscellaneous Report no. 28.

Preusser, F. (1999) Lumineszenzdatierung fluviatiler Sedimente; Fallbeispiele aus der Schweiz und Norddeutschland. *Kölner Forum Geologie Paläontologie* **3**, 1–62.

Preusser, F. (2010) Stratigraphische Gliederung des Eiszeitalters in der Schweiz (Exkursion E am 8. April 2010). *Jahresberichte und Mitteilungen des Oberrheinischen Geologischen Vereins, NF* **92**, 83–98.

Preusser, F., Drescher-Schneider, R., Fiebig, M. & Schlüchter, C. (2005) Re-interpretation of the Meikirch pollen record, Swiss Alpine Foreland, and implications for Middle Pleistocene chronostratigraphy. *Journal of Quaternary Science* **20**, 607–620.

Price, R. J. (1969) Moraines, sandar, kames and eskers near Breiðamerkurjökull, Iceland. *Transactions of the Institute of British Geographers* **46**, 17–43.

Price, R. J. (1973) *Glacial and Fluvioglacial Landforms*. London: Longman.

Pringle, J & George, T.N. (1961) *British Regional Geology, South Wales*. London: Her Majesty's Stationary Office.

Putnam, A., Schäfer, J., Barrell, D.J.A., Vadergoes, M., Denton, G.H., Kaplan, M.R., Finkel, R.C., Scwartz, R., Goehring, B.M. & Kelley, S.E. (2010) In situ cosmogenic [10]Be production-rate calibration from the Southern Alps, New Zealand. *Quaternary Geochronology* **5**, 392–409.

Putnam, A. E., Schaefer, J. M., Denton, G. H., Barrell, D. J. A., Birkel, S.A., Andersen, B.G., Kaplan, M.R., Finkel, R.C., Schwartz, R. & Doughty, A.M. (2013) The Last Glacial Maximum at 44°S documented by a moraine chronology at Lake Ohau, Southern Alps of New Zealand. *Quaternary Science Reviews* **62**, 114–141.

Quinlan, G. & Beaumont, C. (1982) The deglaciation of Atlantic Canada as reconstructed from the post-glacial relative sea-level record. *Canadian Journal of Earth Sciences* **19**, 2232–2246.

Rabassa, J. (2008) Late Cenozoic glaciations in Patagonia and Tierra del Fuego. In: Rabassa, J. (ed.). *Late Cenozoic of Patagonia and Tierra del Fuego*. Amsterdam: Elsevier, Developments in Quaternary Science **11**, 151–204.

Raden, U. J. van, Colombaroli, D., Gilli, A., Schwander, J., Bernasconi, S. M., van Leeuwen, J., Leuenberger, M. & Eicher, U. (2013) High resolution late-glacial chronology for the Gerzensee lake record (Switzerland): δ18O correlation between a Gerzensee-stack and NGRIP. *Palaeogeography, Palaeoclimatology, Palaeoecology* **391**, 13–24.

Rahmstorf, S. (2003) Timing of abrupt climate change: A precise clock. *Geophysical Research Letters* **30**, 17-1–17-4.

Ramm, F. & Topf, J. (2008) *OpenStreetMap. Die freie Weltkarte nutzen und mitgestalten*. Berlin: Lehmanns Media.

Ramsay, W. (1924) On relations between crustal movements and variations of sea-level during the late Quaternary time, especially in Fennoscandia. *Fennia* **44**(5), 39.

Rappol, M. (1983) *Glacigenic properties of till - Studies in glacial sedimentology from the Allgäu Alps and The Netherlands.* Thesis, Universiteit van Amsterdam.

Rappol, M., Haldorsen, S., Jørgensen, P., van der Meer, J. J. M. & Stoltenberg, H. M. P. (1989) Composition and origin of petrographically-stratified thick till in the Northern Netherlands and a Saalian glaciation model for the North Sea basin. *Mededelingen van de Werkgroep voor Tertiaire en Kwartaire Geologie* **26**, 31–64.

Rasmussen, S. O., Andersen, K. K., Svensson, A. M., Steffensen, J. P., Vinther, B. M., Clausen, H. B., Siggaard-Andersen, M.-L., Johnsen, S. J., Larsen, L. B., Dahl-Jensen, D., Bigler, M., Röthlisberger, R., Fischer, H., Goto-Azuma, K., Hansson, M. E. & Ruth, U. (2006) A new Greenland ice core chronology for the last glacial termination. *Journal of Geophysical Research* **111**, D06102, doi:10.1029/2005JD006079.

Raymo, M. E., Ruddiman, W. F. & Froelich, P. N. (1988) Influence of Late Cenozoic mountain building on ocean geochemical cycles. *Geology* **16**, 649–653.

Rice, R. J. (1968) The Quaternary deposits of the central Leicestershire. *Philosophical Transactions of the Royal Society of London A* **262**, 459–509.

Rice, R. J. (1981) The Pleistocene deposits of the area around Croft in south Leicestershire. *Philosophical Transactions of the Royal Society of London B* **293**, 385–418.

Rice, R. J. & Douglas, T. (1991) Wolstonian glacial deposits and glaciation in Britain. In: Ehlers, J., Gibbard, P.L. & Rose, J. (eds) *Glacial Deposits in Great Britain and Ireland.* Rotterdam: Balkema, pp. 25–36.

Richards, A. E. (1998) Re-evaluation of the Middle Pleistocene stratigraphy of Herefordshire. *Journal of Quaternary Science* **13**, 115–136.

Richmond, G.M. (1986) Tentative correlation of deposits of the Cordilleran ice-sheet in the northern Rocky Mountains. In: Richmond, G.M. & Fullerton, D.S. (eds) Quaternary Glaciations in the United States of America. *Quaternary Science Reviews* **5**, 129–144.

Richmond, G.M. & Fullerton, D.S. (1986) Introduction to Quaternary glaciations in the United States. In: Šibrava, V., Bowen, D.Q. & Richmond, G.M. (eds) Quaternary Glaciations in the Northern Hemisphere. *Quaternary Science Reviews* **5**, 3–10.

Richter, K. (1932) Die Bewegungsrichtung des Inlandeises, rekonstruiert aus den Kritzen und Längsachsen der Geschiebe. *Zeitschrift für Geschiebeforschung* **8**, 62–66.

Richter, K. (1936) Ergebnisse und Aussichten der Gefügeforschung im Pommerschen Diluvium. *Geologische Rundschau* **32**, 196–206.

Richthofen, F. von 1877. China–Ergebnisse eigener Reisen und darauf gegründeter Studien, Bd. I. Berlin: Reimer.

Ricken, W. (1983) Mittel- und jungpleistozäne Lößdecken im südwestlichen Harzvorland. Stratigraphie, Paläopedologie, fazielle Differenzierung und Konnektierung in Flußterrassen. *Catena Supplement* **3**, 95–138.

Rinterknecht, V. R., Marks, L., Piotrowski, J. A., Raisbeck, G. M., Yiou, F., Brook, E. J. & Clark, P. U. (2005) Cosmogenic [10]Be ages on the Pommeranian Moraine, Poland. *Boreas* **34**, 186–191.

Rinterknecht, V. R., Pavlovskaya, I. E., Clark, P. U., Raisbeck, G. M., Yiou, F. & Brook, E. J. (2007) Timing of the last deglaciation in Belarus. *Boreas* **36**, 307–313.

Rinterknecht, V. R., Bitinas, A., Clark, P. U., Raisbeck, G. M., Yiou, F. & Brook, E. J. (2008) Timing of the last deglaciation in Lithuania. *Boreas* **37**, 426–433.

Roberts, M. & Parfitt, S. (eds) (1999) *Boxgrove: A Middle Pleistocene Hominid Site at Eartham Quarry, Boxgrove, West Sussex*. English Heritage Archaeological Report **17**, 456 pp.

Robertsson, A.-M. (2000) The Eemian interglacial in Sweden, and comparison with Finland. *Geologie en Mijnbouw–Netherlands Journal of Geosciences* **79**, 325–333.

Rognon, P. (1967) Le massif de l'Atakor et ses bordures (Sahara Central). Etude géomorphologique. Thèse Doctorat d'Etat ès lettres. Paris: Editions du CNRS.

Roland, N. W. (2009) *Antarktis–Forschung im ewigen Eis*. Heidelberg: Spektrum.

Rolfe, C. J., Hughes, P. D., Fenton, C. R., Schnabel, C., Xu, S., Brown, A. G. (2012) Paired [10]Be and [26]Al exposure ages from Lundy: new evidence for the extent and timing of Devensian glaciation in the southern British Isles. *Quaternary Science Reviews* **43**, 61–73.

Romanovsky, N. H. (1985) Distribution of recently active ice and soil wedges in the USSR. In: Church, M. & Slaymaker, S. (eds) *Field and Theory: Lectures in Geocryology*. University of British Columbia, 154–165.

Rose, J. (1987) Status of the Wolstonian glaciation in the British Quaternary. *Quaternary Newsletter* **53**, 1–9.

Rose, J. (1994) Major river systems of central and southern Britain during the Early and Middle Pleistocene. *Terra Nova* **6**, 435–443.

Rose, J. (1995) Lateglacial and Holocene river activity in lowland Britain. *Palaoklimaforschung*, Special Issue **9**, 51– 74.

Rose, J. (2009) Early and Middle Pleistocene landscapes of eastern England. *Proceedings of the Geologists' Association* **120**, 3–33.

Rose, J. & Allen, P. (1977) Middle Pleistocene stratigraphy in southeastern Suffolk. *Journal of the Geological Society of London* **133**, 83–102.

Rose, J., Allen, P. & Hey, R.W. (1976) Middle Pleistocene stratigraphy in southern East Anglia. *Nature* **236**, 492–494.

Rose, J., Whiteman, C.A., Allen, P. & Kemp, R.A. (1999) The Kesgrave sands and Gravels: 'pre-glacial' Quaternary deposits of the River Thames in East Anglia and the Thames valley. *Proceedings of the Geologists' Association* **110**, 93–116.

Röthlisberger, F. (1976) Gletscher- und Klimaschwankungen im Raum Zermatt, Ferpècle und Arolla. *Die Alpen* **52**(3/4), 59–152.

Röthlisberger, H. & Lang, H. (1987) Glacial hydrology. In: Gurnell, A. M. & Clark, M. J. (eds) *Glacio-fluvial Sediment Transfer*. John Wiley & Sons, pp. 207–284.

Rousseau, D. D. (2001) Loess biostratigraphy: New advances and approaches in mollusk studies. *Earth Science Reviews* **54**, 157–171.

Rowan, A. V., Brocklehurst, S.H., Schultz, D.M., Plummer, M.A., Anderson, L.S. & Glasser, N.F. (2014) Late Quaternary glacier sensitivity to temperature and precipitation distribution in the Southern Alps of New Zealand. *Journal of Geophysical Research: Earth Surface* **119**, 1064–1081.

Rowlands, B. M. (1971) Radiocarbon evidence of the age of an Irish Sea glaciation in the Vale of Clwyd. *Nature-Physical Science* **230**, 9–11.

Rózycki, S. Z. (1965) Die stratigraphische Stellung des Warthe-Stadiums in Polen. *Eiszeitalter und Gegenwart* **16**, 189–201.

Ruddiman, W. F. (2003) The anthropogenic greenhouse era began thousands of years ago. *Climatic Change* **61**, 261–293.

Ruddiman, W. F. & Kutzbach, J. E. (1990) Late Cenozoic plateau uplift and climate change. *Transactions of the Royal Society of Edinburgh, Earth Sciences* **81**, 301–314.

Rudoy, A. N. (2002) Glacier-dammed lakes and geological work of glacial superfloods in the Late Pleistocene, Southern Siberia, Altai Mountains. *Quaternary International* **87**, 119–140.

Russell, A. J., Gregory, A. G., Large, A. R. G., Fleisher, P. J. & Harris, T. (2007) Tunnel channel formation during the November 1996 jökulhlaup, Skeiðarárjökull, Iceland. *Annals of Glaciology* **45**, 95–103.

Ruth, U., Bigler, M., Rothlisberger, R., Siggaard-Andersen, M.L., Kipfstuhl, S., Goto- Azuma, K., Hansson, M.E., Johnsen, S.J., Lu, H.Y. & Steffensen, J.P. (2007) Ice core evidence for a very tight link between North Atlantic and east Asian glacial climate. *Geophysical Research Letters* **34**(3), L03706.

Sadiq, A. M. & Nasir, S. J. (2002) Middle Pleistocene karst evolution in the State of Qatar, Arabian Gulf. *Journal of Cave and Karst Studies* **64**(2), 132–139.

Salisbury, R. D. (1892) A Preliminary Paper on Drift or Pleistocene Formations of New Jersey. Annual Report of the State Geologist of New Jersey, 1891, 35–108.

Salisbury, R. D. (1894) Surface Geology–Report of Progress, 1892. New Jersey Geological Survey, Annual Report of the State Geologist for the year 1893, 35–328.

Salisbury, R. D. (1902) The Glacial Geology of New Jersey. Volume V of the Final Report of the State Geologist. Geological Survey of New Jersey, Trenton, NJ. 802 pp.

Saltzman, B. (2001) *Dynamical Paleoclimatology*. Academic Press,

Saltzman, B. & Maasch, K. A. (1990) A first-order global model of late Cenozoic climatic change. *Transactions of the Royal Society of Edinburgh, Earth Sciences* **81**, 315–325.

Sarıkaya, M. A., Zreda, M. & Çiner, A. (2009) Glaciations and paleoclimate of Mount Erciyes, central Turkey, since the Last Glacial Maximum, inferred from 36Cl dating and glacier modeling. *Quaternary Science Reviews* **23–24**, 2326–2341.

Sarıkaya, M. A., Çiner, A. & Zreda, M. (2011) Quaternary Glaciations of Turkey. In: Ehlers, J., Gibbard, P.L. & Hughes, P.D. (eds) *Quaternary Glaciations - Extent and Chronology - A Closer Look*. Elsevier, Developments in Quaternary Science **15**, 393–404.

Sauramo, M. (1958) Die Geschichte der Ostsee. *Annales Academiae Scientiarum Fennicae A* **III 51**, 522.

Sawagaki, T. & Aoki, T. (2011) Late Quaternary Glaciations in Japan. In: Ehlers, J., Gibbard, P.L. & Hughes, P.D. (eds) *Quaternary Glaciations - Extent and Chronology - A Closer Look*. Elsevier, Developments in Quaternary Science **15**, 1013–1022.

Schaefer, I. (1956) Sur la division du Quaternaire dans l'avant-pays des Alpes en Allemagne. *Actes IV Congres INQUA*, Rome/Pise 1953, vol. **2**, 910–914.

Scheffer, F. & Schachtschabel, P. (2010) *Lehrbuch der Bodenkunde*, 16. Auflage. Stuttgart: Spektrum.

Schellmann, G. & Radtke, U. (2004) The Marine Quaternary of Barbados. *Kölner Geographische Arbeiten* **81**, 137.

Schilt, A., Baumgartner, M., Blunier, T., Schwander, J., Spahni, R., Fischer, H. & Stocker, T. F. (2010) Glacial–interglacial and millennial scale variations in the atmospheric nitrous oxide concentration during the last 800,000 years. *Quaternary Science Reviews* **29**, 182–192.

Schindler, C., Röthlisberger, H. & Gyger, M. (1978) Glaziale Stauchungen in den Niederterrassen-Schottern des Aadorfer Feldes und ihre Deutung. *Eclogae Geologicae Helvetiae* **71**, 159–174.

Schirmer, W. (ed.) (1990) Rheingeschichte zwischen Mosel und Maas. *Deuqua-Führer* **1**, 295.

Schlüchter, C. (1980a) Die fazielle Gliederung der Sedimente eines Ufermoränenkomplexes–Form und Inhalt. *Verhandlungen des Naturwissenschaftlichen Vereins in Hamburg, NF* **23**, 101–117.

Schlüchter, C. (1980b) Bemerkungen zu einigen Grundmoränenvorkommen in den Schweizer Alpen. *Zeitschrift für Gletscherkunde und Glazialgeologie* **16**(2), 203–212.

Schlüchter, C. (1981) Bemerkungen zur Korngrößenanalyse der Lockergesteine, insbesondere der Moränen, in der Schweiz. *Verhandlungen des Naturwissenschaftlichen Vereins in Hamburg, NF* **24**(2), 155–160.

Schlüchter, C. (1988) Exkursion vom 11. Oktober 1987 der Schweizerischen Geologischen Gesellschaft im Rahmen der SNG-Jahrestagung in Luzern: Ein eiszeitgeologischer Überblick von Luzern zum Rhein - unter besonderer Berücksichtigung der Deckenschotter. *Eclogae Geologicae Helvetiae* **81**, 249–258.

Schlüchter, C. (1989a) A non-classical summary of the Quaternary stratigraphy in the northern Alpine Foreland of Switzerland. *Bulletin de la Société Neuchâteloise de Géographie* **32–33**, 143–157.

Schlüchter, C. (1989b) The most complete Quaternary record of the Swiss Alpine Foreland. *Palaeogeography, Palaeoclimatology, Palaeoecology* **72**, 141–146.

Schlüchter, C. (2004) The Swiss glacial record–a schematic summary. In: Ehlers, J. & Gibbard, P. L. (eds) *Quaternary Glaciations–Extent and Chronology. Part I, Europe*. Amsterdam: Elsevier, Developments in Quaternary Science **2**, 413–418.

Schmincke, H.-U. (2010) *Vulkanismus*, 3. Auflage. Darmstadt: Primus.

Schnütgen, A. (2003) Die Petrographie und Verbreitung tertiärer Schotter der Vallendar-Fazies im Rheinischen Schiefergebirge, ihre paläoklimatische und–geographische Bedeutung. In: Schirmer, W. (ed.) *Landschaftsgeschichte im Europäischen Rheinland*. GeoArcheoRhein **4**, 155–191.

Schreiner, A. (1989) Zur Stratigraphie der Rißeiszeit im östlichen Rheingletschergebiet (Baden-Württemberg). *Jahreshefte des Geologischen Landesamtes Baden-Württemberg* **31**, 183–196.

Schreiner, A. (1992) *Einführung in die Quartärgeologie*. Stuttgart: Schweizerbart.

Schreiner, A. (1997) *Einführung in die Quartärgeologie*, 2. Auflage. Stuttgart: Schweizerbart.

Schreiner, A. & Ebel, R. (1981) Quartärgeologische Untersuchungen in der Umgebung von Interglazialvorkommen im östlichen Rheingletschergebiet (Baden-Württemberg). *Geologisches Jahrbuch A* **59**, 3–64.

Schröder, P. (1988) Aufbau und Untergliederung des Niederterrassenkörpers der Unterelbe. *Mitteilungen aus dem Geologischen Institut der Universität Hannover* **27**, 119.

Schulz, W. (1998) *Streifzüge durch die Geologie des Landes Mecklenburg-Vorpommern*. c/w Verlagsgruppe: Schwerin.

Schumacher, R. (1981) *Untersuchungen zur Entwicklung des Gewässernetzes seit dem Würmmaximum im Bereich des Isar-Loisach-Vorlandgletschers*. Dissertation, Universität München.

Schuster, M., Roquin, C., Duringer, P., Brunet, M., Caugy, M., Fontugne, M., Macaye, H. T., Vignaud, P. & Ghienne, J.-F. (2005) Holocene lake Mega-Chad palaeoshorelines from space. *Quaternary Science Reviews* **24**, 1821–1827.

Schwan, J. (1988) *Sedimentology of coversands in northwestern Europe*. Thesis, Vrije Universiteit te Amsterdam.

Schwarzbach, M. (1978) Glazigene Sichelmarken als Klimazeugen. *Eiszeitalter und Gegenwart* **28**, 109–118.

Schwarzbach, M. (1993) *Das Klima der Vorzeit*, 5. Auflage. Stuttgart: Enke.

Scotese, C. (2008) Plate tectonic and paleogeographic mapping: State of the art. *Search and Discovery* Article #40312.

Scourse, J. D. (ed.) (2006) *The Isles of Scilly. Field Guide*. Quaternary Research Association.

Scourse, J. D., Haapaniemi, A.I., Colmenero-Hidalgo, E., Peck, V.L., Hall, I.R., Austin, W.E.N., Knutz, P.C. & Zahn, R. (2009) Growth, dynamics and deglaciation of the last British-Irish ice sheet: the deep-sea ice-rafted detritis record. *Quaternary Science Reviews* **28**, 3066–3084.

Sefström, N. G. (1836) Ueber die Spuren einer sehr großen urweltlichen Fluth. *Poggendorfs Annalen der Physik und Chemie* **38**, 614–618.

Sefström, N. G. (1838) Untersuchung über die auf den Felsen Skandinaviens in bestimmter Richtung vorhandenen Furchen und deren wahrscheinliche Entstehung. *Poggendorfs Annalen der Physik und Chemie* **43**, 533–567.

Sejrup, H. P., Haflidason, H., Aarseth, I., King, E., Forsberg, C.F., Long, D. & Rokoengen, K. (1994) Late Weichselian glaciation history of the northern North Sea. *Boreas* **23**, 1–13.

Sejrup, H. P., Larsen, E., Landvik, J., King, E., Haflidason, H. & Nesje, A. (2000) Quaternary glaciations in southern Fennoscandia: evidence from southwestern Norway and the northern North Sea region. *Quaternary Science Reviews* **19**, 667–685.

Sejrup, H. P., Hjelstuen, B.O., Dahlgren, K.I.T., Haflidason, H., Kuijpers, A., Nygard, A., Praeg, D., Stoker, M.S. & Vorren, T.O. (2005) Pleistocene glacial history of the NW European continental margin. *Marine and Petroleum Geology* **22**, 111–1129.

Sejrup, H. P., Nygard, A., Hall, A.M. & Haflidason, H. (2009) Middle and Late Weichselian (Devensian) glaciation history of south-western Norway, North Sea and eastern UK. *Quaternary Science Reviews* **28**, 370–380.

Seppälä, M. (2005a) Dating of palsas. In: Ojala, A. E. K. (ed.) *Quaternary Studies in the Northern and Arctic Regions of Finland*. Geological Survey of Finland, Special Paper **40**, 79–84.

Seppälä, M. (ed.) (2005b) *The Physical Geography of Fennoscandia*. Oxford: Oxford University Press.

Sernander, R. (1894) *Studier öfver den Gotländska vegetationens utvecklingshistoria*. Dissertation, Uppsala.

Serrano, E., González-Trueba, J.J., Pellitero, R., González-García, M. & Gómez-Lende, M. (2013) Quaternary glacial evolution in the Central Cantabrian Mountains (Northern Spain). *Geomorphology* **196**, 65–82.

Shackleton, N. J. (1967) Oxygen isotope analyses and pleistocene temperatures re-assessed. *Nature* **215**, 15–17.

Shackleton, N. J. (1977) Oxygen isotope stratigraphy of the Middle Pleistocene. In: Shotton, F.W. (ed.) *British Quaternary Studies, Recent Advances*. Oxford: Oxford University Press, 1–16.

Shackleton, N. J. (1987) Oxygen isotopes, ice volume and sea lavel. *Quaternary Science Reviews* **6**, 183–190.

Shackleton, N. J. (2000) The 100,000-year ice-age cycle identified and found to lag temperature, carbon dioxide, and orbital eccentricity. *Science* **289**, 1897–1902.

Shackleton, N. J. & Opdyke, N. D. (1973) Oxygen isotope and palaeomagnetic stratigraphy of equatorial Pacific core V28-238: oxygen isotope temperatures and ice volumes on a 105 and 106 year scale. *Quaternary Research* **3**, 39–55.

Shackleton, N. J., Berger, A. & Peltier, W. A. (1990) An alternative astronomical calibration of the lower Pleistocene timescale based on ODP Site 677. *Transactions of the Royal Society of Edinburgh, Earth Sciences* **81**, 251–261.

Shangzhe, Z., Jijun, L., Jingdong, Z., Jie, W. & Jingxiong, Z. (2011) Quaternary Glaciations: Extent and Chronology in China. In: Ehlers, J., Gibbard, P.L. & Hughes, P.D. (eds) *Quaternary Glaciations - Extent and Chronology - A Closer Look*. Elsevier, Developments in Quaternary Science **15**, 981–1002.

Sharp, M. J. (1985) Sedimentation and stratigraphy at Eyjabakkajøkull: an icelandic surging glacier. *Quaternary Research* **24**, 268–284.

Shaw, J. (2002) The meltwater hypothesis for subglacial landforms. *Quaternary International* **90**, 5–22.

Shaw, J. (2010) In defence of the meltwater (megaflood) hypothesis for the formation of subglacial bedform fields. *Journal of Quaternary Science* **25**, 249–260.

Shennan, I. & Horton, B. (2002) Holocene land- and sea-level changes in Great Britain. *Journal of Quaternary Science* **17**, 511–526.

Shi, Y. (scientific advisor), Li, B., Li, J. (chief eds), Cui, Z., Zheng, B., Zhang, Q., Wang, F., Zhou, S., Shi, Z., Jiao, K. & Kang, J. (eds) (1991) *Quaternary Glacial Distribution Map of Qinghai-Xizang (Tibet) Plateau, Scale 1:3,000,000*. Beijing: Science Press.

Shilts, W. W. (1984) Quaternary events, Hudson Bay Lowland and southern District of Keewatin. In: Fulton, R.J. (ed.) *Quaternary Stratigraphy of Canada - A Canadian Contribution to IGCP Project 24*. Geological Survey of Canada, Paper no. 84-10, pp. 117–126.

Shimek, B. (1909) Aftonian sands and gravels in western Iowa. *Bulleton of the Geological Society of America* **20**, 399–408.

Shotton, F. W. (1953) The Pleistocene deposits of the area between Coventry, Rugby and Leamington, and their bearing on the topographic development of the Midlands. *Philosophical Transactions of the Royal Society of London B* **237**, 209–260.

Shotton, F. W. (1968) The Pleistocene succession around Brandon, Warwickshire. *Philosophical Transactions of the Royal Society of London B* **254**, 387–400.

Shotton, F. W. (1976) Amplification of the Wolstonian Stage of the British Pleistocene. *Geological Magazine* **113**, 241–250.

Shotton, F. W. (1983a) The Wolstonian Stage of the British Pleistocene in and around its type area of the English Midlands. *Quaternary Science Reviews* **2**, 261–280.

Shotton, F. W. (1983b) Observations on the type Wolstonian glacial sequence. *Quaternary Newsletter* **40**, 28–36.

Shreve, R. L. (1984) Glacier sliding at subfreezing temperatures. *Journal of Glaciology* **30**, 341–347.

Sissons, J. B. (1967) *The Evolution of Scotland's Scenery*. Edinburgh & London: Oliver and Boyd.

Sissons, J. B. (1979a) The Loch Lomond Stadial in the British Isles. *Nature* **280**, 199–203.

Sissons, J. B. (1979b) The Loch Lomond Readvance in the Cairngorm mountains. *Scottish Geographical Magazine* **96**, 18–19.

Sissons, J. B. (1978) The parallel roads of Glen Roy and adjacent glens, Scotland. *Boreas* **7**, 229–244.

Sjørring, S. (1983) The glacial history of Denmark. In: Ehlers, J. (ed.) *Glacial Deposits in North-West Europe*. Rotterdam: Balkema, pp. 163–179.

Sjørring, S., Nielsen, P. E., Frederiksen, J., Hegner, J., Hyde, G., Jensen, J. B., Mogensen, A. & Vortisch, W. (1982) Observationer fra Ristinge Klint, felt- og laboratorieundersøgelser. *Dansk Geologisk Forening*, Årsskrift for 1981, 135–149.

Skinner, L. C. & Shackleton, N.J. (2005) An Atlantic lead over over Pacific deep-water change across termination I: implications for the application of the marine isotope stage stratigraphy. *Quaternary Science Reviews* **24**, 571–580.

Skupin, K., Speetzen, E. & Zandstra, J. G. (1993) *Die Eiszeit in Nordwestdeutschland: Zur Vereisungsgeschichte der Westfälischen Bucht und angrenzender Gebiete*. Geologisches Landesamt Nordrhein-Westfalen, Krefeld.

Sladen, J. A. & Wrigley, W. (1983) Geotechnical properties of Lodgement Till: A review. In: Evenson, E.B., Schlüchter, C. & Rabassa, J. (eds) *Tills and Related Deposits*. Rotterdam: Balkema, pp. 184–186.

Smed, P. (1962) Studier over den fynske øgruppes glaciale landskabsformer. *Meddelelser fra Dansk Geologisk Forening* **15**, 1–74.

Smed, P. (2002) *Steine aus dem Norden: Geschiebe als Zeugen der Eiszeit in Norddeutschland*, 2. Aufl. Berlin: Borntraeger.

Smiley, T. L., Bryson, R.A., King, J.E., Kukla, G.J. & Smith, G.I. (1991) Quaternary palaeoclimates. In: Morrison, R.B. (ed.) *Quaternary Nonglacial Geology: Conterminous US*. Geological Society of America, Geology of North America Vol. K-2, pp. 13–44.

Smith, A. G. & Pickering, K. T. (2003) Oceanic gateways as a critical factor to initiate icehouse Earth. *Journal of the Geological Society* **160**, 337–340.

Smith, A. M., Murray, T., Nicholls, K. W., Makinson, K., Adalgeirsdóttir, G., Behar, A. E. & Vaughan, D. G. (2007) Rapid erosion, drumlin formation, and changing hydrology beneath an Antarctic ice stream. *Geology* **35**, 127–130.

Smith, B. E., Fricker, H. A., Joughin, I. R. & Tulaczyk, S. (2009) An inventory of active subglacial lakes in Antarctica detected by ICESat 2003–2008. *Journal of Glaciology* **55**, 573–595.

Smith, D. & Lewis, D. (2007) Dendrochronology. In: Elias, S. A. (ed.) *Encyclopaedia of Quaternary Science*. Amsterdam: Elsevier, pp. 459–465.

Smith, G. A. (1993) Missoula flood dynamics and magnitudes inferred from sedimentology of slackwater deposits on the Columbia Plateau, Washington. *Geological Society of America Bulletin* **105**, 77–100.

Smith, L. C., Sheng, Y., MacDonald, G. M. & Hinzman, L. D. (2005) Disappearing arctic lakes. *Science* **308**, 1429.

Solger, F., Graebner, P., Thienemann, J., Speiser, P. & Schulze, F. W. O. (1910) *Dünenbuch. Werden und Wandern der Dünen. Pflanzen- und Tierleben auf den Dünen. Dünenbau*. Stuttgart: Enke.

Sollid, J. L. & Sørbel, L. (1998) Palsa bogs as climate indicator–examples from Dovrefjell, Southern Norway. *Ambio* **27**, 287–291.

Solomon, S., Plattner, G.-K., Knutti, R. & Friedlingstein, P. (2009) Irreversible climate change due to carbon dioxide emissions. *PNAS* **106**, 1704–1709.

Sparks, B. W. & West, R. G. (1964) The drift landforms around Holt, Norfolk. *Transactions of the Institute of British Geographers* **35**, 27–35.

Sparks, B. W., Williams, R. G. B. & Bell, F. G. (1972) Presumed ground-ice depressions in East Anglia. *Proceedings of the Royal Society of London, A* **327**, 329–343.

Speetzen, E. & Zandstra, J. G. (2009) Elster- und Saale-Vereisung im Weser-Ems-Gebiet und ihre kristallinen Leitgeschiebegesellschaften. *Münstersche Forschungen zur Geologie und Paläontologie* **103**, 128.

Stackebrandt, W. (2009) Subglacial channels of Northern Germany–a brief review. *Zeitschrift der Deutschen Gesellschaft für Geowissenschaften* **160**, 203–210.

Stampfli, G. M. & Borel, G. D. (2002) A Plate Tectonic Model for the Paleozoic and Mesozoic. *Earth and Planetary Science Letters* **196**, 17–33.

Stampfli, G. M. & Borel, G. (2004) The TRANSMED transects in time and space. In: Cavazza, W., Roure, F., Spakman, W., Stampfli, G. M. & Ziegler, P. A. (eds) *The TRANSMED Atlas: The Mediterranean Region from Crust to Mantle*. Heidelberg: Springer-Verlag, pp. 53–80.

Starnberger, R., Terhorst, B., Rähle, W., Peticzka, R. & Haas, J. N. (2009) Palaeoecology of Quaternary periglacial environments during OIS-2 in the forefields of the Salzach Glacier (Upper Austria). *Quaternary International* **198**, 51–61.

Stea, R. R., Seaman, A. A., Tronk, T., Parkhill, M. A., Allard, S. & Utting, D. (2011) The Appalachian Glacier Complex in Maritime Canada. In: Ehlers, J., Gibbard, P.L. & Hughes, P.D. (eds) *Quaternary Glaciations - Extent and Chronology - A Closer Look*. Elsevier, Developments in Quaternary Science **15**, 631–659.

Steffen, W., Sanderson, R.A., Tyson, P.D., Jäger, J., Matson, P.A., Moore III, B., Oldfield, F., Richardson, K., Schellnhuber, H.J., Turner, B.L. & Wasson, R.J. (2004) *Global Change and the Earth System: A Planet Under Pressure*. New York: Springer-Verlag.

Steffen, W., Grinevald, J., Crutzen, P. & McNeil, J. (2011) The Anthropocene: conceptual and historical perspectives. *Philosophical Transactions of the Royal Society of London A* **369**, 842–867.

Stephan, H.-J. (1981) Eemzeitliche Verwitterungshorizonte im Jungmoränengebiet Schleswig-Holsteins. *Verhandlungen des Naturwissenschaftlichen Vereins in Hamburg (NF)* **24**(2), 161–175.

Stephan, H.-J. (1987) Moraine stratigraphy in Schleswig-Holstein and adjacent areas. In: Meer, J. J. M. van der (ed.) *Tills and Glaciotectonics*. Rotterdam, Boston: Balkema, pp. 23–30.

Stephan, H.-J. (1998) Geschiebemergel als stratigraphische Leithorizonte in Schleswig-Holstein: Ein Überblick. *Meyniana* **50**, 113–135.

Stephan, H.-J. & Ehlers, J. (1983) North German till types. In: Ehlers, J. (ed.) *Glacial Deposits in North-West Europe*. Rotterdam: Balkema, pp. 239–247.

Stephan, H.-J., Kabel, Ch. & Schlüter, G. (1983) Stratigraphical problems in the glacial deposits of Schleswig-Holstein. In: Ehlers, J. (ed.) *Glacial Deposits in North-West Europe*. Rotterdam: Balkema, pp. 305–320.

Stock, C. (1929) Significance of abraded and weathered mammalian remains from Rancho La Brea. *Bulletin of Southern California Academy of Sciences* **28**, 1–5.

Stockhausen, H. (1998) Geomagnetic palaeosecular variation (0–13000 year BP) as recorded in sediments from three maar lakes from the West Eifel (Germany). *Geophysical Journal International* **135**, 898–910.

Straw, A. (1991) Glacial deposits of Lincolnshire and adjoining areas. In: Ehlers, J., Gibbard, P.L. & Rose, J. (eds) *Glacial Deposits in Great Britain and Ireland.* Rotterdam: Balkema, pp. 213–221.

Straw, A. (2005) *Glacial and Pre-glacial Deposits at Welton-le-Wold, Lincolnshire.* Exeter: Straw.

Streif, H. (1986) Zur Altersstellung und Entwicklung der Ostfriesischen Inseln. *Offa* **43**, 29–44.

Streif, H. (1990) *Das ostfriesische Küstengebiet. Nordsee, Inseln, Watten und Marschen.* 2nd edition. Gebrüder Borntraeger Verlagsbuchhandlung, Sammlung Geologischer Führer no. 57.

Streif, H. (2004) Sedimentary record of Pleistocene and Holocene marine inundations along the North Sea coast of Lower Saxony, Germany. *Quaternary International* **112**, 3–28.

Strelin, J. & Malagnino, E. (2009) Charles Darwin and the oldest Glacial Events in Patagonia: The Erratic Blocks of the Río Santa Cruz Valley. *Revista de la Asociación Geológica Argentina* **64**(1), 101–108.

Stremme, H. (1979) Böden, Relief und Landschaftsgeschichte im Norwestdeutschen Raum. *Zeitschrift für Geomorphologie NF,* Supplement-Band **33**, 216–222.

Strömberg, B. (1983) The Swedish varve chronology. In: Ehlers, J. (ed.) *Glacial Deposits in North-West Europe.* Rotterdam: Balkema, pp. 97–105.

Strömberg, B. (1989) *Late Weichselian Deglaciation and Clay Varve Chronology in East-Central Sweden.* Sveriges Geologiska Undersökning C **73**, 70.

Stuart, A. J. (ed.) (1995) Insularity and Quaternary vertebrate faunas in Britain and Ireland. In: Preece, R.C. (ed.) *Island Britain: A Quaternary Perspective.* Geological Society, London, Special Publication no. **96**, 111–125.

Stuiver, M., Heusser, C. T. & Yang, I. C. (1978) North American glacial history back to 75,000 years B.P. *Science* **200**, 16–21.

Sugden, D. E. & John, B. S. (1976) *Glaciers and Landscape–A Geomorphological Approach.* London: Edward Arnold.

Sugden, D. E., Marchant, D. R., Potter Jr, N., Souchez, R. A., Denton, G. H., Swisher III, C. C. & Tison, J. -L. (1995) Preservation of Miocene glacier ice in East Antarctica. *Nature* **376**, 412–414.

Sutherland, R., Kim, K., Zondervan, A. & McSaveney, M. (2007) Orbital forcing of midlatitude Southern Hemisphere glaciation since 100 ka inferred from cosmogenic nuclide ages of moraine boulders from the Cascade Plateau, southwest New Zealand. *Geological Society of America Bulletin* **119**, 443–451.

Svendsen, J. I. & Mangerud, J. (1987) Late Weichselian and Holocene sea-level history for a cross-section of western Norway. *Journal of Quaternary Science* **2**, 113–132.

Svendsen, J. I., Gataullin, V., Mangerud, J. & Polyak, L. (2004) The glacial history of the Barents and Kara Sea Region. In: Ehlers, J. & Gibbard, P. L. (eds) *Quaternary Glaciations– Extent and Chronology, Part I: Europe.* Amsterdam: Elsevier, Developments in Quaternary Science **2**, 369–378.

Svensson, H. (1984) The periglacial form group of Southwestern Denmark. *Geografisk Tidsskrift* **84**, 25–34.

Szabo, J. P., Angle, M. P. & Eddy, A. M. (2011) Pleistocene Glaciation of Ohio, USA. In: Ehlers, J., Gibbard, P.L. & Hughes, P.D. (eds) *Quaternary Glaciations - Extent and Chronology - A Closer Look*. Elsevier, Developments in Quaternary Science **15**, 513–519.

Taramelli, T. (1898) Del deposito lignitico di Leffe in provincia di Bergamo. *Bollettino Società Geologica Italiana* **XVII**, 202–218.

Tarasov, L. & Peltier, W. R. (2003) Greenland glacial history, borehole constraints, and Eemian extent. *Journal of Geophysical Research* **108(B3)**, 2124–2143.

Taylor, K. (2007) Ice cores: History of research, Greenland and Antarctica. In: Elias, S. A. (ed.) *Encyclopaedia of Quaternary Science*. Amsterdam: Elsevier, pp. 1284–1288.

Teed, R. (2000) A >130,000-Year-Long Pollen Record from Pittsburg Basin, Illinois. *Quaternary Research* **54**: 264–274.

Teeuw, R. M. & Rhodes, E. J. (2004) Aeolian activity in northern Amazonia: optical dating of Late Pleistocene and Holocene palaeodunes. *Journal of Quaternary Science* **19**, 49–54.

Teller, J. T. (1985) Glacial Lake Agassiz and its influence on the Great Lakes. In: Karrow, P. F. & Calkin, P. E. (eds) *Quaternary Evolution of the Great Lakes*. Geological Association of Canada, Special Paper **30**, 1–16.

Thackray, G. D., Shulmeister, J. & Fink, D. (2009) Evidence for expanded middle and late Pleistocene glacier extent in northwest Nelson, New Zealand. *Geografiska Annaler* **91A**, 291–311.

Thieme, H. (1999) Altpaläolithische Holzgeräte aus Schönigen, Lkr. Helmstedt. Bedeutsame Funde zur Kulturentwicklung des frühen Menschen. *Germania* **77**, 451–487.

Thieme, H. & Veil, S. (1985) Neuere Untersuchungen zum eemzeitlichen Elefantenjagdplatz Lehringen, Landkreis Verden. *Die Kunde NF* **36**, 11–58.

Thome, K. N. (1980) Der Vorstoß des nordeuropäischen Inlandeises in das Münsterland in der Elster- und Saale-Kaltzeit - Strukturelle, mechanische und morphologische Zusammenhänge. *Westfälische Geographische Studien* **36**, 21–40.

Thompson, L. G., Mosley-Thompson, E., Davis, M. E., Henderson, K. A., Brecher, H. H., Zagorodnov, V. S., Mashiotta, T. A., Lin, P.-N., Mikhalenko, V. N., Hardy, D. R. & Beer, J. (2006) Kilimanjaro Ice Core Records: Evidence of Holocene Climate Change in Tropical Africa. *Science* **298**, 589–593.

Thompson, W. G. & Goldstein, S. L. (2006) A radiometric calibration of the SPECMAP timescale. *Quaternary Science Reviews* **25**, 3207–3215.

Thorarinsson, S. (1944) Tefrokronologiska studier pa Island; Thjorsardalur och dess foeroedelse; Tephrochronological studies in Iceland. *Geografiska Annaler* **26**(1–2), 217.

Thorarinsson, S. (1969) Glacier surges in Iceland, with special reference to the surges of Brúarjökull. *Canadian Journal of Earth Sciences* **6**, 875–882.

Torell, O. (1858) Bref om Island. Öfversigt af Kongl. *Vetenskaps-Akademiens Forhandlingar* **XIV**, 325–332.

Torell, O. (1865) Inledning. In: Holmström. L. P. (ed.) *Märken efter Istiden Iakttagna I Skåne*. Malmö: Cronholmska Boktrykeriet.

Torell, O. (1875) Schliff-Flächen und Schrammen auf der Oberfläche des Muschelkalkes von Rüdersdorf. *Zeitschrift der Deutschen Geologischen Gesellschaft* **27**, 961.

Toucanne, S., Zaragosi, S., Bourillet, J. F., Cremer, M., Eynaud, F., Turon, J. L., Fontanier, C., Van Vliet Lanoë, B. & Gibbard, P. (2009a) Timing of massive 'Fleuve Manche' discharges over the last 350 kyr: insights into the European Ice Sheet oscillations and the European drainage network from MIS 10 to 2. *Quaternary Science Reviews* **28**, 1238–1256.

Toucanne, S., Zaragosi, S., Bourillet, J. F., Gibbard, P., Eynaud, F., Turon, J. L., Cremer, M., Cortijo E., Martinez, P. & Rossignol, L. (2009b) A 1.2 Ma record of glaciation and fluvial discharge from the West European Atlantic margin. *Quaternary Science Reviews* **28**, 2974–2981.

Toucanne, S., Zaragosi, S., Bourillet, J. F., Marieu, V.,Cremer, M., Kageyama, M.,Van Vliet Lanoë, B., Eynaud, F., Turon, J. L. & Gibbard, P. (2010) The first estimation of Fleuve Manche palaeoriver discharge during the last deglaciation: Evidence for Fennoscandian ice sheet meltwater flow in the English Channel ca 20–18 ka ago. *Earth and Planetary Science Letters* **290**, 459–473.

Tripati, A., Backman, J., Elderfield, H. & Ferretti, P. (2005) Eocene bipolar glaciation associated with global carbon cycle changes. *Nature* **436**, 341–346.

Troll, C. (1924) Der diluviale Inn-Chiemseegletscher - Das geographische Bild eines typischen Alpenvorlandgletschers. *Forschungen zur deutschen Landes- und Volkskunde* **23**, 1–121.

Turner, C. (1970) Middle Pleistocene deposits at Marks Tey, Essex. *Philosophical Transactions of the Royal Society of London B* **257**, 373–440.

Turner, C. (1996) A brief survey of the early Middle Pleistocene in Europe. In: Turner, C. (ed.) *The Early Middle Pleistocene in Europe.* Rotterdam: Balkema, pp. 295–317.

Tyrrell, J. B. (1898) The glaciation of north central Canada. *Journal of Geology* **6**, 147–160.

Urban, B. (1995) Vegetations- und Klimaentwicklung des Quartärs im Tagebau Schöningen. In: *Archäologische Ausgrabungen im Braunkohlentagebau Schöningen.* Hahnsche Buchhandlung, Hannover, 44–56.

Urban, B. (2007a) Interglacial Pollen Records from Schöningen, North Germany. In: Sirocko, F., Claussen, M., Sanchez Goñi, M. F. & Litt, T. (eds) *The Climate of Past Interglacials.* Amsterdam: Elsevier, Developments in Quaternary Science **7**, 417–444.

Urban, B. (2007b) Quartäre Vegetations- und Klimaentwicklung im Tagebau Schöningen. In: H. Thieme (ed.) *Die Schöninger Speere – Mensch und Jagd vor 400.000 Jahren.* Theiss, Stuttgart, 66–75.

Urban, B., Lenhard, R., Mania, D. & Albrecht, B. (1991) Mittelpleistozän im Tagebau Schöningen, Ldkr. Helmstedt. *Zeitschrift der Deutschen Geologischen Gesellschaft* **142**, 351–372.

Urdea, P., Onaca, A., Ardelean, F. & Ardelean, M. (2011) New evidence on the Quaternary glaciation in the Romanian Carpathians. In: Ehlers, J., Gibbard, P.L. & Hughes, P.D. (eds) *Quaternary Glaciations - Extent and Chronology - A Closer Look.* Elsevier, Developments in Quaternary Science **15**, 305–322.

Ussing, N. V. (1903) Om Jyllands hedesletter og teorierne for deres dannelse. *Oversigt over Det Kongelige Danske Videnskabernes Selskabs Forhandlingar* **1903**(2), 1–152.

Vandenberghe, J. (2007) Cryoturbation structures. In: Elias, S. A. (ed.) *Encyclopaedia of Quaternary Science.* Amsterdam: Elsevier, pp. 2147–2153.

Vásquez-Selem, L. & Heine, K. (2011) Late Quaternary Glaciation in Mexico. In: Ehlers, J., Gibbard, P.L. & Hughes, P.D. (eds) *Quaternary Glaciations - Extent and Chronology - A Closer Look.* Elsevier, Developments in Quaternary Science **15**, 849–862.

Vegt, P. van der, Janszen, A., Moreau, J., Gibbard, P.L., Huuse, M. & Moscariello, A. (2009) Glacial sedimentary systems and tunnel valleys of East Anglia, England. In *Glaciogenic Reservoirs and Hydrocarbon Systems*, Petroleum Group Geological Society, December 2009, Abstracts 1–2.

Velichko, A. A., Faustova, M.A., Pisareva, V.V., Gribchenko, Y.N., Sudakova, N.G. & Lavrentiev, N.V. (2011) Glaciations of the East European Plain: Distribution and Chronology. In: Ehlers, J., Gibbard, P.L. & Hughes, P.D. (eds) *Quaternary Glaciations - Extent and Chronology - A Closer Look*. Elsevier, Developments in Quaternary Science **15**, 337–360.

Ventris, P. A. (1985) Pleistocene environmental history of the Nar Valley, Norfolk. Unpublished Ph.D. thesis, University of Cambridge.

Ventris, P. A. (1986) The Nar Valley. In: West R.G. & Whiteman, C.A. (eds) *The Nar Valley and North Norfolk, Field Guide*. Coventry: Quaternary Research Association, pp. 6–55.

Ventris, P. A. (1996) Hoxnian Interglacial freshwater and marine deposits in northwest Norfolk, England and their implications for sea-level reconstruction. *Quaternary Science Reviews* **15**, 437–450.

Verne, J. (1870) *Vingt Mille Lieues sous les Mers*. Paris: Hetzel.

Villinger, E. (2003) Zur Paläogeographie von Alpenrhein und oberer Donau. *Zeitschrift der Deutschen Geologischen Gesellschaft* **154**, 193–253.

Vincent, J.-S. (1989) Quaternary geology of the southeastern Canadian Shield. In: Fulton, R.J. (ed.) *Quaternary Geology of Canada and Greenland*. Geological Society of America, Geology of North America no. K-1, pp. 249–275.

Vincent, J.-S. & Prest, V.K. (1987) The Early Wisconsinan history of the Laurentide Ice Sheet. *Geographie Physique et Quaternaire* **XLI**, 199–213.

Vink, A., Steffen, H., Reinhardt, L. & Kaufmann, G. (2007) Holocene relative sea-level change, isostatic subsidence and the radial viscosity structure of the mantel of northwest Europe (Belgium, the Netherlands, Germany, southern North Sea). *Quaternary Science Reviews* **26**, 3249–3275.

Vlerk, I. M. van der & Florschütz, F. (1950) *Nederland in het Ijstijdvak. De Geschiedenis van Flora, Fauna en Klimaat, Toen Aap en Mammoet ons Land Bewoonden*. Utrecht: de Haan.

Vliet-Lanoë, B. van, Bourgeois, O. & Dauteuil, O. (1998) Thufur formation in northern Iceland and its relation to holocene climate change. *Permafrost and Periglacial Processes* **9**, 347–365.

Vorren, T. O., Landvik, J. Y., Andreassen, K. & Sverre Landberg, J. (2011) Glacial History of the Barents Sea Region. In: Ehlers, J., Gibbard, P.L. & Hughes, P.D. (eds) *Quaternary Glaciations - Extent and Chronology - A Closer Look*. Elsevier, Developments in Quaternary Science **15**, 361–372.

Wade, N. (2009) Scientists in Germany draft Neanderthal genome. *New York Times*, 12 February 2009.

Waelbroeck, C., Labeyrie, L., Michel, E., Duplessy, J.C., McManus, J.F., Lambeck, K., Balbon, E. & Labracherie, M. (2002) Sea-level and deep water temperature changes derived from benthic foraminifera isotopic records. *Quaternary Science Reviews* **21**, 295–305.

Wagenbreth, O. (1978) Die Feuersteinlinie in der DDR, ihre Geschichte und Popularisierung. *Schriftenreihe für Geologische Wissenschaften* **9**, 339–368.

Wagner, G. A. Rieder, H., Zöller, L. & Mick, E. (eds) (2007) *Homo heidelbergensis. Schlüsselfund der Menschheitsgeschichte*. Stuttgart: Theiss.

Walcott, R. I. (1970) Isostatic response to loading of the crust in Canada. *Canadian Journal of Earth Sciences* **7**, 716–726.

Walder, J. S. & Hallet, B. (1986) The physical basis of frost weathering: toward a more fundamental and unified perspective. *Arctic and Alpine Research* **18**, 27–32.

Walker, M. J. C., Berkelhammer, M., Björck, S., Cwynar, L. C., Fisher, D. A., Long, A. J., Lowe, J. J., Newnham, R. M., Rasmussen, S. O. & Weiss, H., (2012) Formal subdivision of the Holocene Series/Epoch: a discussion paper by a Working Group of INTIMATE (Integration of ice-core marine and terrestrial records) and the Subcommission on Quaternary Stratigraphy (International Commission on Stratigraphy). *Journal of Quaternary Science* **27**, 649–659.

Walter, K. M., Zimov, S. A., Chanton, J. P., Verbyla, D. & Chapin III, F. S. (2006) Methane bubbling from Siberian thaw lakes as a positive feedback to climate warming. *Nature* **443**, 71–75.

Wang, Y. J., Cheng, H., Edwards, R.L., An, Z.S., Wu, J.Y., Shen, C.-C. & Dorale, J.A. (2001) A high-resolution absolute-dated Late Pleistocene monsoon record from Hulu Cave, China. *Science* **294**, 2345–2348.

Wansa, S. (1991) Lithologie und Stratigraphie der Tills bei Gräfenhainichen. *Mauritiana* **13**, 189–211.

Wansa, S. (1994) Zur Lithologie und Genese der Elster-Grundmoränen und der Haupt Dren the-Grundmoräne im westlichen Elbe-Weser-Dreieck. *Mitteilungen aus dem Geologischen Institut der Universität Hannover* **34**, 77.

Washburn, A. L. (1979) *Geocryology: A Survey of Periglacial Processes and Environments*. London: Edward Arnold.

Washburn, A. L., Burrows, C. & Rein, R. Jr (1978) Soil deformation resulting from some laboratory freeze-thaw experiments. In: *Proceedings of Third International Conference on Permafrost*, Edmonton, Alberta, **1**, 756–762.

Wateren, F. M. van der (1985) A model of glaciotectonics, applied to the ice-pushed ridges in the Central Netherlands. *Bulletin of the Geological Society of Denmark* **34**, 55–74.

Wateren, F. M. van der (1987) Structural geology and sedimentology of the Dammer Berge push moraine, FRG. In: Meer, J. J. M van der (ed.) *Tills and Glaciotectonics*. Rotterdam: Balkema, pp. 157–182.

Wateren, F. M. van der (1992) *Structural Geology and Sedimentology of Push Moraines. Processes of soft sediment deformation in a glacial environment and the distribution of glaciotectonic styles*. Dissertation, Amsterdam.

Wayne, W. J. (1991) Ice-wedge Casts of Wisconsinan Age in Eastern Nebraska. *Permafrost and Periglacial Processes* **2**, 211–223.

Weddle, T. K. (1992) Late Wisconsinan stratigraphy in the lower Sandy River valley, New Sharon, Maine. *Geological Society of America Bulletin* **104**, 1350–1363.

Weertman, J. (1964) The theory of glacier sliding. *Journal of Glaciology* **5**, 287–303.

Weerts, H. J. T., Westerhoff, W. E., Cleveringa, P., Bierkens, M. F. P., Veldkamp, J. G. & Rijsdijk, K. F. (2005) Quaternary geological mapping of the lowlands of The Netherlands, a 21st century perspective. *Quaternary International* **133–134**, 159–178.

Weidenfeller, M. & Knipping, M. (2008) Correlation of Pleistocene sediments from boreholes in the Ludwigshafen area, western Heidelberg Basin. *Eiszeitalter und Gegenwart* **57**(3/4), 270–285.

Weinberger, L. (1950) Gliederung der Altmoränen des Salzach-Gletschers östlich der Salzach. *Zeitschrift für Gletscherkunde und Glazialgeologie* **I**, 176–186.

Weiss, E. N. (1958) Bau und Entstehung der Sander vor der Grenze der Würmvereisung im Norden Schleswig-Holsteins. *Meyniana* **7**, 5–60.

Welten, M. (1988) Neue pollenanalytische Ergebnisse über das Jüngere Quartär des nördlichen Alpenvorlandes der Schweiz (Mittel- und Jungpleistozän). *Beiträge zur Geologischen Karte der Schweiz NF* **162**, 40.

West, R. G. (1957) Interglacial deposits at Bobbitshole, Ipswich. *Philosophical Transactions of the Royal Society of London B* **241**, 1–31.

West, R. G. (1980) *The Pre-Glacial Pleistocene of the Norfolk and Suffolk Coasts*. Cambridge: Cambridge University Press.

West, R. G. (2011) *Plant Life of the Quaternary Cold Stages: Evidence from the British Isles*. Cambridge: Cambridge University Press.

West, R. G. & Donner, J. J. (1956) The glaciations of East Anglia and the East Midlands: adifferentiation based on stone-orientation measurements of the tills. *Quarterly Journal of the Geological Society of London* **112**, 69–91.

Westerhoff, W. (2009) *Stratigraphy and sedimentary evolution. The lower Rhine-Meuse system during the Late Pliocene and Early Pleistocene (southern North Sea Basin)* Dissertation, Vrije Universiteit Amsterdam.

Whiteman, C. A. (1992) The palaeogeography and correlation of pre-Anglian-Glaciation terraces of the River Thames in Essex and the London Basin. *Proceedings of the Geologists' Association* **103**, 37–66.

Whiteman, C. A. & Kemp, R. (1990) Pleistocene sediments, soils and landscape evolution at Stebbing, Essex. *Journal of Quaternary Science* **5**, 145–161.

Whiteman, C. A. & Rose, J. (1992) Thames river sediments of the British Early and Middle Pleistocene. *Quaternary Science Reviews* **11**, 363–375.

Wickham, S. S., Johnson, W.H. & Glass, H.D. (1988) Regional geology of the Tiskilwa Till Member, Wedron Formation, northeastern Illinois. *Illinois State Geological Survey Circular* **543**, 35 pp.

Willman, H. B. & Frye, J.C. (1970) Pleistocene stratigraphy of Illinois. *Illinois State Geological Survey Bulletin* **94**, 204 pp.

Wilson, L. J., Austin, W.E.N. & Jansen, E. (2002) The last British Ice Sheet: growth, maximum extent and deglaciation. *Polar Research* **21**, 243–250.

Wimmenauer, W. (2003) *Geologische Karte von Baden-Württemberg 1:25. 000, Erläuterungen zum Blatt Kaiserstuhl*, 5. Auflage. Landesamt für Geologie, Rohstoffe und Bergbau Baden-Württemberg, Freiburg im Breisgau.

Winkler, S. (2009a) *Gletscher und ihre Landschaften*. Darmstadt: Primus.

Winkler, S. (2009b) Glacier fluctuations of Jostedalsbreen, western Norway, during the past 20 years: the sensitive response of maritime mountain glaciers. *The Holocene* **19**, 395–414.

Winsemann, J., Asprion, U., Meyer, T., Schultz, H. & Victor, P. (2003) Evidence of iceberg ploughing in a subaqueous ice-contact fan, glacial Lake Rinteln, Northwest Germany. *Boreas* **32**, 386–398.

Winsemann, J., Asprion, U., Meyer, T. & Schramm, C. (2007) Facies characteristics of Middle Pleistocene (Saalian) ice-margin subaqueous fan and delta deposits, glacial Lake Leine, NW Germany. *Sedimentary Geology* **193**, 105–129.

Wintges, T. (1984) Untersuchungen an gletschergeformten Felsflächen im Zemmgrund/Zillertal (Tirol) und in Südskandinavien. *Salzburger Geographische Arbeiten* **11**, 209.

Wintges, T. & Heuberger, H. (1982) Untersuchungen an Parabelrissen und Sichelbrüchen im Zemmgrund (Zillertal) und über die damit verbundene Abtragung. *Zeitschrift für Gletscherkunde und Glazialgeologie* **16**(2), 157–170.

Wintle, A. G. (1991) Thermoluminescence dating. In: Smart, P. L. & Frances, P. D. (eds) *Quaternary Dating Methods–A User's Guide*. Quaternary Research Association, Technical Guide **4**, 108–127.

Wintle A. G. (2008) Fifty years of luminescence dating. *Archaeometry* **50**, 276–312.

Wintle, A. G. & Huntley, D. J. (1979) Thermoluminescence dating of sediments. *PACT* **3**, 374–380.

Wintle, A. G. & Catt, J.A. (1985) Thermoluminescence dating of Dimlington Stadial deposits in eastern England. *Boreas* **14**, 231–234.

Woldstedt, P. (1926) Probleme der Seenbildung. *Zeitschrift der Gesellschaft für Erdkunde zu Berlin* **1926**, 103–124.

Woldstedt, P. (1927) Die Gliederung des Jüngeren Diluviums in Norddeutschland und seine Parallelisierung mit anderen Glazialgebieten. *Zeitschrift der Deutschen Geologischen Gesellschaft, Monatsberichte* **1927**(3/4), 51–52.

Woldstedt, P. (1929) *Das Eiszeitalter. Grundlinien einer Geologie des Diluviums*. Stuttgart: Enke.

Woldstedt, P. (1935) *Geologisch-morphologische Übersichtskarte des norddeutschen Vereisungsgebietes*. Berlin: Preußische Geologische Landesanstalt.

Woldstedt, P. (1938) Über Vorstoss- und Rückzugsfronten des Inlandeises in Norddeutschland. *Geologische Rundschau* **29**, 481–490.

Woldstedt, P. (1939) Vergleichende Untersuchungen an isländischen Gletschern. *Jahrbuch der Preußischen Geologischen Landesanstalt für 1938* **59**, 249–271.

Woldstedt, P. (1950) Quartärforschung: Einleitende Worte. *Eiszeitalter und Gegenwart* **1**, 9–15.

Woldstedt, P. (1954) Saaleeiszeit, Warthestadium und Weichseleiszeit in Norddeutschland. *Eiszeitalter und Gegenwart* **4/5**, 34–48.

Woldstedt, P. (1961) *Das Eiszeitalter. Grundlinien einer Geologie des Quartärs, Band 1*, 2. Auflage. Stuttgart: Enke.

Wolf, L. (1980) Die elster- und präelsterzeitlichen Terrassen der Elbe. *Zeitschrift für Geologische Wissenschaften* **8**, 1267–1280.

Wolf, L. & Alexowsky, W. (1994) Fluviatile und glaziäre Ablagerungen am äußersten Rand der Elster- und Saale-Vereisung; die spättertiäre und quartäre Geschichte des sächsischen Elbgebietes (Exkursion A2). In: Eißmann, L. & Litt, T. (eds) *Das Quartär Mitteldeutschlands*. Altenburger Naturwissenschaftliche Forschungen 7, pp. 190–235.

Woodland, A. W. (1970) The buried tunnel-valleys of East Anglia. *Proceedings of the Yorkshire Geological Society* **37**, 521–578.

Woodroffe, C. D. (2007) Coral records. In: Elias, S. A. (ed.) *Encyclopaedia of Quaternary Science*. Amsterdam: Elsevier, 3006–3015.

Woodward, J. C. (2014) *The Ice Age: A Very Short Introduction*. Oxford: Oxford University Press.

Woodward, J.C. & Hughes, P.D. (2011) Glaciation in Greece: A new record of cold stage environments in the Mediterranean. In: Ehlers, J., Gibbard, P.L. & Hughes, P.D. (editors) *Quaternary Glaciations - Extent and Chronology - A Closer Look*. Amsterdam: Elsevier, Developments in Quaternary Science no. 15, pp. 175–198.

Wooldridge, S.W. & Linton, D.L. (1955) *Structure, Surface and Drainage in South-East England*. London: George Philip.

Young, N. E., Schäfer, J.M., Briner, J.P. & Goehring, B.M. (2013) A ^{10}Be production-rate for the Arctic. *Journal of Quaternary Science* **28**, 515–526.

Young, R. R., Burns, J.A., Smith, D.G., Arnold, L.D. & Rains, R.B. (1994) A single, late Wisconsin, Laurentide glaciation, Edmonton area and southwestern Alberta. *Geology* **22**, 683–686.

Zagwijn, W. H. (1957) Vegetation, climate and time-correlations in the Early Pleistocene of Europe. *Geologie en Mijnbouw* **19**, 233–244.

Zagwijn, W. H. (1984) The formation of the Younger Dunes on the West coast of The Netherlands (AD 1000–1600). *Geologie en Mijnbouw* **63**(3), 259–268.

Zagwijn, W. H. (1989) Vegetation and climate during warmer intervals in the Late Pleistocene of western and central Europe. *Quaternary International* **3/4**, 57–67.

Zalasiewicz, J. (2009) *Die Erde nach uns. Der Mensch als Fossil der fernen Zukunft*. Stuttgart: Spektrum.

Zalasiewicz, J., Williams, M., Smith, A.G., Barry, T.L., Coe, A.L., Bown, P.R., Brenchley, P., Cantrill, D., Gale, A., Gibbard, P.L., Gregory, F.J., Hounslow, M.W., Kerr, A.C., Pearson, P., Knox, R., Powell, J., Waters, C., Marshall, J., Oates, M., Rawson, P. & Stone, P. (2008) Are we now living in the Anthropocene. *GSA Today* **18**, 4–8.

Zalasiewicz, J., Williams, M., Steffen, W. & Crutzen, P. (2010) The new world of the Anthropocene. *Environmental Science and Technology* **44**, 2228–2231.

Zalasiewicz, J., Waters, C. N., Williams, M., Barnosky, A. D., Cearreta, A., Crutzen, P., Ellis, E., Ellis, M. A., Fairchild, I. J., Grinevald, J., Haff, P. K., Hajdas, I., Leinfelder, R., McNeill, J., Odada, E. O., Poirier, C., Richter, R., Steffen, W., Summerhayes, C., Syvitski, J. P. M., Vidas, D., Wagreich, M., Wing, S. L., Wolfe, A. P. & Zhisheng, A. (2015) When did the Anthropocene begin? A mid-twentieth century boundary level is stratigraphically optimal. *Quaternary International*, doi: 10.1016/j.quaint.2014.11.045.

Zandstra, J. G. (1983) Fine gravel, heavy mineral and grain-size analyses of Pleistocene, mainly glacigenic deposits in the Netherlands. In: Ehlers, J. (ed.) *Glacial Deposits in North-West Europe*. Rotterdam: Balkema, pp. 361–377.

Zandstra, J. G. (1988) *Noordelijke Kristallijne Gidsgesteenten*. Leiden: Brill.

Zech, W. & Wölfel, U. (1974) Untersuchungen zur Genese der Buckelwiesen im Kloaschautal. *Fortwirtschaftliches Centralblatt* **93**, 137–155.

Zech, W., Bäumler, R., Sovoskul, O. & Sauer, G. (1996) Zur Problematik der pleistozänen und holozänen Vergletscherung des Westlichen Tienshan. *Eiszeitalter und Gegenwart* **46**, 144–151.

Zeeberg, J. (2002) *Climate and Glacial History of the Nowaja Semlja Archipelago, Russian Arctic: With Notes on the Region's History of Exploration*. Amsterdam: Rozenberg.

Zeeberg, J. & Forman, S. L. (2001) Changes in glacier extent on north Nowaja Semlja in the twentieth century. *The Holocene* **11**(2), 161–175.

Zelčs, V. & Markots, A. (2004) Deglaciation history of Latvia. In: Ehlers, J. & Gibbard, P. L. (eds) *Quaternary Glaciations–Extent and Chronology, Part I, Europe*. Amsterdam: Elsevier, Developments in Quaternary Science **2**, 225–243.

Zelčs, V., Markots, A., Nartišs, M. & Saks, T. (2011) Pleistocene Glaciations in Latvia. In: Ehlers, J., Gibbard, P.L. & Hughes, P.D. (eds) *Quaternary Glaciations - Extent and Chronology - A Closer Look*. Elsevier, Developments in Quaternary Science **15**, 221–230.

Zens, J. (1998) Nachtmarsch in Tibet. *Berliner Zeitung*, 7 October 1998.

Zeuner, F. (1952) Pleistocene shorelines. *Geologische Rundschau* **40**, 39–50.

Ziegler, J. H. (1983) Verbreitung und Stratigraphie des Jungpleistozäns im voralpinen Gebiet des Salzachgletschers in Bayern. *Geologica Bavarica* **84**, 153–176.

Ziegler, P. A. (1982) *Geological Atlas of Western and Central Europe*. The Hague: Shell Internationale Petroleum Maatschappij BV.

Zimmermann, W. F. A. (1885) *Die Wunder der Urwelt. Eine populäre Darstellung der Geschichte der Schöpfung und des Urzustandes unseres Weltkörpers so wie der verschiedenen Entwickelungsperioden seiner Oberfläche, seiner Vegetation und seiner Bewohner bis auf die Jetztzeit*. Berlin: Gustav Hempel.

Zolitschka, B., Brauer, A., Negendank, J. F. W., Stockhausen, H. & Lang, A. (2000) Annually dated late Weichselian continental paleoclimate record from the Eifel, Germany. *Geology* **28**, 783–786.

Index

ablation zone of a glacier 112, 130
accelerator mass spectrometry (AMS) 363
accumulation zone of a glacier 112
active layer 278
Adige River 116
Advanced Spaceborne Thermal Emission and Reflection (ASTER) maps 233, 234
aeolian processes 356
   dunes 369–78
   loess 378–81
   sands 378
Afghanistan 260
Africa 272
   ice-age glaciations 46
Agassiz, Louis 3, 4, 5–6, 172, 265–7
age determination using crystal lattice defects 73–4

agriculture 453–4
Airy ellipsoid 240
$^{26}Al/^{10}Be$ exposure ages 118
albedo 19, 130
Albers cone projection 240
Allerød Interstadial 360
Alpine glaciations 35, 47
   ice-stream networks 116–18
   meltwater 129–30
   Würmian Glaciation maximum 55
Alpine Ice Sheet 269–70
Alpine Quaternary stratigraphy 46–9
Altai Flood 218
Ancylus Lake 416, 421
Angamma Delta 338–9
Anglian Stage 78–81
Antarctic Ice Sheet 274

Antarctica 21, 273–4
antediluvian man 14
Anthropocene 465–7
anthropogenic effects 466
anti-syngenetic ice wedges 299
aquifers 130
ArcInfo 230
Arctic exploration by airship 243–4
Arctic glacial extent 255
Arduino, Giovanni 30
Argentina 269
Asia
   High 20–1
   ice age glaciations 46
asthenosphere 251
atomic mass spectrometry (AMS) 75
Australia 21, 272
   ice-age glaciations 46
axis of rotation tilt 19, 20

The Ice Age, First Edition. Jürgen Ehlers, Philip D. Hughes and Philip L. Gibbard.
© 2016 John Wiley & Sons, Ltd. Published 2016 by John Wiley & Sons, Ltd.
Companion website: www.wiley.com/go/ehlers/iceage

Baden–Württemberg 52
Baffin Island 202, 203
Baginton-Lillington Gravel 82
Baikal Lake 263, 264
Baltic River system 58–9
Baltic Sea 414–22
barchans 372
Barents Sea glaciers 244–6
basal meltout till 140, 141
basal sliding glacial flow 120
Bavelian Stage Complex 50
Belchen plateau 50
benthic foraminifera 38
Bergamo 49
Berwyn Mountains 76
Bessel ellipsoid 240
Biber glacial period 50
Biblical Flood 4
Biharian superzone 314
bioturbation 34
Blanc, Mont 258
bleached loam 334–6
B-maximum orientation 156
bog eyes 327
bogs 327
Bølling Interstadial 360
Bolders Bank Formation 410
boulder clay 145
Bouvet Island 275
Braderup section 60
Brazil 345–6
Bretz, J. Harland 217–18
British Isles 76–8
    Anglian Stage 78–81
    Devensian Ice Sheet 84–6
    East Anglia 198
    glaciation limits 79
    Hoxnian Interglacial Stage 81
    Ipswichian Stage 84
    location map 77
    retreat timing 82
    tills 145
    Wolstonian Stage Glaciations
        81–4
Bronze Age 454–5
brown coal diapirs 293
Brúarjökull 114

Brückner, Eduard 46
Brunhes Chron 34, 393
Buch, Leopold von 4
Buckland, Reverend William 4, 7

Caithness 410
Calabrian Stage 28
calcite shells 38
Caldenius, Carl 267–8
Cambisol 326
carbon dioxide (CO$_2$)
    Holocene increase 363
    recent increase 464–5
    variation in atmosphere 19, 21
Carboniferous glaciations 21
causes of ice ages 18–24
Chad Basin 343
Chad Lake 338–43
channels 193–200
Channelled Scablands 218
Charcot Seamount 63
Chari-Palaeodelta 340, 342
Chernozem 327
Cheshire Plain 85, 86
Chihuahua Desert 336
Chile 269
China 102
chronostratigraphic charts of
        Quaternary Period 31–2
clasts
    erratics 148–9, 159, 160, 219
    indicator clasts 159–63
    orientation 155–8
    transport 147–55
clays 35
    varved 207
cliff erosion 413
climate models and reconstructions
    degree-day modelling 444–5
    empirical, curved or linear
        relationships 443–4
    energy-balance modelling 445
    glacier modelling and climate 442–5
    ice cores 427–9
    ice-sheet modelling 431–42
    marine circulation 429–31
climate zone shifts 336–45

climatic fluctuations 18–24
    ice ages 43
coastal erosion 413
Columbia Basin 217
compaction 38
compressive flow 113
coral reefs 248–9
    Barbados 250
Cordilleran Ice Sheet 98, 217
cosmic ray dating 74–5
cosmogenic isotopes 49
Cossonay 55
Cotswold Hills 82
Cotta, Bernhard 7
Courbassière, Glacier de 9
course of the Ice Age 27
    British Pleistocene succession 76–86
    deep-sea traces 35–42
    global view 100–4
    morphostratigraphy 46–9
    North America 86–100
    Northern Germany and adjacent
        areas 58–74
    Quaternary Period beginning 27–8
    stratigraphic record 33–5
    systematics 43–6
    traces of older glaciations 49–58
crag-and-tail features 127
crevasses 130
Cromer Ridge 408
crust of the Earth 251
cryopediments 288
cryoplanation 286–8
cryoturbation 278, 291–2
crystal lattice defect age determination
        73–4
Cultural Landmark point features
        (CLPOINT) 230
Cuvier, Georges 4
Cvijić, Jovan 271

Dansgaard–Oeschger events 43
Danube Glaciation 50
Danube River 48
Darcy's law 130
Darwin, Charles 4, 12–13, 265
Debeli namet glacier 110

declination 33
deep-sea evidence 35–42
deformable bed glacial movement 120
deformation till 139
deformed subglacial sediments 142
deglaciation
   ice decay 350–4
   kettle holes 354–7
   landforms, contributions to 349
   Little Ice Age (LIA) 363–7
   pressure release 357–8
   sudden transition 359–63
degree-day modelling 444–5
deltas 207
Deluge I 255
Deluge II 255
dendrochronology 364–6
Denmark microfossils 171–2
Denton, George 245
deserts, current global location 337
Devensian Ice Sheet 84–6
Devil's Hole 408
diamictons 137
diapirs 293
Digital Chart of the World (DCW)
   229–30, 231
digital maps 229–35
Digital Terrain Elevation Data (DTED)
   230
Dimlington 86
Dionne, Jean-Claude 5
Dömnitz Interglacial 67
Doha 345
Don Formation 94
Donau glacial period 50
Donau gravel 50
Donegal Bay drumlins 126
Dowdeswell, Julian 274
draas 371
Drava River 116
drift-ice transport 4–5
Drift Theory 10
dropstones 156
drumlins 125–6
dry valleys 386
drying lakes 459–65
Djurdjura Mountains 272

dunes 369–78
   generations 374
Durmitor massif 111, 174

Earth
   area and extent of snow and ice 108
   dating the surface 74–5
   forebulge 251
   geoid shape 226, 240
   ice ages 14–16
   magnetic field changes 33
   orbit 19, 20
   sea-level rise from glacial melting 107
   structure 251
earth pillars 142
Earth Resources Technology Satellite
   (LandSat) 237
Earth Topographic map (ETOPO) 233
East Anglia 198
East-Greenland-type pingos 302, 303
eccentricity of Earth's orbit 19, 20
Eem River 69
Eemian interglacial 27, 54, 69
   diatomite sedimentation rate 322
   temperatures 325
   vegetational succession 320–1
Elbe River 396–400
   particle-size distribution 399
electron spin resonance (ESR) 74
ellipsoids for mapping 240
Elsterian Glaciation 60, 62–6
   European river system 63
   extent in Germany 65
Elsterian till 164
end moraines 172–3, 174
   search for 176–7
energy-balance modelling 445
English Channel 62–3
engorged eskers 201
Enhanced Thermic Mapper (ETM)
   images 237–9
Enns River 53
Enns valley 56
epigenetic ice wedges 299
equilibrium-line altitude (ELA) 104,
   112, 443, 444
Erd, Klaus 66

Eridanos System 58–9
erosion at sea margins 413
erosion from glacial flow 115
erratics 148–9, 159, 160, 219
eskers 127, 200–2
European glaciations 44, 101–2
European river systems
   Elsterian Glaciation 63
   Reuverian 59
European Terrestrial Reference System
   (ETRS89) 233
Euseigne 142
eustasy 246–52
eustatic fluctuations 249
Eutric Regosol 324
Evenlode river 400
extending flow 113
Eyjafjallajökull volcano 91, 133
Eyre Lake 344–5

fatigue 283
firn 112
fjords 191–3
flow till 139, 140, 143
flow velocity of glaciers 113, 114
flow zones of glaciers 113–14
flutes 125
Folldal 201, 212
foraminifera 172, 315–16
forebulge 251
fossil soils 329–31
Freedom of Information Act (FOIA) 237
freeze–thaw cycles 278, 286, 291
frost cracks 297–301
frost fracturing 284–5
frost shattering 284–5, 287
frost weathering 283–6

Geer, Gerard de 209
Geikie, James 12, 13
geoid shape 226, 240
geological timescale 15
Germany 58–52
   Eemian Interglacial 69
   Elsterian Glaciation 62–6
   Holsteinian Interglacial 66
   QEMSCAN till analysis 166–71

Germany *(continued)*
  Rhine area 28
  Saalian Complex Stage 66–9
  till types 145, 146
  Weichselian Glaciation 70–4
  youngest volcanoes 393–4
Girard, Heinrich 220
glacial polish 118
glacial series 46–8
glacial valleys 220
glacial–interglacial cycles 19
glaciation
  extensive 21
  Last Glacial Maximum (LGM)
    39–42, 104, 262, 336
glaciers
  ablation zone 112, 130
  accumulation zone 112
  dynamics 182–8
  equilibrium-line altitude (ELA) 112,
    443, 444
  erosion, characteristic 115
  flow surges 114
  flow velocity 113, 114
  flow zones 113–14
  global area 107
  ice-sheet development 118–19, 121
  ice-sheet dynamics 121–9
  ice-stream networks 116–18
  ice streams 114
  melting 459–65
  meltwater 129–34
  modelling 442–5
  movement 113–14, 120
  origin 107–12
  sea-level rise 107
  striae 127–8
  volcanic eruptions 133–4
  water-filled cavities 113
glaciofluvial deposits 127, 205
glaciomarine sediments 36–7
Glazialtheorie 7, 10
global perspective in ice ages 100
  Early and Middle Pleistocene
    Glaciations 101–3
  Late Pleistocene Glaciations 103–4
  Plio-Pleistocene Glaciations 100–1

Global Topographic map (GTOPO)
  230, 231
  errors 232
Goethe, Johann Wolfgang von 5
gold deposits 154–5
Gondwana 21
Google Earth 239
Google Maps 235
Gorleben 61, 62
Gorner glacier 9
Gottsche, Christian Carl 176–7, 193
Götzinger, Gustav 22
Gower Pensinsula 81
Goz–Kerki spit system 339
Grahmann, Rudolf 22
grain size 35, 36
  phi scale 148
  Wentworth classification 148
gravel 35
  fine 164–6
gravel terraces 202–7
Great Patagonian Glaciation (GPG) 268
green Sahara 337
Greenland Ice Core Project (GRIP) 428
Greenland Ice Sheet 108, 109
Grimes Graves 297
Gripp, Karl 295–6, 351
Grosswald, Mikhail 245–6, 252, 257
Gulf Stream 279–80
Günz Glaciation 47, 50, 51
Gürich, Georg 22

Haplic Luvisol 326
Harden Profile Development Index 328
Haslach Glaciation 51–2
Haslach Gravel 52
hatbakker 213–16
Heinrich Events 43
helicopter electronic measurements
  (HEM) 197
Helmholtz, Hermann von 4
Hermans, Willem Frederik 356
High Asia 20–1
hill–hole pairs 177
Hillington 199
Himalayas 258
Hindu Kush 258

Hirnantian 14
historical background to the Ice Age
    Theory 3
  causes of ice ages 18–24
  difficulties with religious scripture
    4–13
  ice ages of the Earth 14–16
Histosols 327
Holocene series 18
Holocene 362–3
  geobotanical subdivision 362
  North America 99–100
Holsteinian Interglacial 66
  temperatures 325
homo sapiens 452
Hoxnian Interglacial Stage 81
Hudson Bay 99-100, 119
Hughes, Terence J. 245, 254–7
human interference 447–8
  agriculture 453–4
  Anthropocene 465–7
  Bronze Age 454–5
  drying lakes, melting glaciers and
    other problems 459–65
  Iron Age 454–5
  land grab, recent 457–9
  Middle Ages 457
  Middle Stone Age 452
  Neanderthals and *homo sapiens* 452
  Neolithic period 453–4
  Roman civilisation 455–6
  spreading out from Africa 448–51
Humboldt, Alexander von 4, 6
hummocky moraines 181–2
Hüyyenberg Tephra 394

ice-age glaciations
  Europe 44
  North America 45
  South America, Africa, Asia and
    Australia 46
ice ages
  definition 43
  duration 43
  timing 21
ice cores 427–9
ice decay 349, 350–4

ice-sheet modelling 431–3
    Early–Middle Weichselian 435–42
    Saalian Glaciation–Early Weichselian
        433–5
ice sheets
    development 118–19, 121
    dynamics 121–9
ice streams 114
    networks 116–18
ice wedges 297–301
ice-contact deltas 37
ice-dammed lakes 207–13
ice-rafted debris (IRD) 36–7, 84
    icebergs 43
Ichthyosaurus 29
Iller-Lech region 50
Illinoian Glaciation 93
    Pre-Illinoian Glacial Stages 92
    Pre-Illinoian Warm Stages 92
    Pre-Illinoian–Illinoian Interglacial
        92–3
inclination 33
Independence Formation 92
indicator clasts 159–63
infrared radiofluorescence (IR RF) 384
infrared radioluminescence (IR RL) 384
infrared-stimulated luminescence
        (IRSL) 73
Inn valley 48, 56
Innsbruck 48
insolation 18
instantaneous glacierization 121
intensity of magnetic fields 33
interglacials
    beetles, snails and foraminifera
        314–16
    climate zone shifts 336–45
    fauna 312–14
    rainforests 345–7
    soil formation 324–36
    soils and palaeosols 328–36
    tar pits 311–12
    vegetation 316–24
    weathering and soil formation 324–7
internal deformation glacial flow 120
International Committee on
        Stratigraphy (ICS) 30

International Union for Quaternary
        Science (INQUA) 21, 22–3, 28,
        30
International Union of Geological
        Science (IUGS) 28, 30
involutions 291–4
Ipswichian Stage 84
Irish Sea Glaciation 81
Iron Age 454–5
isostasy 246–52
isostatic adjustments 251
isostatic depression 252

Jaberg 55
Jäkel, Dieter 263
Jaramillo Subchron 34
Jerusalem, Johann Friedrich Wilhelm 4
jökulhlaup 132–3
Jura Mountains 48

kame-and-kettle topography 350–1
kames 213–20
Kaolin Sands 60
K–Ar dating 34
Kara Ice Sheet 434
Kara Sea 437
Karakoram 258
Kenya Mount 272
Kesgrave Formation 76, 400
Keskadale 174, 175
kettle-and-kame topography 350
kettle holes 354–7
Kieselooliths 387
Kilimanjaro Mount 272
Kinne Diabase 162, 163
Krems valley 53
Kryižu kalnas 216-7
Kuhle, Matthias 20–1, 258–63
Kun Lun Shan 108, 258
Kurzeme 424
Kverkjökull 131

Laacher See system 394
Lake Agassiz 208
lake drying 459–65
land grab, recent 457–9
landforms resulting from deglaciation 349

landmass location 21
LandSat 237
Lang, Heinrich Otto 10
Langeland Island 213–16
Last Glacial Maximum (LGM) 39–42,
        104, 262, 336
Lauenburg Clay 65–6, 143, 144
Laurentide Ice Sheet 95, 96, 99–100,
        119, 121
Lech River 48, 52
Leffe basin 49
lichenometry 366–7
Lieth 61
Linnaeus 405
lithosphere 251
lithostratigraphy 34
Litorina Sea 416, 422
Little Ice Age (LIA) 271, 363–7, 457
Llanddewi Formation 81
Loch Etive 216-7
lodgement till 139, 140, 141, 156
loess 378–81
loess–loam-rich solifuction sheet 331
Loess Plateau 380
Lomonosov Ridge 252
Loosen Gravels 58
Luvisol 326, 334
Lyell, Charles 4, 7, 246–7

Mackenzie Delta 301, 304, 305
Mackenzie-type pingos 301–2
macrofabric of tills 157
Madsen, Victor 22
magnetic field changes 33
magnetic pole reversals 33
Maidenhead Formation 402
Malvern Hills 81
Mangerud, Jan 27, 245
mantle of the Earth 251
Manych Depresssion 435
mappability of deposits 34
maps 227–9
    digital maps 229–35
    ellipsoids 240
    projections 240
marine circulation 429–31
marine isotope stages (MIS) 18

Marks Tey 324
Martonne, Emmanuel de 271
Matuyama Chron 34, 52
Medieval Warm Period 457
Mediterranean glaciations 269–71
Meiendorf Interstadial 360
Meikirch 53-55
meltwater from glaciers 129–34, 191
    channels 193–200
    eskers 200 2
    fjords 191–3
    gravel terraces 202–7
    ice-dammed lakes 207–13
    kames 213–20
    outwash plains 202–7
Memmingen 48, 52, 57
Menapian Cold Stage 60
Menke, Burkhard 66
Mercator projections 240
Mercer, John Hainsworth 268
Mesolithic era 452
microfabric of tills 157
microfossils 171–2
Middle Ages 457
Middle Stone Age 452
Milankovitch, Milutin 19
Mindel Glaciation 47, 52
Mindel–Riss Interglacial 52–3
Mindel valley 52
minerals, magnetic 34
Missoula Flood 217-8
Mondsee 54, 55
moraines 115
    end moraines 172–3, 174, 176–7
    glacier dynamics 182–8
    hummocky moraines 181–2
    push moraines 173–81
Moravian Gate 64
morphological reversal 50
morphostratigraphy 46–8
    soils 328
moulins 130
Münsterländer Bucht 64
Murzuk Basin 371

Nagelfluh 207
Neanderthals 451, 452

Neogene–Quaternary boundary 28
Neolithic period 453–4
Netherlands 60, 408
Nigardsbreen glacier 131, 132
North America 86–7
    glaciation extent 88
    Holocene 99–100
    ice-age glaciations 45
    Illinoian Glaciation 93
    older glaciations 87–9
    Pleistocene, Early 92
    Pre-Illinoian Glacial Stages 92
    Pre-Illinoian Warm Stages 92
    Pre-Illinoian–Illinoian Interglacial
        92–3
    Sangamonian Interglacial 93–4
    Wisconsinan Glaciation 94–9
North Sea 62–3, 406–14
    tsunami 411
North Sea Drift Formation 78–9
Northern Sea Route (Northeast Passage)
    252
NorthGRIP (NGRIP) core 18
Norway 15
    Mjøsa, Lake 16
    Varanger Peninsula 17
Novaya Zemlya 244, 246
nunataks 140, 283

Ob River 279
Obergünzburg 52
ocean circulation 429–31
ocean deposits 35–42
Oceania 272
'Old Swede' erratic, origin of 148–9, 151
Oldenswort borehole 61
Older Dryas 360
older glaciations 49–58
    Günz Glaciation 51
    Haslach Glaciation 51–2
    Mindel Glaciation 52
    Mindel–Riss Interglacial 52–3
    Riss Glaciation 53–4
    Riss–Würm Interglacial (Eemian) 54
    Würm Glacial 55–8
Older Saalian Drenthe Till 334
Olduvai Subchron 34

Oligocene 21
Open Street Map 235
Operational Navigation Charts (ONC)
    229–30
optically stimulated luminescence
    (OSL) 49, 73, 374, 384
Ordnance Survey maps 237, 240
Ordovician glaciation 21
oscillations in climate 18–24
outwash plains 202–7
oxygen isotope curves 428
oxygen isotope-ratio 38–9
    calculation of 38
oxygen-isotope stratigraphy 38–9
ozone hole 461

Palaeogene–Neogene boundary 27–8
Palaeolithic era 450–1
palaeomagnetism 33–4
palaeosols 328–36, 381
palsas 304–7
Pamir 258
Panama Straits 21
Paris Basin 30
particle-size distribution 145
    phi scale 148
        Wentworth classification 148
Pebble Gravel Formation 400
Peoria Loess 96
Penck, Albrecht 35, 46, 47, 271
periglacial areas
    definition and distributions 277–80
periglacial soil stripes 296–7, 298
periglacial solifluction 294–6, 410
permafrost 277
    global distribution 278
    temperature changes, recent 464
Petermann, August 274
petridelaunic flood 4
phi scale for grain size 148
pillars of earth 142
pingos 301–4, 356
pipkrake 291
pitted outwash 350
plate tectonics 21
Pleistocene extent of frozen ground
    281–2

Pleistocene ice age 23, 24
Pleistocene series 18
Pleniglacial 360
Pliocene–Pleistocene boundary 28
Podzol 326
Polar Ural-type glaciers 109
pollen analysis 316–24
porphyry 161–3
postglacial uplift 252
potassium–argon (K–Ar) dating 34
Precambrian glaciations 21
precession of Earth's axis 19
precipitation and ice sheet formation
    109
pressure release 357–8
projections for mapping 240
Ptolomy Lake 337–8
push moraines 173–81
    cross-section 178–80
Pyrenees Weichselian Glaciation
    maximum 56

QEMSCAN analysis 166–71
Qinghai–Tibetan Plateau 261–2
Quaternary Environment of the
    Eurasian North (QUEEN) 245
Quaternary Period 18
    beginning 27–8
    chronostratigraphic charts 31–2
    stratigraphy 29–30
quicksand 373

radiocarbon dating 209, 380
rainforests 345–7
    current global location 347
    decreasing area 460
Rancho La Brea 310-2
Rapakivi granite 159–61, 162
Raukar 405–6
Rauno Formation 396
Red Baltic Porphyry 161–3
red till 152–3
Regosol 324, 327
relict soils 329
resistance measurements 197
Rhät sandstone 11
Rhine glacier 53, 56

Rhine River 48, 52-3, 56-7, 387–96
Rhinog mountains 124
rhythmites 212–13
ripples on dunes 370–1
Risbury Formation 81
Riss Glaciation 47, 53–4
Riss River 48
Riss–Würm Interglacial (Eemian) 54
Ristinge Klint 178–80, 183
rivers
    dry valleys 386
    processes and landforms 383–5
rock glaciers 288–91
Roman civilisation 455–6
Rosa, Monte 8
Rubified Valley Farm Palaeosol 332, 333

Saalian Complex Stage 66
    early Saalian 66–7
    extent 68
    Saalian Glaciations 67–9
Saalian Glaciation 27
    Palaeogeographical reconstruction
        432, 433, 438, 440
    to Early Weichselian 433–5
Sahara Desert 272, 336–8
    dunes 376
Sahara glaciation 14, 15
Sahel Desert 376
Salvigsen, Otto 245
salt domes 195
Salzach glacier 56
Salzach River 54
Samerberg 52, 54, 55
sand 35
sander 202
sands, aeolian 378
sandur 127, 202
Sangamon Soil 92-4, 336
Sangamonian Interglacial 93–4
Sannox, Glen 175
satellite images 236–9
Scandinavian Ice Sheet 56, 185–8,
    221–2, 392
Scarborough Bluffs 143, 144
Scheffel, Joseph Victor von 29
Schieferkohle 49, 54

Schmalsee Lake 194
Schwansen peninsula 351
Scotland tillite series 15, 17
sea ice in Last Glacial Maximum (LGM)
    42
sea-level rise 459, 462
    from glacial melting 107
Sefström, Nils Gabriel 4, 11
segregation ice 278
Severnaya Zemlya 242-4
Shuttle Radar Topography Mission
    (SRTM) 233, 234
Siberia 252–8
Sicily Terrace 247
Siegsdorf 313
Snowball Earth 15
Sognefjord 191–3
sohlmoräne 120, 142
soil formation 324–7
    palaeosols 328–36
soil stripes 296–7, 298
solifluction 294–6
South America 21, 103
    glaciers 265–9
    ice-age glaciations 46
    volcanoes 265–9
spy satellite images 236–7
Stagnosol 326, 334
Steyr River 53
Storegga Slide 411
stratigraphy 29–30
    Alpine Quaternary 46–9
    chronostratigraphic charts of
        Quaternary Period 31–2
    components 33–5
    oxygen isotopes 38–9
striae, glacial 127–8
string mires 307–8
subaquatic outwash fans 37
subaquatic till 143
Subcommission on Quaternary
    Stratigraphy (SQS) 466
subglacial deformation 120
subglacial meltout till 139, 140, 141
subglacial transport 152
sublimation till 139
supraglacial meltout till 139, 140, 143

supraglacial transport 152
surges in glacial flow 114
Sweden till clast composition 150
Swiss Plateau 48
Switzerland older glaciations 49–50, 55
Sylt Island 60
syngenetic ice wedges 299

Taylor Glacier 152
tephra 36, 87
    dating 90–1
terraces 202–7, 383–5
Tertiary boundary 27–8
Thalgut exposure 50, 51, 53-5
Thames River 400–3
    Quaternary geology 403
thermokarst processes 356
thermoluminescence (TL) 73, 380
Thoringian superzone 314
thufurs 308–9
Thungschneit 54, 55
Thursaston 149
Tibet 258–63
    ice distribution 259
Tibetan Plateau 20–1
Tiglian Stage 59-61
tills
    clast orientation 155–8
    clast transport 147–55
    clasts, indicator 159–63
    definition 137–43
    gravel 164–6
    macrofabric 157
    microfabric 157
    microfossils 171–2
    QEMSCAN analysis 166–71
    red till 152–3
    types 139
    variability 145
tilt of Earth's axis of rotation 19, 20

timing of ice ages 21
tongue basin 115, 177
Torrell, Otto 11
Tottenhill 83
Trans-Himalayas 258
transient-electromagnetic system
        (SkyTEM) 197
Traun valley 56
Travemünde 144
tree rings 364–6
Treene Interglacial 336
tsunami in the North Sea 411
tunnel valleys 193
Turgay Depression 433, 436, 437

Uhlenberg interglacial deposits 50
Ulm 51
Universal Transverse Mercator (UTM)
        system 230
Unterpfauzenwald 52
uranium–thorium (U–Th) dating
        249
urstromtäler 205, 220–4

Vallendar layers 387
Valley Farm Palaeosol 332, 333
Varanger ice age 15
varves 207, 209–11
    thickness measurements 211
Vatnajökull 132
Vector Maps (VMAP1 & VMAP2) 230,
        231, 232
Vector Product Format (VPF)
        230
Vendian 14
ventifacts 372
Villafranchian superzone 314
volcanic eruption under ice 133–4
von Hoff, Karl, Ernst Adolf 4
V-shaped valleys 115

Waardenburg 62
Wahnschaffe, Felix 11
wanderblöcke 49
Warthe Till 334
water-filled cavities in a glacier 113
Wehr Tephra 394
Weichelian glacial maximum 24
Weichselian Glaciation 70–4
    extent 71
    glacier dynamics 182–8
    original limit 70
    stratigraphy 360
Wentworth classification for grain size
        148
West Eifel volcanic field 393–4
Westerhoven 62
White Sands 376, 377
Wildhaus 54
Wind River Mountains 114-5
Wisconsinan Glaciation 94–9
Woldstedt, Paul 22, 196
Wolstonian Stage Glaciations 81–4
Woodward, Jamie 13
World Geodesic Systems (WGS84)
        233
Wrangel Island 253–4, 257
Würm Glaciation 47, 55–8
Württemberg Rottal 51

yardangs 372
Yarmouth Soil 87, 92-3
yedoma 297
Yellow River 379
Yellowstone 89
Yoldia Sea 416, 420
Younger Cover Gravels 52
Younger Dryas 360–1

Zermatt glacier 9
Zimmermann, W.E.A. 13